Insect Pest Management

2nd Edition

D. Dent

CABI Bioscience
UK Centre
Ascot
UK

CABI *Publishing*

CABI *Publishing* is a division of CAB *International*

CABI Publishing
CAB International
Wallingford
Oxon OX10 8DE
UK

CABI Publishing
10 E 40th Street
Suite 3203
New York, NY 10016
USA

Tel: +44 (0)1491 832111
Fax: +44 (0)1491 833508
Email: cabi@cabi.org
Web site: http://www.cabi.org

Tel: +1 212 481 7018
Fax: +1 212 686 7993
Email: cabi-nao@cabi.org

A catalogue record for this book is available from the British Library, London, UK.

Library of Congress Cataloging-in-Publication Data
Dent, David, Ph.D.
  Insect pest management/David Dent.--2nd ed.
    p. cm.
  Includes bibliographical references.
  ISBN 0-85199-340-0 (hc. : alk. paper)--ISBN 0-85199-341-9
    (pbk. : alk. paper)
  1. Insect pests--Integrated control.   I. Title.
  SB931.D45   2000
  632.7--dc21

                                                    99-087126

ISBN 0 85199 340 0 (HB)
      0 85199 341 9 (PB)

Typeset in 9/11pt Melior by Columns Design Ltd, Reading
Printed and bound in the UK at the University Press, Cambridge

Contents

Preface

The subject of pest control is rarely discussed without reference to the concept of integrated pest management, or IPM as it is more commonly known. IPM is essentially an holistic approach to pest control that seeks to optimize the use of a combination of methods to manage a whole spectrum of pests within a particular cropping system. Insect pest management is often thought to be synonymous with IPM, mainly because the concept was developed by entomologists, but in its widest sense IPM refers to the management of weeds, pathogens as well as insects. Insect pest management is a subsystem of IPM but for the sake of simplicity the two are considered synonymous throughout the book – unless stated otherwise.

Pest management involves a number of stakeholders ranging from scientists to farmers and agribusiness to consumers. As a textbook for undergraduate/postgraduate students and for researchers of insect pest management emphasis is placed on the underlying principles and experimental approaches to the science that underpins the development of working IPM systems. However, the different types of stakeholder have an impact on the way pest management is viewed, studied and practised and hence, these different perspectives have been included wherever possible.

The aim of this book is to provide an overview of insect pest management, highlighting the major problem areas and contentious issues and where possible attempting to identify promising lines and directions for future research and implementation. The book includes descriptions and explanations of the different control measures and their use, categorized as insecticides, host plant resistance, biological control, cultural control and interference methods, and quarantine legislation. In addition, there are chapters on sampling, monitoring and forecasting, yield loss assessment, programme design, management and implementation. The introduction provides a brief history of pest management, the causes of pest outbreaks and an overview of the different stakeholders involved in pest management. The final chapter addresses the driving forces that are shaping the future of IPM, including three case studies of working IPM systems.

The book includes case studies and cites examples throughout from a range of entomological disciplines, veterinary, medical, agricultural, forestry and postharvest systems. I have tried in this second edition to reduce the bias towards agricultural systems evident in the first. The use of examples from both temperate and tropical countries and from a wide chronological range are retained – the latter because it is necessary to retain a perspective on current science in relation to what has been done in the past. There is too often a tendency to assume that only the most current research is applicable to today's problems – the best research withstands the test of time and provides the basis for that which is possible now.

A great deal of work is carried out with the objective of developing or implementing IPM programmes but even after nearly 40 years of progress in the science and socio-economics of pest management there are too few opportunities to take an holistic approach to the subject. Ten years ago there was a need for individuals who were able to address the subject of IPM in an interdisciplinary way – for people who can really coordinate, manage and deal with the inter-disciplinary nature of IPM. Although there are now more people who work with this ability there is still a great need for others who can contribute to the science of pest management at this level. I hope now, as I hoped ten years ago when I first wrote this book, that this second edition will contribute something to this change of perspective and encourage others to consider the subject of pest management as a whole rather than as isolated disciplines. For it is still at this holistic level that the greatest and most exciting advances are to be made.

Acknowledgements

I am grateful to a large number of people for their assistance and encouragement during the writing of both the first and second editions of this book. My ideas and views have been shaped in the intervening years between editions by the many discussions I have had with colleagues around the world. Those ideas and views are now reflected in the changes I have made to this second edition. My thanks to my colleagues for their input.

For those who took the time to work on the original draft, my renewed thanks: Tom Coaker, Dereck Campion, Martin Dean, Peter Haskell, Eric Hislop, Mark Jervis, Neil Kidd, Bill Reed, Raoul Robinson, Peter Walker and Tristram Wyatt. My appreciation also to all those who took the time to search for, copy and send me reprints of their papers for this latest edition: Lindsay Baggen, Rod Blackshaw, Colin Bower, Micheal Cohen, Joanne Daly, Ian Denholm, Alan Dewar, T. Drinkwater, Bob Ellis, Annie Enkegaard, Peter Follett, Dawn Gouge, Jean-Claude Gregoire, David Hagstrum, Jane Hayes, Rick Hodges, John Holt, Casey Hoy, Marcos Kogan, Simon Leather, Joop van Lenteren, Graham Matthews, Peter McEwen, Geoff Norton, Larry Pedigo, Les Penrose, John Pickett, Wilf Powell, Martin Shapiro, Les Shipp, Ken Schoenly, Nan-Yao Su, Charlie Summers, Graham Thwaite, Lester Wadhams, Richard Wall, John Westbrook, Peter Witzgall, Steve Wratten, Vicki Yokoyama, Frank Zalom. My apologies if I have left anyone off the list who should have been included.

The typing of this edition has been divided between my colleague and friend Jeremy Harris and my wife Sue – my sincere thanks to you both. To Jeremy also my appreciation for the time and skills at the computer to produce the ten new figures for the book. I remain forever grateful to Sue for having made the first edition possible through all her support and encouragement, and the typing and drawing. Also for too many years when it dominated all of our spare time. And now, my apologies, love and thanks to Sue, Catherine and Thomas for allowing me to spend so much time closeted away writing when there were sunny days outside and many more fun things to be doing.

David Dent
1999

'Pest control constitutes an ancient war, waged by man for 4000 years or more against a great variety of often small and remarkably persistent enemies. Suprisingly, although the war is old, its dynamics and the nature of the principal protagonists seem poorly understood. Even the objectives, at least of man, are ill-defined. It is as though man, in the heat of battle, has not had the time to analyse in any sophisticated fashion the conflict in which he finds himself. Battles have been won and lost but lessons have been learned slowly and painfully. It is only in recent years that people have begun to ask the fundamental questions of principle and to raise doubts about implicit beliefs and objectives.'

Conway 1976

1
Introduction

A pest insect is one that is judged by man to cause harm to himself, his crops, animals or his property. In farming an insect may be classified as a pest if the damage it causes to a crop or livestock is sufficient to reduce the yield and/or quality of the 'harvested product' by an amount that is unacceptable to the farmer. Insects may be classed as pests because they cause damage directly to harvestable products, e.g. codling moth larval damage to apples, or because they cause indirect damage or harm in other ways, e.g. by causing a nuisance to livestock or humans or as vectors of plant or livestock diseases. There are a myriad of ways in which insects can cause harm and they have done so for the thousands of years that man has occupied the earth. Likewise man's attempts to control or manage the harm caused by insects has a long and varied history. A knowledge of this history adds an important dimension to the study of pest management because it can provide insights into the driving forces (technical, economic, social) that have forged current pest management practices which in turn will provide some idea of the forces likely to be acting in the future (Norton, 1993).

1.1 A Brief History of Pest Management

The history of pest management dates back to the beginnings of agriculture and from that time it is a history that combines important events (discoveries and defining moments), influential people, institutions, organizations and governments in ways that have led us to the current concept of integrated pest management (IPM). IPM has been defined as a 'pest management system that in the context of the associated environment and population dynamics of the pest species, utilizes as suitable techniques and methods in as compatible manner as possible and maintains the pest population levels below those causing economic injury' (Smith and Reynolds, 1966). The concept of utilizing a number of techniques in as compatible a manner as possible is, of course, not new and even the different control techniques available to us today were utilized in some shape or form many years BC. For instance, the use of insecticide by the Sumerians who applied sulphur compounds to control insects and mites was first recorded from 2500 BC onwards. Botanical insecticides were used as seed treatments around 1200 BC by the Chinese who also applied mercury and arsenical compounds to control body lice at that time.

First descriptions of cultural controls, especially manipulation of planting dates, were recorded around 1500 BC, while burning as a cultural control method was first described in 950 BC. Evidence of biological control, of manipulating natural enemies, comes from both China and Yemen where

1

colonies of predatory ants (*Oecophylla smaragdina*) were set up in citrus groves, moving between trees on bamboo bridges to control caterpillar and beetle pests (Coulsen *et al.*, 1982). These practices date back over 1000 years. Genetic resistance is one of the oldest recognized bases of plant pest control (Panda and Khush, 1995). Theophrastus recognized the differences in disease susceptibility among crop cultivars as early as 3 BC (Allard, 1960) and, of course, farmers will have selected for resistance to local pests by the practice of saving seed from healthy plants for sowing the next season. Over many generations of selection resistant land races developed, which provided sustainable levels of yield for average local conditions. Thus by 500 AD all the general types of control measure available today – insecticides, host plant resistance, biological and cultural control – had already been developed and used by one civilization or another.

Reference to developments in pest management seem to be few and far between for the period between AD and the 18th century. A number of bizarre events such as the excommunication of cutworms in Berne, Switzerland, in 1476 and the banishing in 1485 of caterpillars by the High Vicar of Valence perhaps depict the lack of understanding of the causes of pest problems at this time. It was not until 1685 that the first correct interpretation of insect parasitism was published by the British physician Martin Lister who noted that the ichneumon wasps emerging from caterpillars were the result of eggs laid by adult female ichneumonids (van Driesche and Bellows, 1996). This was followed by a similar interpretation of parasitism in *Aphidius* species by the Dutchman Antoni van Leeuwenhock recorded in 1700. The work of Linnaeus burgeoned an interest in insect descriptions which promoted greater observation and study of insect biology. It was Linnaeus who first noted in 1752 that 'Every insect has its predator which follows and destroys it' and then commented on how such 'Predatory insects should be caught and used for disinfecting crops'

(Hörstadius, 1974). The earliest documentation on host plant resistance occurs in 1762 with the description of hessian fly (*Mayetiola destructor*) resistant wheat cultivar, Underhill, in the US by Haven (Panda and Khush, 1995).

From these beginnings the scientific discipline of entomology began to take shape at the turn of the 19th century. Pioneers such as John Curtis in the UK, Thomas Say and C.V. Riley in the US created the basis for applied entomology with their publications (e.g. *Farm Insects*, J. Curtis, printed 1841–57 – an illustrated account of pest insects arranged by crops for ease of use by farmers; *American Entomology*, 1824 by T. Say) and their innovative approaches to pest control which formed the foundation for modern pest control. In the US the Rocky Mountain locust plagues of the 1870s initiated the formation of the US Entomological Commission (Sheppard and Smith, 1997). In 1881 C.V. Riley was appointed Entomologist of the Department of Agriculture and the Division of Entomology was created within the USDA. The Entomological Society of Canada was founded by W. Saunders and C. Bethuse in 1863 with the first issue of the journal *The Canadian Entomologist* published in 1868 (Sheppard and Smith, 1997). In 1873 C.V. Riley was responsible for the first international movement of a biological control agent *Tyroglyphus phylloxerae* which was introduced into France to control grape phylloxera, which unfortunately failed to exert sufficient levels of control to be useful. The first successful example of classical biological control occurred in 1888 with the importation and release of the Vedalia beetle (*Rodolia* (formerly *Vedalia*) *cardinalis*) for control of the citrus pest of the cottony cushion scale (*Icerya purchasi*) in California (van Driesch and Bellows, 1996). Within two years the beetle had controlled the scale throughout the state. This project has far-reaching effects because it demonstrated the feasibility of utilizing arthropod predators to control serious pest problems and produce large economic benefits.

Other areas of biological control were also forging ahead during the 19th century. Insect pathogens were first considered as biological control agents by Bassi in 1836, when he proposed that liquids from cadavers of diseased insects could be mixed with water and sprayed on plants to kill insects. However, it was not until 1884 when the Russian entomologist Elie Metchnikoff developed a means to mass produce the fungus *Metarhizium anisopliae* for use in the field against the sugar beet curculio *Cleanus puntiventris* that entomopathogens proved their potential as a practical control option.

The end of the 19th century saw a fundamental change in agriculture and urbanization, particularly in the US. Agriculture was the linchpin to urbanization; a rapid growth from a mainly subsistence way of life towards greater mechanization (substituting labour) and commercialization fuelled urbanization. Since insect pests were a limiting factor in agricultural efficiency, the importance of their control increased. Thus, a demand developed for the investigation, identification and control of insect pests. In this way the subject of entomology became identified as a distinctive scientific discipline with the increase in training and employment of State and Federal entomologists. The numbers increased from less than 20 in the mid-1800s to more than 800 by the 1870s (Sorenson, 1995). In 1889 the first US national professional group, the American Association for Economic Entomology was founded, followed by the publication of the *Journal of Economic Entomology*, in 1908 (Howard, 1930).

The change from subsistence to commercial agriculture had a major impact on farmers' attitudes to pest control, largely because commercialization introduced the need to borrow or make cash investments which had not been necessary in subsistence agriculture. Insect pest problems caused losses of yield in subsistence agriculture but unless they were catastrophic they did not risk the farmer's livelihood, whereas in commercial agriculture, insect problems threatened the safety of other cash investments and thereby threatened the farmer's continued ability to stay in business (Benedict, 1953). This meant that the standards for acceptable levels of insect control increased from those that had been acceptable in subsistence agriculture (Morse and Buhler, 1997). Hence, the need to reduce risk by ensuring effective pest control became a prime consideration among commercial farmers. This in turn provided an ideal opening for chemical insecticides which could provide reliable, demonstrable levels of insect control for a relatively low cost.

The new era, involving the use of toxic substances for insect control, was led by the use of the dye 'Paris Green' which was found to be effective against the Colorado beetle in the US and was used with Bordeaux mixture against Grape Phylloxera in French vineyards (1870 and 1890). In the 1890s lead arsenate was also introduced for insect control. By 1910, lead arsenate and Paris Green were the most widely used insecticides sold on a commercial basis with sales reaching £10 million each year. In 1917 calcium arsenate joined lead arsenate as the leading insecticides in addition to the increasing list of available products which included tar oils, plant extracts, derris, nicotine and pyrethrum (Ellis, 1993). Thus, the first 40 years of the 20th century witnessed an increased use and reliance on chemical insecticides, including the introduction of new compounds such as ethylene oxide, thiocyanates and phenothiazine. However, the application of chemical products tended to be haphazard and very imprecise with the technical material often ineffective. Hence at this time, although growing in importance for use in pest control, chemical insecticides did not yet dominate the approaches adopted by farmers and enormous potential was seen to exist for biological control and breeding resistant crop plants. With the former between 1920 and 1930 there were more than 30 cases of natural enemy establishment recorded throughout the world, based on the increasing number of introductions of exotic pest species promulgated by growing international trade, particularly of food and fibre raw materials.

Plant breeding for agronomic characters had become well established in the 19th century utilizing quantitative genetics, but it was not really until the rediscovery of Mendel's Law of Heredity in 1900 that geneticists understood qualitative breeding approaches and stumbled upon disease and later insect pest resistance factors of economic importance (Robinson, 1996). The systematic research of R.H. Painter in the 1920s on the resistance to Hessian fly in wheat cultivars laid the foundations for the development of resistance breeding against insects utilizing qualitative genetics (Panda and Khush, 1995). However, successes were few and far between with breeding for resistance to insects (unlike pathogens); the most notable success at the time being cotton resistant to *Empoasca* bred in South Africa (Parnell, 1935) and India (Husain and Lal, 1940). It was not until the 1960s that the full potential of host plant resistance to insects was fully appreciated but by then the dominance of chemical insecticides for insect control had reached spectacular proportions.

Chemical insecticides came to prominence on the back of the most famous insecticide – DDT (dichlorodiphenyl trichloroethane) – developed by Paul Müller working for the Geigy Chemical Company in 1939. DDT offered persistence, low cost, virtually no plant damage, broad spectrum activity and low acute mammalian toxicity. It was first used in 1941 by Swiss farmers to control Colorado beetle and by 1945 DDT production had reached 140,000 tonnes a year. The success of DDT also stimulated the search for other similar chemicals and the subsequent development of aldrin, HCH, dieldrin, heptachlor and chlordane (Ellis, 1993). Other methods of pest management paled in significance with the success achieved with these impressive chemicals and hence, the use of more biologically oriented approaches declined dramatically in the 1940s and

1950s. This redirection of effort was mitigated by the overwhelming need to test the efficacy of the ever expanding arsenal of new chemical compounds (Kogan and McGrath, 1993). Pesticides became the only method used by many farmers for the control of agricultural pests.

Ironically, at the same time that DDT had obtained notoriety as a panacea for pest control in 1946, the philosophy of IPM came into being in the alfalfa fields of California's San Joaquin Valley (Summers, 1992). In 1946 K.S. Hagen was hired as the first supervised control entomologist in California, where he monitored 10,000 acres of alfalfa for the alfalfa caterpillar *Colias eurythene* and its parasite *Apanteles medicanis* (Hagen *et al.*, 1971). From the experience of combining natural enemies, with host plant resistance and rational use of chemicals in the 1950s against the spotted alfalfa aphid *Therioaphis maculata* emerged the integrated pest control philosophy presented in Stern *et al.* (1959).

However, the context in which the IPM framework would achieve significance was happening around the world in situations where pesticides were used in excess. Insect resistance to chemical insecticides was first reported in 1946 in houseflies in Sweden but in the 1950s it became widespread in many agricultural pests. Resurgence of target pests, upsurges of secondary pests (both caused by the suppression of natural enemies by insecticides), human toxicity and environmental pollution (Metcalf, 1986) caused by pest control programmes relying on the sole use of chemicals reeked havoc in many cropping systems. The ecological and economic impact of chemical pest control came to be known as the 'pesticide treadmill' because once farmers set foot on the treadmill it was virtually impossible to try alternatives and remain in viable business (Clunies-Ross and Hildyard, 1992). In addition, despite problems with the chemicals used, the faith in and 'addiction' to the chemical technology by farmers was assured through the continued development and availability of new active ingredients coming onto the market (Table 1.1).

Table 1.1. The availability of chemical insecticides between 1950 and 1987 (from Ellis, 1993).

Insecticide	Period
Organochlorines	
DDT dust	1950–1972
DDT liquid	1959–1984
DDT bait	1967–1979
BHC dust	1950–1953
Aldrin	1959–1964
Dieldrin	1959–1964
Organophosphates	
Fenitrothion	1977–1979
Tiazophos	1979–present
Bromophos	1983–present
Chlorpyrifos	1983–present
Trichlorfon	1983–present
Carbamates	
Carbaryl	1985–present
Methrocarp	1970–1972
Synthetic pyrethroids	
Cypermethrin	1987–present
Alphacypermethrin	1987–present

A landmark event in the history of pest management was the publication in 1962 of the book *Silent Spring* by Rachel Carson, which was not really important from the point of view of its technical content but rather from its impact on the general public. *Silent Spring* brought pest management practices into the public domain for the first time and provided the spring board for an increase in the awareness of the general public of the problems associated with chemical insecticide use. This growing environmental concern, combined with the philosophy of integrated control advocated by an increasing vocal band of scientists provided openings for the funding and development of alternative, more environmentally friendly approaches such as insect pheromones, sterile insect techniques, microbial insecticides and host plant resistance. The gypsy moth pheromone was isolated, identified and synthesized in 1962; by 1967 the screw worm *Cochlomyia hominivorax* was officially declared as eradicated from the US through use of the

male sterile technique (Drummond *et al.*, 1988); and in 1972 the first commercial release of a microbial insecticide of *Bacillus thuringiensis* based on the isolate HD-1 for control of lepidopterous pests occurred (Burgess, 1981). Plant breeding using Mendelian genetics had earlier in the century produced very high yielding dwarf wheat cultivars. The need to improve yield in similar ways with other crops and hence to alleviate 'third world' malnutrition fostered the establishment of the first International Agricultural Research Centre (IARC) sponsored by the Consultative Group for International Agricultural Research (CGIAR), the International Rice Research Institute (IRRI) in Los Baños in the Philippines in 1960. This was followed by seven other centres improving crop reproductivity through development of high yielding varieties (HYV). Part of this development included breeding for insect resistance, of which there was some limited success, e.g. cultivar IR36 was resistant to brown plant hopper (*Nilaparvata lugens*), the green leaf hopper (*Nephotettix virescens*), the yellow stem borer (*Scirpophaga incertulas*) and the gall midge (*Orseolia oryzae*) (Panda and Khush, 1995). This period became known as the 'Green Revolution' alluding to the 'greening' of developing countries with high yielding cultivars, particularly of rice (Panda and Khush, 1995). Generally however, the resistance was based on a form of resistance that could be overcome by the insects and particularly in rice, farmers were advised to spray their high yielding crops with chemical insecticides to protect them. Such measures were still necessary until the 1980s when IPM methods were introduced into rice systems for the first time.

The concept of IPM in the 1960s and 1970s was based on restricting pesticide use through use of economic thresholds and utilization of alternative control options such as biological products or methods, biopesticides, host plant resistance and cultural methods (Thomas and Waage, 1996). IPM was launched in the US and ultimately around the world with a large and influential research project known as the Huffaker Project (1972–1979; which was continued in the 1980s as the Atkinson project) focusing primarily on insect pest management in six crops: cotton, soybean, alfalfa, citrus fruits, pome fruits (apples and pears) and some stone fruits (peaches and plums; Morse and Buhler, 1997). Involving scientists from universities, USDA and private industry the project sought to organize research to transcend the disciplinary, organizational, political, geographical and crop specialism barriers that have usually inhibited collaborative efforts between scientific disciplines. In this effort, however, and others that have succeeded it, IPM was designed to the specification of capital intensive, technologically sophisticated farmers. Ironically this was the same design for which the high tech chemical pest control tradition had been based, something that IPM endeavoured to replace (Perkins, 1982; Thomas and Waage, 1996). A different model for the development of IPM came, not from the intensive, high tech world of agriculture of the west but from the Philippines and South-East Asia. Here the emphasis was on IPM training among farmers (Table 1.2), creating sufficient understanding of the interaction between natural enemies and their pest prey/hosts so that economic thresholds are based on observation of the balance between pest and natural enemy numbers in the rice paddy

(Matteson et al., 1994). In Indonesia, after one season's training, farmers' spraying practices have changed from an average of 2.8 sprays per farmer to less than one per season, with the majority of farmers not spraying at all. There was no difference in yields using IPM methods versus those harvested with nationally recommended technical packages (Matteson et al., 1994). The concept of IPM was officially adopted in Indonesia in 1979, but its implementation was negligible because of massive pesticide subsidies, the limited experience and knowledge of extension personnel and massive promotional campaigns by the pesticide companies (APO, 1993; Morse and Buhler, 1997). IPM adoption escalated however, with the Presidential Decree No. 3, instituted in 1986, which banned 57 broad spectrum insecticides used on rice and endorsed IPM as the official strategy for rice production.

At this time other governments around the world were waking up to the need for endorsing IPM. In 1985 India and Malaysia declared IPM Official Ministerial Policy, as did Germany in 1986. IPM was also implicit in the Presidential Declaration in the Philippines in 1986 and Parliamentary decisions in Denmark and Sweden in 1987. In 1992 at the United Nations Conference on Environment and Development, Agenda 21, Rio de Janeiro, the World's Heads of State endorsed IPM as a sustainable approach to pest management.

Table 1.2. Evolution of rice IPM farmer training in Asia 1979 to present (from Matteson et al., 1994).

Period	Training approach
1978–1984	Pilot scale, Philippines. Elaboration of IPM training principles. Group field training by master trainers using conventional teaching methodology. Weekly training classes over an entire cropping season, follow up for 1–2 seasons
1985–1989	(i) Small-scale implementation of the above by NGOs, Philippines (ii) Large-scale implementation through several national T & V extension systems, sometimes supported with multimedia strategic extension campaigns
1990–1992	Medium scale, Indonesia. Dedicated training system in which master trainers use participatory non-formal education methodology

The political endorsement of IPM on the international stage has come about because of the policy shifts towards environmental concerns that have occurred over the last 30 years. It is this concern for the environment that will remain one of the key factors driving pest management during the next millennium.

1.2 Causes of Pest Outbreaks

The history of pest management is a subset of the history largely of agriculture and while pests have been a chronic problem in agriculture since the beginning, many of today's serious pest problems are the direct consequence of actions taken to improve crop production (Waage, 1993). The intensification of agriculture has created new or greater pest problems in a number of ways:

1. The concentration of a single plant species/variety in ever larger and more extensive monocultures increases its apparency to pests and the number of pest species which colonize it (Strong *et al.*, 1984).
2. Generally, high yielding crop cultivars can provide improved conditions for pest colonization, spread and rapid growth.
3. Reductions of natural enemies around crops means that natural enemies of pests must come to the crop from increasingly small and more distant non-crop reservoirs, entering crops too late or in too little numbers to prevent pest outbreaks.
4. Intensification results in a reduction of intervals between plantings of the same crop, or overlap of crops, which provides a continuous resource to pests.
5. The search for better cultivars and accelerated movement of plant material around the world and with it the movement of pests. Plant breeders, commercial importers, distributors of food aid and general commerce inadvertently introduce pest species.
6. Virtual reliance on chemicals leading to an increase in pest problems particularly for insects.

One of the factors influencing the increased introduction of exotic insect species has been the increase in importation of foreign products and materials. In the absence of their normal natural enemy complex or of environmental constraints these introduced species may become pests and cause extensive damage to crops or livestock. Pests may also be transported to countries in which they are not indigenous by the introduction of new crop types or animal breeds, with similar results. In general, it is such changes in agricultural practices as the introduction of new crop species or enlargement and aggregation of fields, use of monocrops and plant density, that have been held responsible for causing many pest problems. Risch (1987) cited:

1. The changes in crop cultivars relative to those of wild relatives.
2. The simplification of agroecosystems compared with natural ecosystems as being the most important contributing factors.

The former occurred because the successful control of insect pests with insecticides in the 1950s and 1960s allowed plant breeders to concentrate on developing new, high yielding cultivars in the safe knowledge that the insects were taken care of cheaply and effectively with insecticides. Hence, the breeders focused their attention more on attaining outstanding yields than on producing insect resistant cultivars (Ferro, 1987). Insect outbreaks then occurred on cultivars having no natural levels of insect resistance when the insecticide umbrella was removed (brought about by insects resistant to insecticides).

The second important contributing factor, the reduction of diversity in large crop monocultures, has long been associated with reasons for pest outbreaks. The reasons for this are that a monocrop is thought to provide a highly suitable habitat for a pest, but a highly unfavourable one for the pest's natural enemies, thus creating conditions appropriate for outbreaks. However, more recently it has generally been recognized that outbreaks are not an inevitability of such trophic simplicity (Redfearn and Pimm, 1987), an idea long recognized in forest entomology. Monocultures do occur in natural ecosystems, e.g. bracken, heather or natural forest. The establishment of species in monocrop plantations is certainly not a radical departure from this and should not automatically make them more vulnerable to pests (Speight and Wainhouse, 1989). In forest systems in particular it has been shown that cyclical pest outbreaks can occur as a consequence of natural changes in the physiological condition of the host, with weather often playing an important role (Berryman, 1987; Speight and Wainhouse, 1989).

Insect outbreaks, especially of migratory pests, are often associated with particular weather patterns, e.g. outbreaks of the desert locust and *Spodoptera* spp. The weather can also directly affect population development, if temperatures are favourable for population growth at an appropriate period during the insect's life cycle then outbreaks can occur, e.g. mild winters in the UK are associated with outbreaks of cereal aphids. Weather can produce a differential development of pests and their natural enemies causing a decoupling of their association and thereby permitting an unregulated pest population increase.

In effect, the goals of agricultural intensification are being undermined by pest problems that are now inadvertently the result of that very process. Sustainable agriculture and hence, sustainable pest management required finding solutions to these pest problems which protect the goals of intensification (Thomas and Waage, 1996).

Insect outbreaks can be triggered through intervention by man, i.e. through the use of insecticides, irrigation, fertilizer or cultivars lacking resistance to insects. Of these, it is insecticides that have had the most widespread influence on insect pest outbreaks. They can be an indirect cause of insect pest outbreaks by a number of means including reduction of natural enemies, removal of competitive species and secondary pest outbreaks, and

through the development of insecticide resistant insects. The use of broad spectrum insecticides (which is now on the decline) can cause the destruction of both pest and natural enemy populations, but the ability of the pest to then rebound in the absence of the natural enemy can lead to an outbreak. Secondary pests occur when an insecticide differentially affects the major pest relative to minor ones. With the subsequent destruction of the major pest, competition is removed, allowing the minor pests to exploit more effectively the resource, and an outbreak ensues. The situation where insect outbreaks occur because of the development of insecticide resistance is well known. Insecticides that are given repeated widespread application create a situation in which there is intense selection pressure for resistant individuals. These individuals can then proliferate in the insecticide treated area and cause subsequent outbreaks.

In addition to all of these factors the changes in consumer requirements for insect free produce in the USA and Europe have been one driving force behind the need to control insects at lower densities than before. Such shifts in consumer standards have meant that some insects are controlled beyond the point at which they are causing physical losses to the crop or animal. More recently demand for 'organic' or 'green' products has created a market at the other end of the spectrum. These are items of food, etc. that have been produced using only environmentally friendly techniques, excluding the use of chemicals such as insecticides or herbicides. With premium prices being paid for such 'green' or 'blemish-free' products then economics or aesthetics determine the level of infestation at which an insect is regarded as a pest.

Thus, consumerism and the needs of the consumer influence pest management practices. There are a whole range of stakeholders as well as consumers who influence pest management, how it has developed in the past and how it will develop in the future.

1.3 The Stakeholders in Pest Management

Stakeholders are individuals and groups of individuals who have a vested interest in a particular issue, cause or enterprise. Their expectations are built on past experiences, assumptions and beliefs and will reflect specific organizational structures (Collins, 1994). Within pest management there are numerous stakeholders who can include, for instance, shareholders, managers, employees, suppliers, customers and communities who are all linked to different degrees to a commercial company that produces a chemical insecticide. The various stakeholders, both between and within groups, can be placed in a hierarchy where the 'stake' each has in the enterprise is similar and influences the stakeholder group in the tiers below it. In pest management the first tier stakeholders tend to be governments and international agencies, the second tier pest management scientists and extensionists, the third commercial companies, the fourth farmers and growers and the fifth tier is the consumers, customers and communities that are the beneficiaries (or otherwise) of the decisions made and implemented by the tiers above. The lower tier stakeholders may also exert some influence over those stakeholders 'above' them. In order to understand pest management, how it has reached where it is today and how to identify where it is going in the future it is necessary to understand something of the motives and interests of each stakeholder group.

1.3.1 Governments, politics and funding agencies

The stakeholders in government need to establish policies that will work to the benefit of all other stakeholders. Since other stakeholders in pest management are so diverse in their interests and needs, governments perform a balancing act, selecting policies that will often reflect a compromise so that no one group of lower tier stakeholders wholly benefits and none lose out completely. Governments are responsible for funding public research interests in its institutes and universities; they are also responsible for ensuring that commercial companies generate new products which need to be manufactured, employing people, generating wealth, paying taxes. Governments also need to ensure that food is available at an appropriate price and quality and that the means by which it is generated does not degrade our environment. Hence stakeholders in government seek to look after the interests of farmers, consumers and communities, through funding research and supporting industry where possible.

During the industrialization of agriculture at the end of the 19th century the government policies sought to provide cheap food for a growing urban population, hence agriculture was targeted for investment and support. However, this process tended to benefit larger farmers through the economies of scale and reduced labour costs from mechanization, at the expense of small holders. The arrival of pesticides was greeted with euphoria since it allowed far more control of the system and assisted the process of agricultural industrialization (Perkins, 1982; Morse and Buhler, 1997). Government policies including tax incentives made the production and use of pesticides attractive to business and farmers. Pesticide development, production and use became institutionalized and farmers became increasingly dependent (Zalom, 1993). For many years after World War II governments needed to demonstrate the availability of surplus food as a political tool. Even though in the late 1950s and early 1960s the problems with pesticides became increasingly apparent, politically agricultural production had to be maintained. Rich farmers, the agrochemical industry and food consumers all wanted the benefits of the system in spite of the public concern over the environmental impact (Morse and Buhler, 1997). The problem for the stakeholders and government was how to maintain production, protect the environment and maintain an economically viable agribusiness sector.

However, with increasingly obvious environmental concerns raised about pesticides, the increased awareness of the general public of these issues, they became politically more important. Hence, a change of policy which allowed for environmental issues to be addressed but that did not impact too heavily on the agribusiness was required. The solution to the problem has been IPM.

1.3.2 The research scientists

The development of an IPM programme requires detailed knowledge of an agroecosystem, its component parts and how they interact; pest management is knowledge intensive. Generation of this knowledge is the job of scientists. It has been argued that IPM is the creation of scientists, and it is scientists who have largely controlled its evolution, albeit subject to various pressures (Morse and Buhler, 1997), and hence, the IPM approach should be seen primarily in terms of their desires and agenda.

It has been argued, often with justification, that IPM and research into IPM are not driven by principles but by the need to solve emerging problems. Certainly this would often seem to be the case. IPM was itself developed in response to emerging problems, mainly those associated with the misuse of insecticides, and alternatives and solutions to many pest problems are continually being sought. The principles of IPM are rarely applied because few scientists take on a management programme at the level at which they can be applied. Usually, because of human nature, a failing strategy will not be replaced by a new one (IPM) but different ways of adapting the existing strategy will be tried in order to minimize the extent of the failure. Hence, much of the research in IPM has been a response to a changing problem simply because scientists are usually reluctant to admit failure or to give up on an approach or idea. It is often much easier to advocate the need for a new, 'state of the science' technique with which to shore-up a failing approach, than it is to admit failure and start again.

There is a tendency within IPM to develop and employ control measures that can provide only a short-term solution to the problem. Such measures can work effectively only over a limited time scale because as soon as their use becomes widespread pests will adapt them and render them useless, e.g. prolonged use of a single insecticide, vertical resistance in crop plants and the use of genetically engineered crop plants. All of these examples provide only short-term answers to pest problems, but each also produces a research/development treadmill from which there is no escape. This may provide work for researchers and short-term economic gains for commercial companies, but it does not ultimately solve pest problems, or contribute a great deal to the development of sound insect pest management strategies. Despite this, many techniques gain acceptance within the framework of IPM because each new technique could be claimed to provide another weapon to be added to the pest control armoury. Diversity is fundamental to pest management, hence any technique can be justified and proclaimed under the IPM banner.

Control measures usually pass through a period during which they represent the 'new', 'in vogue' approach. The development of transgenic crop plants is one such example currently receiving a great deal of interest and, of course, funding. While it cannot be denied that transgenic crops will have a major influence on crop protection in the future there is also the serious possibility that insects will quickly develop means of circumventing engineered resistance mechanisms. It should be remembered that we have been through this situation before, where a product held immense promise and received extensive funding, only to find after years of research that the approach was applicable only to certain situations, e.g. male sterile techniques, mass trapping with pheromones, juvenile hormones and monitoring with pheromones.

The changes in seasonal abundance of a pest are easily described but much less easily explained. The research commitment required to describe a process and that necessary to explain it involve different orders of magnitude. Processes that influence population growth are rarely easily explained. Often, the more research that is carried out, the more questions arise, and the complexity of the problem increases. The understanding that is central to the philosophy of IPM necessitates an in-depth enquiry by scientists into the complexities and subtleties of insect biology and ecology. Such studies can easily become further and further removed from the original question that prompted the research. While it is necessary to study a subject in detail in order to understand how it functions, there is a danger that the research can get so far removed from its original objectives that the final results are inapplicable.

In each case, whether there is the need to shore up a failing strategy, the development of new solutions to pest problems (often products for the agribusinesses which are unsustainable when in widespread use), the need for a greater understanding of a particular pest/crop system, it is scientists who benefit, it is science that is required to provide the answers. All this is in the context of a sustained, consistent erosion of the base budget for agricultural research and extension over the last 20 years in the US (Zalom, 1993) and in the UK (Lewis, 1998). Hence, there has been tremendous pressure on research budgets but the public sector scientists have managed to diversify the number of options studied under the umbrella of IPM. The downside has been that with dwindling resources funding agencies at national and international levels have called for collaborative multidisciplinary research programmes where the idea has been to make more effective use of limited resources (Dent, 1992). In doing so, however, sponsors have taken little account of the constraints of the specialist nature of research in relation to developing multidisciplinary programmes, the appraisal and reward systems (Zalom, 1993), organizational structures and management systems (Dent, 1995), all of which are geared towards specialism and not interdisciplinary approaches. Despite the obvious role for interdisciplinary research in integrating control measures at a research level the statement made by Pimentel in 1985 still remains largely true today that: 'most remain *ad hoc* efforts by individual pest control specialists, each developing so-called integrated pest management programmes independently of one another'.

1.3.3 Commercial companies

Commercial enterprises generate income through the provision of services, products or a combination of the two. Within agribusiness there is a greater emphasis on manufacturing and sale of products rather than the service side of the industry. Growers expect to budget for tangible items such as machinery, pesticides and fertilizer but the concept of purchasing, for example advice, is less acceptable (Zalom, 1993). Hence, product inputs tend to dominate agribusiness in general and in pest management control products have gained in importance since the turn of this century. Whereas chemical pesticides were the predominate type of control product in the 1960s, since that time there has been a proliferation of different types of pest management products including: monitoring devices (e.g. insect traps); biopesticides (e.g. Bt); semiochemicals (e.g. for mating disruption); insect parasitoids (e.g. *Trichogramma* spp.) and predators (e.g. *Chrysoperla carnea*) and most recently genetically manipulated crop plants (e.g. Bt cotton). All of these products are purchased as off-farm inputs which generate an income for the commercial company and reduce the risk for the farmer. Some products are more successful in this than others so that a number of products service large, generally international markets while others meet the demands for smaller, more local and specific markets. In general, however, commercial companies are searching for products which have a range of characteristics (Box 1.1) that will ensure their uptake and sustained use over many years and provide an adequate return of the investment necessary to develop, produce, market and sell them. Hence, provided a product life cycle is sufficiently long to generate a suitable return on the investment, then a company will have achieved its objective. There is no need for a 'sustainable' product *per se*; all income generated above the required return is a bonus, which is why the concept of a 'product treadmill' is not such an anathema to the commercial stakeholders, as perhaps it is to others. Commercial companies are not in the business of alleviating the world's pest problems, but rather providing solutions that will generate a viable income and maintain the longer term prospects of the individual companies. The pest control business is worth billions of dollars worldwide each year, its presence influences the whole philosophy of pest management, continually driving for new technologies which can be sold as off-farm inputs, feeding and maintaining the demand for 'its' products. The commercial company stakeholders are major players in pest management affecting agricultural policy, R&D and farmers' expectations and needs. The wealth and taxes, the employment and the assurance they generate provide a powerful incentive for their continued role in the future.

1.3.4 Farmers and growers

Users of insect pest control technologies are not a uniform group. Even among farmers and growers there is a tremendous diversity, so that pest management means different things to different individuals, farmer groups and communities. The awareness and level of understanding of IPM among growers in the UK differs among the horticultural and agricultural arable sectors (Bradshaw *et al.*, 1996). This is not surprising since IPM has been practised in the horticultural sector for over 20 years, when it was introduced due to problems with insecticide resistance. The same types of problems have not occurred in the agricultural sector and hence IPM has been slower to catch on.

Farmers' objectives may vary. They may, for example, be interested in the maximization of profit or alternatively the minimization of risk (Zadoks, 1991). In subsistence farming food security will be the primary objective whereas reduction of labour may be secondary. In general a hierarchy of objectives may be expected (Norton and Mumford, 1983). A grower's perception of risk can be a function of loan and contractual commitments, which may require them to follow 'prudent' practices to meet yield or quality objectives (van den Bosch, 1978; Zalom, 1993). The practices they adopt for control of insects may also be market driven and if that means reducing pesticide residues through adopting the principles of IPM then this is what the farmers will do. Farmers will adopt practices that are expedient in the face of risk, are available and secure them a satisfactory livelihood.

Farmers have often been viewed as passive recipients of pest management technologies, however, this view is changing and farmers tend now to be seen as an integral part of the pest management stakeholder network, with a role in defining pest control needs, evaluating their effectiveness and influencing their wider adoption. Farmers, more than any other group are sensitive to customer needs and the more competitive and intensive farming becomes the more consumers will dictate the pest control practices adopted by farmers.

Box 1.1. Characteristics of products sought by commercial pest control product companies (from Dent, 1993).

Commercial value
Broad spectrum effects
Generally applicable
Easily marketable
High performance
Reliable
Visibly effective
Low hazard and/or toxicity to humans

1.3.5 Customers and consumers

Consumers in developed countries have increasingly high expectations concerning food quality. It is now unacceptable for insects, their body parts or frass to be found on fresh produce or packaged products. In addition, there is increasing concern about pesticide residues on food despite, paradoxically, the fact that the use of pesticides provided the means by which pest free produce first became possible, and now continues to make high cosmetic standards realizable. High quality standards have largely been imposed by government agencies on producers, processors, packers and retailers in response to consumer concerns (Zalom, 1993). It will be the need to maintain consumer confidence in the food industry that will continue to drive other stakeholders to invest in 'safe' technologies. This approach is being mirrored in developing countries wherever they serve developed country markets that demand high quality standards.

The views of the general public on the perceived environmental hazard posed by some pest control measures, particularly pesticides, continues to have an impact on pest management at all stakeholder levels. The concerns first expressed in *Silent Spring* have been maintained in the public arena by vociferous groups committed to environmentalism. These groups which initially campaigned successfully to maintain a high profile on the problems with pesticide use are now equally vigilant and vocal concerning the potential hazard posed by genetically manipulated crop plants. Public concern may yet significantly influence the widespread use of these and other novel control measures.

1.3.6 Balancing costs and benefits

Pest management that may be good for individual farmers is not necessarily good for farmers in aggregate, or society at large (Mumford and Norton, 1991). Different conflicts of interest occur between different segments of society and hence value judgements must be made about the relative claims of consumers, farmers, commercial companies and scientists. The responsibility ultimately lies with politicians and hence on political judgements rather than economic, scientific, environmental facts. Policy makers face difficult decisions on what constitutes success in a pest management policy. The traditional view that what is good for individual farmers must be good for society (Perkins, 1982) no longer holds true. The political importance of farmers is waning and that of the consumer increasing. The general public have an increasing scepticism about science and with concern for the environment, food safety and occupational health growing political issues, there will be an increasing number of conflicts of interests between consumers and scientists, agribusiness and farmers. Resolution of these conflicts through the development of acceptable pest management strategies is only likely to occur if the policy makers understand the objectives of each group and can work towards compromise policies that attempt to balance costs and benefits among all stakeholders (Mumford and Norton, 1991).

2

Sampling, Monitoring and Forecasting

2.1 Introduction

Monitoring in insect pest management can be used to determine the geographical distribution of pests or to assess the effectiveness of control measures, but in its widest sense monitoring is the process of measuring the variables required for the development and use of forecasts to predict pest outbreaks (Conway, 1984b). Such forecasts are an important component of pest management strategies because a warning of the timing and extent of pest attack can improve the efficiency of control measures. A forecast may take the form of a simple warning of when to spray insecticides, based on the occurrence of the pest, or a more complicated forecast by prediction (Hill and Waller, 1982). Inevitably the type of forecast that can be provided will depend on the insect pest and the effectiveness of the monitoring programme, but without reliable forecasts of outbreaks the outlook for judicial use of insecticides remains poor.

Central to any insect pest monitoring programme is the sampling technique that is used to measure changes in insect abundance. Although supplementary information about the insect's life history and the influence of weather may be needed to produce a pest forecast (Hill and Waller, 1982), the pest sampling techniques provide the basic estimate by which the state of the system is assessed. The estimate of pest abundance or change in numbers provides the essential measure by which control decisions are often made. Hence, it is important that the sampling technique used in any monitoring programme is appropriate and robust.

A great deal of research effort is directed at developing sampling techniques and devices, particularly insect traps, that may be used for monitoring. Given the amount of research that is carried out there are surprisingly few working monitoring systems that provide reliable forecasts. Research on insect traps provides an easy means by which large amounts of apparently useful data can be obtained; the continuing proliferation of papers on pheromone trapping provide ample evidence of this. A great deal of trap catch data has been collected and analysed for many insect species, and various trap designs, placed at different positions, using different lures and doses, but in general few scientists have seriously considered why and how these variables influence trap catch, or how traps may be improved on the basis of such an understanding. Often this is because such work requires painstakingly detailed experimentation or extensive direct observational studies (e.g. Vale, 1982a). Most insect traps provide only relative estimates (see below) and as such are particularly prone to changes in environmental conditions, with the consequence that data collected on different occasions are not strictly comparable unless conditions have remained static. This provides a major drawback in their use but, surprisingly, since the 1960s (Johnson, 1950; Taylor, 1962a, 1963) there have been few serious attempts to explore and evaluate the extent of this problem. Instead the effects have been either ignored or just described by linear regression analysis of trap catch against environmental variable. Insect traps provide the most popular form of monitoring device for entomologists, but if the development of monitoring and forecasting systems is to be given the serious consideration it deserves then there is a need for a move away from empirical field studies towards a more critical evaluation of trap development and a greater realism about the potential of traps to provide robust and reliable monitoring and forecasting systems (Wall, 1990).

One of the problems that has plagued the development of effective monitoring systems is that, initially, sampling often only takes place as part of a study of the ecology and biology of a pest insect. As data accrue there is an increased temptation to make as much use of the data as possible. The data may then be analysed to determine whether or not they can provide forecasts of pest abundance. If useful relationships are not immediately apparent then further studies may be conducted to obtain more information on factors affecting trap efficiency, such as the weather, crop phenology, moonlight, etc. If the relationships prove to be useful then a forecast may be devised, but all too often it is only at this late stage that the practicalities of implementation are considered. The objectives of a sampling programme must be clearly defined at the start and if the objectives are those other than the development of a monitoring and forecasting system, they may need to be carefully scrutinized and redefined at a later stage. It is only by carefully defining the objectives of the research that the extent of the complexity and difficulties of devising an appropriate monitoring and forecasting system can be appreciated and the research directed in a meaningful way. There is a great need with

monitoring to get away from the 'look and see' approach which has dominated the subject in the past.

A good understanding of sampling theory is essential for research on insect monitoring as well as every other component of insect pest management. The basic concepts of sampling theory are dealt with in this chapter simply because of their obvious direct relation to monitoring. Aspects of sampling relevant to other components of insect pest management are considered in appropriate chapters. The aim of the chapter is to provide both a basic framework of information on sampling, monitoring and forecasting and an insight into some of the problems confronting scientists working in this field.

2.2 Sampling

Reliable estimates of pest incidence must be based on samples that are representative of the range of insect abundance encountered in the area of interest. Bias, due to an individual's preference for certain sampling sites, should not be introduced. The influence of this kind of bias can be avoided by the use of an appropriate sampling pattern. The most commonly used patterns involve random and stratified random sampling. Systematic sampling patterns are less widely used and in general, data obtained in this way cannot be analysed statistically (Southwood, 1978).

2.2.1 Random, stratified random and systematic sampling

Random sampling involves selecting a number of samples from a population such that every sample has an equal chance of selection. The simplest method for field or tree crop studies is to base the sampling pattern on coordinates selected from random number tables.

Stratified random sampling differs from random sampling only in the division of the population into different strata from which random samples are then taken. The strata are subdivisions of the samples based on knowledge of the distribution of the population; for example, insects may exhibit different preferences for particular plant parts. The strata reflect real differences in population levels while each stratum consists of a more homogeneous subpopulation. The number of samples taken in each stratum may be varied according to the variance encountered within that stratum. Hence, stratified random sampling may produce a gain in precision and efficiency in the estimate of the size of the whole population. However, these advantages can be obtained only when the proportions of the population in the different strata are known. The sample size would be less than that indicated for a simple random sample only when the standard deviation within strata is known to be less than that for the population as a whole (Church, 1971).

Systematic sampling involves taking samples at fixed intervals. The size of the fixed interval and the reference or starting point for the intervals are chosen, within defined limits, from random number tables. The first sample is taken at the reference point and subsequent samples at successive intervals beyond that, e.g. for a fixed interval of 10 units and a reference unit of 15, samples would be taken at intervals of 15, 25, 25, 14 ..., etc.

The systematic sample and the stratified random sample patterns may appear similar but they differ in that the systematic sample units occur at the same position in each stratum, whereas with stratified random sampling the position in the stratum is determined at random (Cochran, 1977) (Fig. 2.1). The systematic sample is spread more evenly over the population, and in some circumstances this can mean that it is a more precise sampling method than random stratified sampling (Cochran, 1977). Researchers should always be aware that if an insect attack is distributed systematically then a systematic sampling pattern, which did not coincide with the pattern of distribution of the insect, may lead to error (Walker, 1981).

The sampling procedures used by research workers are usually carried out with due regard for statistical requirements with the aim of obtaining reliable and accurate population estimates. These estimates are then often used to develop monitoring and forecasting programmes for use by farmers and extension workers. A monitoring programme based on estimates derived from sampling techniques used by scientists that requires similar sample inputs from farmer or extension workers is unlikely to be appropriate. Farmers and extension workers cannot be expected to obtain a large number of random samples; sampling at this level needs to be adapted to the farmers' needs, i.e. the sampling method should be uncomplicated and not too intensive. This may mean adopting a pseudo-random or systematic sampling pattern. That is, a pattern that mimics randomization or systemization as closely as is practicable. This leaves scientists with two choices, they can either develop a monitoring programme using sampling patterns and techniques that are appropriate for farmers, or after developing a programme modify it for farmer use. To date, scientists more commonly adopt the latter approach but in the end leave it to the extension staff to modify the technique for farmer use!

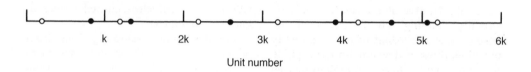

Fig. 2.1. Systematic (○) and stratified random (●) sampling within six sampling units (k) (after Cochran, 1977).

2.2.2 Sample size

There is a point in any sampling programme when a decision has to be made about the number of samples that should be used. Since the number of samples taken will influence the precision of the final estimate the importance of this should not be underestimated. Too few samples will reduce the value of the estimate (e.g. Vlug and Paul, 1986) and too many will increase the cost of the programme, where the cost may be measured in terms of time, labour, equipment or financial outlay (Blackshaw *et al.*, 1994). Occasionally, a compromise will have to be made between precision and sampling costs due to a lack of resources. On such occasions it may be appropriate to leave the sampling programme until adequate resources become available.

A decision about suitable sample sizes can only be made with some prior knowledge of the sampling variability likely to be encountered in the proposed programme. Such information can be obtained from evaluation of similar experiments and/or by carrying out a sampling pre-test. In reality the former is rarely possible, not because they are technically difficulty but because the information on which they rely is seldom available (Perry, 1997). However, given the importance of such estimates for experiments or decision making, the required level of precision can be estimated by use of a pre-test which determines the level of sample variance in the proposed experiment and permits the most appropriate sample size to be defined.

The precision required, and hence the amount of error that can be tolerated in the sample estimate, will depend on the use to which the estimates will be put. The generally accepted maximum level of error is 5% of the mean value, but the error should be chosen according to the likely degree of separation of events or treatments and the required certainty of differentiation between them. For instance, a trap catch population estimate with a 25% error would only allow a distinction between a doubling or halving of the population size.

The number of samples chosen has a great impact on the precision of the estimate obtained but large increases in sample size cause only small decreases in error.

The formula for the number of samples required when dealing with a sample of proportions (e.g. the proportion of plants or plant parts infested by insects) assumes random sampling and an *a priori* knowledge of the value P, the probability of occurrence. For example, approximately 30% of the plants in a pre-test were infested with a pest insect, hence $P = 0.30$. The number of samples (n) that needs to be taken, assuming only a 5% error (c, expressed as 0.05) is:

$$n = \frac{t^2 pq}{c^2} \qquad (2.1)$$

where:
t = 'Student's t' from statistical tables, for c = 0.05 is 1.96
$q = 1 - p$
p = the proportion of uninfested plants or plant parts.

The formula for determining 'n' for continuous data assumes that simple random sampling is used. A pre-test is needed to determine the values for the standard deviation of the sample (s) and the mean of the sample estimate (y).

$$n = \frac{s^2}{cy} \qquad (2.2)$$

where:
c = the pre-determined level of error, as above.

Direct counts of insects on crop plants provide accurate estimates of insect numbers, provided a large enough sample size is used. However, direct counting may become prohibitively time consuming if the pest is abundant or patchily distributed (Ward *et al.*, 1985). In certain situations a measure of 'incidence levels' (any number > 0 or proportion of plant parts or plants infested by the insect) could be used instead of direct counts provided the incidence measures were accurate enough.

Since sample size has a large influence over the accuracy of an estimate, the size of sample of an incidence count required to give an equivalent accuracy to direct counts can be determined. Ward *et al.* (1985) considered this problem and used two different models for determining sample size with incidence counts.

1. The Probit model

$$\log \mu = a + b \cdot \text{probit}(P) \qquad (2.3)$$

where:
μ = the mean density
P = the proportion of plants or plant parts infected
a and b are the intercept and slope obtained from regression analysis.

2. The Nachman model

$$\log \mu = a + b \cdot \log\left[\ln \frac{1}{1-P}\right] \qquad (2.4)$$

where:
ln = the natural logarithm
log = the logarithm to base 10.

The required sample size in each case can be given by:

1. For the Probit model, where c is the required sample accuracy and z is the ordinate of the normal curve (from statistical tables):

$$n > \frac{b^2 \cdot P \cdot (1-P)(\ln_{10})^2}{c^2 \cdot z^2} \qquad (2.5)$$

2. For the Nachman model:

$$n > \frac{b^2 \cdot P}{c^2[\ln(1-P)]^2(1-P)} \qquad (2.6)$$

The number of samples for the direct counts can be estimated as above or from the formula for Taylor's Power Law (Taylor, 1961) which gives:

$$n > \frac{a \cdot \mu^{b-2}}{c^2} \qquad (2.7)$$

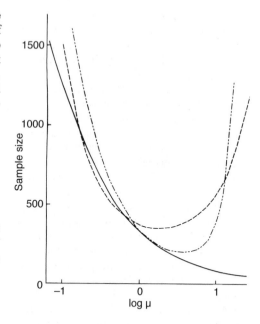

Fig. 2.2. Minimum sample sizes for population estimates (μ) with a required level of accuracy of CV = 0.1; using direct counts (——) and incidence counts assuming the Probit (– – – –) and Nachman (–··–··) models (after Ward *et al.*, 1985).

where:
a = a sampling factor
b = index of aggregation.

Ward *et al.* (1985) compared the different models using incidence counts and mean densities of cereal aphids on winter wheat (Fig. 2.2). At very low densities (μ = 0.1 aphids/tiller) n was very large for both incidence and direct counts. As the density increased, the sample size for direct counts declined steadily whereas the sample size from both incidence models declined but then increased again. At mean densities of less than one aphid per tiller there is little difference in the sample sizes required for incidence and direct counts, but at mean densities between 1 and 10 aphids per tiller, then up to three times the number of samples need to be taken to give an equivalent accuracy of incidence counts compared with direct counts ($c = 0.1$ in this case). At higher densities the minimum sample size for incidence counts increases markedly, but because incidence counts are made easily and take little time, the use of a large sample size may more than compensate for the time it would take for direct counts at the same densities. Obviously in selecting a monitoring procedure attention should be paid to both the accuracy required and the time involved in the sampling.

2.2.3 Sample independence and interaction

One of the fundamental principles of sampling theory is that the measurement of units in one sample does not influence or bias the estimate of a second sample, so that the samples remain independent. Relative estimates are particularly prone to problems associated with sample interaction. The size of the 'active area' (e.g. the distance for which the pheromone is observed to initiate a response) of a trap is often variable and difficult to measure. Sampling involving a number of traps may be subject to unquantified interaction effects if they are not spaced at a distance sufficient to prevent interaction (van der Kraan and van Deventer, 1982). The interaction may increase or decrease the size of a trap's catch above that of a solitary unbiased trap. Studies of pheromone trap catches of *Cydia nigricana* have shown that when interaction does occur, both the number and the position in a line of traps affect the size of each trap catch, and that these interactions are further modified by the orientation of the line to the mean wind direction (Perry *et al.*, 1980). Pheromone traps are difficult to work with in this context because their active areas are suspected to be large (e.g. 300 m for *Trichoplusia ni*; Kishaba *et al.*, 1970; Toba *et al.*, 1970) and particularly susceptible to the influence of the direction and speed of the wind.

Case Study: The effect of trap spacing on trap catch of *Erioishia brassicae* (Finch and Skinner, 1974)

A study of the influence of trap spacing on catch in various types of water trap was carried out with the cabbage root fly (now *Delia radicum*). Finch and Skinner (1974) considered the influence of different trap densities (one, four, nine, 16 and 25 traps) on the mean catch of four types of trap: non-fluorescent traps; fluorescent traps; water traps and allyl isothiocyanate (ANCS) traps (Fig. 2.3). The mean catch of traps that are not interacting should remain the same irrespective of trap density (the non-fluorescent water trap, Fig. 2.3) since each trap performance is unimpaired by other traps. Previous experiments had shown that the non-fluorescent water traps were only attractive at a range of 0.75 m. At the closest spacing in this experiment of 2.2 m there was no overlapping of trap active areas and trap catch did not increase with increasing distance between the traps. Some overlapping was apparent with the fluorescent yellow trap at trap spacings below 5 m (Fig. 2.3) and indicated a trap active distance of approximately 2.5 m. The mean trap catches of the two traps using the attractant, ANCS, were still increasing at the widest trap spacing of 8 m (Fig. 2.3) indicating that there was still some trap active area overlap. The attractant increased the active area of the traps, and hence they would need to be spaced further than 8 m apart before they could even be considered as acting as independent samples.

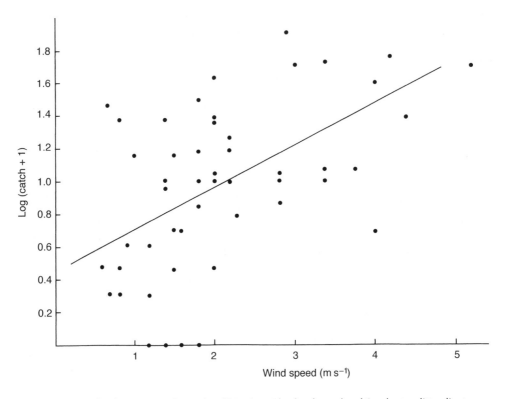

Fig. 2.5. Relationship between wind speed and hourly catch of male moths of *Spodoptera littoralis* at a pheromone trap during the period of peak catches (1800–2200 h), 21 October to 3 November 1972 (after Campion *et al.*, 1974).

Traps are one of two types, those that:

1. Influence an insect's behaviour in order to attract it to the trap, e.g. light and pheromone traps.
2. Those that have some form of mechanical device for catching insects, either passively, e.g. Malaise traps and pitfall traps or actively, e.g. suction traps.

Each trap can also be described by the area around it from which it samples the insects. This area may be referred to either as an active area or a capture area, depending on the type of trap and its mode of action. The active area is the area around a trap within which the insects may respond adequately to the trap stimulus and be attracted towards it (Nakamura, 1976; Shorey, 1976; van der Kraan and van Deventer, 1982), e.g. the area of a pheromone plume. The capture area is the area around the trap towards which it is probable that the insect will orient and move into the active area, and thence to the trap. A pheromone trap and light trap have both active and capture areas, with the capture area always slightly larger than the active area, whereas a Malaise trap that does not attract insects has a very small active area, the size of the trap aperture. The capture area alone represents the area from which these traps sample.

It is a characteristic of most traps that the size of the active and/or capture area varies according to the environmental conditions. This can mean that trap catches are so variable that it becomes difficult to interpret the trap data. In order to better

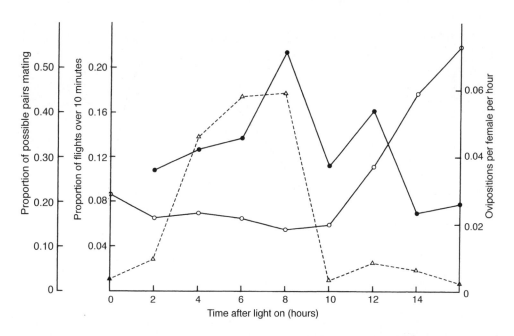

Fig. 2.6. Daily periodicities in flight (●—●), mating (○—○) and oviposition (△– – –△) in *Oncopeltus fasciatus* (after Dingle, 1972).

understand trap catch data a great deal of research has been carried out in an attempt to quantify the effects of various factors on trap catch: for instance, moonlight on light trap catches (Bowden, 1973, 1982; Bowden and Church, 1973; Bowden and Morris, 1975) (Fig. 2.4), wind speed on pheromone trap catch (Campion *et al.*, 1974; Lewis and Macauley, 1976) (Fig. 2.5), wind direction on sticky trap catch (Schoneveld and Ester, 1994) or a combination of environmental factors on light and pheromone traps (Dent and Pawar, 1988). However, very few studies have actually managed to define the limitations of a particular trapping technique or to provide correction factors in order to improve the precision and reliability of the results (Taylor, 1962a,b; Bowden and Morris, 1975).

Relative and absolute measures are both affected by changes in specific behavioural or physiological states of insects. This effect is referred to as an insect phase effect (Southwood, 1978). For example,

attractant traps depend for their effectiveness on the response of the insect to a particular trap stimulus. The insect may only be receptive to the stimulus at a specific age on specific behavioural or physiological states during its life cycle. Similarly, non-attractant traps may depend on insect activity that occurs only at specific developmental stages. When an insect is susceptible to a trap stimulus or it is in a particular mode of activity that makes it susceptible to capture then its behavioural or physiological state is considered to be in phase with the trap (Southwood, 1978).

Many insects exhibit circadian rhythms of activity which may include a daily sequence of feeding, mating and oviposition (Fig. 2.6). A specific mode of activity may make an insect more susceptible to capture than another mode, e.g. pheromone traps elicit a sexually appetitive response from insects, a response that may only be elicited during the period of a day when an insect would normally respond to the stimulus.

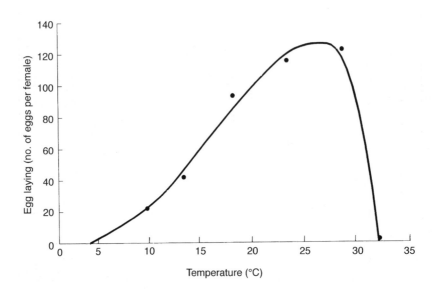

Fig. 2.7. Total number of eggs laid by *Encarsia tricolor* females from days 3 to 20 after adult emergence at different constant temperatures (after Artigues *et al.*, 1992).

2.3.3 Natality

Natality is the rate at which new individuals are added to the population by reproduction. The vast majority of studies addressing reproduction in insects deal with fecundity rather than fertility due to the difficulties of measuring the latter (Barlow, 1961). Fecundity is the total number of eggs produced or laid during the lifetime of the female insect (Jervis and Copeland, 1996). A range of biotic and abiotic factors influence insect fecundity. Among the biotic factors there are intrinsic factors such as insect size or clone and extrinsic factors such as host plant effects which may include plant species, cultivar

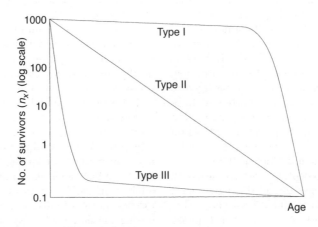

Fig. 2.8. Hypothetical survivorship curves (after Pearl, 1928).

Table 2.1. Ways of expressing population change and age-specific mortality when stages of the insect do not overlap (from Varley et al., 1973).

Line	Eggs (N_E)		Small larvae (N_{L1})		Large larvae (N_{L2})		Pupae (N_P)	Adults (N_A)	
1 Population	1000		100		50		20	10	
2 Number dying in interval	900	+	50	+	30	+	10		Sum: 990 dead
3 % mortality	90	+	5	+	3	+	1		Sum: 99% mortality
4 Successive % mortality	90		50		60		50		
5 Successive % survival	10		50		40		50		
6 Fraction surviving	0.1	×	0.5	×	0.4	×	0.5		Product = 0.01 survival
7 Log population	3.0		2.0		1.7		1.3	1.0	
8 k-value	1.0	+	0.3	+	0.4	+	0.3		Sum k = 2.0

or growth stage (e.g. Leather et al., 1985; Soroka and Mackay, 1991). The main abiotic factor influencing fecundity in insects is temperature. The number of offspring produced increases with temperature up to a critical value above which fecundity is reduced and declines to zero (Fig. 2.7). The nature of the relationship between temperature and fecundity is rarely linear, but if it can be defined then it can prove useful in prediction of population growth.

2.3.4 Development and growth

Insect growth refers to the increases in size and weight of an insect as it passes through different developmental stages from egg, larva/nymph, pupa to adult. The total development time of an insect is defined as the period between birth and the production of first offspring by the adult female (Bonnemaison, 1951). Development times can also be determined for the period between each instar, for pupal development and for a preoviposition period. The rate of development of insect eggs and pupae is strongly influenced by temperature while the development of larval or nymphal stages is dependent on both abiotic (particularly temperature; see Section 2.5.2) and biotic

(particularly host plant and/or diet) factors. Humidity and photoperiod have also been shown to influence insect development (Subramanyan and Hagstrum, 1991; Brodsgaard, 1994). Insect diet, host plant species, cultivar and growth stage can all have an influence on insect development times (Dent and Wratten, 1986; Yang et al., 1994). Host/diet quality can be influenced by temperature and humidity and thereby have an indirect impact on development (Subramanyan and Hagstrum, 1991).

2.3.5 Survival and mortality

Mortality can be expressed in a number of ways (e.g. percentage mortality, fraction surviving; Table 2.1) or graphically as survivorship curves (Fig. 2.8). Such survivorship curves are useful as a means of depicting theoretical forms of mortality in relation to age but they have little practical value. A number of mathematical functions have been used for this purpose, particularly the exponential model which assumes that insect mortality rates are independent of age (Clements and Paterson, 1981) and the Gompertz and Weibull functions which assume that the probability of dying in the next interval increases over time (Dent, 1997a; e.g.

Readshaw and Van Gerwen, 1983; Bartlett and Murray, 1986).

Mortality in insects can be measured either directly using mark–recapture experiments or indirectly by estimating the numbers of individuals in successive developmental stages. Mark–recapture experiments which involve various means of marking individuals, releasing them, and recapturing them after an interval, have been reviewed by Reynolds *et al.* (1997). The data generated by such experiments can be analysed by a number of different models each of which makes certain assumptions that affect their interpretation and value. For instance, the equation of Fisher and Ford (1947):

$$N_t = \frac{n_t a_i \theta_{i-t}}{r_{ti}} \qquad (2.8)$$

where:

N_t = the population size estimated
n_t = the total sample time at t
a_i = the total of marked insects released at time i
θ_{i-t} = the survival rate over the period, $i - t$
r_{ti} = the recapture at time t of insects marked at time i.

This method assumes that the sampling is random, survival rates and probabilities of capture are unaffected by marking, survival rates are independent of age and are approximately constant (Manly, 1974). In contrast to this the Manly and Parr (1968) equation takes account of age dependent mortality:

$$N_i = \frac{an_i}{r_i} \qquad (2.9)$$

where:

N_i = the population estimate on day i
n_i = the total number captured on day i
r_i = the total number of marked insects recaptured on day i
a = the total number of insects marked.

The techniques for measuring stage-frequency data, that is the survival rate and number entering each stage, are numerous (see Dent, 1997a). For an extensive review of this topic see Manly (1990).

The factors that cause mortality in insects can, of course, be natural (environmental, predators, parasitoids, age) or contrived by man for that purpose (chemicals, biopesticides, etc.). The most common abiotic factors affecting insect survival include temperature (e.g. Shanower *et al.*, 1993), humidity (e.g. Kfir, 1981), rainfall (e.g. Chang and Morimoto, 1988) and wind speed (dislodging insects from their host plants; e.g. Cannon, 1986). The influence of parasitism and predation on insect populations are considered in Chapter 6 and those of the host plant in Chapter 5 dealing with host plant resistance. In the latter, survival is usually measured as insect numbers or proportions of a population surviving to adults on different host plant species or cultivars (e.g. Bintcliffe and Wratten, 1982).

2.3.6 Migration

A generally accepted definition of migration is: 'persistent and straightened-out movement affected by the animal's own locomotory exertions or by its active embarkation on a vehicle (wind or water)'. It depends on some temporary inhibition of station-keeping responses, but promotes their eventual distribution and recurrence (Kennedy, 1985). Migration contrasts with on-going vegetative movements which include those for feeding, reproduction and accidental displacement. Evidence for migration of insects has been obtained by observation such as the conspicuous arrival of insects in places where they cannot be accounted for by local breeding (e.g. Pedgley *et al.*, 1995) or experimentally by a variety of techniques. These include various mark–recapture methods where marking may be achieved with paints and inks (e.g. Ives, 1981), dusts and dyes (e.g. Bennet *et al.*, 1981), and radio isotopes (Service, 1993). In addition, it has been possible to link insect specimens to their source location by natural markings (e.g. phenotypic variation of size and colour; Hill, 1993), their elemental composition, genetic markers, the presence of pollen and mites (e.g. Drake, 1990; Lingren *et al.*, 1994).

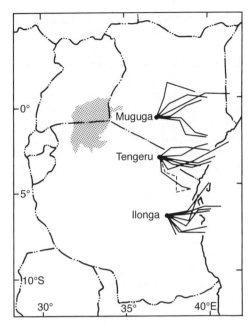

Fig. 2.9. Back-tracks calculated assuming downwind movement at wind speed plus 10 km h^{-1} on 1.5 nights for nights in December to March of 1973–74 and 1974–75 when there were sudden increases (as defined in the text) of *Spodoptera exempta* moths in the light-traps at Muguga, Tengeru and Ilonga, East Africa (after Tucker *et al.*, 1982).

The arrival of a migrating population of insects such as locusts can have a significant impact on the area in which they alight, however, quantifying migration itself, identifying when it will occur and in what direction is incredibly difficult. A number of techniques have been developed for this purpose which include at its most basic visual observation (e.g. Turchin *et al.*, 1991 who used a team of observers to quantify the movements of butterflies) to the more sophisticated use of video (e.g. Riley, 1993) and radar methods (e.g. Riley, 1974; Wolf *et al.*, 1993; Riley and Reynolds, 1995). The use of these methods and laboratory techniques (e.g. static tethering, flight mills and the use of wind tunnels) have been reviewed in Reynolds *et al.* (1997).

Weather has a very strong influence on insect migration, particularly atmospheric processes associated with wind fields. Wind speed decreases near to the ground due to the friction with the earth's surface. This layer of slower air speeds is known as the flight boundary layer (Taylor, 1974) and is where, when the wind speed is lower than the insects' flight speed, the insects can control the direction (track) of their flight. Above the flight boundary layer, wind speeds exceed the insects' flight speed and they will be carried downwind, irrespective of the direction that they attempt to fly.

Migration may occur on dominant winds or on temporary atmospheric disturbances. Generalized seasonal movement may be related to the prevailing wind fields (Tucker *et al.*, 1982) and produce characteristic seasonal patterns of infestation. A technique known as trajectory analysis can be used to back-track the movements of insects based on prevailing wind fields and synoptic maps of the wind field.

The synoptic wind field charts are prepared by meteorological stations and may be available for reference. A back-track can be constructed if a line segment is drawn up-wind from the trap position of the insect (using the wind, the flight speed × the flight time × map scale). The flight speed can be taken as either the wind speed alone or the wind speed plus the insect's flight speed. The flight time for each line segment should be divided into a number of periods that correspond to the times of the measurements used to prepare the synoptic charts. The charts are usually prepared every 6 hours and occasionally every 3 hours. A 12 hour flight period would use two or four different synoptic charts and hence produce two or four adjoining line segments (Fig. 2.9). A forward-track can be constructed in the same way with the line segments added in the up-wind direction.

Trajectory analyses of this kind have been undertaken for *Spodoptera exempta* (Tucker, 1984a), *Spodoptera frugiperda* and *Helicoverpa zea* (Westbrook *et al.*, 1995a, 1997). Trajectories can be assessed directly through the use of tetroons which are specially designed balloons that maintain a constant volume and drift passively at altitudes of constant air density. Tetroons released at the time of migration can act as surrogate markers tracked by ground based radar (Westbrook *et al.*, 1994, 1995b).

2.4 Monitoring Strategies and Objectives

Monitoring strategies can be roughly classified as:

1. Surveys.
2. Field based monitoring strategies.
3. Fixed position monitoring strategies.

The particular sampling technique used in each of these monitoring strategies

depends largely on the precise objectives of the proposed programme (Fig. 2.10).

2.4.1 Surveys, field based and fixed position monitoring

A survey may be carried out to study the distribution of a pest, or it may involve a study of both the distribution and the abundance of a pest species. The aim of a distribution survey is to locate and map the geographical distribution of a pest species. This may be used to assess the pest status of a particular insect, the spread of an introduced species or the spread of a mobile endemic pest that is extending its range.

Surveys measuring both the distribution and abundance of a pest can be used to assess the relative level of pest infestation and pest migration. A survey can identify areas of relatively high infestation and may show up seasonal patterns of occurrence in

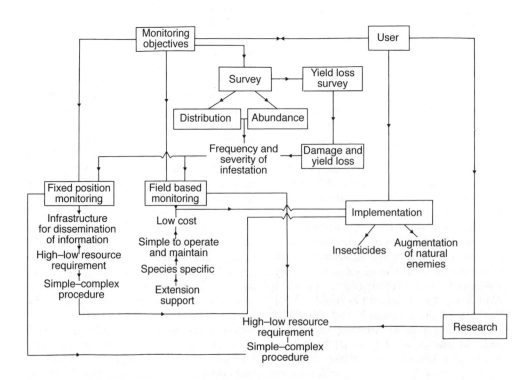

Fig. 2.10. The different monitoring strategies and factors affecting their use.

different locations. Such seasonal patterns may be related to differences in environmental conditions and may provide some understanding of factors influencing pest population dynamics. Levels of infestation and/or environmental factors in particular regions may be shown to be indicative of impending pest outbreaks and hence used in regional forecasts.

Survey techniques can also be used to establish the presence of the seasonal or sporadic movements/migration of some pest species. A knowledge of likely source populations and pest movement from these areas may have potential uses in pest forecasting. Pest migration may also be related to weather systems.

The aim of a field based monitoring strategy is to provide the farmer with a decision-making tool. The field based sampling technique, which could be an insect trap or a method of counting eggs or larval stages, is used to collect data on local pest population changes in a particular crop or field. This information is then used by the farmer to make decisions about the implementation of control measures. Preceding this, in research terms, field based monitoring may form part of a study of pest population ecology. Information gained in this way may then be used to produce a practical monitoring strategy on which management practices can be based.

Sequential monitoring is carried out with the objective of measuring the way in which pest numbers change over time. It can be done in different crops, different crop growth stages and over different seasons, and may identify the development times, mortality and generation times of pest populations under different conditions. The data may be used to determine rates of population development which in turn may be used to produce pest forecasts. The damage caused may be related to levels of infestation and used to determine damage thresholds.

Field based monitoring can also be carried out to follow the progress of population development up to a pre-defined number of insects or action threshold. This may be any number of insects from one, and may indicate that insecticide application is necessary, or it may be used in developmental models which predict the most appropriate time to apply insecticide.

Fixed position monitoring is usually restricted to research stations where a trap, e.g. a suction or light trap, is maintained in a fixed position and is used to sample insects over a number of seasons and years. The method can be used to identify the pest insects present at a given location, and to compare relative changes in pest numbers between seasons. It can provide a general overview of pest abundance and may be useful, if related to other pest estimates, for forecasting outbreaks on a regional basis.

2.4.2 Biology and nature of attack

The biology of a pest and the nature of its attack and damage it causes will influence the type of monitoring/forecasting strategy that is eventually adopted. The level of damage that can be tolerated, the number of insects that cause this and the rate of population increase are all important concerns. However, the most important biological criteria for determining which strategy to adopt will be the frequency and distribution of pest outbreaks.

Infrequent but widespread outbreaks would be most effectively dealt with using a centralized regional monitoring/forecasting strategy since this would not involve farmers and agricultural extension workers in a large amount of unnecessary sampling during non-outbreak years. Regional centres (normally research institutes) can continuously monitor pest population levels and issue warnings to farmers when an outbreak is predicted. Infrequent outbreaks may be caused by influxes of highly mobile migrant pests on weather fronts or sporadic disturbances, bringing insects from source areas that may be located in other regions, or countries. Forecasts of such outbreaks would depend on coordination and cooperation between monitoring

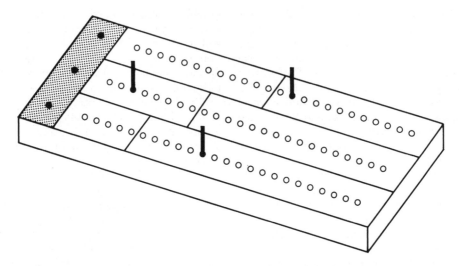

Fig. 2.11. A pegboard for counting insect numbers and producing a recommendation on insecticide application. The pegboard is divided into three parallel strips painted different colours, each of which is drilled with 24 holes. At one end is a black band with holes for pegs prior to use. One of the painted strips is used for counting the number of plants sampled on a diagonal transect across the field. When the line across the strip is reached the farmer starts sampling along the second diagonal. The other two strips are used for counting the number of eggs or larvae sampled for two insect species. The line across the strip indicates the action threshold for insecticide application (after Beeden, 1972).

centres and this would be best achieved through research institutes.

For pests that reach outbreak levels regularly each season, and for outbreak years of 'infrequent pests' then a field level monitoring/forecasting strategy, or at least a farm level strategy, will be most appropriate. Farm level monitoring will be satisfactory where there is little variability in infestation between fields or perhaps crops. Where infestation is highly variable in space then field by field monitoring would be most appropriate for optimizing insecticide use between fields, provided that the costs of extensive monitoring did not outweigh their usefulness.

In situations where pest outbreaks are infrequent and patchily distributed then the chances of developing any form of monitoring/forecasting strategy is remote. In such cases the level of damage and hence the economics of crop production would be critical factors in assessing the potential value of a monitoring pro-

gramme. Where outbreaks were frequent but patchily distributed across a region then a farm level monitoring/forecasting strategy or a combination of a farm and regional centre strategy would be most appropriate.

2.4.3 Availability and suitability of monitoring techniques

Monitoring techniques fall into two groups: those that are only suitable for use at research centres (where there are the necessary resources available) and those potentially suitable for use by farmers. Monitoring techniques appropriate only to research centres are techniques such as light and suction traps that require an electrical supply, or other traps that also require trained personnel to sort and identify the insects caught. Insect-specific traps and crop counts of insect stages that are easily identified represent potential monitoring techniques for use by farmers, provided that the technique (which will primarily have been used by scientists developing the monitoring and forecasting scheme) can be modified appropriately. Monitoring techniques to be used by farmers must be simple to operate or carry out and the data easily interpreted, traps must cost little to run and be readily maintained. Monitoring and forecasting methods that involve the graphing and charting of the progress of infestation are to be avoided unless there are opportunities for extensive farmer training.

Techniques that require training and supervision stretch the resources of extension workers. It cannot be stressed enough that the monitoring must be appropriate for the situation to which it is to be applied. Farmers in developing countries may be illiterate and/or enumerate so that even simple counts of insect numbers may provide a problem. Beeden (1972) devised a simple pegboard for counting insect numbers and for making an insecticide recommendation, a method which involves no reading or writing (Fig. 2.11). The pegboard provides a simple practical application of a quite sophisticated monitoring and decision making technique that is appropriate for farmers (Beeden, 1972); it represents a simple application that has potential for development and extensive use in insect pest management.

The distribution of an actual or potential insect pest species may be determined through the use of an appropriate survey programme. The sampling technique used in a survey programme will depend on the biology and life cycle of the insect in question. Since the distribution of an insect species is usually dependent on the mobile winged stage of the life cycle, various aerial trapping techniques are often used in distribution surveys. Traps or other techniques that have a high degree of pest specificity and can be readily checked and maintained are helpful in reducing the costs of sample surveys. Pheromone traps are particularly useful in this respect and have been used successfully in monitoring the spread of a number of pests in Europe (e.g. Bathan and Glas, 1982).

The survey should adequately cover the area of interest and individual sample locations should be chosen on the basis of what is known about insect host preference and general biology. The data collected from a trap network or a field count distribution survey can often be used to determine the relative levels of abundance in different locations. Such surveys may provide some understanding of the driving variables behind regional population changes.

The amount and type of information that can be gained from surveys carried out at different locations often depends on the type of supplementary data that is collected or available. On their own, data of relative levels of infestation can provide information about seasonal changes in abundance in different locations and may identify areas and time of high infestation. However, this information has limited value unless it can be related to other variables.

If the sampling technique used is a trap that catches mobile adult insects, a relationship between this catch and the levels of damage, or abundance of the insect, in adjacent crops or the surrounding area may be required. The method of collection of additional information, such as crop counts, damage estimates, crop types and growth stages must be standardized between locations. The sampling may need to be continued for a number of years before it can be used in any regional forecasting programme. The importance of correctly evaluating the type of information that is going to be required in a forecasting programme at the start of the sampling survey cannot be over-emphasized. Time and effort can easily be wasted in collecting inappropriate data. It is not unheard of, after a few years of a survey, for it to be realized that relevant data have not been collected.

The identification of an appropriate sampling and monitoring technique in order that the data generated can be used for forecasting requires clear objectives and planning. Various types of trapping device are used extensively for monitoring, but rarely are the ultimate uses of the data to

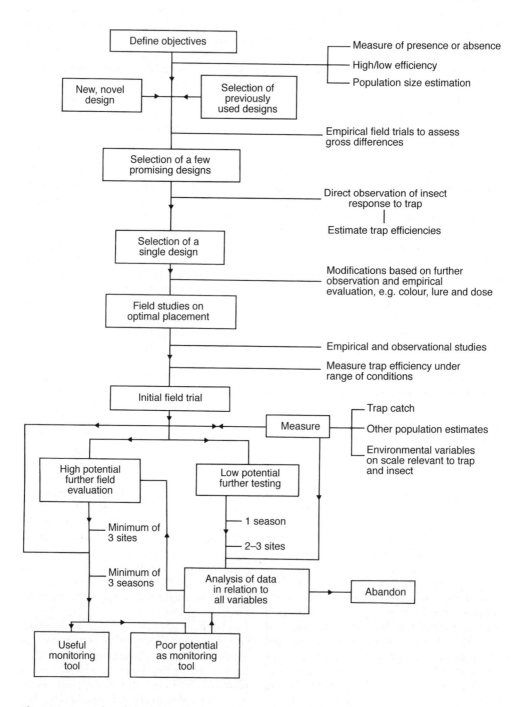

Fig. 2.12. Research pathway for developing traps as monitoring devices.

be generated considered at the outset. The use of traps for the monitoring of insects is fraught with difficulties mainly caused by the variability of trap catch data. This variability makes interpretation of the data difficult and this in turn reduces the value of such traps as monitoring devices. It is important if traps are to be used for monitoring that they are developed for the specific purpose and conditions for which they are needed. In an ideal situation a trap would be developed and evaluated along the lines of the scheme in Fig. 2.12. The most critical phase of the development occurs after the initial field test. If successful then full scale trials over a number of sites and seasons could take place. This could not be justified if the initial trial were not successful, but if results were partially successful the temptation always exists to expand the trial with the hope that more data will clarify the situation. A successful monitoring procedure will provide consistent and reliable forecasts; if the data are variable then the value of the procedure will be diminished.

2.5 Forecasting

Forecasting systems are becoming increasingly sophisticated. Even the simplest are built on statistical models, whereas the more sophisticated require computer models and complicated specialist software.

Forecast methods are classified here according to whether they are based on threshold counts, physiological time, predictive models or geographic information systems.

2.5.1 Action thresholds

An action threshold may be defined as the pest density that warrants the initiation of a control strategy, usually insecticide application. The application of insecticides according to the level of infestation, rather than on a fixed schedule basis, may reduce the number of insecticide spray applications, or at least ensure that the number of applications is economically viable. Replacing the fixed schedule system with a 'spray when necessary' strategy requires there to be an appropriate method of monitoring pest abundance and a reliable action threshold. The monitoring technique must not involve intensive sampling, especially if it is to be carried out at regular intervals by the farmer. Traps should be inexpensive and easily maintained, and catch data easily interpreted. Action thresholds have been developed for *Cydia pomonella* and *Archips podana* in UK apple orchards (Alford *et al.*, 1979), *Bactrocera oleae* in olives (Delrio, 1992), for *Keiferia lycopersicella* in tomato crops (van Steenwyk *et al.*, 1983), for *Psila rosae* in carrots (Schoneveld and Ester, 1994) and for *Thrips tabaci* and *Acrolepiopsis assectella* in leeks (Hommes *et al.*, 1994).

Case Study: A pheromone trap capture threshold of pink bollworm moths in hirsutum cotton (Taneja and Jayaswal, 1981)

Taneja and Jayaswal (1981) carried out a field trial during the 1979–80 cotton growing seasons at Hisar, Haryana (India) to determine the capture threshold of male pink bollworm moths in pheromone traps used for timing of insecticide applications. The pheromone traps were omnidirectional with a diameter of 30 cm, a height of 7 cm with eight holes (diameter 2 cm). They were positioned in each plot and the number of males counted and removed daily. The plot size was 60 × 60 m and the experiment was laid out as a simple randomized block design having six treatments and four replications. The six treatments involved the application of insecticide to the cotton plots according to four different pheromone trap catch thresholds, a fixed spray schedule and a no spray control. The four threshold treatments were 4, 8, 12 and 16 moths per trap-night and spray application was made within 24–48 hours of the observed threshold. The fixed spray schedule was the application of insecticide at an interval of 13–14 days.

The incidence of pink bollworm during the experiment was recorded, in the squares, flowers, green bolls and opened and unopened bolls. Only the data from the opened and unopened bolls are discussed here.

Twenty plants were tagged in each plot and all the opened and unopened bolls from these plants were picked and the incidence of larvae recorded (Table 2.2).

The highest incidence of larvae in open and unopened bolls in each season occurred in the control plots where no insecticide was applied. In 1978–79 the percentage incidence in both opened and unopened bolls increased from treatment 1 to treatment 5, in the 1979–80 season the differences between treatments were less pronounced (Table 2.2). The first treatment recorded a significantly lower incidence of larvae in the opened and unopened bolls than other treatments in both years. Hence, the application of insecticides, when the number of moths averaged four or eight moths/trap/night, proved to be a better strategy than the fixed spray schedule. However, in this particular case, the spray programme based on a threshold of four moths/trap/night resulted in an increased number of insecticide applications. A study of the economics of this programme still indicated that even with a greater number of sprays than the fixed schedule the threshold programme produced a higher profit.

Table 2.2. Incidence of pink bollworm on open and unopened bolls in plots treated with insecticides according to four different pheromone trap catch thresholds (T1–T4), a fixed schedule treatment (T5) and an untreated control (T6) (after Taneja and Jayaswal, 1981).

Treatment threshold Number of moths/ trap/night		Total number open bolls	Percentage increase	Total number unopen bolls	Percentage increase
1978–1979					
T1	4	758.5	19.4	164.0	42.9
T2	8	769.7	25.8	168.7	45.9
T3	12	768.0	34.0	153.5	56.0
T4	16	765.0	47.3	175.7	66.9
T5	Fixed schedule	778.5	48.9	169.5	72.8
T6	Control	724.2	84.9	172.5	91.0
F-test		NS	S	NS	S
1979–1980					
T1	4	751.5	25.6	73.5	34.0
T2	8	770.2	32.7	72.5	43.6
T3	12	794.7	48.2	78.2	59.0
T4	16	811.5	47.8	84.0	56.4
T5	Fixed schedule	812.5	44.2	90.2	49.9
T6	Control	821.5	72.4	92.5	78.6
F-test		NS	S	NS	S

2.5.2 Temperature and physiological time

Temperature has a major influence on insect development and, hence, can be used to predict emergence of a particular life stage. The use of age to define the development stage of insects or plants has been replaced by the idea of physiological time. Physiological time is a measure of the amount of heat required over time for an organism to complete development, or a stage of development (Campbell *et al.*, 1974). Physiological time, which is the cumulative product of total time × temperature above a developmental threshold (Southwood, 1978), is measured in day degrees and is considered to be a thermal constant (Andrewartha and Birch, 1954). Hence, if it is assumed that development is a linear function of temperature (above a threshold) then the insect develops in proportion to the accumulated area under the temperature vs. time graph (Allen, 1976). Before physiological time can be computed an estimate of the insect's threshold temperature for development needs to be obtained and the form of the relationship between temperature and the insect's development rate established.

The simplest estimate of the threshold for development is the point at which a regression line crosses the x-axis (where the development rate = 0) in a development rate vs. temperature relationship (Fig. 2.13). The development rates are determined for each individual as the reciprocal of the time to development (1/development time). They are normally obtained by measuring the development times for a number of individuals at constant temperatures in an environmentally controlled cabinet or growth chamber. Fluctuating temperatures may influence the time to complete insect development (Siddiqui *et al.*, 1973; Foley, 1981; Sengonca *et al.*, 1994) although Campbell *et al.* (1974) found no difference when constant and fluctuating temperatures were used, provided that the fluctuations were not extreme, i.e. they did not extend below the threshold and the average temperature was not in the upper threshold region. The insect's diet or host may also influence the development time, hence the host plants used in experiments should reflect those used by the insect in the field (Campbell *et al.*, 1974; Williams and McDonald, 1982).

The use of the linear regression x-axis method for calculating a developmental threshold is sufficiently accurate for most applications; however, techniques for more

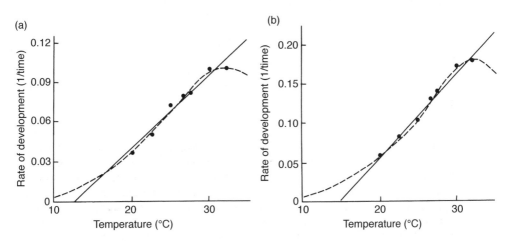

Fig. 2.13. A comparison of linear (——) and sigmoid (– – – –) approximations of the relationship between rate of development (1/time) and temperature for (a) larvae and (b) pupae of *Trichoplusia ni* (after Stinner *et al.*, 1974).

accurate threshold estimates are possible. If care is taken to obtain data at very low and high temperatures the relationship between development rate and temperature may be found to be sigmoidal (Fig. 2.13). The sigmoid relationship shown in Fig. 2.13 simulates the lower temperature relationship more accurately than the linear relationship (Stinner *et al.*, 1974). The function used to produce the sigmoid relationship is as follows:

$$R_T = \frac{C}{\left(1 + e^{k1=k2T'}\right)} \qquad (2.10)$$

where:

R_T = rate of development (1/time) at temperature T

C = maximum development rate $\times$ e^{k1+k2}, i.e. the asymptote

T_{opt} = temperature at which maximum development rate occurs

$k1$ = intercept (a constant)

$k2$ = slope (a constant)

T' = T (temperature), for $T < T_{opt}$

$T'= 2 \cdot T_{opt} - T$, for $T > T_{opt}$

Stinners sigmoid model has been widely used (e.g. Whalon and Smilowitz, 1979; Allsopp, 1981).

The daily temperature cycle needed in any day-degree calculation can be obtained from a daily temperature trace or it can be simulated using a sine wave approximation. Using only daily maximum and minimum temperatures a sine wave can be generated that closely simulates a daily temperature cycle. A formula for calculating half day, 'day-degrees' in situations where the temperature cycle is intercepted by a lower threshold temperature for development is provided by Allen (1976); a similar formula is also provided by Frazer and Gilbert (1976).

The accuracy with which a day-degree model predicts a biological event will depend on the accuracy of the developmental threshold, the type of temperature measure used and the precise time that the day-degree accumulation begins (Collier and Finch, 1985). Agronomists working

with annual crops commonly use sowing date as the time to start an accumulation (Pruess, 1983), i.e. a clearly defined starting time, with an obvious biological significance. However, entomologists using day-degree models find it more difficult to define a single discrete time/event that can be used to initiate an accumulation. Where insect growth stages are carefully monitored and where one growth stage is used to predict the occurrence of a future stage (e.g. Jay and Cross, 1998), for example adult catches in traps used to predict egg hatch, then models may have a starting time that can be clearly defined, and are usually quite successful (Pruess, 1983; Garguillo *et al.*, 1984). However, in situations where models are used to predict insect development after a period of dormancy or diapause, initiation times for an accumulation may not be easily defined. Often arbitrary dates are chosen (Baker *et al.*, 1982; Collier and Finch, 1985). The justification for this is usually that day-degree accumulation before that date is likely to be negligible. Use of such arbitrary starting times is likely to become a major source of error in any day-degree accumulation: an error that cannot be corrected by mathematically precise day-degree calculations (Pruess, 1983). Where possible, starting dates for any day-degree calculation should be based on meaningful biological criteria.

Day-degrees can be most effectively used in combination with a monitoring technique to predict the onset of a particular life stage of an insect, e.g. egg hatch. This information can then be used to ensure timely application of a control measure. Most commonly it is used with insecticides (Minks and de Jong, 1975; Glen and Brain, 1982; Potter and Timmons, 1983; Garguillo *et al.*, 1984; Pitcairn *et al.*, 1992); however, the principle applies to other control methods as well e.g. crop covers used to control *Penphigus bursarius* on lettuce (Collier *et al.*, 1994), which is especially important when insects are only vulnerable to insecticides for very short periods of time, for example with boring insects.

Case Study: Forecasting the time of emergence of sheep blowfly (Wall *et al.*, 1992)

Adult female *Lucilia sericata* lay batches of up to 200 eggs in the wool of sheep, close to the skin surface often choosing areas soiled by faeces and urine (Craig, 1955). After hatching, the larvae pass through three instars, feeding on the epidermal and skin secretions (Evans, 1936). After completing feeding the third instar larvae drop to the ground where they disperse (the larval wandering stage) and eventually burrow into the soil to pupate. The first generation of adults emerge in spring; the females seek a suitable oviposition site on a host sheep. In late August and early September in temperate areas female adults produce larvae that burrow into the ground and diapause. In mid-March these larvae migrate to the soil surface and pupate and they emerge as first generation adults in the spring.

Sheep blowfly strike is a significant economic and husbandry problem throughout the northern hemisphere and in New Zealand, parts of South Africa and Australia. Control is usually obtained through use of insecticide sheep dips when the sheep strike levels reach economic thresholds. Information on the effects of temperature on the rates of development of the different stages of the blowfly life cycle were obtained (Table 2.3) and have been used in a simulation model to identify the impact of different control measures.

The simulation model shows that mortality of >90% would need to be achieved on each generation to reduce population growth, which is clearly unrealistic. Other simulations indicated that a reduction in the first generation was the most crucial to suppressing population growth and that a significant kill at this stage in combination with spot treatments as necessary during the rest of the season could provide sustainable control. However, the use of such strategic treatments against blowfly populations requires prediction of their spring emergence. Spring emergence could be predicted based on information that the overwintering larvae require 30 day-degrees to complete larval development and that pupation required 12.5 day-degrees above 9°C following which adults emerge (Table 2.3). Hence, extrapolation from temperature measurements made from previous years can be used to give advanced warning of blowfly spring emergence and the need or not for sheep dipping.

Table 2.3. Threshold base-temperatures and day-degree requirements (± SE) for completion of pre-adult life-cycle stages and adult egg batch maturation (data from Wall *et al.*, 1992).

	Base temperature (°C)	Day-degrees (SE)
Larval wandering stage	9.5	45.6 (5.1)
Pupation	8.8	126.1 (3.7)
Post-diapause larvae	9.2	27.7 (0.3)
First egg batch	11.2	62.0 (2.3)
Second egg batch	11.1	29.7 (0.9)

2.5.3 Predictive models

A model is a simplified representation of a system (Holt and Cheke, 1997). Within the context of forecasting, models are the relationships that are quantified in some way that enable prediction of likely incidence at some future time. The temporal and spatial scales of such forecasts tend to be dependent on the type of monitoring that is undertaken. Where monitoring is based at the farmer's field level, forecasts are usually short term and restricted to a particular crop within a restricted area. They usually employ easy to use monitoring systems, e.g. pheromone or coloured traps that provide simple population counts and a means by which a forecast can be determined.

The models commonly involve regression or multiple regression analyses of insect number of one stage against another (adults vs. larvae) or against damage.

Case Study: Pheromone trap catches of male pink bollworm, *Pectinophora gossypiella*, and cotton boll infestation in Barbados (Ingram, 1980)

Pink bollworm, *Pectinophora gossypiella*, is a major pest in almost all of the cotton growing areas of the world, causing both quantitative and qualitative losses (Taneja and Jayaswal, 1981). Control of the pink bollworm is normally achieved by enforcement of a close season (Hill, 1983) when no cotton is grown and emerging adults can find no hosts on which to oviposit – a suicide emergence (Ingram, 1980). However, in situations where control is not obtained through the use of a close season then chemical control may be necessary. Chemical control against the larvae is usually ineffective because they feed within the fruiting bodies where they are protected. Conventional insecticides are applied mainly to control the number of adult moths and the larvae not yet bored into the tissue (Taneja and Jayaswal, 1981).

The cost of chemical control must be more than compensated for by the increased yield obtained from controlling the pest. Hence, some measure of the level of infestation that causes damage or yield loss must be determined. With the pink bollworm there is a direct relationship between cotton boll damage and the level of infestation. The presence of a larva in a cotton boll may be taken as a direct indication of boll damage, hence the percentage of bolls infested in a sample will provide an estimate of overall percentage damage in a field. The number of male adults can be easily monitored using a pheromone trap. What is needed is to establish if there is a relationship between adult numbers (represented by pheromone trap catches) and the boll damage as measured by the percentage of infested bolls.

Sampling of adult male populations was carried out using pheromone traps and related to larval boll infestations in the 1976–77 and 1977–78 seasons. A single pheromone trap (omnidirectional water trap) was set up in a field of cotton 11 weeks old and daily catches were recorded until the plants were uprooted at 28 weeks. The trap was positioned just below canopy height and was raised as the plants grew taller. The pheromone attractant septum was replaced every two weeks. Samples of 100 large green bolls were selected randomly and opened so that the percentage mined by pink bollworm could be determined. Boll sampling was carried out at weekly intervals, from week 13 until week 26. The data from the 1976–77 cotton season are shown in Fig. 2.14.

The mean daily catch of the total number of moths trapped for the week ending on the day on which the boll samples were taken was regressed against the weekly percentage of damaged bolls (Fig. 2.15). The relationship was shown to be significant for both years and an analysis of variance showed there were no differences between slopes ($F = 0.58$) or elevation ($F = 0.28$).

Spraying is usually advised when 10% of the bolls are infested, but this level should be adjusted each season when the cost of inputs can be weighed against the expected increase in pickable cotton and the current price obtainable for the lint (Ingram, 1980). The price of a product and the cost of producing it are going to vary between seasons. Hence, the level of damage that can be tolerated before it becomes economically viable to control the pest is also going to fluctuate. This in turn will influence the threshold value of the catch data used to monitor population change. In this context a relationship similar in form to Fig. 2.15 will be useful since then new catch thresholds can be easily estimated once a particular seasonal damage threshold has been identified. However, this does depend on there being a consistent relationship between the catch and field data, in both time and space. The existence of relationships can depend on many of the factors already discussed in this chapter, as well as the life history strategy and population phenology of each particular pest. Each pest must be assessed according to its particular circumstances, but it would be advisable to study population development in a number of fields and over a number of seasons, to validate any relationship to be used in field forecasting.

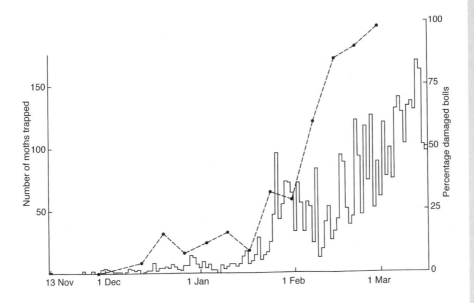

Fig. 2.14. Weekly boll infestations (●) and nightly pheromone trap catches of the pink bollworm, *Pectinophora gossypiella*, during the 1976–77 cotton season, Barbados (after Ingram, 1980).

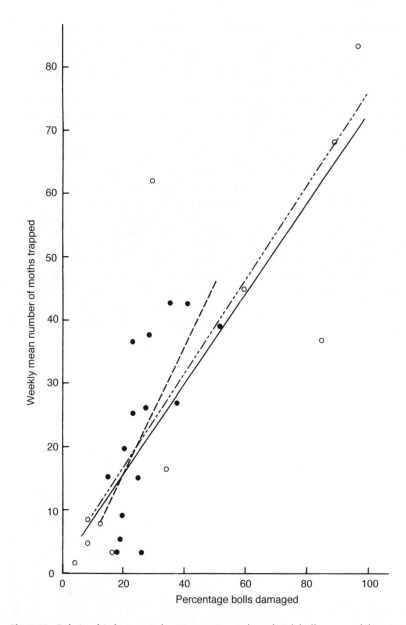

Fig. 2.15. Relationship between pheromone trap catches of pink bollworm and the percentage damaged bolls in the 1976–77 (○; ——) and 1977–78 (●; – – – –) seasons and for combined data (–·–·–·) (after Ingram, 1980).

Regional forecasts tend to be based on the long term collection of population data, normally trap catch, at a specific fixed location. Traps that sample aerial populations such as light traps and suction traps are most commonly used. The catches are often collected on a daily basis and are characteristically collected for a number of years in order to establish the presence and form of seasonal population trends. Often it is hoped that the trap estimates may reflect regional population changes and be used to forecast levels of field infestations and provide spray warning for farmers.

The advantages of such fixed position monitoring systems used for regional forecasts are:

1. It can be used in areas where personnel for field inspections are limited.
2. Few traps may be required, hence limited resources can be centralized (perhaps at agricultural research stations) where there are the trained personnel necessary to collect, sort, analyse and interpret the trap data.

The efficient dissemination of the forecast information and complementary field inspection may also be necessary.

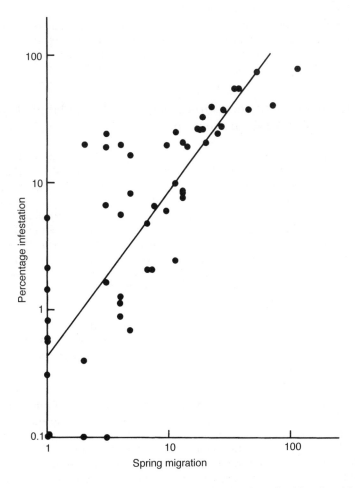

Fig. 2.16. Regression predicting the expected level of crop infestation by *Aphis fabae* (log % bean plants infested). From the spring aerial migration sample (Way *et al.*, 1981).

There are a number of disadvantages with a fixed position monitoring system for use as a regional forecasting programme. One of the biggest disadvantages is that data in such a system need to be collected for at least 10 years in order to validate the forecasting system (Way et al., 1981), especially if only one or a few traps are used, since then replication and validation is only assessed between years. This means the use of a fixed position monitoring system in order to provide a regional forecasting programme is a long-term project, requiring a large investment of resources but from which there is no guarantee of a successful forecasting system. The cost of the monitoring technique may be reduced and the probability of obtaining a successful forecasting system is enhanced if the data for a number of important pests can be collected and studied at the same time.

The important components of a fixed position monitoring and forecasting system include:

1. A thorough knowledge of the biology and seasonal cycle of each pest (Lewis, 1981).
2. An appropriate monitoring technique for both aerial and field populations.
3. A consistent relationship between catch estimates and field infestation.
4. A relationship between the level of field infestation and damage and if possible between trap estimates and crop damage.
5. Sufficient replication in time and space for validation of the forecasting programme.

A number of such regional forecasting systems exist (Way et al., 1981 (Fig. 2.16); Pickup and Brewer, 1994; Linblad and Solbreck, 1998).

Case Study: Short-term forecasting of peak population density of the grain aphid *Sitobion avenae* on wheat (Entwistle and Dixon, 1986)

The grain aphid, *Sitobion avenae*, is a major pest of wheat in the UK and Europe. A number of severe outbreaks have prompted many studies of the biology and damage relationships of this and other cereal aphids (Vickerman and Wratten, 1979). *S. avenae* feeds on the ear of the wheat and can cause substantial yield loss. It is most effectively controlled through the use of insecticides. An ability to forecast the abundance of this aphid would provide a warning of when insecticide applications may be necessary. Entwistle and Dixon (1986) have produced a short term forecasting system based on two counts on the crop. The derivation of this forecasting system is related here. The growth stages (GS) of cereals have been classified according to a numerical scale (Zadoks et al., 1974). The growth stages important in the context of this work are:

Growth stage	Description
53	One-quarter of ear emerged
59	End of ear emergence
65	Mid-anthesis
69	End of anthesis

A total of 32 sites were sampled from 1975 to 1983 and a further nine sites were sampled in 1984. Wheat tillers at each of these sites were examined weekly from mid-May until anthesis (GS 65) after which they were sampled twice weekly (Leather et al., 1984; Entwistle and Dixon, 1986). A range of between 50 and 100 tillers were examined, depending on the sampling site (Entwistle and Dixon, 1986) and on the aphid density. The number of tillers sampled at some sites was reduced as aphid densities rose, maintaining a standard error of 10% of the mean (Leather et al., 1984). The mean number of *S. avenae* per tiller was calculated on the day closest to the growth stages GS 53, 59, 65 and 69. The aphid density each day was estimated with the assumption of a constant rate of increase in population size between monitoring dates. The crop growth stage each day was also estimated assuming equal daily increments on the Zadoks *et al.* (1974) scale between monitoring dates.

The log peak density of aphids was regressed against the log density of aphids at growth stages 59, 65 and 69. The relationship was positive for each growth stage, although the strongest relationship occurred at growth stage 69 (Fig. 2.17). The percentage variance explained by the regression line increased from GS 59, 29%; GS 65, 50.4% to GS 69, 70.5%.

The observed rates of increase of aphids per tiller were calculated between growth stages 53(i)–59(ii), 59(i)–65(ii) and 65(i)–69(ii) using the following formula:

$$y = \frac{\log_e\left[\text{aphids} + 0.01 \text{ at GS(ii)}\right] - \log_e\left[\text{aphids} + 0.01 \text{ at GS(i)}\right]}{\text{number of days between (i) and (ii)}} \tag{2.11}$$

The log peak density of aphids was then regressed against the rate of increase at GS 53–59, GS 59–65 and GS 65–69. The relationship was positive for each

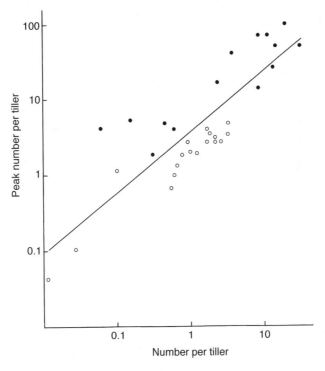

Fig. 2.17. The peak mean number of *Sitobion avenae* per tiller (*y*) in relation to the mean number of *S. avenae* per tiller (*x*): at the end of anthesis (GS 69).
$\log_e (y) = 1.27 + 0.779 \log_e (x)$, $r + 0.84$, d.f. = 30, $P < 0.001$.
● = results from 1976, 1977, 1978, 1979 and 1980 in Norfolk.
○ = results from 1980 outside Norfolk, 1981, 1983.
Numbers are shown on log scale (after Entwistle and Dixon, 1986).

growth stage interval, and similarly to above, the strongest relationship occurred at the later growth stage (Fig. 2.18). The percentage variance explained by the regression line again increased with increased growth stage from GS 53–59, 36%; GS 59–65, 38% to GS 65–69, 46%.

A multiple regression, having the following equation was computed:

$$\log_e\left[(\text{peak no. aphids / tiller}) + 0.01\right]$$
$$= K_1 + K_2 \log_e\left[(\text{current no. aphids / tiller}) + 0.01\right]$$
$$+ K_3 (\text{observed rate of increase}) \tag{2.12}$$

The incorporation of the rate of increase considerably improved the relationships between peak density and densities at the end of ear emergence (percentage variance explained = 45.7%) and during anthesis (percentage variance explained 66.2%) with the multiple regression for the end of anthesis explaining 86.7% of the variance in the relationship.

The regression equations were used to obtain predicted values for peak population densities. Plots of the peak population densities against predicted values were drawn (Fig. 2.19 shows the relationship for GS 69). The accuracy of the prediction increased with advancing growth stage (Fig. 2.19). The forecast model now needs to be related to a damage model before this method can be used to develop a monitoring procedure for practical use.

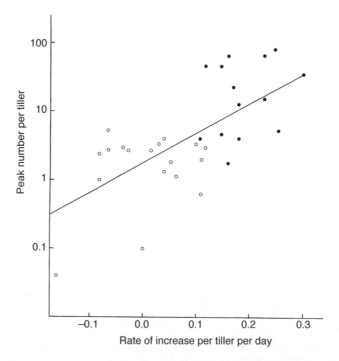

Fig. 2.18. The peak mean number of *Sitobion avenae* per tiller (y) in relation to the rate of increase in mean number of *S. avenae* per tiller per day (x): from GS 65 to the end of anthesis (GS 69). $\log_e (y) = 0.528 + 10.5 (x)$, r + 0.68, d.f. = 30, $P < 0.001$. Symbols as Fig. 2.17, y-axis on log scale (after Entwistle and Dixon, 1986).

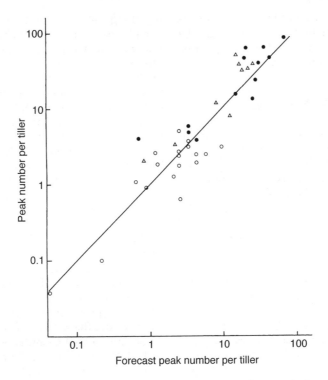

Fig. 2.19. The peak mean number of *Sitobion avenae* per tiller (*y*) in relation to the forecast from the multiple regression equation (see text) at the end of anthesis (GS 69). Symbols as Fig. 2.17, △ results from 1984 for model validation. Numbers shown on log scale (after Entwistle and Dixon, 1986).

2.5.4 Geographic Information Systems

The quantification of the spatial dynamics of pest populations especially in relation to associated environmental changes, e.g. wind direction (Tucker *et al.*, 1982) or rainfall (Tucker and Pedgeley, 1983; Tucker, 1984a, b) is not new; regional and national density distribution maps of pest species constructed from trap catch records using computer graphics have been possible since the 1970s (Taylor, 1974; Campion *et al.*, 1977). However, the more recent increase in computer power and sophistication of software has made the analysis of geographic variables in relation to the spatial aspects of the dynamics of pest populations more widely available. This type of analysis based on Geographic Information Systems (GIS) utilizes data referenced by spatial or geographic coordinates (Star and Estes, 1990). Geographical information such as altitude, temperature, soil type and distribution of crop varieties can be stored as a spatial map layer. When a number of map layers are overlaid, valuable insights such as pest zones or areas prone to pest outbreaks may be identified. Information generated in this way can be fed into models that use the maps as variables and analyse these to determine whether correlations exist and to construct forecast maps. GIS has been successfully used in pest management to forecast grasshopper incidence in Canada (Johnson and Worsbec, 1988), dispersal of rice insects in South Korea (Song *et al.*, 1994) and to identify the risk posed by exotic insect pests in the UK (Baker, 1994; Head *et al.*, 1998). It is likely that GIS analyses will figure more commonly in the understanding of pest populations and the ability to forecast susceptible outbreak regions in the future, as this powerful technology gains more widespread use.

2.6 Discussion

Sampling using relative measures of insect abundance tends to be popular among entomologists mainly because the various types of trap used provide quick and simple methods for acquiring a large amount of population data. However, the value of such data, for the purposes of monitoring and forecasting, may be limited if the trap catch cannot be related to other appropriate population estimates. The type of monitoring system employed will be dependent on the user, the pest, the resources available and whether or not the monitoring is being carried out for research purposes or as a practical aid in insecticide application. More attention needs to be given at the start of a sampling programme to the eventual use of the data and whether or not a monitoring/forecasting programme could result from the work (Wall, 1990). If this is the case, then it is important to identify the potential users, their needs and abilities so that an appropriate sampling and monitoring programme can be devised. The tendency to embark on long term monitoring programmes with the hope that 'something will come out of it' needs to stop and be replaced by a more rational approach where specific goals, time limits and available resources are identified. The type of sampling techniques used in a monitoring programme dictates the type of forecasting system that can be developed, hence both the monitoring programme and the perceived forecasting system must be appropriate for the situation and conditions to which it will be applied.

In the absence of any monitoring/forecasting system a farmer would either apply an insecticide regardless of the level of pest attack (prophylaxis) or not apply an insecticide at all. The no control option would mean that in non-outbreak years (or seasons) the farmer would maximize his profits with respect to that particular pest/crop system; however, in an outbreak year total yield and monetary loss may occur. The prophylaxis option in non-outbreak years would reduce the farmer's profits by the cost of the unnecessary insecticide application, whereas in outbreak years the farmer would control the pest but the prophylaxis strategy may mean that the farmer uses too many (or too few) insecticide applications for effective pest control. An ineffective control due to prophylaxis could reduce the farmer's profits. Hence, neither strategy will always provide the optimal solution in terms of maximizing farmers' profits. The common factor in the two strategies that will influence overall farmer profit is the probability of an outbreak and the severity of attack. A monitoring and forecasting system must improve on both the no control and the prophylaxis option if it is to be adopted by farmers, i.e. a monitoring and forecasting system must correctly predict outbreak and non-outbreak years so that insecticide is applied only in outbreak years, and only when it is cost-effective to do so. Hence, the accuracy and consistency of the forecast will directly affect the yield and profits of the farmer. The value of a forecast scheme and the importance of its accuracy can be assessed in economic terms (Watt, 1983).

The accuracy of a forecast will depend on the variability encountered in the forecast model. Forecast models are often based on regression models which are prone to a number of potentially serious defects. Firstly, the regression model should explain a large proportion of the variance if it is to be at all reliable. Secondly, regression models are really only reliable within the range of data from which they were originally derived. They may work for many years but then a combination of circumstances outside the original conditions arises, causing a model to fail (Conway, 1984b). These considerations must be taken into account when developing forecast schemes, so that the reliability that can be placed on forecasts can be evaluated. A farmer would soon stop using a monitoring and forecasting system if it were shown to be unreliable.

3

Yield Loss

3.1 Introduction

The amount of useful product that is obtained from crop plants or livestock is commonly referred to as 'yield'. Estimates of yield may be: (i) quantitative, e.g. weight of grain, fruit or tubers per unit of land; or (ii) qualitative, e.g. the percentage of a product meeting certain cosmetic standards which will vary for a given crop or livestock system according to weather, the levels and types of input and pest incidence. If all factors are optimal then the highest attainable yield is obtainable. However, since conditions are rarely optimal, actual yields are normally well below those that are theoretically obtainable. The best a farmer can usually hope to achieve is a yield that provides the highest possible return on his investment of inputs.

Yield loss assessments attempt to account for the difference between actual and attainable yield as a result of the inputs, weather and pests in a given cropping system. Hence, the losses due to pests constitute only one of a number of complex production constraints. Even so yield losses due to insects alone provide a multitude of problems for the scientist to unravel and evaluate.

3.2 Approaches and Objectives

Information on yield losses caused by pest insects is of great interest and value to

decision makers whether these are farmers, extension agents or government officials. Farmers require yield loss information to assist in management decisions concerning selection and timing of control measures while governments require yield loss data for food and crop production planning and to assist in the process of resource allocation for research, extension and control operation (Walker, 1987).

There are a number of common approaches or frameworks that have been used for assessing economic losses due to pests (MacLeod and Norton, 1996). These include:

1. Estimation of expenditure on control or eradication (e.g. historical annual average or most recent year).
2. Estimation of the income losses or an opportunity cost given the threat of pest damage (e.g. consequences of utilizing one control option relative to another).
3. Economic threshold models.
4. Assessment of impact of actions on other parties (e.g. drift from chemical insecticides on neighbouring farmers and the wider community).
5. Estimates of the impact in terms of the economic value of investing public resources into particular areas (e.g. public research on Bt insecticides or Bt transgenic crops).
6. Investment models which involve a relatively large initial investment that provide

a stream of benefits over a longer term (e.g. investment in a plant breeding programme at a national level which requires a stream of outlays and benefits to the farmer accrued over a period of time).

The last two methods require the skills and knowledge of economists to implement and hence, are generally less applicable. The other methods, with the exception of (1) the expenditure method, rely on some quantitative assessment of a yield/damage function. The generation of the data necessary to define the yield/damage function, especially in economic terms, underpins the majority of experimental approaches dealing with yield loss. This is especially true of the economic threshold concept (Stern *et al.*, 1959) which has formed the basis of many pest management systems and is considered by some a central tenant of IPM. In addition to experimental approaches, surveys may be undertaken in order to understand and quantify farmers' perceptions of yield loss due to pests.

The scale on which yield loss assessments are undertaken represents a further dimension to take into account. An assessment of crop losses on a regional level may be required to allow policy decisions to be made, perhaps concerning priorities for research (which pests and crops to study), to assess the need for control and to identify the regions, farmers and communities most in need of assistance. Regional evaluation of crop loss can be carried out by use of either survey or experimental methods or a combination of both (Rai, 1977; Nwanze, 1989).

On individual farms, yield loss assessments are normally carried out to establish criteria on which to base crop and pest management decisions. Crops are normally considered in isolation and the effect of a single major pest on yield is evaluated. Detailed observation and experimentation are required in order to assess the impact of the pest on crop yield under a variety of conditions, such as the timing of infestation in relation to crop growth stage, the weather and the use of different crop production practices, e.g. fertilizer and pesti-

cide use. In the latter situation the impact of pesticide inputs on the yield of a crop may be used to determine the value of such a practice to the farmer.

3.3 Measurement of Yield Loss

The primary aim of a yield loss assessment is to determine the type of relationship that may exist between pest infestation and yield. Initially experiments or surveys may simply attempt to establish that losses *per se* occur but more detailed information is usually required in order to determine the way in which pest infestation influences yield loss.

3.3.1 Pest intensity

The intensity of pest attack can be described as the product of three effects:

1. The numbers of the pest present.
2. Their development stage.
3. The duration of the pest attack.

It is the combination of these three factors in relation to the crop that influences crop yield.

Estimates of insect numbers or density are usually made through actual counts of the insect on the crop, or by measuring the proportion of those plants or plant parts that are infested. Occasionally, relative sampling methods may also be employed. Another sampling procedure that is used in yield loss assessment studies is that of a scale of damage or infestation and the classification of field samples by a visual rating. This is a technique commonly used in breeding trials that assess the effect of plant resistance on insect/pest numbers (Chapter 5).

When a count of insects is made and used as a measure of pest intensity, it is assumed that each individual insect contributes an equal amount to the total yield loss of the plant or crop. However, different insect developmental stages may have a differential effect on plant yield. Hence, in order to assess accurately the effect of insect intensity on yield loss some account should be taken of the population structure of the infesting pest population. The population structure of the pest can be determined through the use of a more refined count procedure, so that the individual insects are classified according to developmental stage or by the use of an index that reflects the developmental stage. The yield loss caused by each developmental stage or index level can be related to that caused by the most damaging developmental stage of the insect.

Developmental stages having a similar effect can be clumped together; for example, in the assessment of yield loss caused by aphids on cereals, adults and fourth instar nymphs were given an index of one and nymphs younger than this one-third, so that three nymphs were required to equal one adult (Wratten et al., 1979). Aphid counts are thus adjusted to 'adult equivalents' which are then used as the measure of pest intensity. Alternatively the damage or area consumed by the immature stages required to complete their development can be determined, e.g. the larvae of the green clover worm (Plathypena scabra) consume on average 54 cm^2 of soybean leaves in order to complete development (Hammond et al., 1979; Browde et al., 1994a). The length of time for which a pest infestation is present on a plant or crop will also influence the extent of yield loss. Hence, any index of the size of an infestation needs a temporal component that can take this into account. The level of attack can then be expressed as insect days, which is the area beneath a graph of insect numbers (or adult equivalents) plotted against time (e.g. Smelser and Pedigo, 1992; Annan et al., 1996).

3.3.2 Types of pest damage

The presence of an insect pest in a crop is usually characterized by a particular type of damage. The damage may take the form of injuries caused by insect feeding, the presence of contaminants, such as frass, that reduce the market quality of the harvestable product or indirect insect damage caused by the presence of bacteria or viruses transmitted by the insect. The type

of pest damage will in turn influence both the likelihood and the extent of yield loss.

Insects feed and consume plant tissue or plant sap by chewing, sucking or boring. Chewing insects that consume leaf tissue will reduce the area of photosynthetic material available to the plant. Although the damage caused may be obvious, the loss of leaf area does not necessarily result in a concomitant loss in plant yield simply because plants can often compensate for damaged tissue by enhanced growth (Section 3.9.1).

However, where chewing insects feed directly on flowering or fruiting structures then substantial yield loss can occur. For instance, a single adult bean leaf beetle (Ceratoma trifurcata) will feed on average on 0.494 soybean pods per day in a 'normal' outbreak year (Smelser and Pedigo, 1992).

Insects that bore into plant tissue include leaf miners, shootflies, stem borers and those insects that bore into fruits and grains. The last group may cause direct yield loss due to consumption of crop grains, while others may reduce the value of the product by causing a decrease in quality. Stem borers can be particularly destructive because their larvae bore into the developing stems, often killing them and causing a yield loss by reducing the number of grain bearing shoots or by weakening stems to the extent that they lodge and cannot be harvested.

Insects that imbibe plant sap use their mouth parts to pierce and probe within the plant tissue until they locate a phloem vessel from which they take up the sap. The presence of sucking insects acts as a sink for the phloem, redirecting a large part of it away from the tissue for which it was intended and into the insect gut. In this way an infestation of phloem feeding insects may interfere with the normal partition of photosynthates between plant organs (Bardner and Fletcher, 1974).

The extent of the yield loss will often depend on the feeding sites of the sucking insects. For example, there are marked differences in the feeding sites of cereal aphid species in the UK and these are important in relation to the amount of damage which they cause (Vickerman and Wratten, 1979). The aphid *Metopolophium dirhodum* is mainly a leaf feeder on wheat, where it intercepts the nitrogen and carbohydrates in the flag leaf that are allocated to the developing ear. *M. dirhodum* has been shown to reduce overall grain weight by as much as 7%. The grain aphid, *Sitobion avenae*, however, feeds at the glume bases and hence directly reduces the supply of assimilates in the developing grain reducing the yield by 14% (Wratten, 1975). The differing effects of the two aphid species resulted from the degree of nutrient drain imposed at the particular feeding sites, combined with a reduction in the leaf area duration of the flag leaf.

The contamination of the harvested product with frass, exuviae or the insect itself, while not directly affecting yield, can be considered as damaging to the crop since it can reduce the market value of the product. This is a factor that greatly affects the market value of food products in developed countries where extensive grading systems for food quality exist. For instance, in California in the USA, processing tomatoes are rejected if 2% or more of the tomatoes by weight have a larva or excreta of insects in the flesh of the tomato. Open holes that are clean and contain no larvae are not subject to the 2% tolerance. However, if a hole penetrates into the tomato so that the seed pocket is visible, the tomato is scored 'as limited use' and may be subject to a quality deduction by the processor. Between 1988 and 1990 an average 62% of loads were scored as having a trace or more of damage but rarely were loads graded as exceeding the 2% tolerance (0.5%) (Zalom and Jones, 1994).

Insect vectors rarely cause direct losses to a crop, rather it is the diseases they transmit that cause the major problem. However, often the most appropriate means of managing the disease is to control the insect vector. Virus yellows disease of sugar beet is one of the most important diseases to affect the crop in Europe causing reduction in sugar yields by up to 50% in some years (Smith and Hallsworth 1990; Dewar, 1992b). Virus yellows diseases are caused by two viruses, the beet mild yellowing luteovirus (BMYV) and the beet yellows closterovirus (BYV). Their control relies on insecticides either applied at drilling or as a foliar application to prevent build up of their aphid vectors, the peach-potato aphid *Myzus persicae* and the potato aphid *Macrosiphum euphorbiae* (Stevens *et al.*, 1994).

Case Study: The effects of feeding by larvae of *Plutella xylostella* and *Phaedon cochleariae* on the yield of turnip and radish plants (Taylor and Bardner, 1968)

A laboratory experiment was carried out to assess the effects of larval density on the defoliation of radish and turnip leaves and the consequent effect of this on root yield. Seedlings of radish and turnip were infested with five, ten, 15 or 20 larvae. There were four replicates of each density and four controls for both turnip and radish. The insects fed until they pupated and then the plants were harvested, their remaining leaf area measured and the dry weight of roots determined. Table 3.1 shows that with increasing larval density the leaf area remaining decreased. The decrease in leaf area had a significant and detrimental effect on root yield with the two insect species decreasing yield of radish similarly. However, in turnip the relationship between the larval density of *P. xylostella* and leaf area is less straightforward than with radish. The leaf area of the infested plants decreases compared with uninfested plants, but the changing larval density has no apparent effect on leaf area or ultimately on root yield.

This can be explained by the compensatory growth of the turnip after insect attack. The differing response of turnip to attack by the two insects can be explained by the way in which the two insects differ in both their method of feeding and where they feed on the plant. *P. xylostella* larvae feed indiscriminately on all leaves, but only damage the area actually eaten, leaving the veins intact; yield was not affected because attacked plants retained their older leaves longer than unattacked plants. These older leaves grew large after *P. xylostella* had stopped feeding; the plants also grew side shoots and produced more leaves than unattacked plants. In contrast, *P. cochleariae* fed mainly on older leaves and caused greater damage because the larvae sever leaf veins and rasp the leaf surface thereby killing more tissue by desiccation than they actually consumed.

Table 3.1. The average leaf area defoliated by the larvae of *Plutella xylostella* and *Phaedon cochleariae* and its effect on root yield of radish and turnip plants (from Taylor and Bardner, 1968).

Larvae per plant	Plutella xylostella		Phaedon cochleariae	
	Leaf area (cm^2)	Root dry weight (g)	Leaf area (cm^2)	Root dry weight (g)
Turnip plants				
0	500	1.85	355	2.13
5	414	1.99	282	2.26
10	426	1.91	205	1.68
15	429	1.71	217	1.55
20	409	2.83	163	1.19
Radish plants				
0	283	2.96	287	3.83
5	191	1.85	217	2.09
10	125	0.73	155	1.45
15	100	0.80	94	0.74
20	56	0.39	93	0.98

3.3.3 Measures of yield and yield loss

Crop yield is usually measured in terms of kilograms of harvested product per hectare. As a standard measure this allows easy comparison between fields and trials. However, on an individual plant level, yield will be influenced by the number of plant parts bearing the harvestable product (grains, fruits and tubers), the actual weight or size of these products and the number per plant part. For instance, in cereals yield loss due to aphids may result from a reduction in the number of ear-heads, the grain weight (usually measured as 1000-grain weight) or the number of grains per ear head (Vickerman and Wratten, 1979). These different yield components are not always stated, but they are useful for indicating the ways in which yield may be improved. It is of little value to quote yield reduction as a percentage loss unless this figure is supported by an absolute measure of the yield or yield loss. In the same way, a value for yield based on kilograms per unit length of row has little meaning unless the number of plants or the plant spacing is also given. Care must be exercised to ensure that yield loss data are quoted in a form that makes them meaningful to others.

The losses caused by insects in postharvest storage, such as *Sitophilus* spp. and *Prostephanus truncatus*, are usually measured by reference to grain weight. The loss may be measured as a change in the weight of samples over a particular period, by comparative weights of damaged and undamaged kernels or by the determination of the percentage of insect damaged grain which is then converted into a measure of weight loss. In fruit tree crops losses may be measured by fruit damage, decreased yield or mortality. The losses incurred due to tree mortality may be assessed in terms of the costs of replanting and the delay in production, for example in young apple and pear trees attacked by the leopard moth, *Zeuzera pyrina* (Audemard, 1971). Losses in forest trees caused by defoliating pests such as sawflies (e.g. *Neodiprion sertifer*), the jack pine budworm, *Choristoneura pinus*, and the spruce budworm, *C. fumiferana,* may be measured by incremental loss in terms of shoot elongation, radial increment, area of annual rings, tree height, top and root mortality and ultimately by the lost volume of wood (Kulman, 1971; Day and Leather, 1997).

There are some circumstances where the effect of a pest on the quality of a product can be considered as a yield loss. In horticultural crops where grading and sorting are carried out, pest damage may render a certain percentage of the crop unfit for sale, in which case the percentage of discarded crop will provide a measure of yield loss (Southwood and Norton, 1973). Where a lower price will be obtained for a poorer quality product the loss can be expressed as the proportion of the crop falling within each grade/category.

Insect pests can also have a direct effect on the quality of the harvested product. Infestations of the aphid *Sitobion avenae* have been shown to affect significantly the bread making quality of wheat by reducing the percentage flour extraction, increasing the colour, nicotinic acid and thiamine (vitamin B_{12}) content and reducing the baking value of the flour (Lee *et al.*, 1981). However, different aspects of the grain quality do not change in parallel with one another, or with yield changes

and thus damage thresholds have to vary according to the yield/quality measure under consideration.

3.4 Crop Loss Surveys

Crop loss surveys may be undertaken simply to determine the types of losses occurring and their main causes, to determine the distribution of losses in different areas or to actually evaluate losses with a view to forecasting crop production or justifying control measures (Walker, 1987). In addition, surveys may be undertaken to meet more specific objectives such as defining farmers' perception of yield losses (Mulaa, 1995) or to establish baseline data by which to evaluate area-wide control measures (e.g. classical biocontrol introductions; Farrell *et al.*, 1996). Such surveys may be conducted as face-to-face interviews with individual or groups of farmers (e.g. Mulaa, 1995) or by use of postal questionnaires (e.g. French *et al.*, 1992, 1995). The question/interview approach can be supplemented by on-site evaluation of actual yield losses. In the study conducted by Mulaa (1995) the interviews were conducted and then the numbers of maize plants damaged by the stalkborer *Busseola fusca* were counted on each farm from five areas (10 m × 10 m) selected at random. The percentage of damaged plants per hectare was then determined using a simple formula, taking into account plant spacing, which allowed comparison of the distribution and extent of plants damaged by stalkborers in the Trans Nzoia District, Kenya. Surveys provide useful information on yield loss that it is often impossible to collect by any other means. However, to carry out surveys correctly requires a great deal of resources. In the *B. fusca* example above, 300 farmers were interviewed in five different agro-ecological zones (Mulaa, 1995).

Case Study: Prevalence and regional distribution of sheep blowfly strike in England and Wales (French *et al.*, 1992)

Sheep blowfly strike (ovine myiasis) is the infestation of the skin of sheep by the larvae of blowflies. The problem occurs in many sheep producing countries of the world but in the United Kingdom the common myiasis fly is *Lucilia sericata*.

A self-administered, two page questionnaire was used to collect data on the prevalence and distribution of blowfly strike in England and Wales in 1988 and 1989. These were important years for the control of blowfly strike in the UK because the number of compulsory dips (total immersion in insecticide) for the control of sheep scab was reduced from two to one in 1989. The questionnaire was sent to a random sample of 2451 sheep farmers with a minimum flock size of 50 sheep. The sample was stratified by region and contained 4.6% of the holdings in England and Wales with more than 50 sheep. Of the 2451 sheep farmers, 1819 returned the questionnaires giving an overall response rate of 74.2%. Of these 1638 were usable (66.8%).

The analysis of the questionnaire indicated that a large proportion of farms in England and Wales reported one or more cases of strike in their flocks. The proportion in both years was similar (77.5% in 1988; 80.0% in 1989) even between regions (Fig. 3.1). However, there were significant differences between regions. In the north of England the proportion was significantly lower (58.6% in 1988 and 1989) than in the four other regions. The proportion of affected farms was similar in central England and Wales (76.2–79% in 1988; 80.2–82.5%

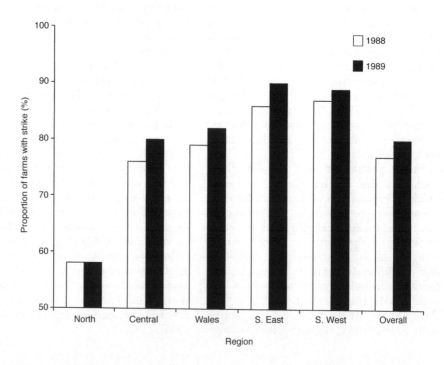

Fig. 3.1. Proportions of farmers reporting at least one case of strike in 1988 and 1989, in each region and overall (showing 95% confidence intervals; after French *et al.*, 1992).

in 1989) and significantly different from the other three regions (χ^2 analysis at $P < 0.05$). Similarly the south west and south east regions were similar (87.9% and 86.9% in 1988; 89.5% and 89.9% in 1989) but significantly different from the other three regions ($P < 0.05$; Fig. 3.1).

Until recently the control of blowfly strike has been helped by statutory requirements to dip sheep in the summer. With this removed there was concern that the prevalence of sheep strike would increase. Although this study did reveal a consistent small increase in strike in 1989 there are many factors that could have contributed to this and the importance of the increase of 0.1% is questionable. Despite this, by providing estimates of the frequency of blowfly strike, the survey provides useful information for the future assessment of strike and its control.

3.5 Plant Growth Analysis and Modelling

The process of infestation and its effects on plant yield involves a complex of interactions between the insect and the plant, a process that cannot be fully understood until some attention is paid to the effects of the insect or plant growth (Bardner and Fletcher, 1974). Insect infestation is just one of the factors that can affect a plant's growth during its normal life cycle, the main influencing factor being the environment. One of the methods used by plant physiologists to study the effect of the environment on plant growth is called plant growth analysis. This is a powerful method of estimating net photosynthetic production and analysing the physiological adaptations of different plant species (Jones, 1983); it can be readily applied to estimating the effect of insect infestation on plant growth.

Plant growth analysis is based on readily available measurements of plant dry weights and leaf dimensions, obtained from regular harvests at different times during a plant's life cycle. The analyses can be directed at changes in the growth of individual plants, plant parts or plant stands. Plant growth can be followed through a series of relatively infrequent large harvests (the classical approach to plant growth analysis) or by smaller more frequent harvests (the functional approach; Hunt, 1982). Both these approaches produce a series of sequential measurements from which up to four principal types of derived quantity can be constructed.

1. Simple rate of change.
2. Simple ratios between two quantities.
3. Compounded rates of change (rates involving more than one variable).
4. Integral durations (the area beneath a time series of primary or derived quantities; Hunt, 1982).

Also a number of important relationships exist between these derived quantities.

Most photosynthesis occurs in plant leaves, hence the leaf area will influence the amount of photosynthesis that can take place. The leaf area ratio (LAR) is used as an index of plant leafiness. The LAR is the ratio of the total leaf area to whole plant dry weight and in a broad sense represents the ratio of photosynthesizing to respiring material within the plant (Hunt, 1978).

$$\text{LAR} = \frac{LA}{W} \tag{3.1}$$

where:
LAR is the leaf area ratio
LA is the leaf area of the plant
W is the dry weight of the plant.

Plants simultaneously depend upon their leafiness and on the efficiency of their leaves to produce new material for growth. An index of leaf production efficiency is supplied by the unit leaf rate (E) which is the net gain in weight of the plant per unit leaf area (Hunt, 1978):

$$E = \frac{W_2 - W_1}{T_2 - T_1} \times \frac{\log_e LA_2 - \log_e LA_1}{LA_2 - LA_1} \quad (3.2)$$

where:
W is the dry weight of the plant
LA is the leaf area at times T_2 and T_1.

The estimates of plant leafiness (LAR) and of the assimilatory capacity of its leaves (E) enable a calculation of overall relative growth rate (RGR) to be made. The RGR is simultaneously dependent on both LAR and E, and can be simply expressed as:

$$RGR = LAR \times E \quad (3.3)$$

The RGR at the whole plant level is expressed in the classical approach to plant growth analysis as the mean RGR:

$$\text{Mean RGR} = R =$$
$$\frac{\log_e W_2 - \log_e W_1}{T_2 - T_1} \quad (3.4)$$

and in the functional approach as an instantaneous:

$$R = \frac{d(\log_e W)}{dt} \quad (3.5)$$

The RGR of the whole plant may be useful when there is a need to compare species and treatment differences on a uniform basis; further useful information may be obtained by determining relative growth rates of various plant structures such as leaves, shoots or roots. For instance, Nicholson (1992) revealed through growth analysis that potato plants defoliated by the Colorado beetle had higher relative growth rates for leaf expansion than non-defoliated plants.

Crop growth, like individual plant growth, is influenced by both leaf production efficiency and by the leafiness of the crop. The leaf production efficiency expressed as the unit leaf rate (ULR) provides exactly the same information at the crop level as at the individual plant level. However, the leaf area per plant is an inappropriate measure of leafiness of the whole crop since it does not take into account variation due to plant density. Since the number of plants per unit area of land will influence the leaf area present per unit area of land, a measure is required that will take plant spacing into account. Hence the use of the leaf area index (L), expressed as:

$$L = \frac{La}{P} \quad (3.6)$$

where:
P is the land area
La is the total leaf area above the land area P.

The value L is dimensionless and represents the mean crop leafiness. Leaf area indices have been shown to be influenced by infestations of *Empoasca fabae* on lucerne (Hutchins and Pedigo, 1989), *Sogatella furcifera* on rice (Watanabe *et al.*, 1994) and *Sitobion avenae* on wheat (Rossing and Wiel, 1990). The changing value of the leaf area index during crop development can be plotted on a time series, and the area under the time curve calculated. This integral, referred to as the leaf area duration (LAD), is a measure of the total crop leafiness during its growth period and represents the crop's whole opportunity for assimilation (Hunt, 1978). Annan *et al.* (1996) used the LAD as one measure by which they compared the impact of *Aphis craccivora* on the growth and yield of susceptible and resistant cultivars. Analysis revealed that the susceptible cultivar ICV-1 had a greatly reduced LAD value for given levels of infestation compared with the resistant cultivar ICV-12. Infested ICV-1 seedlings also showed stunting and other growth deformities which were not observed in ICV-12 plants.

As plants grow, photosynthetic materials are distributed to particular organs according to the developmental stage of the plant. During the initial stages of growth plants produce large leaf areas and hence materials are directed towards leaf development. However, when it is necessary to produce other structures such as flowers or fruits assimilates are switched away from other organs to accommodate the growth of these new structures. Since there is a limit to the amount of new assimilates that can be produced, an increase in the requirements for materials in one organ would imply a concomitant decrease in the amount distributed elsewhere. The sequence in which this partitioning occurs and the impact that insect infestation can have on this process is of great interest, particularly in defining the relationship between crop yield and the timing and intensity of insect attack.

If the dry weight of plant parts is measured at regular intervals during crop development then the distribution of these weights can be plotted to provide an indication of how assimilates are partitioned (Fig. 3.2). Alternatively, the distribution of weights as a percentage of total dry weight can be plotted which will give an indication of the magnitude of any switch in the partitioning of assimilates. The changes in

weight gain by individual structures can be determined by calculating the dry weight increment between one sample and the next. Pests can have an impact on the dry matter partitioning in plants, which is particularly important where it affects the allocation of assimilates to the harvested product. Mites (*Tetranychus urticae*) markedly affect both the shoot nitrogen accumulation and partitioning in cotton (Sadras and Wilson, 1997c). Lint yield losses were significant and were greatest for early mite infestation. Reduced shoot dry matter and reduced harvest index (lint yield/shoot dry weight) both contributed to this yield reduction (Sadras and Wilson, 1997b). Reductions in both fruit number per unit shoot dry matter and seed cotton mass per fruit also contribute to a reduction in the harvest index of damaged crops.

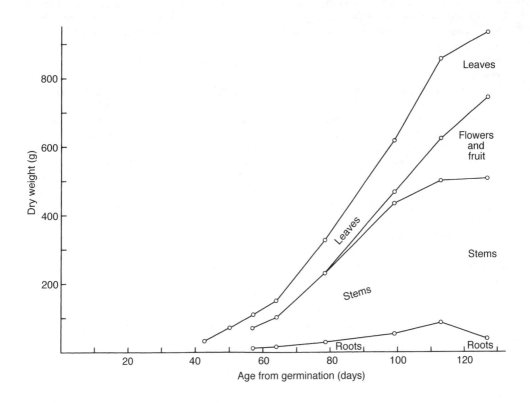

Fig. 3.2. The relative distributions of mean dry weight at harvest of the various types of plant organ, in relation to age for plants of *Helianthus annuus* (after Evans, 1972).

The analyses of partitioning of assimilates in relation to pest attack can provide a greater understanding of the process of yield loss. However, it remains an area that has not yet been adequately addressed by entomologists.

3.5.1 Crop growth models

Crop growth can be modelled at different levels of physiological detail (Spitters, 1990). At the simplest level crop growth and yield are modelled as direct responses to weather, climate and the environment (Pace and MacKenzie, 1987). Regression models are normally used to predict yield against one or a number of variables such as rainfall or temperature. As with all such empirical approaches the models are of little use outside the environmental ranges for which they were developed (France and Thornley, 1984). In general, empirical growth models have little direct application to pest loss estimation although they are often components of mechanistic models, which attempt to explain the effects of the environment on lower level plant physiological processes. Among those most often considered are light interception and photosynthesis, root activity and nutrient uptake, leaf area expansion, transpiration, carbon and nutrient partitioning, growth, development and senescence of organs (Pace and MacKenzie, 1987). This modelling approach can be used to simulate pest damage and crop interactions. Various crop models have been developed: GLYCIM a plant growth model for soybean (Acock et al., 1983); SUBSTOR predicts yield in potatoes (Travasso et al., 1986); OILCROP-SUN a process orientated model for sunflower crops (Villaloboss et al., 1996); a number also simulate pest damage effects, for instance CERES a rice crop growth model (Pinnschmidt et al., 1995) and GOSSYM a cotton pest model (Baker et al., 1983).

Above, the use of leaf area index (L) and leaf area duration (LAD) were introduced in relation to plant dry matter production. The leaf area index is a measure of the leafiness of a crop and leaf area duration is an integral of its changes over a season. Any insect infestation that defoliates a crop will reduce the amount of leaf area which will in turn influence the L and the LAD and hence subsequently reduce crop yield. The photosynthetic capability of a plant is not just dependent on its leafiness, but also on the ability of the leaves to absorb light. The range of light wavelengths important for photosynthesis is broadly similar to the visible spectrum and is referred to as the photosynthetically active radiation (PAR) (Jones, 1983). The amount of PAR absorbed by the leaves and the efficiency of its use, combined with the area of leaves available for absorption will thus influence the plant's dry matter production. A graph of total biomass against intercepted light will give a straight line through the origin which represents the average crop light use efficiency (CLUE). Pests may reduce the amount of light intercepted, CLUE or both (Rossing and Heong, 1997). When CLUE is unaffected, injury simply causes a reduction in the amount of energy available for crop growth (Waggoner and Berger, 1987; see Case Study). Damage is then proportional to the amount of energy that was not intercepted. When CLUE is decreased, the relation between intercepted light and damage is more complex and would justify a move away from empirical studies to use of more comprehensive crop growth models. A number of studies have utilized CLUE (synonymous with radiation use efficiency RUS; g dry matter MJ^{-1} or PAR intercepted by the canopy) to evaluate effects of pest damage (Sadras, 1996; Sadras and Wilson, 1997a).

Case Study: Impact of *Phytophthora infestans* on potato yield (Waggoner and Berger, 1987)

An insect that defoliates a crop or mines leaves will reduce the leaf area available for light absorption. Waggoner and Berger (1987) when considering the effects of plant diseases (the ideas are equally applicable to insect defoliators) proposed that it would be logical to subtract the area of diseased leaves from the LAD by integrating the size of the healthy leaves $[(1 - x)L]$ (where x is the fraction of disease incidence) over the season. This provides a measure of HAD, the healthy leaf area duration.

Waggoner and Berger (1987) used the data of Rotem *et al.* (1983a, b) on yield losses caused by *Phytophthora infestans* to potato to determine if HAD would provide a good predictor of the yield of a diseased crop. The HAD was calculated as the sum of the healthy haulm area multiplied by the interval of time that constituted a season, for three separate periods, the springs of 1978 and 1979 and the autumn of 1978. The relationship of HAD with yield is shown in Fig. 3.3. The data for spring in the two years differ because the growth occurred over an extended period in 1978. The amount of insolation also changed markedly

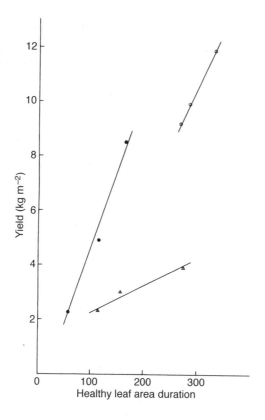

Fig. 3.3. The relationship between the healthy leaf area duration (HAD) and the tuber yield of potatoes during three seasons: spring 1978 (O), autumn 1978 (△) and spring 1979 (●) (after Waggoner and Berger, 1987).

were compounded by higher mealybug populations. A similar situation occurred with the green mite, where the damaging effects of drought on yield were compounded by mite feeding (Gutierrez *et al.*, 1988a,b,c). The model was also used to simulate the population dynamics of a predator and parasitoid of the mealybug, indicating the advantages of such multitrophic models. However, as with the model of Waggoner and Berger (1987), the need for extensive field data makes the approach difficult to utilize in practice. But if yield loss assessment is to develop beyond the use of simple yield/infestation regression then the resources for more studies of the type outlined above may well be necessary.

3.6 Manipulative Techniques

The key to yield loss assessments is the ability to determine the level of pest attack coincident with observed yield loss. Often the variation in levels of pest incidence from field to field and season to season makes it impossible or extremely difficult to plan and carry out meaningful yield loss studies. In order to achieve controlled levels of damage/insect attack in such experiments, entomologists have artificially infested plants with insects, simulated insect damage to plants as well as employing a variety of techniques to manipulate natural levels of infestation in order to more fully evaluate the impact of pests on yield.

3.6.1 Artificial infestation
Artificial infestation involves augmenting natural levels of pest intensity with insects either collected from the wild or with insects that have been reared in the laboratory (Josephson *et al.*, 1966; Kouskolekas and Decker, 1968). The use of insects collected from the field has the disadvantage that the insects may not be available in sufficient quantities when they are needed for the experiment and that the work of collecting adequate numbers may be very

labour intensive. However, collected insects, unlike laboratory reared ones, are acclimatized to field conditions, have a wide genetic base and should respond normally to the presence of suitable host plants. Laboratory reared insects will need to be acclimated before release if their rearing conditions have not been similar to the actual environment. The length of the acclimation period will depend on the difference between the actual and the rearing conditions. Unless wild genes are regularly introduced into laboratory cultures, the reared population can very quickly become inbred, especially when selection for the fittest individuals takes place in small cultures (Dent, 1990). Hence, there is always the risk that the use of reared insects will not truly represent the response of wild insects. This may be particularly the case when insects that have been reared on artificial diets are released onto their normal host plants. However, despite these disadvantages the use of reared insects has many advantages for artificial infestation, including the ready availability of large numbers and particular developmental stages at the most appropriate time for experiment.

One of the simplest methods of infestation with field collected insects, and insects that have been reared on their host, is to place infested cuttings on or around the experimental plants (Wratten *et al.*, 1979). As the excised plant material dies the insects will move off onto the growing plants. The disadvantage of this technique is that it may be difficult to control the level of infestation when the number of insects on each piece of excised plant is highly variable, especially if the insects are small and difficult to count.

Insects can be applied directly to the plant (Bailey, 1986; Kolodny-Hirsch and Harrison, 1986; Mulder and Showers, 1986; Pena *et al.*, 1986) but these techniques can be labour intensive if a large number of plants need to be infested. Some effort has been directed towards the development of simple methods and devices capable of inoculating plants with a specific number of insects. The eggs of Lepidoptera can be attached to plants by the substrate on which they were laid, when the larvae hatch they move from this onto the host (Josephson *et al.*, 1966). A suspension of eggs in agar has been used to infest maize with *Helicoverpa* (Widstrom and Burton, 1970; Wiseman *et al.*, 1974) and with *Drabrotica virgifera* (Palmer *et al.*, 1979; Sutter and Branson, 1980). Larvae of *Spodoptera frugiperda, Diatraea saccharalis, D. lineolata* and *D. grandiosella* have been inoculated on to maize using a manual larval dispenser referred to as a 'bazooka' (Ortega *et al.*, 1980). The bazooka is calibrated to deliver a uniform number of larvae and corn cob grits mixture to each plant to provide a highly uniform infestation of first instar larvae. This technique certainly has a wide potential for artificial infestations and is now commercially available for use not only with Lepidoptera but also with aphids and leafhoppers (Mihm, 1989; Panda and Khush, 1995).

Cages can be used to maintain pest infestations in isolation from the rest of a crop. In this way the levels of attack can be simulated irrespective of the size of the natural pest population. Cages can cover individual plants or large numbers of plants depending on plant size and spacing (Simmons and Yeargan, 1990; Helm *et al.*, 1992; Smelser and Pedigo, 1992). Large cages (e.g. 8 m^3; Wratten, 1975) that cover a larger number of plants have the advantage that the area contained within the cage can be considered as a plot and sampling carried out within it in the same way as other treatment plot experiments. There should be sufficient cages to permit replication of both treatments (insect densities) and controls (cages having no infestation). The growth and yield of the crop inside and outside the control cages should be com-

pared to determine the effect of the cage environment, and the treatment yields should be compared with the yields of plants in the control cages.

Yield loss assessments using artificial infestation techniques are not easy to carry out, however, if consideration is given to careful timing of inoculation (to simulate natural attack) this technique can provide the most effective method for controlled manipulation of conditions.

3.6.2 Simulated damage

If an insect causes damage to the above ground parts of a plant through defoliation or consumption of growing points and flowers then it may be possible to simulate this damage through artificial removal of these plant parts. The effect of damage on yield has been measured in this way in field crops, particularly soybean (Simmons and Yeargan, 1990; Browde *et al.*, 1994a,b,c) and in forestry (Ericsson *et al.*, 1985; Britton, 1988).

This technique appears to provide a simple and practical method of studying the effects of damage but it does have a number of problematic drawbacks. Firstly, it may not be easy to simulate the exact nature of insect damage, since it may not simply be a question of, for instance, leaf area removed. In the case study of Taylor and Bardner (1968; Section 3.3.2) the larvae of the beetle *Phaedon cochleariae* rasped the leaf surface, thus killing leaf tissue by desiccation in addition to that which they ate. The larvae of the moth *Plutella xylostella*, however, ate cleanly through the leaves and caused no secondary damage. Thus the type of defoliation caused by *P. xylostella* would be more easily simulated than that caused by *P. cochleariae*.

Case Study: The effect of artificial defoliation on the yield of pole bean *Phaseolus vulgaris* (van Waddill *et al.*, 1984)

The major defoliating pests of the pole bean in southern Florida are the leaf miners *Liriomyza* spp., cabbage loopers *Trichoplusia ni*, and the bean leafroller, *Urbanus proteus*. The importance of the timing and extent of defoliation on the yield of the pole bean was not known, and information obtained could be used to provide recommendations for the necessity and timing of insecticide application. The study was carried out to determine:

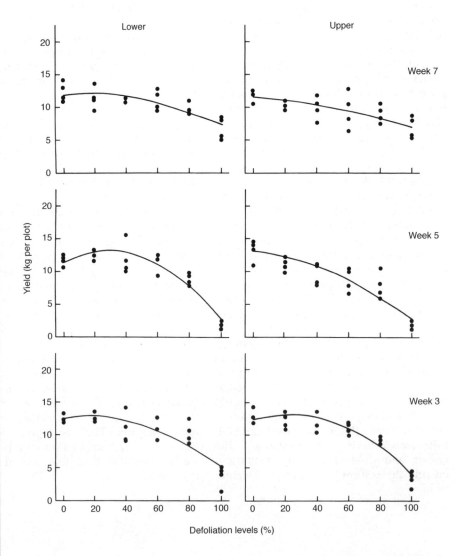

Fig. 3.6. The relationships between pole bean yields and defoliation levels for plants defoliated on a single occasion during weeks 3, 5 and 7 after plant emergence; for plants defoliated starting at the top of the plant (upper) and from ground level (lower) (after van Waddill *et al.*, 1984).

1. The time at which the beans are most susceptible to defoliation; and
2. The relationship between pole bean yield and manual defoliation of beans, when beans were defoliated once, repeatedly and at different levels within the upper and lower halves of the plant.

A split-plot design was used to evaluate the effects of defoliation time (main plots) and various defoliation levels (sub-plots). Treatments were replicated four times. Each sub-plot was defoliated only once, at one of the defoliation levels of 100, 80, 60, 40 and 20% with defoliation starting from either the top of the plant ('upper' treatment) or from ground level ('lower' treatment). Sub-plots which were not defoliated were the controls. The times (main plots) at which defoliation was carried out were 1, 3, 5 and 7 weeks after plant emergence.

For the repeated defoliation experiment, defoliation levels of 0, 10, 20, 30 and 50% were assigned to plots in a randomized complete block design. Foliage was removed weekly so that each plot was maintained at the required defoliation level when compared with the undefoliated control.

The regression analyses of the relationships between yield and levels of defoliation were all significant ($P < 0.005$; Fig. 3.6) except for week 1. Week 5 corresponds to the blooming (flowering) period, as it was at this time that the plants appeared most sensitive to defoliation. As expected in the continuous defoliation experiment, yield decreased with increasing defoliation (Fig. 3.7) so that a continuous 50% reduction in foliage resulted in yield losses of approximately 40%. Van Waddill et al. (1984) concluded from this study that a large proportion of pole bean yield loss could be accounted for by defoliation. Plants were most sensitive to defoliation during the blooming and pod set period.

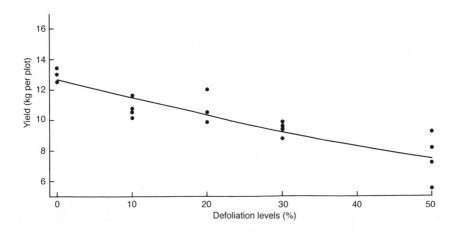

Fig. 3.7. The relationship between pole bean yields and weekly defoliation levels (after van Waddill et al., 1984).

3.6.3 Control of initial levels of infestation

Levels of natural infestation can be controlled to provide differences of pest intensity at the start of a yield loss trial. The most common form of control measure used is insecticides but plant varieties have also been used (Walker, 1987).

Insecticides can be used to create different levels of pest infestation through use of different doses applied at specific times (Egwuatu and Ita, 1982), predetermined application times (Dina, 1976; Kirby and Slosser, 1984) or different insecticides can be used to create and maintain different levels of infestation (Yencho *et al.*, 1986). With natural infestations the absolute level of attack and hence the degree of yield loss between treatment plots cannot be predetermined. In seasons where population levels are low yield loss may be negligible and treatment differences insignificant. In addition, in situations where a number of different pests attack a given crop different pesticides may have to be used to control the different pests in order to separate their effects.

The natural level of pest infestation can also be controlled by the use of plant varieties which exhibit different levels of susceptibility to the pest insect. The range of possible intensities that can be produced depends on the number of varieties having variable but distinct levels of resistance. The biggest drawback with this technique is that ideally in the absence of pest attack or at the same level of attack the different varieties should produce similar yields (Walker, 1981). In practice this is often difficult to achieve.

3.7 Paired Treatment Experiments

The most commonly used and the simplest method of evaluating losses due to insect pests is the paired-treatment experiment (e.g. Fisher and Wright, 1981; Cole *et al.*, 1984; Barnard, 1985). This technique is equally applicable to field crops, orchards,

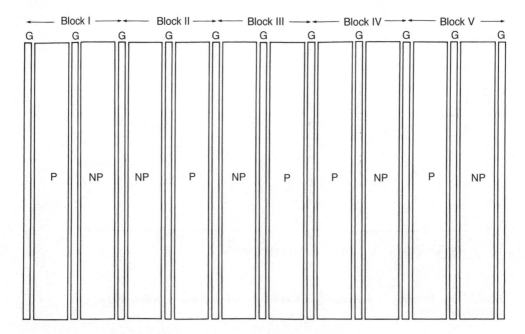

Fig. 3.8. The layout of a randomized complete block strip trial in a paired-treatment experiment where: P = protected plot; NP = unprotected plot and G = guard rows (after LeClerg, 1971).

forests, stored products and livestock systems and in its simplest form involves a comparison of the yield in control plots where there is no infestation (control by insecticides may be necessary) with the yield from plots having natural or controlled levels of infestation. A suitable number of replicates is used. Treatment and control plots should be arranged in pairs, but the allocation of treatment or control to each of the plots should be made at random (Fig. 3.8). In many situations however, it is not possible to follow this simple model, although in principle a comparison is still made between uninfested and infested hosts, but the way this is achieved is dependent on the pest/host system.

In forest systems, a common approach is to compare the growth of trees that are attacked by heavy infestations of defoliators with that of less heavily attacked trees (Thomas and Miller, 1994) or to compare growth of the same trees during periods of high and low pest attack (Seaby and Mowat, 1993; Day and Leather, 1997). Such studies have the problem of having only 'pseudo-controls' but measuring sub-lethal effects of infestation in forest stands is difficult if not impossible to manipulate in any other way. Aerial photography has also provided a means of estimating yield losses from insect pests in forest plantations (Chiang and Wallen, 1971).

Case Study: The effect of the forest tent caterpillar, *Malacosoma disstria*, on the growth and development of aspen, *Populus tremuloides* (Duncan and Hodson, 1958)

The forest tent caterpillar is a defoliating pest of aspens in Minnesota, USA. During an outbreak period between 1953 and 1955, 80 plots were studied and information collected on:

1. the plot type as a whole;
2. individual trees;
3. the growth of the stand.

These data were then used to determine whether the defoliation caused by the forest tent caterpillar reduced aspen growth.

The various intensities of defoliation were categorized as complete, heavy, moderate or light. Individual tree data included a rating of defoliation, a classification of leaf size and increment cores taken from the tree trunk at chest height. Nine to twelve per cent of the trees were estimated to have normal leaf development one year after the last defoliation while between 63 and 88% of all trees had normal leaf development two years after defoliation. The data had been collected from 22 and 14 plots respectively. The average basal area growth (mean of five cores) of the aspen in 1954 was 4.88 cm without defoliation, 4.10 cm (84%) during the first year of light defoliation, 1.38 cm (28%) during the first year of heavy defoliation, 1.02 cm (21%) during the year of heavy following light defoliation and 0.655 cm (13%) during the second year of heavy defoliation.

Overall, the forest tent caterpillar significantly reduced the growth of the aspen. Basal area losses were found to vary with defoliation intensity and history from very little in the first year of light defoliation to nearly 90% of the prospective growth in the second and third years of heavy defoliation. In stands with heavy defoliation a growth reduction of between 14 and 20% occurred after cessation of defoliation.

Estimates of losses in stored grain are dependent on a comparison of damaged grain with either that prior to storage or with undamaged grain sampled at the same time as the damaged. The methods for estimating yield losses using both these approaches have been reviewed by Adams and Schulten (1976). The situation in stored grain systems is complicated by variations in moisture levels during storage which can have a significant effect on results unless these are taken into account during the experiment. The baseline of undamaged pre-storage grain is obtained by the calculation and use of a dry weight/moisture content graph.

To estimate the loss of grain weight due to pests and micro-organisms after a particular period, a standard volume of grain is taken and this is converted to dry grain weight. The dry weight/moisture content graph is then used to find the equivalent dry weight of a sample at the same moisture content so that the weight loss can be calculated:

% dry weight loss =

$$\frac{\text{dry weight from graph} - \text{dry weight in sample}}{\text{dry weight from graph}} \times 100 \qquad (3.7)$$

For example, if the farmer's grain sample with a moisture content of 14% had a dry weight of 500 g then, at 14% moisture content from the graph, the undamaged dry weight is 550 g, and the percentage loss is:

$$\frac{550 - 500}{550} \times 100 = \frac{50 \times 100}{550} = 9.1\% \qquad (3.8)$$

Another method of estimating losses is the count and weight method, although this is mostly inappropriate for either very high or low levels of infestation unless large random samples are taken. A sample is randomly selected and divided into damaged and undamaged grains. The damaged grains can be subdivided according to type of damage or pest. The number of grains in each category and their weights are determined and the percentage weight loss calculated as follows:

% weight loss =

$$\frac{(Wu.Nd) - (Wd.Nu)}{Wu(Nd + Nu)} \times 100 \qquad (3.9)$$

where:
Wu = weight of undamaged grain
Wd = weight of damaged grain
Nu = number of undamaged grains
Nd = number of damaged grains.
The sample sizes for this method must be in the order of 100 to 1000 grains.

In field crops paired treatment experiments carried out in a single field (or a number of paired treatment trials can be carried out in a number of different fields) are more objective and more reliable than comparisons between treated and untreated crops in separate fields. This is because field to field differences in soil type, crop variety and other cultural practices may affect crop responses and yields to an equivalent or to a greater extent than the treatments (Le Clerg, 1971).

In veterinary pest experiments animals that remain uninfested may well exist within the same herd or flock as those that are infested, hence can be considered to be exposed to similar conditions. Any effect due to animal size can be reduced by assessing differences in weight gain between infested and uninfested animals (see Case Study).

Case Study: A comparison of weight gain in sheep attacked and unattacked by the sheep headfly, *Hydrotaea irritans* (Appleyard *et al.*, 1984)

The sheep headfly, *Hydrotaea irritans* is responsible for the development of open wounds on the heads of sheep during the summer in the northern UK. The effects of this damage on the weight gain of 125 Scottish black face lambs was investigated. The weights of the lambs were recorded at the beginning and end of the trial and the weight gain was calculated for each lamb. The lambs' heads were examined weekly for the presence of wounds which were quantified as follows: 0 = no lesions detected; 1 = minor lesions detected only on close inspection; 2 = moderate lesions readily detected; 3 = severe lesions requiring therapy; and 4 = extreme damage. The level of headfly damage during the trial was low but affected animals had damage assessed as lesion score 2 on one or more occasions during the trial. The unaffected animals were free of clinically significant lesions throughout the trial. The effect of headfly damage on weight gain is shown in Table 3.2. Lambs which suffered clinically obvious damage had a mean weight gain of 12.18 ± 2.31 kg compared with 14.05 ± 1.77 kg for those that were unaffected ($P < 0.001$).

Table 3.2. The effect of headfly damage on weight gain of lambs (after Appleyard *et al.*, 1984).

	Affected ($n = 23$) (kg)		Unaffected ($n = 102$) (kg)	
	Mean	SD	Mean	SD
Starting weight	18.45	± 2.38	17.68	± 2.49[NS]
Finishing weight	30.63	± 3.52	31.72	± 3.01[NS]
Weight gain	12.18	± 2.31	14.05	± 1.77[**]

[NS] not significantly different; [**] significant at $P < 0.001$.

The simple paired-treatment experiments are used to determine whether yield losses occur at a naturally occurring level of pest intensity. Such experiments are carried out in a single field over which the pest intensity is often assumed to be of the same order of magnitude between replicate treatments. Multiple treatment experiments take the paired treatment approach one stage further by increasing the number of treatments, with each treatment representing a different intensity of pest insects. A multiple treatment yield loss experiment involves a more complex experimental design, the principles of which are pertinent to many aspects of field trials methodology in pest management.

3.8 Field Trials: Principles

Experimental field trials are common to most components of insect pest management since every type of control option ultimately has to be tested in the field. With small and specific variations field trial techniques that have been developed in relation to agronomic evaluations are applicable to evaluations of yield loss. The coverage presented here provides an overview of the principles of field trials methodology to complement those methods relevant to yield loss assessment already mentioned above. Specific variations of the general methodology are dealt with in more detail in the relevant chapters.

Field experimentation can only be carried out with good knowledge of sampling methodology, experimental design and statistics. Sampling methodology has been dealt with in Chapter 2 and readers requiring further information should refer to Southwood (1978) and Cochran (1977). The advice of statisticians should always be sought at the planning stage of the trials although entomologists should themselves have a good understanding of the principles and statistical techniques to be used in such trials. Most standard statistical textbooks cover both trial design and the statistical analysis of trial data; useful books on the subject include: Bailey (1981), Puntener (1981) and Snedecor and Cochran (1978). An excellent review of principles has also been written by Perry (1997).

Field experimentation should begin with small plot observations and if the results are positive then larger scale trials will be considered. Field trials are costly and may take many months or years to complete, hence it is important that avoidable mistakes are kept to a minimum. This requires a great deal of thought and careful planning, and supervision and monitoring of the experiment. The objectives of the experiment must be carefully defined and the experiment only started after consideration of the crop and the environment in which the pest is to be tackled and of the biology, ecology and epidemiology of the pest (Unterstenhofen, 1976; Reed et al., 1985). Under-estimation of the importance of this understanding can seriously jeopardize the experiment and the usefulness of the results obtained. Reed et al. (1985) warn of anomalies that may occur as a result of conducting experimental trials at research stations. Research stations often have prolonged cropping seasons, irrigation facilities, pesticide free crops, sick plots for maintaining pathogens and weed plots for maintaining weeds. There is a clear need to quantify the differences between crops on research stations and those in typical farmers' fields with regard to the crop itself, the pests and their natural enemies in order to determine just how relevant are results obtained on research stations to the real world (Reed et al., 1985).

The properties of an experimental area are rarely uniform; the land might have a slight gradient in height, soil conditions or exposure to weather such as the number of sunshine hours. To ensure that these effects are distributed over all treatments, i.e. to prevent plots of one treatment experiencing conditions not occurring in plots of other treatments, it is necessary to incorporate randomization and replication into the trial design.

The simplest method of ensuring that particular treatments are not positioned in such a way as to produce any bias is to randomly allocate treatment replicates to spe-

(a)

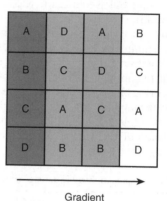

Gradient

(b)

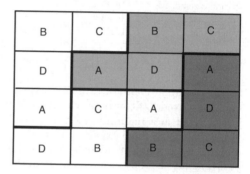

Fig. 3.9. A complete randomized block design: (a) the blocks arranged according to the gradient; and (b) according to various levels of infestation, depicted by the intensity of shading.

cific locations within the trial area. One way of dealing with this is to use a randomized block design, which is based on the principle that patches of ground that are close together tend to be similar, while more distant patches differ. The blocks in the trial (consisting of a small number of plots) and the plots in each block are considered to be experiencing similar soil, pest and environmental conditions. Within each block, each treatment is randomly allocated a plot position. The conditions may differ between blocks (Fig. 3.9a). The layout and shape of blocks will depend on the conditions in each experimental area; although the ideal shape for a block is a square (this reduces the edge length relative to the area occupied) the experimenter should use his knowledge of the differences in soil fertility, yield uniformity, drainage etc. to determine the exact shape of the blocks. A pre-count of pest numbers may reveal differences in the relative size of pest infestation over the experimental area, and blocks could be allocated to take this distribution into account (Fig. 3.9b). The blocks act as replication for each treatment.

Thus, the randomization of treatment plots within a block ensures that localized variability in conditions does not cause bias and the blocking ensures that any differences over the experimental area do not bias treatment differences. However, when variation gradients within a site are not known, the use of blocks may result in their being positioned across gradients, in which case the assumptions made in any subsequent analysis would be incorrect. In such situations a complete randomized design would be more appropriate. The complete randomized block design is fairly common in experiments where the number of treatments varies between five and 20 with fewer replicates than treatments (Simmonds, 1979).

E	C	B	A	D
A	D	C	B	E
B	E	A	D	C
D	A	E	C	B
C	B	D	E	A

Fig. 3.10. A Latin Square trials design, with five treatments: A, B, C, D and E.

A variation of the basic complete randomized block design is the Latin Square design. This is used in experiments where there are only a small number of treatments. The basic property of the Latin Square is that each treatment must appear once in every row and once in every column (Fig. 3.10). Differences in conditions between rows and differences between columns are both eliminated from the comparison of the treatment means, with a resultant increase in the precision of the experiment (Snedecor and Cochran, 1978). The use of the Latin Square is limited to situations where the number of replicates can equal the number of treatments. To construct a Latin Square, write down a systematic arrangement of the letters and rearrange rows and columns at random. Then assign treatments at random to the letters (Snedecor and Cochran, 1978).

When a large number of treatments is used then the size of the blocks in a randomized block design is large and variation in conditions within the block might occur. For large numbers of treatments a range of experimental designs collectively referred to as 'Incomplete Block' designs may be used. They all have in common the use of compact, smallish blocks, any of which contains only a proportion of the total entries (Simmonds, 1979). These designs make comparisons between pairs of treatments ensuring that all pairs of treatments are equally accurate, that differences between blocks can be eliminated. In balanced incomplete block designs every pair of treatments occurs together in the same number of blocks and hence all treatment comparisons are of equal accuracy (John and Quenouillle, 1977). For example, five treatments may be arranged in ten blocks of three treatments; ABC, ABD, ABE, ACD, ACE, ADE, BCD, BCE, BDE and CDE. In this design each treatment occurs six times and each pair of treatments occurs together in three blocks (John and Quenouillle, 1977). However, this design uses every possible combination of three treatments which requires a large number of replications. The number of replications may be

X1	X1	X2	X1	X1	X1
Y3	Y2	Y2	Y2	Y3	Y1
Z1	Z3	Z2	Z1	Z3	Z3
X2	X1	X1	X2	X2	X2
Y3	Y3	Y2	Y1	Y3	Y1
Z1	Z2	Z2	Z2	Z2	Z3
X1	X2	X2	X2	X1	X2
Y1	Y2	Y1	Y2	Y1	Y2
Z1	Z1	Z1	Z2	Z2	Z3

Fig. 3.11. An example of a factorial trial design with three factors. Factor X had two treatments while factors Y and Z have three treatments each. The number of plots is: $2 \times 3 \times 3 = 18$.

decreased by using designs in which every possible combination of treatments are not used. This subject is dealt with further in Chapter 5.

More complex experimental designs will be required if the interaction of a number of treatments needs to be assessed, for example, when testing the effects of different dosage levels of two insecticides on the yield of resistant and susceptible crop cultivars. In this situation, where the interaction of a number of combinations of different variables needs to be studied, a factorial trial design can be used (Fig. 3.11). Factorial experiments compare all treatments that can be formed by combining the different levels of each of the different factors (variables). A variation of the straightforward factorial design is the split-plot or nested design where precise information is required on one factor and on the

interaction of this factor with a second, but less precision is required on the second factor. This type of design is particularly useful where small scale experiments need to be tested on different large scale schemes such as irrigated and non-irrigated land or land cultivated by different means. The irrigated/non-irrigated plots provide the main plots which are then divided into

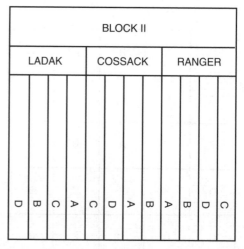

Fig. 3.12. The first two blocks of a split-plot experiment on alfalfa, illustrating the random arrangement of main and subplots (after Snedecor and Cochran, 1978).

smaller subplots, for the smaller scale experiments (Fig. 3.12). The essential feature of a split-plot design is that the subplots are not randomized within each block (Snedecor and Cochran, 1978).

All of the above experimental designs have been used successfully for many years for experiments carried out on research stations. Ultimately trials should be carried out to assess the effectiveness of treatments under conditions experienced by farmers and growers. In the real world it may not be possible or even desirable to set up and carry out experiments using classical experimental design procedures. In the tropics for instance, it can often be difficult to use permanent markers for plots because local people find the materials useful for other purposes and an experiment can easily be ruined if the plot markers are removed.

An alternative to conventional experimental designs that is used in, for instance, mating disruption experiments is the use of whole field experiments or area-wide experiments where plots tend to be large fields or a number of fields and the treatments are not replicated. There are many disadvantages with this approach: systematic errors may arise because treatments are not replicated and uniform trial areas are unlikely on such a large scale; there is a resultant increase in variability and only relative rather than absolute measurements can be obtained (Puntener, 1981). However, if treatments are effective then differences have been demonstrated under realistic practical conditions and can provide tangible evidence of the value of a treatment. Certainly all prospective pest management control options should be tested in on-farm trials as the final proof of their value before being recommended to users.

The choice of plot size for experimental trials will depend on:

1. The type of crop used.
2. The type of equipment needed or used to apply treatments.
3. The amount of plant/pest material required for sampling and evaluation of treatments.

4. The size required to maintain plot variability at a suitable level.

The use of too large a plot will be wasteful of land and resources, while too small a plot will increase the variability of the data. Such variability will be partly dependent on interference or inter-plot effects, when the treatment of one plot interacts with an insect population on an adjacent plot. This effect is thought to be caused by movement of the pest and/or its natural enemies between treatment and control (untreated) plots. In insecticide trials spray drift can also have an effect. The inter-plot effect can potentially influence the results of experimental trials, with yield and insect numbers affected by the proximity of a trial plot to other treated or untreated plots. For instance, the yield of untreated plots of cotton and the size of pest infestation of plots (each 4.2 ha) were affected by the presence of insecticide treated plots 150 m away (Joyce and Roberts, 1959). The insect pests can move from untreated plots where infestation may be high to treated plots where they are killed; thus without equivalent immigration into the untreated plots the infestation is reduced and yield increased to a level greater than would normally be expected in true untreated field conditions. Movement and death of predators and parasites in treated plots would confound the difference in the untreated plots. Joyce and Roberts (1959) proposed an ideal layout of treatments to determine the size of the untreated area that would be required for a plot to approximate to being in an entirely unsprayed environment (Fig. 3.13). Measures to reduce the inter-plot effect have however become commonplace in field trial methodology. These include the use of discarded or disregarded guard rows or border areas and the sampling and harvesting of only the most central area of each plot (Fig. 3.13).

3.9 Economics of Yield Loss

Pest infestations can affect crop/livestock yield which in turn will determine the

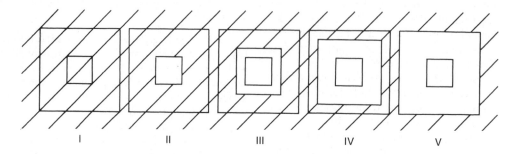

Fig. 3.13. The layout of treatments that would be required to determine the size of the untreated area needed for a plot to approximate an environment in which no insecticides are sprayed. The hatched area is the sprayed area and the central square the actual experimental plot. The margin surrounding the experimental plot gets progressively larger per treatment I–VI (after Joyce and Roberts, 1959).

revenue received by the farmer for sale of the product. If effective pest control techniques are available then farmers can increase their yields and their profits provided they know when it is economic to apply control measures. Thus, for effective and economic control of pests, and for maximizing returns on crop yields information is needed on the relationships between pest infestation and crop yield, crop value, control costs and control effectiveness.

3.9.1 Infestation and yield loss

The type of damage caused by insects varies greatly, due to the confounding effects of the intensity of infestation, duration of attack and plant growth stage. Despite this, it is useful to identify general forms of the relationships between yield and damage caused by insects. These relationships are categorized as: susceptive, tolerant or over-compensatory (Poston *et al.*, 1983).

The susceptive response (Fig. 3.14a) is typical of insects such as seed borers that cause direct damage to their host. With direct damage the yield declines in direct proportion to the number of insects present. For instance, the total number of seeds damaged will be the product of the total number of seeds consumed during the lifetime of a larva and the total number of larvae present. The second response is

referred to as the tolerant response (Fig. 3.14b) and is typical of insects feeding on the plant foliage or roots where a certain level of damage can be tolerated before yield is affected. Above the threshold level of damage, yield declines rapidly with increasing insect intensity, in much the same way as the susceptive response. The third response is the over-compensatory response where the plant initially reacts to the presence of damage in such a way that yield is actually increased above that which would have been achieved in the absence of the pest. This response is usually limited to early infestations and low levels of damage, so that damage greater than that causing over-compensation will reduce plant yield (Fig. 3.14c). Over-compensation is less common than partial or complete compensation for insect damage (Capinera *et al.*, 1986). The ability of a plant to compensate is influenced by several factors, including plant phenology, environmental conditions and the level of injury (Bardner and Fletcher, 1974; McNaughton, 1983).

Even extensive defoliation may not significantly reduce yield (Bennett *et al.*, 1997) and some cultivars have a high tillering capacity and can readily replace tillers damaged by stem borers (Rubia *et al.*, 1996; Bennett *et al.*, 1997) while plant death can induce compensatory growth of surviving plants thereby reducing yield loss (Dewar, 1996). Such compensatory effects need to

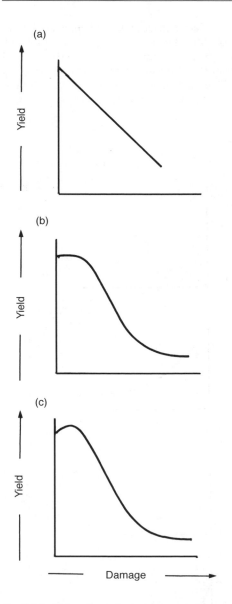

(a)

Yield

(b)

Yield

(c)

Yield

Damage

Fig. 3.14. Generalized plant responses to insect damage: (a) susceptive response; (b) tolerant response; and (c) over-compensatory response (after Poston *et al.*, 1983).

be taken into account in any analysis of crop loss, especially with regard to the economics of control.

The presence of a pest insect infestation

that creates sufficient damage to cause a yield loss will affect the value of the crop and hence, the income of the farmer. The form of the relationship between pest intensity, crop injury, crop yield, crop price and revenue can be generalized for insects attacking foliage/roots or crop products (Fig. 3.15). For pests that attack foliage, roots and crop products (grains, fruits, tubers) the amount of injury is linearly related to the pest intensity. The relationships can differ, however, for the effect of pest injury on yield (Fig. 3.15). When insect pests damage the foliage and roots of plants, and these are not the crop product, then at low pest intensities the price obtained for the crop product will remain high and only begin to tail off when pest intensities are high (Fig. 3.15). The effect of damage to the crop product on price is similar in form to the relationship between pest intensity and yield, while the effect of the pest on revenue, because of the importance of the pest–quality relationship, will result in the curve being a more extreme form of the pest–yield curve (Fig. 3.15). In most cases the pest–revenue curve for foliage and root attacking insects will be identical to that of the pest–yield relationship (Southwood and Norton, 1973).

3.9.2 Economic threshold concept

Historically one of the first major shifts in emphasis away from sole use of chemical insecticides for pest 'control' and towards 'management' of pests and IPM, involved the development and use of economic thresholds. The concept of the economic threshold has remained relatively consistent over time, although there has been a preponderance of terms devised to describe essentially the same thing, which has created some confusion (Morse and Buhler, 1997). The original definition of the economic threshold was coined by Stern *et al.* (1959) as 'the density at which control measures should be determined to prevent an increasing pest population from reaching the economic injury level'. The 'economic injury' level (EIL; sometimes referred to as the 'damage threshold') is the 'lowest population density that will cause economic damage' and 'economic damage' being the 'amount of injury which will justify the cost of artificial control measures' (Stern *et al.*, 1959). What was meant by 'will justify' was not made entirely clear but it has subsequently been accepted to mean 'the density of the pest at which the loss through damage just exceeds the cost of control' (Mumford and Norton, 1984).

The mathematical formulae for calculating economic injury levels are simple

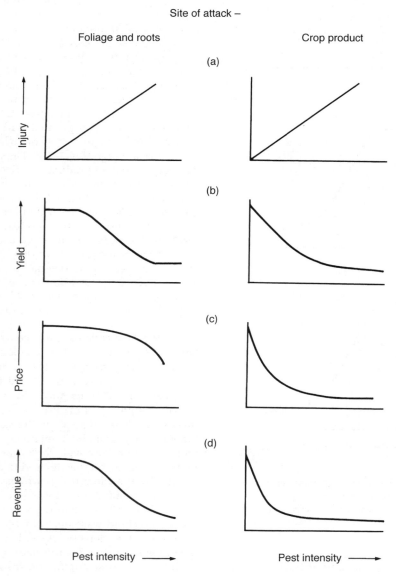

Fig. 3.15. The generalized forms of relationship between pest intensity, injury, crop yield, price and revenue. Effect of: (a) pest on crop injury; (b) pest injury on yield; (c) pest injury on crop price; and (d) pest on crop revenue (after Southwood and Norton, 1973).

enough. A general model for a range of pests that has been widely used is that of Pedigo *et al.* (1986; see below) but other variations on this form are available (e.g. Norton, 1976; Regev *et al.*, 1976).

$$\text{EIL} = C / V I D K \qquad (3.10)$$

where

EIL = economic injury level
C = cost of control ($ ha^{-1})
V = market value of product ($ tonne^{-1})
I = injury per insect per production unit (e.g. % defoliation per insect h^{-1})
D = damage per unit injury (tonnes reduction ha^{-1} = % defoliation)
K = control coefficient (the percentage reduction in pest attack).

However, obtaining the above information to incorporate into the formula is not easy. It is especially difficult where natural enemies are involved in population regulation, where a number of insecticide applications are required during a season or where damage can be caused at different stages of plant growth (Mumford and Norton, 1984). Reichelderfer *et al.* (1984) provide summary guidelines for a method that is probably as generally applicable as any:

1. For a range of pest densities (including zero pests), measure yield and quality of the crop by means of controlled experiments. Pest density measurements should be made by counting or rating the percentage of plant parts damaged. These yield and pest observations should be on plots with no management actions, but should be made early enough in the season so that management actions can be taken.
2. Total crop revenue is computed for each management action at each density by multiplying yield by price per unit of output.
3. Subtract the cost of each management action from the crop revenue for that action at each of the various initial pest densities; call these net revenues.
4. Beginning at very high pest densities and moving to lower densities, compare net revenues for taking a management action with those for taking no action. By

comparing net revenues, find the pest population where net revenue for taking an action is equal to that for not taking an action. The pest density where the net revenues under controlled and uncontrolled circumstances are equal is the action threshold.

The cost of each management action has previously only been viewed in terms of the direct costs (and benefits) to the user but more recently there is a growing appreciation that some management actions also bear an environmental cost. If we can place a monetary value on these environmental costs (e.g. the cost of pollution or of destroying non-target populations with a chemical insecticide) then it is possible to include these costs in the variable C (Pedigo and Higley, 1992). Higley and Wintersteen (1992) estimated the level of risk posed by 32 field crop insecticides to different environmental components (surface water, ground water, aquatic organisms, birds, mammals and beneficial insects) and to human health (acute and chronic toxicity). Producers were then questioned about how much they would be willing to pay (higher pesticide costs or yield losses) to avoid different levels of risk from a pesticide application. By including these environmental costs in C, many of the EILs doubled. Such additions in costs included in EILs may result in less frequent insecticide application or by assigning more realistic costs to insecticide use, alternative methods with lower or negligible environmental costs may become economically feasible (Pedigo and Higley, 1992).

3.9.3 Economic thresholds in practice

Stern *et al.*'s (1959) economic threshold is more commonly referred to as the spray or action threshold and the economic injury level as the economic threshold (Blackshaw, 1995). In practical terms there are different types of economic threshold depending upon their flexibility and how they have been determined (Table 3.3; Poston *et al.*, 1983; Morse and Buhler,

Table 3.3. Types of thresholds employed in crop protection (from Morse and Buhler, 1997; Poston *et al.*, 1983).

Type of threshold	Description
Nominal (e.g. subjective ET)	Based on field experience and logic
	Values remain static
Simple (e.g. subjective ET)	Calculated from crude quantification of the 'average' pest-host relationship in terms of pest damage potential, crop market value, control costs, and potential crop yield
	Generally inflexible to change over time
Comprehensive (e.g. objective ET)	Based on interdisciplinary research incorporating the total production system on a given farm and including factors such as multiple pest and crop stress effects
	Very flexible to change over time
	Fixed ET: set at a fixed percentage of the EIL
	Descriptive ET: includes projections of pest population growth based on simulation models
	Dichotomous ET: based on samples taken over time and classifying the population as economic or non-economic as a result of analysing the sample data

1997). At one end of the scale are the subjective ETs (nominal and simple thresholds), which are more or less fixed figures representing an average of the pest density at which the cost of control is warranted, and at the other end of the scale are objective ETs, which are based on comprehensive research (Morse and Buhler, 1997). The objective ETs are based on estimated ETs and are flexible over time whereas the subjective ones are typically derived by experience and are often no more than 'rules of thumb' or 'guestimates'. In practice, it is the subjective ETs that predominate (Pedigo, 1996) and amongst these the 'action threshold' is very common.

Action thresholds have been calculated for a number of insect species, e.g. *Amblyomma americanum* on cattle (Barnard *et al.*, 1986), *Aeneolamia varia* in sugar cane (Norton and Evans, 1974), *Keiferia lycopersicella* in tomato, *Acyrthosiphon pisum* on green peas (Yencho *et al.*, 1986), *Tipula* species in barley (Norton, 1976; Blackshaw, 1994), *Epiphayas postvittana* in top fruit (Valentine *et al.*, 1996) and aphids in cereals (Elliott *et al.*, 1990). Composite thresholds have been calculated for *Helicoverpa*

armigera, Earias vittelli and *Pectinophora gossypiella* in cotton (Keerthisinghe, 1982) and for *Trichoplusia ni, Plutella xylostella* and *Pieris rapae* in cabbage (Kirby and Slosser, 1984).

Considerable effort has gone into economic thresholds (e.g. Table 3.4 for pests of soybean) but this may have reached its limit for some crop pest complexes. Such a large number of variables makes it extremely difficult to obtain economic thresholds that are generally applicable. Hence, although the economic threshold concept serves as a basis for decision making in insect pest management, the determination of such thresholds has proved to be one of the weakest components in management programmes, with the result that very few research based thresholds have been developed (Poston *et al.*, 1983). Ultimately some situations are just too complex or, irrespective of the suitability of the data or how good the understanding of damage functions, some pest-crop complexes are just inherently uncertain (Table 3.5). Thresholds represent only one way of assisting decision making and should not be seen as a universal solution (Mumford and Knight, 1997).

Table 3.4. Economic injury levels (EILs) for selected foliage- and pod-feeding arthropods on soybean at growth stages R3 to R5. Values based on average recommendations for the major soybean growing regions of the USA, which may differ for different growing regions of the world (Sinclair *et al.*, 1997).

Pest complex	Representative species	Economic injury level
Foliage-feeding arthropods at growth stages R3 to R5		
Lepidopterous	*Anticarsia gemmatalis*	20–25 larvae (>1.2 cm)/row m + 15% defoliation
	Epinotia aporema	30% of growing tips attacked[1]
	Helicoverpa zea	8–10 larvae/m row
	Plathypena scabra	25–30 larvae (>1 cm)/row m + 15% defoliation
	Pseudoplusia includens	20–25 larvae (>1.2 cm)/row m + 15% defoliation
Coleopterous[2]	*Cerotoma* spp.	20 beetles/row m + 15% defoliation
	Epilachna varivestis	15–20 adults + larvae (<0.5 cm)/row m + 15% defoliation
Pod-feeding arthropods at growth stages R5 to R7		
Coleopterous	Cerotoma spp.	20–25 beetles/row m = 8–12% pod injury
Hemipterous	Various species[3]	2–3 large bugs/row m
Lepidopterous	*Helicoverpa zea*	2–3 larvae (>1.5 cm)/m row
	Tortricid and Pyralid spp.	EILs not defined

[1] EIL defined for conditions in southern Brazil.
[2] Other Coleopterous species that could also follow these EILs are: *Aulacophora* sp., *Colaspis brunnea*, *Diabrotica* spp. and various species of *Meloidae*.
[3] Includes: *Acrosternum hilare* and *Nezara viridula*.

3.10 Discussion

Yield loss assessment data are fundamental to insect pest management, since they are the means by which an insect is judged a pest. Yield is also the ultimate criterion by which the efficacy of control measures is assessed, and they form the basis for decision making in insect pest management programmes. Despite these reasons, there are surprisingly few yield loss studies carried out, and this is true for both developed and developing countries (Reed, 1983). To a large extent the reasons for this are associated with the difficulties involved in carrying out yield loss experiments. Even the simplest approach to crop loss assessments, paired-plot comparisons, are fraught with difficulties. Economic entomologists apply insecticides knowing that they provide the largest return if they are applied to dense, well fertilized, high yielding cultivars. Hence, many paired-plot comparisons are carried out under such conditions which can result in massive yield differences, thereby giving high estimates of insect pest losses (Reed, 1983). Given the difficulties of yield loss assessment it is not surprising that yield losses are assumed when large pest infestations occur, or when losses are so obvious that it is considered inappropriate to waste effort quantifying losses when control measures are desperately needed. Even when thresholds have been obtained, they may not be used correctly (Wratten *et al.*, 1990), only

Table 3.5. Conditions that make threshold prediction more or less uncertain (Mumford and Knight, 1997).

	Threshold uncertainty is reduced	Threshold uncertainty is increased
Prices	Fixed	Market values
Type of damage	Direct (i.e. feeding on the part of plant harvested)	Indirect
Crop growth	Consistent (irrigation)	Variable (rainfed)
Pest attack	Short duration Slow reproduction Tied to crop stage Endemic pests Later season	Long duration Fast reproduction Independent of crop stage Immigrant pests Earlier season
Weather conditions	Controlled (greenhouse) Constant (irrigated crops, arid areas)	Temperate areas
Scouting	Cheap Easy to detect damage or presence	Expensive Cryptic stages or damage
Control application	No pesticide resistance Not affected by weather (adjuvants/stickers, or in dry climate)	Variable resistance Affected by weather
Natural enemies	Very high or very low Natural control, stable source nearby	Variable natural control

as a guide (Blackshaw, 1994), they may be inefficient (Waibel, 1987) or uneconomic (Szmedra *et al.*, 1990). Hence, there appear to be few incentives for carrying out experiments to obtain yield loss data and calculate economic thresholds.

It could be that a change of emphasis is required. The methodologies exist, it is just the complex interactions of so many variables over space and time that seem to make the loss assessment approach so unworthwhile. Perhaps longer term experiments with pests in specific ecological zones are necessary to resolve the problem (Judenko, 1972). In the same way that research institutes and experimental sta-tions are prepared to run long term monitoring devices such as suction or light traps (not without substantial investment of time and resources in some instances), so they could run long term field evaluations of crop losses to pests. National and regional research institutes would be obvious locations for such work. These studies should be carried out in tandem with those providing information for crop physiological models (Sections 3.5.1; 3.5.2) and the simulation models (e.g. Allen, 1981). This combined approach should ensure that yield loss assessment receives adequate attention and provides the much needed baseline information for IPM programmes.

4
Insecticides

4.1 Introduction

Chemical insecticides have been considered an essential component of insect pest control since the early 1950s when organochlorine insecticides were first widely introduced. Since that time, however, the problems associated with insecticide misuse and the advent of more ecologically sound IPM approaches have raised doubts about the wholesale use of insecticides as a sole means of pest control. Increasingly the use of chemical insecticides has been considered in terms of judicial applications within the context of a more sustainable IPM approach. Despite this, however, the chemical insecticide market is estimated to be worth US$8 billion annually which ably demonstrates the value placed on insecticides by farmers and other purchasers worldwide. While it may be appropriate to decry insecticides for their poor environmental and safety record, insecticide use remains a cornerstone of pest management and is likely to continue as such for many years to come.

The correct and rational use of insecticides is a complex process that draws on a thorough knowledge of:

- insect population dynamics and the impact of chemical use;
- the active ingredients of the insecticide, its mode of action and formulation;
- delivery of the chemical, its application

and pick-up at the target site;
- ease of use, safety and economics;
- toxicological and ecotoxicological impact and insecticide resistance.

In this chapter, all of these factors are considered in order to provide an overview of the benefits, difficulties and problems associated with chemical insecticide use.

4.2 Objectives and Strategies

In general, if used correctly, chemical insecticides are incredibly effective at killing their target pest. If the most appropriate insecticide is selected, if it is targeted effectively when applied and the timing, rate of application and number of applications is optimized in relation to the application costs and subsequent benefits achieved through increased crop yield, then insecticides remain an efficient and economic means of controlling insect pests. Even when used inappropriately, and applied incorrectly, the perceived benefits of insecticides to the farmer still seem justified in relation to their costs, the yields obtained and the perceived benefits of reduced risk from pest damage. Hence, even in this simplistic way it is easy to understand why insecticides have proved so popular among users as a means of pest control.

One of the objectives of pest management research since the 1950s has been to improve the decisions made in relation to insecticide application, mainly through the use of action or economic thresholds (Sections 2.5.1; 3.9). However, such thresholds should not be seen as a universal solution (Mumford and Knight, 1997); alternative approaches which may be equally valid, depending on the circumstances, include calendar-based applications and standard operating procedures.

A calendar-based or 'scheduled, prophylactic' application means that the insecticide is applied to the crop at regular intervals without information on the level of pest infestation. On each occasion the level of infestation is unknown or assumed to be sufficient to justify application. Where crop value is high, or damage thresholds low, and while the returns from high insecticide input remain high this strategy, at least in the short term, may appear attractive. However, in the longer term the disadvantages can far outweigh the benefits. There are now many well documented examples of how, when insecticides were used prophylactically over a number of years, insecticide resistance developed, secondary pest insurgence occurred, natural enemy populations were destroyed and eventually insecticide use no longer constituted a viable control method. The possible damage to the health of farmers and their families as a result of over-exposure to poisonous chemicals or their residues must also be taken into account.

Standard operating procedures provide an alternative to calendar treatments that substitute a predetermined application schedule (e.g. spraying every two weeks)

for more direct decision making. So, for example, a simple decision rule such as 'spray every time it looks like there will be a three day dry spell at least two weeks after the previous spray' may, if based on sound economic and biological criteria, prove more cost-effective than either calendar or threshold treatments without the need for complex information or management (Mumford and Knight, 1997). The objectives of the farmers will ultimately affect the strategy adopted. Other stakeholders have other objectives, however, and the strategies they propose may be entirely different. Lower dose rates below those recommended to reduce residue levels may be requested by retailers; better means of targeting and application and pesticide specificity will be required by environmentalists wishing to reduce the impact on non-target organisms and the environment. All these and other objectives have an influence over the strategies adopted in the development and use of chemical insecticides. Over time there has been a change in the type of pesticide used (Fig. 4.1), a decline in application rates (Fig. 4.2) and active ingredient, a general increase in specificity and a decrease in persistence combined with reduced mammalian toxicity (Table 4.1) (Geissbuhler, 1981).

Despite these improvements in reducing the disadvantages associated with chemical insecticides they remain hazardous substances that need to be treated as such, but within this context they continue to represent a highly effective and economic (to the user) means of pest control.

Table 4.1. Comparative toxicology and use rates for pyrethroids, carbamates and organophosphates (from Perrin, 1995). a.i., Active ingredient.

Chemical group	Typical range of field use rates (g a.i. ha^{-1})	Acute oral LD$_{50}$ (rat) (mg kg^{-1})	Acute dermal LD$_{50}$ (rat) (mg kg^{-1})
Pyrethroids	5–100	100–5000	1000–5000
Carbamates	125–1000	20–100	1000–5000
Organophosphates	250–1500	10–500	50–3000

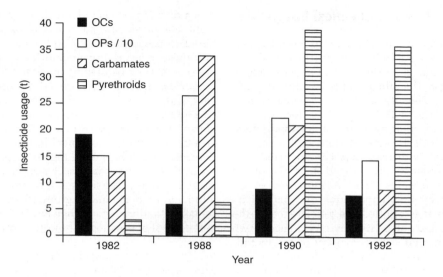

Fig. 4.1. Comparison of insecticide usage on cereals in Great Britain 1982–1992, amount used (tonnes) (after Wilson, 1995).

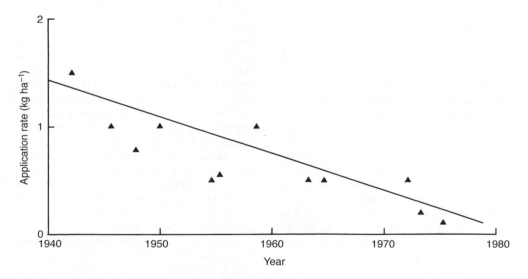

Fig. 4.2. The reduction in application rates of insecticides since 1940. The graph is based on the following insecticides: DDT (1942), parthion (1946), dieldrin (1948), diazinon (1951), azinphos-methyl (1955), dimethoate (1956), carbaryl (1958), chlordimeform (1963), monocrotophos (1965), diflubenzuon (1972), permethrin (1973), deltamethrin (1975) (after Geissbuhler, 1981).

4.3 Classes of Chemical Insecticides

The majority of chemical insecticides consist of an active ingredient (the actual poison) and a variety of additives which improve the efficacy of its application and action. The chemical and physical properties of the technical product, for example, its state, melting point and vapour pressure, will to some extent influence the type of additives that can be used and subsequently the overall formulation of the insecticide. The formulation of an insecticide will influence the method by which it is applied, its persistence in the field and also its toxicity. These two aspects then, the chemistry of the technical product and the insecticide formulation, are central to an understanding of chemical insecticides. Chemical insecticides are usually divided into four major classes, the organochlorines, organophosphates, carbamates and the pyrethroids. In addition to these there are now a number of new compounds representing other 'smaller' classes e.g. insect growth regulatory, imidates, phenylpyrrazoles.

4.3.1 Organochlorines

The best known organochlorine insecticide is DDT, noted for its broad spectrum of activity, its persistence and its accumulation in the body fat of mammals. DDT shares these properties with the other organochlorines, examples of which are aldrin, dieldrin, endosulfan and gamma-HCH or gamma-BHC. The organochlorines can be divided into subgroups according to structural differences but they have in common chemical activity that affects synaptic transmission, their stability, low solubility in water, moderate solubility in organic solvents and a low vapour pressure (Hill and Waller, 1982). The stability and solubility of the organochlorines means that they are highly persistent and this may lead to long term contamination of the environment and gradual accumulation in animals at the higher end of the food chain. If organochlorines present in the body fat reach a high level, when the fat is broken

down during periods of food shortage sufficient chemical can be released into the blood to cause poisoning and even death. For these reasons organochlorine insecticides have been banned by most developed countries, although they are still produced, sold and used in many developing countries where they are usually one of the cheapest insecticides available. The broad spectrum activity of organochlorines, their persistence and hazard to the environment mean that their use in insect pest management is largely considered inappropriate, although as a chemical group there still remain situations where they are an important control option, e.g. mosquito control.

4.3.2 Organophosphates

The organophosphate insecticides (OPs) were originally developed as a by-product of research into nerve gases which was carried out during World War II and they work by inhibition of the respiratory enzyme cholinesterase. They are a large group of insecticides that are prepared from a small number of intermediates by combination with a wide variety of chemicals that are readily available in the chemical industry (Barlow, 1985). Many of the organophosphates are highly toxic to mammals but they are usually non-persistent and hence are less of a threat to the environment than organochlorines (Edwards, 1987). Nevertheless, they should be handled with care since doses may be cumulative (Hill and Waller, 1982). The fast breakdown of organophosphate insecticides is an advantage as far as the principles of IPM are concerned but it also means that the timing of application is critical to ensure an efficient kill. They therefore require an effective monitoring and action threshold strategy in order to ensure timely application and maximum economic return.

A large number of organophosphate insecticides have a systemic action that is particularly effective against phloem feeding insects, many also have effective contact and stomach action. Organophosphate insecticides have been used to control a wide range of pest insects from sheep blowfly (diazinon), locusts and grasshoppers (fenitrothion), nuisance flies such as houseflies (dichlorvos), aphids (dimethoate) and lepidopterous pests (malathion).

4.3.3 Carbamates

These are derivatives of carbamic acid which have been developed more recently than the organophosphates although their mode of action is basically the same as that of OPs, that is, they affect the activity of acetylcholinesterases. However, in the case of carbamates the enzyme inhibition is more easily reversed and the insects can recover if given too low a dose. Carbamates have a broad spectrum of activity and usually act by contact or stomach action, a few possess systemic activity (e.g. aldicarb and carbofuran). Carbamate insecticides are formulated in a similar way to organophosphates with the most toxic ones (i.e. aldicarb and carbofuran) being available only as granules (Walker *et al.*, 1996). The carbamates are used principally to control insect pests of agricultural and horticultural crops.

4.3.4 Pyrethroids

The most effective and safe natural insecticides are the pyrethrins derived from the flowers of *Pyrethrum cinearaefolium* and their synthetic analogues the pyrethroids (Barlow, 1985). Synthetic pyrethroids are, in general, more stable chemically and biochemically than are natural pyrethrins, however they are readily biodegradable and have short persistence. Synthetic pyrethroids have high contact activity and are particularly effective against lepidopterous larvae (King and Saunders, 1984). The effect of pyrethroids is often extended beyond that of other insecticides because of their ability to repel insects (Hammond, 1996). Permethrin and cypermethrin have been shown to offer such extended control through repellence of the Mexican bean beetle *Epilachna varivestis* on soybeans (Dobrin and Hammond, 1983, 1985).

Most, but not all, pyrethroids have a very low mammalian toxicity (Elliot *et al.*, 1978). The hazards they present relate mainly to short term toxicity, and particularly toxicity to fish and non-target invertebrates. Although largely formulated as emulsifiable concentrates for spraying they can be microencapsulated (e.g. Tefluthrin) for control of soil insect pests. They are used to control a wide range of insect pests of agriculture and horticultural crops and for use in the control of insect vectors of disease (e.g. tsetse fly in parts of Africa) (Walker *et al.*, 1996). Pyrethroids have also made substantial inroads into public health, industrial, amenity and household outlets, as well as grain and food stores (Perrin, 1995). Pyrethroids have become the first choice for an insecticide in these different situations because dosages can be kept extremely low. For instance, 5 g of deltamethrin or lambda-cyhalothrin can protect the same area of cereals from aphid damage as 0.5–1 kg of an organophosphate and 15 kg can treat as many houses for mosquito control as one tonne of DDT. Since their introduction in the mid-1970s the pyrethroids have proved a powerful insecticide control tool which has in some cases threatened their longer term viability through rapid development of pest resistance.

4.3.5 Insect growth regulators

Insect growth regulators (IGRs) interfere with embryonic, larval and nymphal development, disrupt metamorphosis and reproduction. They are highly selective to insects and arthropods but because they kill through disruption of growth and development they take more time to reduce insect populations than conventional insecticides. IGRs can be classified as juvenile hormones, chitin synthesis inhibitors and triazine derivatives.

Two hormones are involved in the control of larval and nymphal moulting: the moulting hormones or ecdysones and the juvenile hormones. The ecdysones are necessary for the resorption of the old cuticle, deposition, hardening and tanning of the new cuticle while the juvenile hormones are present during the larval stage at each moult to prevent the insect maturing (Bowers, 1971). The ecdysones have generally proved to be too expensive to synthesize and use as control agents. However, recently a new class of insect growth regulators which are non-steroidal ecdysone antagonists have proven effective against Lepidoptera (Chandler *et al.*, 1992; Cadogan *et al.*, 1997). The product tebufenozide demonstrates activity on neonate larvae of *Helicoverpa zea*, *Spodoptera frugiperda* (Valentine *et al.*, 1996) and *Choristonsura fumiferaria* (Cadogan *et al.*, 1997).

The role of juvenile hormones in meta-morphosis is to perpetuate immature growth and development when they are present and permit maturation when they are not (Bowers, 1971). A number of juve-nile hormones have been identified and structures established. Derivatives of these hormones have also been produced, known as juvenile hormone analogues or mimics (sometimes referred to as juvenoids). Potential utilization of the juvenile hor-mone or its analogues is dependent upon application at a late larval or early pupal stage when it can induce morphogenetic damage, resulting in development of inter-mediate larval/pupal or pupal/adult stages or 'monster' individuals that are unable to mature, but take some time to die. Herein lies a disadvantage of juvenile hormones. In pest species in which the larval stage is the most destructive, hormonal extension of the feeding period may reduce any value from control even if the insect does not manage to reproduce. This has restricted the use of these IGRs to situations where only adult stages are pests, e.g. mosquitoes, ants and fleas, or where it is appropriate to prevent the build-up of small populations causing negligible damage to levels causing eco-nomic loss e.g. stored products, long term control of cockroaches (Menn *et al.*, 1989).

A number of juvenile hormone mimics are available, the oldest, methoprene, is used for control of flies (particularly of livestock), fleas, mosquitoes, stored food and tobacco pests and pharaohs ants. A commonly used newer IGR is fenoxycarb which can be used to control *Cydia*

pomonella and *Epiphyas postvittana* (Valentine *et al.*, 1996) and termites (Su, 1994). Pyriproxifen, another mimic, is active against fleas at very small doses and has recently been released in the USA in spray, collar and wash formulations (Wall and Shearer, 1997).

The insect cuticle presents a potentially vulnerable and specific target for the disrup-tion of its chemistry, structure and function by insecticides (Reynolds, 1989). The amino sugar polysaccharide chitin is a particularly important component of the insect cuticle. If synthesis of chitin is disrupted at crucial times, such as egg hatch or moult, then the insect will die. Among the substances known to inhibit chitin synthesis are the benzoylphenylureas, e.g. diflubenzuron, hexaflumuron and triflumuron. These IGRs have been found to be effective against a range of pest species including termites, lep-idopteran, mites and scarid flies.

The substituted melamine, cryomazine, has IGR effects causing reduced growth and eventually death from integumental lesions. However, it is considered to be in its own class of triazine insect larvicides (Kötze and Reynolds, 1989) since it does not act directly on chitin synthesis. Cryomazine death of insects is character-ized by a rapid stiffening of the cuticle, and it seems to have more specificity than the benzoylphenylureas, affecting mostly lar-vae of Diptera.

4.3.6 New classes and insecticide leads

New insecticides are developed through a process known as biorational design and the use of natural product leads or through a directed synthesis approach. Whichever process is used the majority of the impor-tant insecticides that are developed are neurotoxins which target one of four sites: presynaptic acetylcholine production, the GABA-gated chloride channel, the voltage regulated sodium channel and the post-synaptic acetylcholine receptor (Lund, 1985; Wing and Ramsay, 1989).

The pyrethroids still remain the last major insecticide class that has been devel-oped and new pyrethroids continue to be produced (Hammond, 1996; Jin *et al.*, 1996). Pyrethroids interfere at the sodium channel with axonal transmission of the nerve impulse. A new insecticide DPX-KN128, for broad spectrum control of lepidopteran pests, blocks sodium channels which leads to poor coordination, paralysis and ultimately death of the insect (Harder *et al.*, 1996).

The structural features of pyrethroids have been combined with those of DDT to produce a new broad spectrum class of insecticides known as the 'Imidate insecticides'. The imidates are less persistent than DDT and more stable to enzyme hydrolysis than pyrethroid esters (Fisher *et al.*, 1996). This has been achieved by replacing the ester functionality of pyrethroids with an imidate ester having a more hydrolytically stable linkage. The imidates have significant activity against lepidopteran larvae, have favourable environmental properties and low toxicity to fish.

Another new class of insecticide, the phenylpyrazoles, has similarities to DDT. The phenylpyrazoles class is perhaps best represented by fipronil which acts by inhibiting the neurotransmitter γ-aminobutyric acid (GABA). The effect is highly specific to the invertebrate GABA receptor making it very safe to the mammalian host. The avermectins (Fisher, 1990) are also highly active against the GABA site. The avermectins, ivermectin and abamectin are fermentation metabolites of the Actinomycete fungi *Streptomyces avermytilis* and have a broad spectrum of activity against arthropods and nematodes and low vertebrate toxicity (Wall and Shearer, 1997).

The screening of pathogens and their metabolites is proving a viable approach to identifying novel insecticide leads. The active ingredient of the spinosyn class of insecticide (Larson, 1997) was originally isolated from a soil sample of *Saccharopolyspora sinosa* in the Virgin Islands in 1982. Spinosyns utilize a novel nicotinic site of action giving increased sensitivity to acetylcholine. Spinosyn A was first synthesized in 1988 and this and others identified have both contact and stomach action on Lepidoptera, Diptera, Hymenoptera, Thysanoptera, Isoptera and Homoptera and a favourable toxicological profile (Larson, 1997).

The nitroquanidine insecticides, of which imidacloprid is a member, interrupt the transmission of nerve impulses by blocking the nicotinergic receptors on the postsynaptic membrane of insect nerve cells, as opposed to the acetylcholine production on the presynaptic membrane. In this respect imidacloprid behaves in a similar manner to the old fashioned insecticide nicotine (Dewar *et al.*, 1993) but is representative of a new generation, known as neonicotinoids. A second generation of neonicotinoids, the thianicotinyl insecticides, are now emerging (Senn *et al.*, 1998) which have clear advantages of lower dose rates and a much broader spectrum of control than other neonicotinoid insecticides. They can be used on most agricultural crops and control a wide range of sucking and chewing insects, including some Lepidoptera.

Research continues to search for new leads among insect hormones. Despite considerable research effort aimed at identifying or synthesizing anti-juvenile hormone agents, until recently, with the discovery of naturally occurring allatostatic neuropeptides, this field has not fulfilled its undoubted promise (Schooley and Edwards, 1996). However, novel IGRs are still entering the market. Novaluron is a novel benzoylphenylurea which acts like others of its class by inhibiting chitin formation but is dissimilar from other benzoylphenylureas in that it acts both through ingestion and contact (Ishaaya *et al.*, 1996, 1998). Antiecdysis effects have also been found in leads based on hymenopteran venom. The parasitic wasp *Eulophus pennicornis* produces a venom protein which has activity as a moult inhibitor in the tomato moth *Lacomobia oleracea* (Marris *et al.*, 1996; Weaver *et al.*, 1997). The discovery of low molecular weight toxins, the polyamine amides in the venom of certain spiders (Usherwood *et al.*, 1984) and the demonstration that they are glutamic acid receptor antagonists (Usherwood and Blagborough, 1989) provide considerable potential as novel structures for new classes of insecticide (Quicke, 1988).

The development of novel insecticides and new leads continues apace despite the increasing costs of such development for agrochemical companies. In general, the commercial companies still appear to be concentrating on developing broad spectrum products, but with lower mammalian toxicology and generally more favourable ecotoxicological profiles.

4.4 Formulations

The technical product of a chemical insecticide is rarely suitable for application in its pure form. It is usually necessary to add other non-pesticide substances so that the chemical can be used at the required concentration and in an appropriate form, permitting ease of application, handling, transportation, storage and maximum killing power. Thus, chemical insecticides may be formulated as solutions, emulsion and suspension concentrates, water dispersible powders, baits, dusts, fumigants and granules and pellets.

4.4.1 Solutions
One of the early methods of applying DDT was as a solution of the technical product in an organic solvent applied at high volumes. However, since this is both a dangerous and expensive method of application it was later discontinued. Now solutions of technical products are used where the active ingredient is soluble in water or an organic solvent and applied using equipment suitable for ultra-low volume (ULV) application.

4.4.2 Emulsion concentrates
Technical products that are insoluble in water can first be dissolved in an organic solvent combined with emulsifying agents. This product can then be diluted with water to an appropriate strength in the spray tank before use. The emulsifying

agent or agents cause the technical product and organic solvent solution to disperse evenly in the water when mixed by stirring. It is sometimes necessary to maintain a level of agitation in order to keep the chemical dispersed in the tank. If separation of the liquid into two phases occurs, a process known as creaming, then the emulsion has been broken.

4.4.3 Water dispersible powders
Water dispersible powders are, as the name suggests, powders that are insoluble but can be dispersed in water. The technical product can be either a solid or a liquid. Liquid insecticides can also be absorbed onto an inert solid and then used as a water dispersible powder. Solid formulations are ground to a fine powder and surface active agents added to promote dispersion of the particles when they are added to water. If the particles are not agitated during application then a certain amount of sedimentation might occur, this would result in a gradient in the concentration of the technical product in the tank, giving an uneven application rate.

Water dispersible powders are usually applied at high volume rates with low concentrations of the technical product but there is a tendency, also reflected in the use of emulsion concentrates, to use smaller volumes at high concentrations (Barlow, 1985).

4.4.4 Suspension concentrates
Suspension concentrates differ from water dispersible powders in that they are formulated as a finely ground solid held as a suspension in a non-solvent liquid. This suspension is then diluted with the same solvent (usually water) for application. Sedimentation may be reduced by the addition of polymers that tend to 'fix' the solid in the suspension. However, one disadvantage of suspension concentrates is that they may be unable to withstand tropical conditions of storage (Barlow, 1985).

4.4.5 Baits
Bait formulations combine insect attractants with an insecticide. The attractant is

used to lure the insect to the insecticide which is then transferred by contact or ingestion. The attractant may be a potential food source or, more recently, sex pheromones, which lure specific pest species to the bait. The active ingredient of the insecticide must not deter feeding or repel the insect and in some cases, for instance with termites, the insecticide needs to be slow acting (Su *et al.*, 1982). Three types of toxicant are commonly incorporated into baits: chemical insecticides, biopesticides and insect growth regulators (Su *et al.*, 1995). Bait formulations are used to control insects as diverse as locusts and grasshoppers (Caudwell and Gatehouse, 1994), ants and cockroaches.

4.4.6 Dusts

Insecticide dusts consist of the technical product mixed with an inert carrier which is then ground to a fine powder. They are applied, either as a seed dressing or as a foliage application, with a dust blower machine.

4.4.7 Granules and pellets

Granules are solid particles between 0.1 and 2.5 mm in size (Barlow, 1985) that consist of the technical product combined with an inert carrier. Granules provide a safe, easy to handle product which allows for precise targeting and slow release of the active ingredient. The pelleting process by which seeds are coated with an insecticide varies from country to country, but the pellet ingredients are usually organic in nature, such as wood fibre. Pelleting allows a number of chemicals to be applied to a matrix without necessarily coming into contact with the seed, thus reducing the risk of phytotoxicity (Dewar and Asher, 1994). Products such as imidacloprid have been formulated as either a pellet or a granule for use in control of sugar beet pests (Dewar, 1992a).

4.4.8 Fumigants

Fumigants are chemical formulations that have a relatively high vapour pressure and hence can exist as a gas in sufficient con-centrations to kill pests in soil or enclosed spaces. The toxicity of the fumigant is proportional to its concentration and the exposure time. They are characterized by a good capacity for diffusion and act by penetrating the insect's respiratory system or being absorbed through the insect cuticle. All fumigants are toxic to humans (Benz, 1987) and hence should only be used where the safety of the operators can be assured. The number of chemicals suitable for fumigation is limited; examples include hydrogen phosphide, methyl bromide, ethylene dibromide and dichlorvos (Benz, 1987). Fumigants are applied to control stored product pests under gas proof sheets, in gas sealed rooms or buildings or in special chambers, they are, however, costly to use and require strict supervision.

Fumigants are also used in a wide range of quarantine treatments. For instance, methyl bromide for control of lepidopterous pests in stone fruits (Yokoyama *et al.*, 1992), cabbage and lettuces (Yokoyama and Miller, 1993) and hydrogen phosphide for control of Hessian fly in hay (Yokoyama *et al.*, 1994).

4.4.9 Controlled release formulations

The important feature of controlled release formulations is that they allow much less active ingredients to be used for the same period of activity (Quisumbing and Kydonieus, 1990). The different formulations used for pesticides include: reservoir systems with a rate controlling membrane, e.g. tefluthrin applied to the surface of sugar beet pellets (Asher and Dewar, 1994); monolithic systems, e.g. plasticized polyvinyl chloride (PVC) strips with dichlorvos as an active ingredient used for fly control, as tags or collars for cat flea control or as ear tags for hornfly (*Haimatobia irritans*) control and as laminated structures, e.g. chlorpyrifos used for household and industrial use against cockroaches, ants and wasps.

4.4.10 Spray adjuvants

Spray adjuvants are chemicals that are usually part of the formulated insecticide although in some circumstances they are added to the insecticide tank mix prior to application. These additives may improve mixing with the dilutent, or improve insecticide activity in the field. The former type of additive includes dispersants that are added to wettable powders to prevent sedimentation, and emulsifiers which are used to ensure that stable emulsions occur when the oil soluble form of a technical product is mixed with water, as in emulsion concentrates. The types of additive that improve activity in the field include surfactants, spreaders, penetrants, stickers and humectants. The surfactants and spreaders improve both the contact with and the spreading of the insecticide over the sprayed surface, by reducing the surface tension of the spray droplets. The penetrants are oils that are added to improve the penetration of a contact insecticide through the insect cuticle. These are particularly important if the insect has a very waxy cuticle. Stickers are added to increase the adhesion of the spray to the leaf surfaces to improve the persistence of the insecticide, especially in situations where rainfall is high and a large amount of runoff from the foliage is expected. Humectants are more commonly used with herbicides and they decrease the rate of evaporation of the water dilutent in a spray formulation.

4.5 The Target and Transfer of Insecticide

An insecticide can only control an insect pest if it is suitably toxic and applied in such a way that it reaches its intended target. A large number of interacting factors are involved in the process of ensuring a suitably formulated, toxic active ingredient is applied to provide adequate coverage and subsequent pick-up and mortality of the insect. It is first necessary to ensure that, under optimal conditions, the active ingredient is sufficiently efficacious, then that the application parameters produce droplets or deposits that can be picked-up

and transferred to the targeted stage of the insect life cycle.

4.5.1 Efficacy testing

The techniques used to evaluate insecticide efficacy have been reviewed by Matthews (1984, 1997a) and Busvine (1971). The experimental process starts with laboratory evaluation and then progresses with small plot trials and then large scale testing on research station fields (Reed *et al.*, 1985). Following this trials will be carried out as multi-location experiments on the fields of cooperating farmers.

Laboratory bioassays to establish insecticide efficacy include standardization of insect species, stage, sex, age and physiological and behavioural condition (Dent, 1995), since all of these factors will influence the susceptibility of a pest. Extrinsic factors such as temperature, humidity, feeding and time of treatment, density of treated insects and illumination also have an impact on susceptibility and need to be standardized. The insecticide may be applied topically to the outer surface of the insect using micro-pipettes or special syringes, as a residual film applied to a suitable surface (e.g. glass slides, filter papers, leaves) or systemically through the xylem of treated laboratory plants to evaluate efficacy against sucking pests such as aphids and whitefly.

The results of a bioassay (the mortality recorded over a range of insecticide concentrations) are plotted as dosage mortality curves. These are analysed by Probit analysis (Finney, 1971; Gunning, 1991) (Fig. 4.3). Toxicity, quoted in milligrams of active ingredient for each kilogram of body weight (i.e. parts per million of the test organism), is measured as the dose at which 50% of the test insects are killed, in a specified time (often 24 hours) and is referred to as the LD_{50} (LD = lethal dose) (Busvine, 1971; Finney, 1971).

Field trials can be costly, hence there may be situations where smaller scale investigations are needed to more precisely determine the treatments to be used in larger, more formal field trials (Matthews,

1997a). Field trials methodology is dealt with in Section 3.8 (see also Perry, 1997) but there are a number of aspects of field trials methodology that are specific to the testing of insecticides, particularly the size and shape of plots, methods of preventing spray drift and the use of control untreated plots. The size of plot used in insecticide trials will be dependent on the type of insecticide application equipment used to apply the treatments. Quite small plots can be used for granular insecticides, plots treated with knapsack sprayers need to be at least 10 × 10 m in size, while plots of 30 × 30 m have been used with hand held spinning disc sprayers (Matthews, 1984; 1992). Smaller plots than these need to be shielded in some way to prevent drift contamination between plots. Rectangular plots can go some way to preventing drift contamination if the long axis of the plot coincides with the direction of the wind at the time of application, but if the wind direction is at all variable this approach is unsatisfactory. Square plots have a rela-

tively low perimeter length and reduce the area sacrificed to guard rows (Reed et al., 1985). Small plots can be physically shielded from drift particles with erection of portable screens but these require extra staff to move them between plots as each plot is treated. The most reliable method of preventing drift contamination is the use of sufficiently large plots or spraying systems which minimize drift (Matthews, 1981).

The use of untreated control plots can promote undesirable inter-plot effects. In insecticide trials untreated control plots can sometimes be replaced with a standard check insecticide treatment. The check treatment could be the current recommendation for the insecticide and its rate of application. The experiment would then determine the value of differing dosages, timings of application and different insecticides from those currently used. Such an approach would reduce, although not remove, the inter-plot effects. Reed et al. (1985) suggested the need for maintaining standard unsprayed crop areas well away from insecticide sprayed crops to provide information on seasonal fluctuations in pest numbers, unaffected by insecticide treatments, however, there are probably few research stations that could afford this luxury, despite its obvious value.

The insecticide treatment may be applied using conventional spray equipment or, because field trials are often carried out over relatively small areas, a number of specialized sprayers have been developed as plot sprayers (e.g. the Oxford Precision Sprayer) (Matthews, 1997a).

The evaluation of treatments is ultimately concerned with the measurement of crop yield but measures of insect damage and numbers may also be considered relevant. The ease with which these variables can be measured depends on many things but the simplest trials will involve univoltine pests for which only a single application of insecticide is required to reduce yield loss. Where multivoltine pests reinfest a crop after insecticide application, repeated applications may be needed which may require estimation of the extent of pest re-invasion.

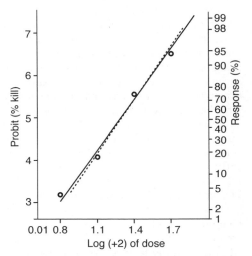

Fig. 4.3. A hypothetical example of a regression line (solid line) used to calculate an LD$_{50}$ through probit analysis. The dotted line is a provisional line from which the expected probits are determined. Response % is the percentage of insects killed; probits are a transformed estimate of the % response (after Matthews, 1984).

Table 4.2. The approximate range of spray volumes applied in field crops and trees and bushes (from Matthews, 1979).

Description of spray	Volume of spray (l ha^{-1})	
	Field crops	Trees and bushes
High volume	>600	>1000
Medium volume	200–600	500–1000
Low volume	50–200	200–500
Very low volume	5–50	50–200
Ultra low volume	<5	<50

The size of an insect infestation can be measured before and after insecticide application to quantify the effectiveness of the treatment and to measure the subsequent rate of reinfestation. Problems may arise with this method with slow acting microbial insecticides because of the time needed to kill insects (e.g. Prior *et al.*, 1996). Counts immediately after application would probably not differ from counts in controls or be greater than those from a check standard chemical treatment. The time taken for insect mortality to occur in the field would need to be known before counts could be usefully employed as a measure of effectiveness. Puntener (1981) provides a number of formulae using insect counts before and after treatment to calculate insecticide efficacy. Two formulae based on counts of surviving insects are included here. The Henderson–Tilton formula should be used when infestation between plots is non-uniform before treatments are applied:

$$\% \text{ efficacy } = 1 - \frac{Ta}{Ca} \times \frac{Cb}{Tb} \times 100 \qquad (4.1)$$

where:

Tb and Ta are the sizes of infestation in the treated plots before and after application.
Cb and Ca are the corresponding infestations in the control or check plot.

If the infestations in plots before treatments are uniform then the Henderson–Tilton formula reduces to Abbott's formula:

$$\% \text{ efficacy } = \frac{Ca - Ta}{Ca} \times 100 \qquad (4.2)$$

since $Tb = Cb = 1$.

Measures of yield are considered in Chapter 3.

4.5.2 Spray characteristics and droplet deposition

Insecticide application aims to provide sufficient insecticide dose and cover to ensure an adequate level of control of the target insect. In the past this was achieved through the use of high volume applications. These used large amounts of the water diluent so that the treated surface became wet and coverage was assured. However, the time required to repeatedly refill large sprayers tanks, and the environmental concern associated with insecticide runoff has led to the use of lower volume applications. These are classified differently for field crops and tree crops (Table 4.2) although the abbreviations HV, MV, LV, VLV and ULV for high, medium, low, very low and ultra low volume are usually accepted in both cases. Reducing the volume that is applied to a surface means that other interacting factors (Fig. 4.4) also have to change in order to ensure adequate coverage. The size of droplets is particularly important as this affects their impaction and retention on the treated surface and, depending on the relative size of the insect, the likelihood that a deposit will transfer

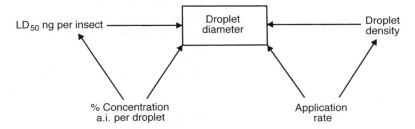

Fig. 4.4. The interaction of droplet diameter with density and application rate, and with LD$_{50}$ ng per insect and the percentage concentration of active ingredient (a.i.) per droplet.

from the surface to the insect. Thus, the type of target must influence the choice of droplet size (Table 4.3) which in turn will influence the volume application rate (Fig. 4.5). The droplet density that is required will depend on the size of the insect and its mobility, higher densities of small droplets would be required for a small, immobile insect egg than for a large, highly mobile caterpillar.

Reducing droplet size but maintaining the same level of active ingredient increases the concentration of the insecticide which is not necessarily desirable. The required concentrations for reduced volumes of spray have to be evaluated by experiment, but the concentration of active ingredient should not normally exceed 10% of a droplet (Matthews, 1984; Fig. 4.6). The influence of differing percentage concentrations of active ingredient for a range of droplet sizes on the toxicity of the acaricide difocol against eggs of the red spider mite, *Tetranychus urticae*, was aptly illustrated by Munthali (1981). This study indicated that the most effective treatment did not occur at the highest concentration

Table 4.3. Optimum droplet size ranges for selected targets (from Matthews, 1979).

Target	Droplet sizes (μm)
Flying insects	10–50
Insects on foliage	30–50
Foliage	40–100
Soil (and avoidance of drift)	250–500

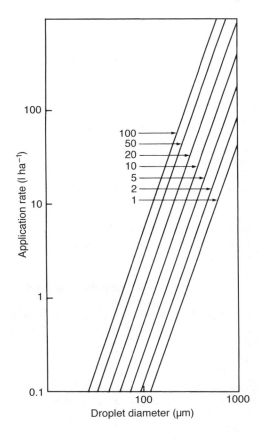

Fig. 4.5. The relationship between the number of droplets (1, 2, 5, 10, 20, 50, 100; number cm^{-2}), their diameter and the volume of spray applied (after Johnstone, 1973).

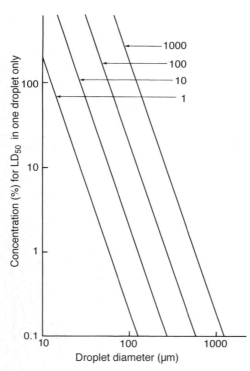

Fig. 4.6. The relationship between toxicity (1, 10, 100, 1000; LD_{50} ng per insect), droplet diameter and the concentration of a.i. for one droplet to contain an LD_{50} dose (Johnstone, 1973).

of active ingredient but at a 1% concentration with an appropriate droplet size and density (Fig. 4.7).

4.5.3 Factors influencing the target and pick-up

An insecticide can only control an insect pest if it is applied in such a way that it reaches its intended insect target. This target can be the eggs, nymphs, larvae, pupae or adults, depending on which are the most harmful stages and the ease with which they can be targeted. The different life cycle stages of an insect vary in behaviour and habitat use which affects their 'transparency' to the insecticide, that is, their presence at a site that can be usefully treated with an insecticide. For instance, boring stages of an insect will be more diffi-

cult to target with an insecticide than foliar feeders. Thus, there is a need to understand the basic biology, resource use and behaviour of a pest insect before an insecticide can be applied effectively (Fig. 4.8).

Insecticides can be transferred to their target insects by direct interception, by gaseous or contact transfer or through ingestion of treated material. Direct interception of droplets by a target insect can occur by either impaction or sedimentation and is dependent on the insect's size and shape as well as droplet size and density. Different parts of an insect will be more likely to intercept droplets than others, so that the surface distribution of droplets will depend on the morphology of the insect. In flying locusts the area on which sedimentation and impaction occurred was calculated using a measurement referred to as the horizontal equivalent area (HEA; Wootten and Sawyer, 1954). When only large droplets and the process of sedimentation on the horizontal plane of the insect is involved, then the HEA was defined as the horizontal plane area which, when passed through a curtain of droplets of a given diameter at the same velocity of a locust, collected the same number of droplets as the locust. However, droplets are also transferred by impaction as the locust flies into a *spray* and hence the HEA by definition is the horizontal area which collects by sedimentation alone the same number of droplets as the locust collects by sedimentation and impaction. An approximate allowance for impaction was made by adding to the area of the locust, the projected horizontal area of the surfaces on which impaction takes place, and then treating the problem as one of sedimentation only (e.g. Fig. 4.9; Wootten and Sawyer, 1954). The HEA was greater for small droplets than for large and was also influenced by the air speed of the locust (Fig. 4.10). A change in air speed from 3 to 5 m s^{-1} (Fig. 4.10) added about 10% to the HEA for larger droplets and about 80% for smaller ones; that is, as the locust increases its air speed the importance of pick-up by impaction of small droplets increases dra-

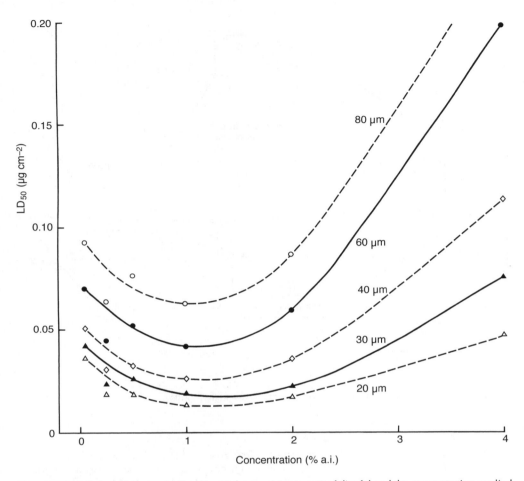

Fig. 4.7. The relationship between the LD_{50} (4 days post-treatment) of dicofol and the concentration applied against the red spider mite *Tetranychus urticae* on the adaxial surface of beans, for a range of droplet sizes (after Munthali and Scopes, 1982).

matically. The impaction of droplets occurred mainly on the wings of the locust and on the head region; this in association with a more even distribution of deposit was thought to be responsible for the observed toxicity response.

Insects such as locusts will readily intercept droplets because they are relatively large and mobile, so that if they are within a spray area their chances of intercepting an insecticide will be increased the more they fly around. The situation with small, sessile life stages of insects is com-

pletely different. In such cases the probability of interception is entirely dependent on the location of the insect and its size in relation to the size and density of droplets. When difocol was applied to the adaxial leaf surface of bean plants with an in-flight diameter of 53 μm and density of 300 drops cm^{-2}, only 10% of eggs of the red spider mite were directly hit by spray droplets (Munthali and Scopes, 1982).

Small insects such as aphids, mites or whitefly may be less susceptible to direct interception of insecticide droplets than

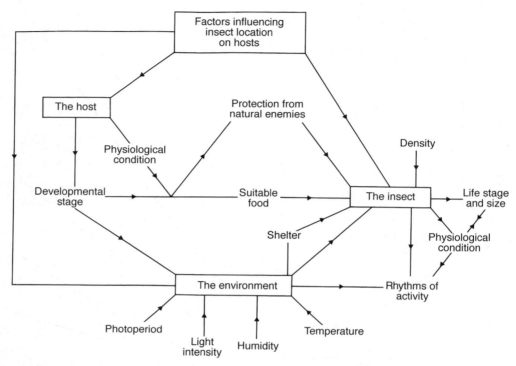

Fig. 4.8. Factors that could influence the location of insects on their hosts.

larger insects but because of their small size they may be more susceptible to a gaseous form of insecticide transfer. Insecticides that have a high vapour pressure ($>10^{-2}$ Pa), such as demeton-S-methyl, dichlorvos and phorate may, in conditions of still air, saturate the air of the boundary layer with toxic vapours. Small insects that are within the boundary layer may receive a lethal dose of an insecticide by gaseous transfer, while larger insects that have a significant portion of their bodies above the boundary layer are less likely to succumb. Ford and Salt (1987) considered the effectiveness of gaseous transfer to be dependent on the insect body size, its location and behaviour (sessile insects would be most susceptible), the air speed at the plant surface, vapour pressure and the toxicity of the active ingredient. However, there are

few insecticides which have high enough vapour pressures at appropriate temperatures for this method of transfer to be generally applicable; the main processes of transfer of insecticides involve direct contact and ingestion by insects.

Contact insecticides are only appropriate against insects that are sufficiently mobile to ensure they come into contact with the insecticide deposit. Contact with the insecticide deposit will depend on droplet size and density. In addition, the proportion of the insect that comes into contact with the deposit and the extent of movement over treated areas will also influence insecticide transfer.

The size and nature (e.g. oils) of the deposit will affect the transfer of insecticide to the insect. While large particles may be more readily dislodged from an

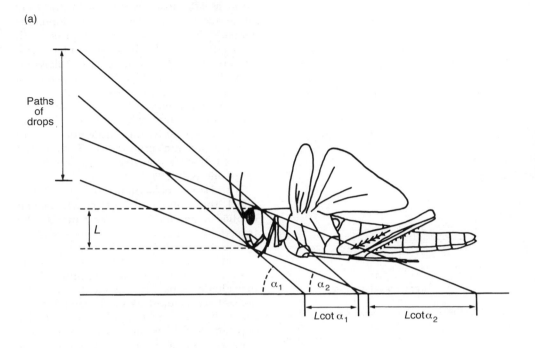

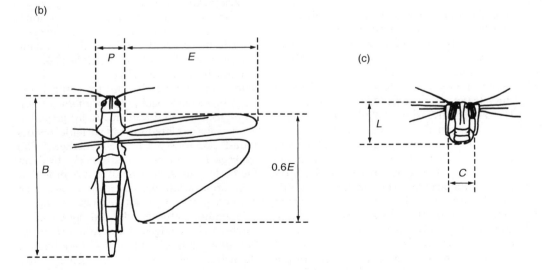

Fig. 4.9. The estimation of the horizontal equivalent (HE) area (HEA) of a locust: (a) variation of HE of a vertical measurement projected at various angles of approach of droplets; (b) horizontally projected area; (c) vertical areas presented to droplets (after Wootten and Sawyer, 1954).

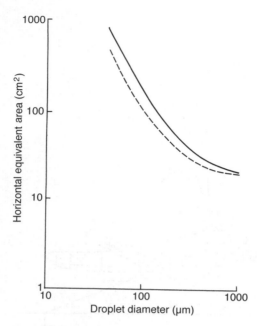

Fig. 4.10. The horizontal equivalent area (HEA) as a function of droplet size at locust flight speeds of 3 m s⁻¹ (– – – –) and 5 m s⁻¹ (———) (after Wootten and Sawyer, 1954).

of time crawling or walking across treated surfaces are most susceptible to this form of insecticide transfer, e.g. lepidopterous larvae. The behaviour of the insect will also affect the likelihood of contact with insecticides. However, insect behaviour may be influenced by the presence of insecticide deposits (Head *et al.*, 1995). *Spodoptera littoralis* moves 45 times faster when walking than when feeding (Salt and Ford, 1984) and this difference could account for the proportion of contacted deposit that is transferred to the insect (Ford and Salt, 1987).

The feeding behaviour of the insect may affect the uptake of a stomach poison applied to the food of the pest insect. The insecticide must be applied to a site on which the insect feeds, the coverage must be sufficient for a lethal dose to be ingested, repellancy and vomiting of the insecticide must be avoided, and the deposit size should be small enough to be eaten by the insect. Wettable powder deposits tend to remain proud of the leaf surfaces and hence are less likely to be ingested than emulsifiable concentrates that sometimes mix with leaf epicuticular waxes (Ford and Salt, 1987).

The behaviour of target insects must always be studied, in some instances it reveals unexpected results and invaluable information. A study of particle size and bait position on the effectiveness of boric acid and sodium borate to control the American cockroach, *Periplaneta americana*, (Scriven and Meloan, 1986) provided a possible explanation for variable results found previously by other workers. The cockroaches do not intentionally feed on the boric acid or borate but if it is provided at an appropriate particle size (< mesh size 80) and placed in positions where normal behavioural patterns take the cockroaches (such as along edges and corners) then the fine powder is picked up by their bodies. However, the insecticidal effect only occurs when the cockroaches ingest the particles as they clean themselves. The results of experiments depend on where the compounds are placed and the particle

insect cuticle once picked up, small particles which can penetrate the insect cuticle more effectively tend to adhere to a surface making transfer to the insect more difficult (Ford and Salt, 1987). The pick-up and retention of DDT crystals was described for the tsetse fly, *Glossina palpalis*, by Hadaway and Barlow (1950). Particles were picked up on the legs and ventral surface of the abdomen. The small particles were transferred from the legs to the antennae, head and other body parts as the flies cleaned themselves. The few larger crystals picked up by the insects were easily removed from the body during cleaning and other movements. Particles were picked up on the legs and ventral aspect of the abdomen as these were the parts of the insect that came into contact with the treated surface. Insects that are not capable of flight or larvae that have a large surface/contact area and spend a large amount

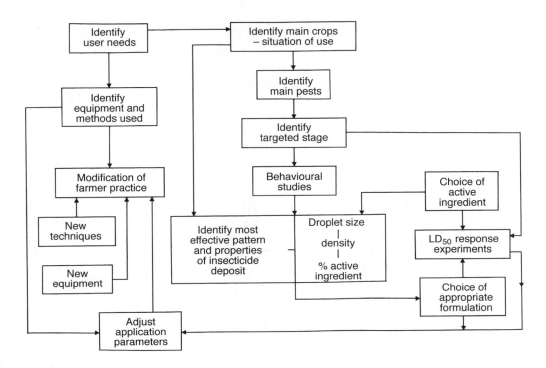

Fig. 4.11. An approach to the development of effective insecticide application procedures based on the properties of insecticide deposits.

size distribution of the powder (Scriven and Meloan, 1986).

A knowledge of insect behaviour is imperative if insecticide targeting is to be improved (Fluckiger, 1989; Adams and Hall, 1990). This knowledge combined with optimal placement, density and droplet size of insecticides provides a means by which more effective transfer can be achieved (Adams and Hall, 1989). An understanding of these interactions can then be used to determine the most effective formulations and to modify spray characteristics and application procedures (Fig. 4.11). This approach is likely to gain favour over empirical observation made during field trials as new techniques lead to a better understanding of processes underlying the field performance of insecticides (Ford and Salt, 1987).

4.6 Application Equipment

A detailed description of the equipment used for the application of insecticides is beyond the scope of this book (see Matthews, 1992) but because spray patterns, droplet size distribution and volume applied are influenced by the type of equipment used, a brief outline of the subject is required.

The application of insecticides is still dominated by techniques initiated over a century ago, that is, as water based sprays applied through hydraulic nozzles (Matthews, 1997a). These sprayers force the liquid insecticide and dilutent, under pressure, through a small aperture nozzle producing a sheet of liquid that becomes unstable, breaking into droplets having a wide range of sizes and volumes. The

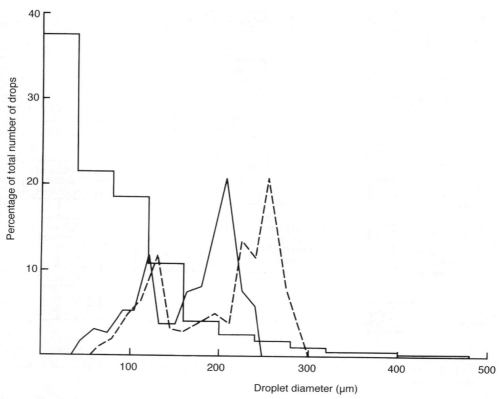

Fig. 4.12. The percentage of the total number of droplets of different diameters produced by a hydraulic nozzle (histogram) and by a centrifugal energy nozzle with a flow rate of 2.0 ml s^{-1} (——) and 55 ml s^{-1} (– – – –) (after Dombrowski and Lloyd, 1974; Matthews, 1979).

volume of the largest droplets can be up to one million times larger than the smallest. Herein lie the disadvantages of hydraulic nozzles. The high proportion of small droplets produced (<100 µm; Fig. 4.12) are prone to drift and evaporation while most of the spray volume is taken up with a few large droplets that tend to bounce or run off the foliage (Matthews, 1985). In arable farming hydraulic nozzles are mounted as a series along a boom which is typically held 50 cm above the ground. In tree and other crops the nozzles are usually mounted in an air stream to project the spray droplets into the crop canopy. Air assisted application is particularly useful in glass houses and stores where very small droplets of < 50 µm diameter are required. Droplets of this size produce fogs which readily penetrate through air to cover surfaces, foliage and stored produce. Use of such small droplet sizes lends itself to rates of insecticide application that are classified as ultra low volume.

Ultra low volume (ULV) application rates became possible through development of equipment that permitted controlled droplet application, i.e. the application of a very narrow range of droplet sizes. The accurate control of droplet size and flow rate was made possible with the development of spinning disc or cup sprayers. These sprayers make use of the centrifugal energy of a spinning disc which produces uniform sized droplets at the edge of the disc. The range of droplet sizes can be varied by the speed of the disc and the flow rate (Fig. 4.12).

A relatively novel form of application involving the electrostatic charging of droplets was developed during the 1980s. The electrodynamic sprayers have no moving parts, they charge the spray as it passes through the nozzle so that the droplets are then deposited on the nearest earthed object, i.e. the crop. Although an excellent idea, in practice the penetration of the crop by droplets was often inadequate to control some pests. Electrostatic charging can be applied to spinning disc and hydraulic nozzles but to date the only commercially successful applicator is a hand held sprayer which is used for spraying cotton in a number of countries in Africa.

Developments to spray equipment in the next few years will be mainly in relation to improved use of existing equipment (Matthews, 1997a). The development of a commercially viable large scale sprayer based on electrostatic charging of droplets would revolutionize insecticide spray equipment, but in the absence of this, efforts will concentrate on better nozzle selection, modifications to improve operator safety and ease of use and reducing wastage of insecticide.

4.7 The Farmer/User Requirements

The decision to adopt insecticide application as a method of insect pest control will not only depend on the type and level of pest infestation but also on a number of other factors which relate to the potential user. Although individual circumstances will differ and the relative importance of each factor will vary, every user will have in common the need for insect control that is easy to use, is safe and economically viable (Fig. 4.13).

4.7.1 Ease of use

The extent to which insecticides have been used in the past tends to lend credence to their effectiveness and to some extent their ease of use, since one would not expect a technique that was ineffective or too difficult to use to gain its present level of acceptance. However, there are still obvious practical problems with insecticide use, such as the hazard during mixing and application and hence the need for protective clothing, difficulties in sprayer calibration, the necessity for appropriate weather conditions, practical thresholds for timing applications and disposal of used containers (Fig. 4.13).

Practical problems associated with insecticide application include the time and difficulty associated with transporting and moving large volumes of water. A large proportion of the time used applying an insecticide can be spent filling a tractor mounted sprayer (ADAS, 1976; Matthews, 1984), a fully loaded knapsack sprayer can weigh 21.5 kg (Pawar, 1986) and it can take up to three to four man days to spray a 1 ha field (Matthews, 1983). The effort required to carry the full sprayer, to refill it (this may be needed over 30 times to spray 1 ha), the continuous pumping required (if the sprayer is a hydraulic energy sprayer) is certainly sufficient to discourage many farmers from using insecticides, especially in a hot climate. Some of these problems have been overcome with the introduction of ULV sprayers, which as the name suggests, use less water than high volume sprayers and hence do not have to be refilled as often. The hand carried versions are light (4 kg; Pawar, 1986) and require less manual effort for application because they are battery powered.

Application equipment needs to be reliable and simple in design so that it requires little maintenance. This simplicity and ease of maintenance has been one of the main advantages of hydraulic knapsack sprayers in developing countries. A further concern for a user is the calibration of the spraying equipment and calculation of application rates. If an insecticide is to be applied at the recommended dose then the user must know the rate at which the insecticide is delivered by the sprayer

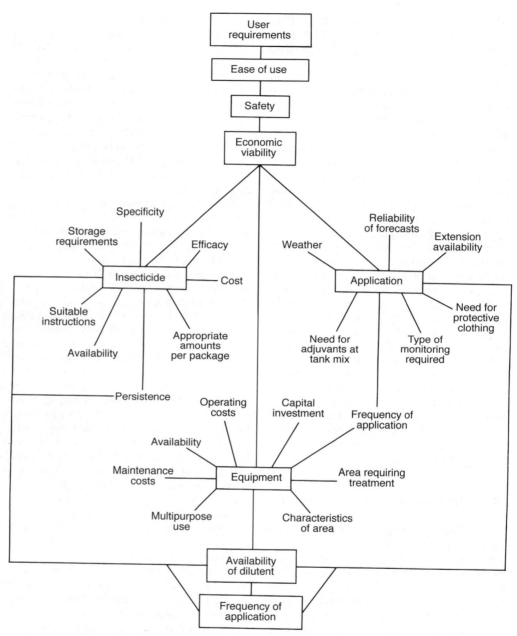

Fig. 4.13. Abiotic factors governing the use of insecticides as a control option.

which in turn requires measures of volume, swath widths and walking speed to be made (Spencer and Dent, 1991). The user then has to calculate the amount of insecticide needed for the total area to be sprayed at the required dose, the number of sprayer loads and the quantity of the insecticide required for each load. Calculations such as this may seriously reduce the chances of an appropriate dose being applied (Dent and Spencer, 1993).

In addition, simple errors in application can also have a significant impact on whether or not the required dose of insecticide is actually applied. For instance, with knapsack sprayers an operator needs to practise walking at a constant speed whilst pumping the hydraulic sprayer, the speed at which he walks will influence the coverage of the crop. For instance, with a swath width of 1 m and a flow rate of 0.5 l min^{-1}, then for a walking speed of 0.4 m s^{-1} the application rate will be:

$$\frac{0.5 \text{ l min}^{-1}}{1 \text{ m} \times 24 \text{ m min}^{-1}}$$
$$= 0.21 \text{ l m}^{-2} \times 10,000$$
$$= 200 \text{ l ha}^{-1} \tag{4.3}$$

whereas if the walking speed is 1 m s^{-1}

$$\frac{0.5 \text{ l min}^{-1}}{1 \text{ m} \times 60 \text{ m min}^{-1}}$$
$$= 0.0083 \text{ l m}^{-2} \times 10,000$$
$$= 83 \text{ l ha}^{-1} \tag{4.4}$$

Even when a farmer has correctly calibrated the sprayer and determined the appropriate amount of insecticide to apply, unless the environmental conditions are suitable, an application should not be made (Table 4.4). Conditions under which insecticides can be applied are referred to as 'weather windows', times when: it is not raining; crop leaves are not wet with dew; the wind speed is appropriate; the plant is not under moisture stress; temperatures are below a certain maximum and sunlight is not too intense (Smith, 1983). Hence, despite their value as crop protection agents insecticides, if applied correctly, are not easy to use.

4.7.2 Safety

Chemical insecticides are inherently toxic substances but they are only a hazard to man if used inappropriately. In developed countries, where users are aware of the need for care and caution when dealing with toxic substances, the hazard from pesticides arises mainly from carelessness and negligence. There is no reason to expect that procedures recommended for the safe handling of insecticides will be followed any more conscientiously than those from other sources of hazard (Barnes, 1976). Provided users in a developed country are aware that they are dealing with a toxic chemical, have a knowledge of the correct procedures and proper protective clothing readily available, then they are totally responsible for their own safety. Accidents occur because users become complacent.

The situation is totally different in developing countries, however, where people may be unused to handling and using toxic chemicals. In some developing countries the use of pesticides may be a community's first contact with toxic chemicals (Barnes, 1976), and they will view the

Table 4.4. Conditions under which pesticides should not be applied (from Smith, 1983).

When it is raining
When leaf is wet with dew
When plant is dry; under moisture stress
Under windy conditions
When temperature is over 18°C
When temperature is below 10°C
In intense sunlight
When mixed with non-compatible materials
On open blossom
To resistant pathogens or pests
To plants to be consumed within 14 days
To untested varieties
To uncleared minor crops
When water supply is hard
Without adding cationic or non-ionic wetter
Before reading instructions on can
After material has been exposed to frost
To rapid growing lush foliage
If unable to convert pints per acre to litres per hectare
If unable to calibrate sprayer
If unable to identify the target
When ground is too wet to carry machine
When unsure of effectiveness of material
If you have no field experience
When customers are looking
If unable to afford cost

problems of hazard and toxicity from a totally different perspective. The safe use of insecticides under these circumstances can only be achieved through education and training. The difficulty is exacerbated because non-users such as family members will inevitably come into contact with insecticide deposits as they wash clothes, prepare food or re-use discarded insecticide containers. Even illness caused by insecticide poisoning may not be diagnosed as such or associated with their use by an unsuspecting community doctor.

The level of hazard associated with different insecticides varies markedly. The World Health Organisation classifies chemicals in terms of hazard according to acute oral and dermal toxicity data for solid and liquid formulations. Thus, good quality granular materials are less hazardous than sprays of the same chemical (Matthews, 1992). The classification in four categories: Ia, Extremely Hazardous, e.g. phorate; Ib, Highly Hazardous, e.g. endosulfan; II, Moderately Hazardous, e.g. deltamethrin; and III, Slightly Hazardous, e.g. malathion, provides a guide to the care that should be taken in dealing with the chemical and the need for a type of protective clothing that should be worn.

For most insecticides it is advisable to wear rubber gloves, overalls, face mask and boots. However, in tropical conditions heavy protective clothing becomes unbearable with high temperatures and humidities. Often no protective clothing is worn (Srivastava, 1974) either because it is not available or it is too expensive for farmers to purchase. At least a light overall or normal clothes should be worn. Many farmers normally walk barefoot, which could seriously increase their exposure to the insecticide; shoes or boots must be worn. A handkerchief tied over the face can be used as a face mask. All clothes used for insecticide application should be washed with soap and detergent after completion of use of the insecticide (Matthews and Clayphon, 1973). Operators should under no circumstances apply insecticides without clothes covering the legs, torso and arms. However,

in reality, this can often occur, thus exposing the spray operator to contamination.

The greatest risk to the user comes from dispensing and mixing the chemical concentrates with the dilutent in the spray tank. A number of solutions have been developed to reduce the operator exposure during mixing. Low level mixing units eliminate the need to climb on a tractor mounted sprayer tank to pour insecticide into the tank; closed chemical transfer systems (CCTS) couples the pesticide container directly to the sprayer or mixing unit, and injection systems mix water with the pesticide concentrate as it is applied (Matthews, 1997a). The agrochemical companies have also agreed to standardize container apertures, which is promoting the development of CCTS.

Other problems with safety aspects of insecticide use concern packaging, labelling and disposal of insecticides and containers. In developing countries adulteration of insecticides can and does take place at repackaging and distribution centres. In Manila's rice bowl 70% of pesticide bottles at local retailers were found to contain chemicals adulterated to more than twice the acceptable standard of deviation (Goodell, 1984). Regulations governing the packaging of insecticides need to be closely controlled, but while monitoring tests for adulteration remain expensive and difficult to carry out this is unlikely to happen, and the dangers from use of such insecticides remain. Labelling of packages needs to be in the local vernacular, using symbols and diagrams meaningful to the user. The label should include the brand name, details of the active ingredient and materials used in the formulation, intended use of the product, full directions for correct and safe use and how to dispose of the container (Matthews and Clayphon, 1973; Matthews, 1992).

The usual recommendation for the disposal of insecticides is to puncture the container and to bury it, preferably in clay soils at a depth of at least 1.5 m (Matthews and Clayphon, 1973). In the tropics disposal of containers in this way is unlikely to occur since the farmer is being asked to dig a pit 1.5 m deep to rid himself of a container that has so many potential and valuable uses to him. The disposal of insecticide containers is something that needs serious re-evaluation; perhaps the container needs to be changed to a type that has no intrinsic value after use.

Case Study: Assessing and quantifying ease of use (ICRISAT, 1989)

The development of a twin spinning disc knapsack sprayer (ICRISAT, 1989) included an assessment of its ease of use, its effectiveness compared with a conventional knapsack sprayer and a comparison with a hand held single disc sprayer of spray droplet distribution on the operator. The swath produced by the twin disc sprayer was 3 m wide, required 15 l of water and about 1.5 man hours to cover 1 ha. A knapsack sprayer needed more than 400 l of water and 200 man hours to spray the same area. Although the twin spinning disc sprayer required less water and less time than the conventional knapsack sprayer to spray 1 ha, the level of control achieved for the leaf miner *Aproaerema modicella*, the thrips *Frankliniella schultzei* and *Scirtothrips dorsalis* and the jassid *Empoasca kerri* on groundnut were equivalent (Table 4.5).

Table 4.5. The number of insects per groundnut plant 36 hours after insecticide application using two types of spraying equipment (from ICRISAT, 1989).

Sprayer type	Post-rainy season		Rainy season		
	Leaf miner	Thrips	Leaf miner	Thrips	Jassids
Conventional knapsack sprayer	0.63	2.00	12.4	0.10	0.02
Twin spinning disc sprayer	0.25	1.72	14.0	0.24	0.02
Control (no spray)	2.48	4.57	43.2	4.20	0.40

When the twin spinning disc sprayer was compared with the conventional hand held single disc sprayer to assess the risk to the operator, there was no significant difference in droplet deposition on the feet, legs and waist of the operator (Table 4.6). However, because the twin spinning disc sprayer has double the work rate, the operator exposure time would only be half that of the single spinning disc sprayer, although the legs of the operator receive a relatively high proportion of the spray droplets when using the twin spinning disc sprayer (Table 4.6). One solution to this problem was to suspend a sheet of polythene from the sprayer to reduce deposition on the operator's legs.

Table 4.6. The deposit of spray (crops cm^{-2}) on the feet, legs and waist of an operator spraying groundnut. The amount of chemical solution falling on to an operator spraying groundnut was measured when the wind direction was at 30° or more across the operator's direction of travel and wind speed was less than 10 km h^{-1} (from ICRISAT, 1989).

Type of spinning disc sprayer	Drops cm^{-2}				FA (%)			
	Feet	Legs	Waist	Total	Feet	Legs	Waist	Total
Single	1.13	0.34	0.10	1.57	16.4	6.4	5.6	28.4
Twin	0.50	0.84	0.04	1.18	11.8	21.2	6.0	31.4
Mean	0.81	0.59	0.07		14.1	13.8	5.8	

FA = area exposed to the chemical as a function of the area of sample paper used for measurement.

4.7.3 Economic viability

Insecticides are without doubt an effective means of killing insects, quickly and on demand. No other control method provides users with an immediate and visibly effective means of response to signs of impending pest outbreaks. This ability has meant that risk-averse farmers spray at the first signs of insect pests, whether the numbers observed are indicative of a serious outbreak or not, and some may even spray insecticides regardless of the possible need as an insurance against a perceived risk. Obviously, there is a cost associated with application of an insecticide. The insecticide itself tends to be inexpensive relative to the value of the crop; for instance in the UK, in apples pesticide costs £747 ha^{-1},

8.3% of total income (Webster and Bowles, 1996); in field beans purchase of Pirimicarb to control black bean aphid costs £12.50 ha^{-1} providing a yield benefit of 3.5% on a crop yielding 5 t ha^{-1} (Parker and Biddle, 1998). Pesticide application costs are estimated to be around £10.00 ha^{-1} (Anon, 1993; Oakley et al., 1998) although the operating costs of the application equipment used will differ (Table 4.7).

In general, the use of pesticides is profitable (e.g. Oakley et al., 1998) or their use is perceived to be profitable through reducing the risk of damage to a crop. Farmers may use pesticides to reduce the risk of pest attack without establishing whether or not control is justified economically. Such prophylactic approaches to control usually

Table 4.7. Operational costs with knapsack and hand-carried CDA sprayers I and II (actual costs of equipment and labour will depend on local conditions; the cost of the chemical, which is also affected by the choice of formulation, is not included) (from Matthews, 1992).

	Manually operated knapsack	Motorized knapsack mistblower	Hand-carried sprayer I	Hand-carried sprayer II
Initial capital cost (£)	60	350	45	45
Area sprayed annually (ha)	20	20	20	20
Tank capacity (litres)	15	10	1	1
Swath width (m)	1	3	1	3
Life in years	3	5	3	3
Hectares h^{-1} spraying[*]	0.36	1.08	0.36	1.08
Overall ha h^{-1} (% efficiency)[*]	0.18 (50)	0.65 (60)	0.31 (85)	0.97 (90)
Use (h annum^{-1})	111	30.8	64.5	20.6
Annual cost of ownership (£)	20	70	15	15
Repairs and maintenance[†] (£)	6	35	4.5	4.5
15% interest on half capital (£)	4.5	26.3	3.4	3.4
Total cost of ownership (£)	30.5	131.3	22.9	22.9
Ownership cost per hour (£)	0.27	4.26	0.36	1.11
Ownership cost per hectare (£)	0.76	3.95	0.99	1.03
Labour costs per hectare[††] (£)	1.38	0.39	0.80	0.25
Operating cost including batteries[**] (£ ha^{-1})	–	0.68	2.2	0.74
Labour costs to collect water[***] (£ ha^{-1})	1.38	0.92	0.13	0.04
Total operating costs per hectare (£)	3.52	5.94	4.12	2.06

[*] Assuming walking speed is m s^{-1}, actual efficiency will depend on how far water supply is from treated area, application rate and other factors.

[†] 10% of capital cost.

[††] Assumes labour in tropical country at £2 per 8 hour day.

[**] Assumes batteries cost 50p each and a set of 8 will operate for 5 h with a fast disc speed. Fuel for mistblower at 44p l h^{-1}.

[***] Water required for washing, even when special formulations are applied at ULV.

Battery consumption is less on some sprayers with a single disc and smaller motors. The 'Electrodyn' sprayer uses only 4 batteries over 50+ hours, so the costs of batteries on a double row swath is 0.6 instead of 2.2.

involve spraying insecticides at set times or intervals during a cropping season, i.e. calendar spraying. In contrast to this insecticides can be applied responsively, according to need in terms of the level of pest attack. This normally depends on the availability of an appropriate monitoring and forecasting system. It is generally considered preferable for farmers to switch from prophylactic calendar based applications to responsive need based applications, simply to improve targeting and to reduce the environmental impact of unnecessary insecticide applications (Dent, 1995). However, the extent to which farmers are willing to make this change is dependent on a number of factors. The difficulties associated with a choice between the two strategies, prophylactic (calendar) spraying and responsive, monitoring and spraying programmes are illustrated by the hypothetical net revenue lines

depicted in Fig. 4.14 (Norton, 1985). In years when the level of pest attack is very low the farmer may benefit from not monitoring or controlling the pest, but with monitoring still providing a more profitable option than prophylactic applications. As the level of pest attack increases the benefits of monitoring and spraying outweigh those of both no control and prophylactic control. However, in situations where pest attack is more frequent the need for the monitoring programme is reduced, and the costs exceed the benefits, making prophylactic control the most financially attractive option (Fig. 4.14; Dent, 1995).

4.8 Insecticide Resistance

Insecticide resistance is the result of the selection of insect strains tolerant to doses of

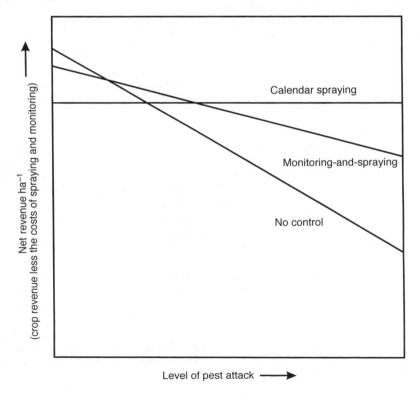

Fig. 4.14. Hypothetical net revenue curves for a prophylactic and a responsive-spray strategy at different levels of pest attack (after Norton, 1985; calendar spraying).

insecticide that would kill the majority of the normal insect population (Cremlyn, 1978). These strains tend to be rare in the normal population, but widespread use of an insecticide can reduce the normal susceptible population thereby providing the resistant individuals with a competitive advantage. The resistant individuals multiply in the absence of intraspecific competition, and over a number of generations quickly become the dominant proportion of the population. Hence, the insecticide is no longer effective and the insects are said to be resistant.

Resistance was recognized as a phenomenon as early as 1911 when citrus scales treated with hydrogen cyanide acquired a certain level of tolerance; however, resistance only really became a concern in the late 1940s with the use of the organochlorine insecticides. The effects were noticed first among pests of medical importance (Busvine, 1976) where the insecticides were being heavily used to control insect vectors of human diseases. The incidence of resistant species of medical importance increased quickly from the late 1950s, and this was followed by a concomitant increase in the numbers of resistant insects of agricultural importance. The number of confirmed resistant insect and mite species continued to rise (Fig. 4.15), to a level of 447 recorded by 1984 (Roush and McKenzie, 1987).

The increasing level of insect resistance to chemical insecticides has been one of the driving forces for change in insect pest management. The development of insect resistance prompted the search for alternative means of control both as a substitute to chemical control and as a means of delaying the establishment of resistance. Faced with the unequivocal fact that no insecticide is immune to resistance, far greater emphasis is being given to insecticide resistance management and to evaluating resistance risk prior to approval of new toxicants (Denholm et al., 1998). The economic costs of resistance to the agrochemical industry can be very high. The cost of a single non-performance complaint can negate the value of 10–1000 individual

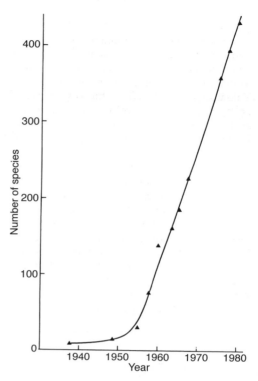

Fig. 4.15. The increase in the number of insect species known to be resistant to at least one insecticide (after Georghiou and Mellon, 1983).

sales (Thompson, 1997). Costs can include sales and technical service staff time, replacement products, legal costs and crop yield replacements as well as loss of future product sales and damage to the reputation of the company. Thus, agrochemical producers have been increasing their commitment to confronting resistance problems through, for instance, evaluating the risk of cross resistance to new insecticides as part of the product development programme. Such cross resistances may occur when a new product shares a target site or common detoxification pathway with compounds already in widespread use against the same pest species. Resistance generally originates through structural alteration of genes encoding target-site proteins or detoxifying enzymes, or through processes (e.g. gene amplification) affecting gene expression (Soderland and Bloomquist, 1990).

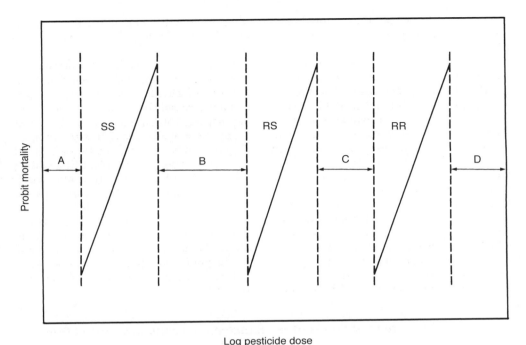

Fig. 4.16. Four ranges of insecticide concentrations: (A) no mortality of any genotype and hence no selection; (B) mortality of SS only (with R resistance dominant); (C) mortality of RS and SS (resistance recessive); and (D) all genotypes killed and hence no selection (after Roush and McKenzie, 1987).

Insecticide resistance can be conferred on individuals by one or a number of genes, but major genes are responsible for the 10–100-fold increases in resistance that are often observed. The major gene inheritance follows the classical Mendelian laws providing individuals with definite genotypic differentiation. Polygenic inheritance, involving cumulative resistance due to many small changes in minor genes, can occur in the field but this type of resistance can be superfluous once a large increase in resistance is conferred on individuals by major genes. Polygenic resistance often occurs in laboratory selection experiments and has been responsible for some confusion over mechanisms of inheritance of resistance (Whitten and McKenzie, 1982). Such laboratory based studies of the inheritance of insecticide resistance therefore have little value, since they are not representative of field induced mechanisms of resistance.

Traditionally resistance studies have identified only two phenotypes, resistant (R) or susceptible (S), and simplifying assumptions have been made about inheritance of resistance (Daly and Trowell, 1996). The process has been seen as one involving simple major gene inheritance where in the absence of selection resistance genes would exist at very low frequencies; the homozygous resistant genes RR would be very rare compared with the heterozygous RS (resistant/susceptible) and the homozygous susceptible SS very common. The presence of the RS gene is maintained purely by a balance between mutation and selection. Once the insecticide is applied the resistant individuals have a selective advantage and their genes rapidly spread through the population (Fig. 4.16; Roush and McKenzie, 1987). In reality, however, the evolution of resistance can involve more than one major gene, the genetic basis of resistance can change over time and the expression of resistance can vary throughout the insect life cycle.

Techniques for detecting insecticide resistance in insects have also advanced greatly in the last ten years. Traditionally resistance was monitored in field populations using bioassays of whole insects but there are many problems associated with the use of such techniques (Brown and Brogdon, 1987). More recently a range of biochemical and molecular methods have become available such as enzyme electrophoresis, enzyme assays and immunoassays (Symondson and Hemingway, 1997) which have revolutionized the means by which resistance can be detected and managed. It seems likely that DNA based detection systems will become available for field detection in the near future. These moni-toring tools will complement the range of tactics available for the management of resistance.

There are relatively few resistance management tactics that have been proposed that are sufficiently risk-free and intuitive to stand a reasonable chance of success in the majority of circumstances. Foremost among these are: (i) restricting the number of applications over time and/or space; (ii) creating or exploiting refugia; (iii) avoiding unnecessary persistence; (iv) alternating between chemicals; and (v) ensure targeting is against the most vulnerable stage(s) in the pest life cycle (Denholm et al., 1998). However, experience would suggest that the most difficult challenge in managing resistance is not the identification of appropriate tactics but ensuring their adoption by growers and pest control operatives.

Case Study: Pyrethroid resistance management against cotton bollworms in Australia

The larvae of the cotton bollworm *Helicoverpa armigera* are major pests of crops in Australia including cotton, grain legumes, coarse grains, oilseeds and many vegetables. The species has 4–6 generations each year, depending on latitude, and has a pupal diapause in winter (Zalucki et al., 1986). Pyrethroid resistance was first detected in natural populations in 1983 after field failures with this insecticide (Daly, 1993). At this time three mechanisms of resistance appeared to be involved (Gunning et al., 1991):

1. A general mechanism of resistance that leads to reduced penetration through the cuticle.
2. A metabolic component inhibited by piperonyl butoxide and so probably associated with the mixed function oxidase (mfo's).
3. Target site insensitivity.

Since then, studies have indicated that the majority of resistance is determined by a single major gene associated with metabolism (Daly and Fisk, 1992); however, the frequency of the resistance phenotype has been shown to fluctuate annually (Forrester et al., 1993). This pattern was shown to be due to the effect of diapause, such that adults that had diapaused had lower estimates of pyrethroid resistance than those adults that had not diapaused which had implications for when pyrethroid insecticides were applied to the summer generation (Daly and Fisk, 1995). In addition, studies of the biology of the field larvae indicated a variation in levels of resistance in different aged larvae such that insecticides needed to target eggs and early larval stages.

A resistance management strategy based on general principles was first introduced in late 1983 (Daly and McKenzie, 1986). The cropping year from spring to autumn was divided into three stages, with the second stage lasting only 5–6 weeks to correspond to a single generation of *H. armigera*. Each major group of insecticide was used in either one or two stages but pyrethroids were limited to the second stage only. This strategy of insecticide rotation proved highly successful, for farmers who were able to keep using pyrethroids as a control option and to the scientists who understood that the strategy moderated the resistance frequency by restricting to one generation the period during which pyrethroid resistant individuals were at a selective advantage relative to susceptible ones.

4.9 Ecotoxicology

The short term benefits from the use of insecticides are immense, with reductions in disease transmission by insect vectors and in losses from field crops, stored products, orchard crops etc. As a result of these benefits insecticides have proved popular as a means of pest control, but there are also indirect costs associated with their use. The evaluation of these effects is dealt with in the context of ecotoxicology, which is 'the study of harmful effects of chemicals upon ecosystems' (Walker *et al.*, 1996). Ecotoxicology involves obtaining data for risk assessment and environmental management, meeting the legal requirements for the development and release of new pesticides, and developing principles to improve knowledge of the behaviour and effects of chemicals (Forbes and Forbes, 1994). This is achieved through study of the distribution of pesticides in the environment and their effects on living organisms at the individual, population and community levels. Since the subject of ecotoxicology largely arose from toxicology and the need to regulate the introduction of toxic chemicals into the environment, the use of cheap, reliable and easy to perform tests on the toxicity of chemicals to individuals has for many years dominated ecotoxicology. However, there is now a move towards evaluation of the impact of pesticides that relates more closely to 'real life' situations. Although the concept of the most sensitive species is still used to relate the results of toxicity tests to the real world there is now increasing emphasis on measurement of effects rather than just death and the use of microcosms, mesocosms and macrocosms (small, medium and large multispecies systems) to

assess the ecotoxicological impact of the use of chemicals (Edwards *et al.*, 1998).

4.9.1 Non-target arthropods

Many non-target arthropods will come into contact with insecticides when they are applied to control pest insects. With only 1% of the applied insecticide actually reaching its insect target (Graham-Bryce, 1977) a large proportion of insecticide is available in the environment to affect non-target species. Unfortunately parasitoids and predators of pests are often susceptible to insecticides and hence can be severely affected (Aveling, 1977, 1981; Jepson *et al.*, 1975; Vickerman, 1988). Since natural enemies are often responsible for regulating pest population levels their destruction can exacerbate the pest problem which leads to further applications of insecticide. In some cases natural enemies that normally keep minor pests in check are also affected and this can result in secondary pest outbreaks. Hence, it is essential, where possible, to use only those insecticides that have a low toxicity to natural enemies and to ensure that they are applied in such a way that the likelihood of natural enemy mortality is reduced, e.g. decreasing plant runoff would decrease the encounters of ground dwelling natural enemies with the insecticide. The general sequence of tests to evaluate the likely impact of insecticides on non-target invertebrates should include laboratory, semi-field and field studies (Fig. 4.17).

Laboratory tests on glass or sand and extended laboratory tests on plant substrates are designed to screen out the harmless insecticides (Brown, 1998). They are arguably the most critical from a regulatory point of view since they are routinely applied to all chemicals and identify those

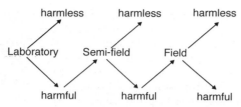

Fig. 4.17. A sequential procedure for testing the side-effects of pesticides on beneficial arthropods (after Hassan, 1986).

which require further testing. Semi-field studies expose the individual species outside but within cages or enclosures incorporating the crop, thereby providing more realistic residue conditions. Field tests for the side-effects on non-target arthropods are designed to investigate potential effects on naturally occurring populations of arthropods. Hence, while the laboratory test will always represent a worst case scenario, with artificial exposure of laboratory reared organisms, the field study offers

realism. Despite this, however, field studies are often difficult to interpret and provide little information on the mechanism of any effects observed (e.g. Jepson, 1987). Ultimately, well designed and carefully conducted field studies will be the final test of acceptability for new and existing pesticides (Brown, 1998). In order for this to occur there is a need for a series of comprehensive field guidelines.

A number of guidelines for sequential testing schemes have been devised (Hassan, 1989; Oomen, 1998); however, for non-target organisms other than honey bees, the exact sequence of tests required, the criteria for moving between them and the interpretation of data obtained at each stage, are still subject to much debate (Denholm *et al.*, 1998). Appropriate test procedures are incorporated into risk assessment schemes (e.g. Table 4.8) which ultimately define the methodology for regulatory risk management.

Table 4.8. The step-by-step development of the risk assessment scheme for honey bees (after Oomen, 1998).

Step	Development procedure	Honey bees
1	Define objective and target	Protection of bees from killing by pesticide use
2	Define acceptability of effects, i.e. define risk classes	Statistically insignificant mortality in exposed bee hives is acceptable
3	Select or develop laboratory testing methods	Dose-response tests LD_{50} (Smart and Stevenson, 1982)
4	Collect laboratory test results and collect field information	Publications, registration data, experiments, incidents
5	Compare and analyse laboratory and field data	
6	Apply acceptability criterion to field information	Discern known harmless (o) and harmful (+) uses
7	Correlate criterion to laboratory data	Find typical hazard ratio of safe cases
8	Set trigger value for laboratory data with safety margin	Hazard ratio < 50
9	Develop sequential decision making scheme, using laboratory information in first tier	EPPO/CoE Scheme 10 (1993) lab–cage–field
10	Test, improve and validate scheme	EPPO/CoE exercise (Aldridge and Hart, 1993)
11	Use scheme and classify risks	EU countries widely used
12	Apply risk management by regulating uses with high risks	High: use not allowed where exposure expected
13	Examples of risk management	No spraying during crop flowering

4.9.2 Other organisms

The most publicized and hence well known example of the negative impact of insecticides on higher organisms has been that of the population decline in raptorial birds due to DDT. The mechanism of decline was the reduced reproductive success due to eggshell thinning which was first discovered in the UK in 1967 (Ratcliffe, 1967). It was found that thinning in excess of 18% was associated with population decline (Peakall, 1993). The link between egg shell thinning and DDT was made in the early 1970s and was shown to be widespread on a global basis. With the decline in use of DDT and other organochlorine insecticides in the early 1970s the extent of the problem was reduced and many populations have recovered even though residues still persist (Edwards, 1987).

Birds are also at risk from insecticides other than the organochlorines. Carbamate and organophosphate insecticides have been recorded as causing mortality to birds and other vertebrates via a number of different sources (Table 4.9; Hunter, 1995).

In recent years, the rigorous approval process for pesticides has meant that problems following the proper use of pesticides are rare (Barnett and Fletcher, 1998; Fletcher and Grave, 1992). The majority of poisoning incidents continue to be the intentional illegal use of pesticides. Often the negligent use of a pesticide is accidental and in some incidences it may reveal a lack of clarity in label instructions (Barnett and Fletcher, 1998).

The pyrethroid insecticides are known to have an impact on fish when entering freshwater rivers and lakes. Pimental *et al.* (1993) quote a USEPA estimate that between 6 and 14 million fish were killed annually between 1977 and 1987 by pesticides in the USA. With organophosphates, however, the hazard is greatly reduced since this group of insecticides is much less toxic to fish (Kips, 1985). Carbamate insecticides are highly toxic to earthworms and some organophosphates have also reduced earthworm numbers (Edwards, 1987). Reducing earthworm numbers can influence soil structure and composition which can have knock-on effects for breakdown of organic matter and drainage.

4.9.3 Fate of insecticides in air, soil and water

The fate of insecticides depends on their physiochemical properties and the medium to which they are applied and ultimately adsorbed.

The presence of insecticides in air can be caused by: spray drift and volatization from the treated surface; the extent of drift is dependent on: droplet size and wind speed; and volatization is dependent on: time after treatment, the surface on which the insecticide settles, the temperature, humidity and wind speed and the vapour pressure of the active ingredient. Generally it may be expected that the ecological risks

Table 4.9. Vertebrate incidents involving carbamate and organophosphate insectides (UK Data 1990–94) excluding abuse (from Hunter, 1995).

Source	Animals affected
Granular applications	Blackbird, sparrow, pheasants, gulls
Secondary poisoning	Buzzard
Seed treatments	Corvids, duck, dog, geese, pigeons
Improper disposal	Gulls, dog, cattle
Poor storage	Dog
Veterinary medicines	Buzzard, corvid, parrot, poultry, dog

associated with insecticides in the atmosphere are probably very small when viewed in the context of atmospheric pollution in general (Kips, 1985) and certainly the impacts are negligible compared with that from direct agricultural applications. For most compounds aerial deposition in remote, natural or semi-natural ecosystems also results in negligible environmental exposure, although for some highly volatile, persistent compounds there appears to be the potential for prolonged exposure, albeit at low concentrations (Dubus *et al.*, 1998).

The sources of insecticides in aquatic systems can be quite diverse, from the chemicals being applied directly to control various pests, from runoff from treated agricultural land, from the atmosphere during precipitation, leaks, spillages and improper use of pesticides. Drinking water supplies in the EU should contain no more than 0.1 µg l^{-1} of a single pesticide and 0.5 µg l^{-1} total pesticides while in the USA and Canada maximum limits for pesticides are based on toxicological assessments (Carter and Heather, 1995). Pesticides are present in water which is treated to produce drinking water at concentrations above regulatory limits and conventional water treatment processes have a limited capacity for pesticide removal (Croll, 1995). Soluble insecticides tend to be washed through aquatic systems and cause little harm but less soluble chemicals can bind to suspended particles and accumulate in sediments and become a source of possible future contamination of water (Edwards, 1973, 1987).

Edwards (1987) placed the various factors that influence the persistence of insecticides in soil in the following order of importance:

1. The chemical structure of the insecticide; the less volatile an insecticide the longer it persists. Granules persist longer than emulsions which persist longer than miscible liquids. Wettable powders and dusts have a very low persistence.
2. The type of soil to which the insecticide is applied; the organic content is the single most important factor influencing persistence. Agricultural soil contains bacteria, fungi, actinomycetes, algae and protozoa and the total and relative numbers are influenced by type of soil, seasonal changes, tillage, crop type, fertilization levels and aeration (Kips, 1985). The amount of soil moisture and rainfall is also important.
3. The microbial population of the treated soil, depth of cultivation of the insecticide, mean temperature of soil (increased temperatures increase degradation).
4. The mineral content and acidity of the soil, clay soils retaining insecticides longer than sandy soils.
5. The amount of plant cover and formulation of insecticide and its concentration.

Pesticide persistence in the soil is also influenced by the phenomenon of enhanced biodegradation following repeated application at the same site. Biodegradation, i.e. primarily decomposition by micro-organisms, is by far the most important mechanism preventing the accumulation of some insecticides in the environment (Römbke and Moltman, 1996). Biodegradation has been shown to be responsible for the failure of a number of soil insecticides (e.g. Suscon Blue).

4.9.4 Insecticide residues

Traces of insecticide in a final product that is to be used as a foodstuff are generally considered to be unwanted contaminants. Published information would suggest that between 60 and 80% of food samples taken do not contain detectable residues and only 1–2% of samples contain any residues in excess of maximum recommended levels (MRLs; Smith, 1995). When residues are detected they can usually be explained by:

1. Insecticide used on the growing crop for control of insects.
2. Insecticides used postharvest for the preservation of food during storage.
3. The carryover of residues from past use of organochlorines.

4. The current use or abuse of organo-chlorine insecticides.

4.10 Rational Insecticide Use

Since the prolific growth of use of insecticides in the 1950s their utilization has been modified as a result of pressure exerted by various stakeholders: the agrochemical companies, the regulatory authorities, users and the general public (Haskell, 1987). At present there is a great deal of research effort devoted to the optimization of insecticide use in order to reduce environmental contamination while maximizing their effectiveness against the target insect. This has led to the consideration of rational insecticide use, the physiological and ecological selectivity of insecticides, where physiological selectivity is characterized by differential toxicity between taxa for a given insecticide and ecological selectivity refers to the modification of operational procedures in order to reduce environmental contamination and unnecessary destruction of non-target organisms.

4.10.1 Timing of insecticide application

The timing of insecticide application primarily depends on the availability of suitable weather windows, the time at which the pest can be best controlled and when least damage will be caused to the environment, especially to non-target organisms. For instance, insecticides should not be applied when the wind speed is above 2.5 m s^{-1} (Matthews, 1981) since this will cause excessive insecticide drift and result in the unnecessary contamination of adjacent areas. Spraying should not proceed if the wind is blowing towards grazing livestock or regularly used pastures, and should always start near the downwind edge of the field and proceed upwind so that the operator continually moves into unsprayed areas (Matthews and Clayphon, 1973).

The timing of insecticide application should coincide with the most appropriate time to control the pest. This may be dependent on an economic threshold value or the time when the most vulnerable stage of the insect life cycle is apparent and susceptible to control. The most judicial use of insecticides against crop pests can be made if pest numbers can be monitored and the insecticide applied as the numbers reach an economically damaging level (Chapters 2 and 3). However, there are still many insect pests for which monitoring systems have not been devised and economic thresholds are not available. In these cases either a scheduled spray regime can be adopted, starting at the period when damage is thought likely to occur (usually a specific crop growth stage) or experiments can be carried out to determine the most appropriate time to apply the insecticide (Hull and Starner, 1983; Richter and Fuxa, 1984).

The choice between whether a scheduled spraying or a specific growth stage application is undertaken will usually depend on the pest population dynamics or whether control is required for a pest complex. Where a pest complex is involved or where overlapping multiple generations occur, the periods during an outbreak when it would be most appropriate for application are less easily defined. Hence, spraying on a scheduled basis can sometimes be the only option until a greater understanding of the pest population dynamics can be obtained. Where discrete, single species and single generation, outbreaks occur and a monitoring technique has not been devised, the above approach for defining the crop growth period appropriate for maximizing yield should be used.

The time of day or the particular period during a season may make beneficial insects more susceptible to insecticides. Bees are particularly prone to insecticide contamination during flowering periods in crops and during the middle of the day, hence applications should not be made at these times. The behaviour and population development of natural enemies should also be taken into account. If natural enemies account for high pest mortality during certain stages of the pest life cycle, insecticides should not be applied at these times, since it could be wasteful of the insecticide and reduce build-up of natural enemies. Since insecticides can have a devastating effect on beneficial insects, research needs to be conducted on optimizing insecticide application in relation to maximizing the potential effects of beneficials.

Case Study: Pre-flowering and post-flowering insecticide applications to control *Aphis fabae* on field beans (Bardner *et al.,* 1978)

The application of a demeton-S-methyl spray and a granular application of phorate were tested for their effectiveness against the black bean aphid, *A. fabae*, on pre-flowering and post-flowering field beans. The granules were compared with the spray formulation in order to determine whether the granules, which are less harmful to bees, provide adequate control of the aphid. The pre- and post-flowering periods were also chosen since there is less bee activity in the crop at these stages. The spray was applied at 0.25 kg a.i. in 370–562 l of water per hectare and 10% phorate granules at 11.2 kg ha^{-1} to randomized plots replicated five or six times and repeated over six consecutive years. The value of the treatments was considered in terms of the size of the aphid infestation, the yield and the economic viability.

In general, the early treatments significantly decreased the number of aphids compared with the untreated plots, while early spray and granule applications appeared equally effective. The early spray and granule applications were better than the late treatments for controlling the aphids and plots sprayed both early and late had the greatest reduction in aphid numbers (Table 4.10).

Table 4.10. The effects of insecticide treatment on aphid infestations (*Aphis fabae*) on field beans. The values in the table are the means of all weekly counts between the end of the primary migration in June and the decline in numbers in late July or early August. The counts are expressed as infestation categories on a logarithmic scale: 0 = no aphids, 1 = 1–10, 2 = 11–100 etc. (from Bardner *et al.*, 1978)

Treatment	Aphid infestation					
	1968	1969	1970	1971	1972	1973
Sprayed early	0.08	0.41	0.16	0.05	0.41	0.23
Sprayed late	0.39	0.74	0.49	0.08	0.53	0.41
Sprayed early and late	0.06	0.23	0.06	0.04	0.27	0.18
Early granules	–	0.45	0.12	0.05	0.37	0.23
Late granules	–	–	0.14	0.07	0.73	0.48
Untreated	0.76	1.09	1.60	0.04	0.78	0.60
SE of differences	±0.188	±0.109	±0.072	±0.188	±0.225	±0.237

There were significant increases in yield in treated plots in all years except 1971 and 1972, although there were large fluctuations within treatments. This was probably caused by variability in aphid abundance and the inherent variability in bean yields. The largest yield increases were the result of early granule application while the combined early and late spray treatment and the early spray treatments were often as good. Lower yield increases relative to untreated plots occurred with both late spray and granule application (Table 4.11).

Table 4.11. The effects of insecticide treatments on the yield of field beans (t ha^{-1}).

Treatment	Yield of field beans					
	1968	1969	1970	1971	1972	1973
Sprayed early	2.13	2.79	1.48	2.16	3.11	4.21
Sprayed late	1.99	2.79	1.38	2.30	2.97	3.90
Sprayed early and late	2.23	3.00	1.39	2.26	3.15	4.27
Early granules	–	2.76	1.60	2.42	3.19	4.38
Late granules	–	–	1.48	1.94	2.90	3.65
Untreated	1.67	2.56	1.07	2.15	3.16	3.70
SE of differences	±0.188	±0.109	±0.072	±0.188	±0.225	±0.237

An economic analysis of these experimental results shows that two sprays were more effective than a single early spray but the extra spray decreased the overall profit (Table 4.12). The application of early granules was the most profitable treatment (£32 t^{-1}) even though the application of a spray costs less than granules.

Table 4.12. The value of mean increase in yield of field beans with insecticide treatments.

Treatment	Mean yield increase (t ha^{-1})	Value of increase at £130 t^{-1}, less cost of treatment (£)	Value of increase, less cost of treatment, if treatments applied only in 1968, 1970 and 1973 (£)
Sprayed early	+0.26	23.32	24.55
Sprayed late	+0.17	11.41	12.63
Sprayed early and late	+0.33	21.73	20.72
Early granules	+0.34	32.01	26.48
Late granules	−0.03	15.90	5.60

These results show that pre-flowering treatments were the most effective and pre-flowering granular applications provided the greatest profit while minimizing the risk to foraging bees.

A great deal of research effort is now directed towards the screening of insecticides for their selectivity so that chemicals which are less toxic to natural enemies can be recommended for use. The physiological selectivity of an insecticide depends on either a decreased sensitivity in the natural enemy at the target site or an enhanced rate of detoxification compared with the pest insect. If the natural enemies have different detoxification pathways from the pest, then this could be exploited by using insecticides that can be detoxified more readily via the pathway used by natural enemies. Insect herbivores and omnivores rely heavily on oxidative detoxification pathways in developing resistance to insecticides (Georghiou and Saito, 1983) whereas entomophagous arthropods appear to use esterase and transferase activity to detoxify insecticides. Thus, the design of insecticides that are primarily detoxified non-oxidatively or activated oxidatively should produce chemicals that are favourably selective for natural enemies (Mullin and Croft, 1985).

There is a great need for research in this subject area to ascertain the biological, physiological and biochemical differences between pests and natural enemies, so that screening for physiological selective insecticides can proceed at a greater pace.

4.10.2 Dosage and persistence
The insecticide dose applied should be sufficient, but no greater than the level required, to provide satisfactory control. The insecticide manufacturer will set the dosage level to an amount that ensures an acceptable level of control, produces acceptable levels of residue and maximizes the return per unit of formulated insecticide. Insecticide doses reduced from recommended rates have been shown to provide adequate levels of control (Hull and Beers, 1985). There are a number of advantages in reducing the dosage level of an insecticide that, both directly and indirectly, benefit the user. The direct benefit would be the lower costs due to the reduced level of chemicals required, while the indirect benefits would be the extra control exerted by natural enemies that may obviate the need for further applications, and a slowing in the rate of development of insecticide resistance if pest populations are only reduced rather than decimated.

The benefits from natural enemies might be diminished, even with lower doses, if a persistent insecticide is used. Mobile natural enemies may avoid an insecticide application or move into a previously treated area, or if present during an application they may survive as a resistant stage or within sheltered refuges. However, if an insecticide is persistent the chances of contamination of natural enemies by insecticide residues are greatly enhanced. Since persistent insecticides are more likely to contribute to insecticide resistance and have a greater effect on natural enemy populations, the advantages of persistence in terms of improved pest control are short lived.

4.10.3 Selective placement
Few insecticide application procedures distribute the insecticide efficiently. With less than 1% reaching the intended insect target a large amount of the applied insecticide is wasted. It is vitally important that the target for the insecticide be precisely defined, and with an understanding of insect biology and behaviour the properties of the insecticide deposit at different application and dosage rates can be optimized (Section 4.5). Selective placement for a particular target can reduce the amount of insecticide applied so that only those surfaces most likely to mediate transfer of a deposit to the insect are treated.

Case Study: Spray deposition in crop canopy and the deposition of insecticide on non-target insects (Cilgi *et al.*, 1988)

The two main cereal aphid pests in the UK, *Sitobion avenae* and *Metopolophium dirhodum*, prefer to feed on the ears or the upper leaves of wheat, respectively (Vickerman and Wratten, 1979), hence insecticide applied against these species would be most effectively targeted if confined to these plant parts. Cilgi *et al.* (1988) sought to determine the distribution of spray within the crop canopy and then to ascertain whether natural enemies at different positions in the crop were susceptible to direct contact spraying.

The pattern of spray deposition within a mature cereal crop was determined by applying Fluorescene, a fluorescent tracer (0.05% w/v + 0.7% wetting agent), to the crop using conventional spraying equipment and a suitable application rate. After application samples of 25 ears, 25 flag leaves and 25 first leaves were taken together with 25 sections of flag leaves that had been placed across 10 cm^2 glazed tiles at ground level. The leaf samples were individually placed in vials

Table 4.13. Deposition levels ($\mu l\ cm^{-2}$) of fluorescent tracer applied at a rate of 200 l ha^{-1} to winter wheat. Values sharing the same letter are not significantly different ($P > 0.05$; from Cilgi *et al.*, 1988).

Crop stratum	Mean deposition rate	SE
Ear	0.360	0.029 a
Flag leaf	0.495	0.053 b
First leaf	0.298	0.021 a
Ground level	0.336	0.017 a

containing 10 ml of phosphate buffer for 8 hours. The amount of tracer released into the buffer was then determined by comparison with a standard calibration curve obtained from a known amount of original spray solution in known volume of buffer measured in a fluorescent spectrophotometer. The data were then converted to volume of tracer per cm^2 from area measurements of the foliage and ear samples. The major difference occurred between deposition on the flag leaf and other sprayed surfaces (Table 4.13) with a large amount of spray reaching ground level.

To determine whether beneficial insects were exposed to direct contact with insecticides, dead individuals of four species were placed at various positions within the plant canopy. Each insect was pinned to a 1.0 × 7.5 cm piece of glazed photographic paper. The tracer deposition on each insect was determined as above with the amount landing per unit insect area, calculated from estimates of the mean ground area coverage of each species.

Insects on the undersides of leaves and at ground level had lower deposition than those insects on the upper surface of leaves and the ears (Table 4.14) hence there will be a different level of risk to beneficial insects in different strata of the crop. The aphid specific predators and parasitoids that inhabit the upper levels of the crop are most at risk, while the mostly nocturnal, ground dwelling polyphagous beneficials are at least risk. Data of this type linked with laboratory LD$_{50}$ toxicity data can provide important insights into the probable levels of mortality to beneficial insects caused by insecticide application (Section 4.9.1).

Table 4.14. Deposition levels ($\mu l\ cm^{-2}$) of fluorescent tracer on insect species placed at different positions in the canopy of winter wheat. Values sharing the same letter are not significantly different ($P > 0.05$; from Cilgi *et al.*, 1988).

Crop position	Insect	Mean deposition rate	SE
Ear	*Coccinella septempunctata*	1.827	0.208 a
Flag leaf (dorsal side)	*Coccinella septempunctata*	0.901	0.169 bc
Ground	*Pterostichus melanarius*	0.412	0.066 cde
First leaf (dorsal side)	*Coccinella septempunctata*	0.309	0.116 de
Ground	*Harpalus rufipes*	0.288	0.070 e
Ground	*Hebria brevicollis*	0.267	0.082 e
Flag leaf (ventral side)	*Coccinella septempunctata*	0.044	0.044 e

4.11 Discussion

Chemical insecticides have provided the principal means of insect pest control since 1945 with the advent of DDT and gamma-HCH. The status which they have achieved provides ample evidence of their value and the confidence placed in their effectiveness by the user, but one might question why insecticides are still used in such large quantities given the well publicized drawbacks associated with their use, such as insecticide resistance, destruction of beneficial insects and environmental contamination. There is of course no simple answer but among other things insecticides do provide users with a means of making an immediate response to an impending pest outbreak, and because of their fast rate of kill they also present the user with tangible evidence of their effectiveness.

Insecticides are also relatively cheap, compared with the potential loss if not applied, and, at the start of the insecticide revolution, they were much cheaper, e.g. dieldrin and DDT, than the more complex insecticides developed more recently, e.g. the synthetic pyrethroids. Insecticides were initially, and to a large extent remain, broad spectrum in activity and as a means of control they can be used against a diverse range of pests in extremely varied conditions and environments, e.g. field crops, stored products, medical and veterinary pests. No other single control method offers this versatility and common assurance of success. As far as farmers or other users are concerned insect pests remain a potential, if not a real, threat to their livelihood.

Insecticides provide an adequate means of controlling pests of their livestock and crops. A farmer may be aware of the problems that insecticides can cause but many of these concerns appear long term and unrelated to the immediate need to reduce the uncertainty associated with pest control. Individual farmers may feel they have little to benefit from changing their practices while neighbours continue in traditional style, an example of Hardin's (1962) 'Tragedy of the Commons'. Also, to date,

any problems with insecticide use such as the development of resistance and environmental contamination with persistent organochlorines have always led to the production of new insecticides as replacements and improvements, so users have had few reasons for changing their practice or seeking alternative methods of control. Insecticide companies are, however, now finding it more costly to develop new insecticides.

In 1956, 198,00 compounds had to be screened to produce a new pesticide product whereas in 1972 10,000 were needed (Johnson and Blair, 1972). The newer insecticides such as the synthetic pyrethroids are more sophisticated than the earlier insecticides such as DDT, requiring many more synthetic steps in the production process, hence production costs are greater. The costs of research and development of a new insecticide from discovery to marketing are estimated at between £10 and 15 million (Haskell, 1987) the recovery of which must be made over 10–15 years. With the rapid development of resistance to many insecticides that occurs through widespread use there is a growing interest within agrochemical companies in prolonging the life of their insecticides. If resistance to the insecticides develops at too fast a rate (especially if cross resistance occurs) an insecticide may become unusable before a satisfactory return on investment can be made. Agrochemical companies are thus interested in ways in which the life of their insecticide products can be prolonged, after all, commercial companies have to be concerned with making a profit. Insecticide companies also have to take into account the concerns of users, special interest/lobby groups and the general public. The powerful influence of such groups on the development and use of insecticides over the past 40 years should not be underestimated. As an example of the influence of special interest groups, Haskell (1987) identified the 'bird lobby' who have effectively nullified progress towards efficient and practical control measures against birds, one of the most damaging groups of pests in world agriculture. Other groups and the pressure of media and public opinion have gradually influenced the course of development of the insecticide industry over 40 years, some of it indirectly through changes in legislation from government and some directly by the need to improve their public image.

The image of the agrochemical industry as the 'bad boys' of agriculture came into being after the publication of *Silent Spring* by Rachel Carson in 1962, heralding the 'Era of Doubt' for insecticide and pesticide use (Metcalf, 1980). The agrochemical industry then, and still today, do their case little justice by presenting too optimistic a picture of their conduct and interests. However, in general such agrochemical companies have reacted positively to criticism and pressure and have contributed to a rationalization of insecticide use through a cautious contribution to the objectiveness of insect pest management. Geissbuhler (1981) considers the agrochemical industry has contributed in five ways:

1. By increasing the biological activity of protection agents against the target organism, which has led to a continual reduction in application rates.
2. By improved selectivity of recently produced compounds.
3. With the development of new products there has been a reduction in properties detrimental to the environment, e.g. a decrease in persistent chemicals.
4. Safety evaluation and risk assessments have been extended.
5. Improved application procedures have aided the targeting and efficiency of insecticides.

It has of course been in the self-interest of the agrochemical industry to make these changes, but their development in this way aids progress towards a more rational use of insecticides within the context of IPM.

The concept of rational insecticide use within IPM has taken a long time to develop, mainly because of the apparent complexity of pest management relative to the more readily identifiable goal of pest eradication. The former requiring a greater understanding of the agroecosystem and pest ecology before implementation is possible in contrast to the quick fix philosophy possible with insecticide use. However, in Europe and the USA the environmental costs associated with insecticide use are now being considered to be too high and there is a general movement in public opinion towards environmentally safe means of pest control and crop production.

The public and environmental pressure groups can and do support alternatives to the use of chemical insecticides. Organically produced fruit, vegetables and meat are gaining in popularity but this will only cater for the higher priced markets. Most people are concerned about what they consider to be excessive and unnecessary use of insecticides; they are not interested in barring them altogether because costs of produce could then be too high. IPM offers the possibility of a rational and reduced use of insecticides which suits the public demands for change.

The extra hazard that insecticides pose in developing countries, due to differences in the perception of toxic substances, the level of education and the availability of trained extension staff for advice, is very real and cannot be emphasized enough. One of the major problems is that the complexity of insecticide use makes it unlikely that they will be used properly. If insecticides are not used correctly then the hazard they pose to the user and anyone else involved with their application or dealing with or consuming treated material will be markedly increased. The safe disposal of unused chemicals and insecticide containers remains a largely unsolved problem. While it is extreme for responsible policy makers to refer to the agrochemical industry as exporters of death (Hessayon, 1983) when discussing trade of chemical insecticides with developing countries, the chemical companies must be aware, or should be made aware, of the extent of the hazard insecticides pose in these countries. Ultimately though it must be the responsibility of the governments of importing countries to decide what constitutes a hazard in relation to their own people, culture and society and to legislate accordingly. One of the reasons that chemical insecticides are readily accepted in developing countries relates to their governments' short term overriding requirement to improve and secure food production. When countries struggle to produce sufficient food to meet their needs, environmental concerns about chemical insecticides on the problems of insecticide resistance seem almost irrelevant. The short term requirements for increased crop production tend to outweigh the need for a longer term rational policy on insecticide use and insecticide companies are not going to ignore a major marketing and sales opportunity. Unless developing countries devise alternative strategies for the development of agriculture to those used in developed countries then the misuse of chemical insecticides, mimicking that in Europe and the USA, is almost inevitable. IPM, with its rational insecticide use philosophy, offers an alternative but the development of such an approach must begin now. It is better not to start the treadmill than to try to stop it once it has started moving.

The development of strategies for the rational use of insecticides within the framework of IPM requires a great deal of research. There is a tendency within IPM research to emphasize the alternative non-insecticide methods of control rather than concentrate on insecticides, perhaps because they represent the old approach, and there is always a tendency for everyone to jump on the latest bandwagon. There is at present a great need for independent work to identify reduced dosage levels that provide adequate control for optimizing application rates, droplet size and density and improving targeting, as well as studies on the influence of insecticides on beneficial insects and the timing of application. There is also a great need for continued research into the development of suitable packaging and disposal procedures as well as refining application equipment and techniques, all of which will rationalize the use of insecticides so that they are used in a more generally acceptable way.

5

Host Plant Resistance

5.1 Introduction

In the 1960s, the so-called 'Green Revolution' led to the availability of high yielding varieties of wheat and rice which heralded the possibility of stable food production for developing countries. At the end of the 1990s and into the 21st century, we are facing a second revolution in crop plant development as a result of genetic manipulation techniques. These techniques and the transgenic crops they produce will undoubtedly have a major impact on the future of food production. Whereas the Green Revolution led to problems of crop susceptibility to pests (since most were bred under a pesticide umbrella), the revolution in transgenic crops should provide a number of different solutions for pest management, mainly through introducing novel genes for insect resistance in crop plants.

The impact of such approaches to pest management are yet to become apparent but the opportunity offered by this technology is certainly tremendous. Whether, in the long term, the utilization and implementation of the technology will be judged to have been beneficial is a matter very much open to debate.

5.2 Objectives and Strategies

Host plant resistance may be defined as the collective heritable characteristics by which a plant species, race, clone or individual may reduce the possibility of successful utilization of that plant as a host by an insect species, race, biotype or individual (Beck, 1965). Reiterated in terms appropriate to crop production, host plant resistance represents the inherent ability of crop plants to restrict, retard or overcome pest infestations (Kumar, 1984) and thereby to improve the yield and/or quality of the harvestable crop product. From the point of view of the farmer, horticulturist and others, the use of resistant cultivars represents one of the simplest and most convenient methods of insect pest control, provided that the cultivars do not require expensive inputs of fertilizer in order to guarantee high yields.

The approaches utilized to obtain host plant resistance can be categorized by the methods used to manipulate (in the broadest sense) plant genetics: the conventional approach utilizing Mendelian genetics, the biometrician's approach and more recently the approaches of the biotechnologists.

The Mendelians utilize the inheritance of characteristics from plant germplasm that are qualitatively variable and can be transferred from a source plant(s) to the recipient plants by a process of breeding and selection. The Mendelians deal with single-gene characters that are readily identifiable and easily transferred during breeding. In contrast to this, the biometricians utilize the inheritance of characters that are quantitatively variable and controlled by many genes (i.e. polygenic characters). The biometricians developed methods of plant breeding (known as population breeding) that involve changes in gene frequency for a particular character (Robinson, 1996). Differences tend to be statistical, hence the name biometricians.

The biotechnologists are more akin to the Mendelians but the approach is based on techniques to transfer single genes from unrelated sources in crop plants. Both the Mendelians and the biometricians manipulate only the primary and secondary gene pools of the cultivated species for crop improvement while the biotechnologists, through advances in tissue culture and molecular biology, have made it possible to introduce genes from diverse sources such as bacteria, viruses, animals and unrelated plants into crop plants (Panda and Khush, 1995). This development provides the opportunity to develop transgenic crops with novel genes for resistance.

Each of the above approaches aims to achieve a useful expression of resistance in farmers' crops in a range of different cropping environments and against a range of insect 'races' and for such resistance to remain viable for a considerable period of time, even when widely planted. However, none of the approaches are likely to fulfil all of the above criteria; the Mendelian and biotechnologists' approaches are capable of developing crop plants that are applicable to a range of environments and insect 'races' but these resistances are liable to break down under heavy selection pressure. The biometricians on the other hand can produce resistance that is 'permanent' and expressed against all insect 'races' for a given locality. The crop plants tend to be locally adapted and hence the resistance may not be transferable across regions. The relative merits of the different approaches will be addressed throughout the chapter.

5.3 Genetics of Virulence and Resistance

Host plant resistance to pests is ubiquitous but there exists a great deal of variation in the levels expressed by plants. The level of resistance will obviously depend on the specific morphological and biochemical defences utilized by the plant, but ultimately the expression and stability of the resistance characters depend on the plant genotype, the pest genotype and the genetic interactions between the plant and the pest (Gallun and Khush, 1980). The genetics of pathogenicity combined with the study of epidemiology have played an important role in the development of pathogen/host plant resistance studies. It is largely within this context that the major differences between pathogen and insect plant resistance studies have arisen.

Van der Plank (1963) postulated that resistance to diseases in plants could be placed in one of two categories, vertical or horizontal resistance. The absolute definition of vertical resistance is that it involves a gene-for-gene relationship (Flor, 1942). For a pathogen to colonize a plant successfully the gene or genes for resistance must be matched by a corresponding gene or genes in the pathogen. If the pathogen lacks one or more corresponding genes then the plant is resistant to that individual and to all other individuals or races likewise lacking in that, or other combinations of genes. The plant will be susceptible to all individuals and races of pathogen having the appropriate corresponding genes or more than all the genes. Vertical resistance is thus characterized by a differential interaction – there is either colonization or no colonization, there is a matching gene-for-gene relationship or the genes do not match – there is no intermediate situation. Vertical resistance is controlled by various combinations of major genes which are highly heritable and which show clear cut, discrete segregation in crosses between resistant and susceptible plants, i.e. the effects are qualitative (Gallun and Khush, 1980). Because vertical resistance is controlled by major genes, it can only be a temporary resistance, it is within the micro-evolutionary change of the pathogen because simple gene mutations can produce individuals having appropriate combinations of major resistance genes (Robinson, 1980a,b).

Horizontal resistance is considered to be a more permanent form of resistance. The absolute definition of horizontal resistance is that it does not involve a gene-for-gene relationship (Robinson, 1980a). All polygenetically inherited resistance is horizontal; any major gene resistance that does not involve a gene-for-gene relationship is also horizontal resistance. Generally a large number of minor genes are involved, with each making a small additive contribution to the resistance (Gallun and Khush, 1980). The ability of the pathogen to parasitize the host is independent of the host's resistance capability because the effect of the sum of many minor resistance genes is independent and unassociated with any comparable sum of minor parasite genes in the pathogen. The resistance is quantitatively inherited and involves a constant ranking relationship, and operates equally well across all races of the pathogen.

The term vertical and horizontal resistance have no literal meaning. Their derivation lies in two figures produced by van der Plank (1963). Figures 5.1a and 5.1b are modified versions of these, depicting the infestation of two potato cultivars by 16 races of the blight, *Phytophthora infestans*. The cultivar in Fig. 5.1a possesses a single resistance gene R_1 and is susceptible to all blight races having a pathenogenicity gene V_1 and completely resistant to the blight races lacking this gene. Figure 5.1b is similar to Fig. 5.1a except that the resistance gene R_2 in this cultivar is different and is only susceptible to those blight races having a V_2 virulence gene. The term vertical resistance is derived from the differing response exhibited by each cultivar. Comparing Figs 5.1a and 5.1b, it can be seen that the difference in resistance response is parallel to the vertical axis and hence this kind of resistance that differentiates between races is called vertical resistance. The origin of the term horizontal resistance can be explained with reference to Fig. 5.2. This figure depicts three cultivars that do not possess any R genes but vary in their degree of resistance to the blight. The difference between these cultivars are not influenced by the races of blight and are parallel to the horizontal axis of the diagram, hence the term horizontal resistance. These terms, vertical and horizontal resistance, are by necessity non-descriptive, abstract terms that describe a concept (Robinson, 1976). It must be remembered that they have no literal meaning and they transcend specific descriptive terms used within a discipline. Vertical and horizontal resistance are thus multidisciplinary terms that are used at the level of the system.

5.3.1 The pathosystem concept

A system may be defined as a pattern of patterns, and a pattern within a pattern is called a subsystem (Robinson, 1976). A pathosystem is a subsystem of an ecosystem and is defined by the phenomenon of parasitism (Robinson, 1980a,b). As with an ecosystem the geographical, biological, conceptual and other boundaries of a pathosystem may be specified as convenient (Robinson, 1980b). The plant pathosystem involves plants used as hosts by parasites which includes fungi, bacteria, viruses, insects, mites and nematodes. In a natural pathosystem the evolutionary survival of wild plants is not impaired by their parasites. If the parasitic ability of the parasites were too great then the host might become extinct and consequently so would the parasite, hence there must be an upper limit to the parasitic ability of parasites (Robinson, 1980a). Balance between hosts and their parasites is maintained, their existence testifies to this.

Before addressing the different types of pathosystem, it is first necessary to understand some of the terminology associated with epidemiology. In epidemiology, there are two types of infection that correspond to two subdivisions of the epidemic. Alloinfection of a pathogen is an infection obtained from another host individual and is responsible for that part of the epidemic referred to as the exodemic. Autoinfection occurs by reinfection of the

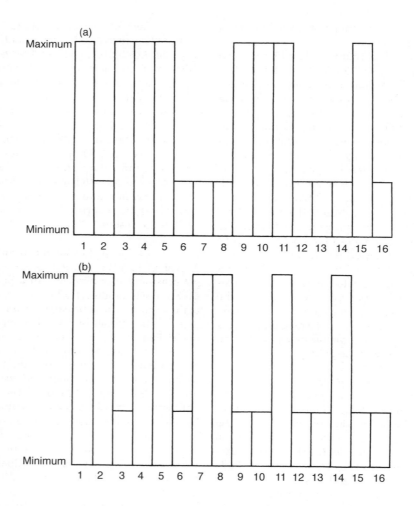

Fig. 5.1. The origin of the term vertical resistance. The amounts of blight suffered by various potato cultivars when exposed to 16 different races of *Phytophthora infestans*. (a) A cultivar having an R_1 resistance gene; (b) a cultivar with an R_2 resistance gene (see text for further explanation) (after van der Plank, 1963).

same individual host, i.e. it was produced on the same host, and this corresponds to that part of the epidemic referred to as the esodemic. Vertical resistance can only prevent infection and it can only prevent allo-infection that occurs during the exodemic. Horizontal resistance on the other hand operates during the esodemic, after a matching allo-infection has occurred, to reduce auto-infection.

There are two types of wild pathosystem, the discontinuous and the continuous. There is also the crop pathosystem under the deterministic control of man. The discontinuous wild pathosystem is autonomous and combines both the vertical and horizontal pathosystems. The vertical pathosystem reduces the effective parasite population to the spores that happen to allo-infect a matching host individual. In a

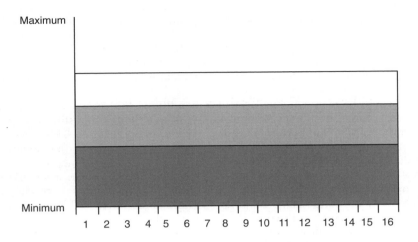

Fig. 5.2. The origin of the term horizontal resistance. The amounts of blight suffered by three potato cultivars having no *R* genes that were exposed to 16 different races of *Phytophthora infestans* (see text for further explanation) (after van der Plank, 1963).

pathosystem where a number of pest generations can occur in a single season, the vertical subsystem can have a major influence on the form of the epidemic and its effect on individuals within the pest population. If a fungal pathogen has four generations per season this results in four separate exodemics, the primary, secondary, tertiary and quaternary exodemics. The vertical pathosystem will influence how often an individual experiences an exodemic (a spatial effect), how often that exodemic is primary, secondary, tertiary or quaternary and because the amount of damage carries with the sequence of exodemic, the extent of the damage received. The primary exodemic is the most damaging, but only a few individuals that had matching allo-infections will experience this. The horizontal subsystem will then influence the development of the esodemic which will influence the population size of the subsequent pathogen generation that initiates the secondary exodemic. The number of individuals present in the secondary exodemic will be dependent on the reproductive rate of the fungi which will

itself have been determined by the level of horizontal resistance in each of the infected hosts. All the vertical genotypes will again be equally represented and allo-infection of further individuals will occur in the secondary exodemic but damage will be less than in the primary exodemic. This process continues through succeeding generations after which it is broken by leaf fall in deciduous trees, dieback in herbaceous perennials or seed set in annuals. The parasite survives the sequential discontinuity as a resting stage.

The continuous wild pathosystem has, as the name implies, no sequential discontinuity of host tissue, hence vertical resistance is made redundant. Continuous wild pathosystems are controlled exclusively by the horizontal subsystem. There is spatial and sequential continuity of host tissue and as a consequence of this, the esodemic is continuous. Evergreen perennial plant populations provide typical examples, having genetic uniformity because of their vegetative reproduction and sequential continuity because of their evergreen habit. Continuity of host tissue is more likely to occur in the tropics and is more common among perennials than annuals, especially of stems and roots.

Vertical resistance may supplement horizontal resistance (which is ubiquitous among plants) in more discontinuous situations (but not necessarily so) so that vertical resistance would be more common in temperate zones, among deciduous perennials and among annuals. But horizontal resistance remains the basis for control of the pathosystem that is supplemented in some situations by vertical resistance. Vertical subsystems probably occur only in angiosperm hosts and possibly in a few deciduous gymnosperms and evolved as a means of conferring stability in a fluctuating environment (Robinson, 1980b). The horizontal subsystem is subject to the changes in positive and negative selection pressures that occur over many generations but remain relatively unbuffered against short-term changes, e.g. marked fluctuations in weather. The parasite may be so favoured by a short-term change in weather that a host is decimated. The vertical subsystem can buffer against such vagaries and hence provide a supplementary level of control.

The deterministic control of crop pathosystems by man has, in the majority of crop pathosystems, promoted an imbalance at the subsystems level so that vertical resistance has been arbitrarily promoted at the expense of horizontal resistance. The identification of vertical resistance provided breeders with a simple and convenient means of developing pathogen resistant cultivars. The simple major gene inheritance of vertical resistance made it easy to manipulate and incorporate into agronomically superior varieties and it was also easily identified under field screening conditions because it conferred complete resistance. The widespread adoption of the resistant cultivar creates a selection pressure for virulent pathogen individuals. Any individual pathogen that has a virulent gene at the locus corresponding to the resistance gene of the new crop cultivar has a selective advantage and can multiply rapidly

(Gallun and Khush, 1980). Since crop host plants are usually members of a homogeneous crop population that have identical genes for resistance, then the descendants of the pathogen individuals having the matching gene will successfully colonize and damage the whole crop. Under these conditions the resistance is said to have broken; although it is still present it is just operationally inactive. This cycle of events has become known as the 'boom and bust cycle of cultivar production'. Plant breeders have not, however, been discouraged because some vertical resistance lasts longer than others and breeders hope they have discovered and utilized one of the long lasting forms.

The length of time that a vertical resistance remains active is dependent on the frequency of the matching virulence gene in the pathogen population and the extent to which the cultivar is grown. The rate at which resistance breaking biotypes develop will be partly dependent on the initial frequency of the gene within the pathogen population. In wild pathosystems all vertical genomes occur with equal frequency and hence they are of equal strength. But with crop pathosystems where the resistance gene may be derived from a wild host or ancestor, or from a different environment, then the strength of the gene is liable to change so that some are strong (a low frequency) or weak (a high initial frequency) (Robinson, 1980a). A strong vertical resistance is of more benefit to the breeder and farmer/grower alike since this will confer resistance for a longer period than weak vertical resistance.

Imbalance within the vertical subsystem has also occurred in crop pathosystems because often only one gene for resistance is incorporated into a cultivar, whereas in their wild counterparts there are secondary vertical genomes in the plant and pathogen population. This enforced simplicity increases the homogeneity of the vertical subsystems which only serves to render it ineffective. The incorporation of a number of major resistance genes is being used as a means of prolonging vertical resistance.

Further imbalance in the crop pathosystem has been introduced through the erosion of the horizontal subsystem due to the emphasis placed on selecting for vertical resistance. During the breeding for vertical resistance, there is a negative selection pressure for horizontal resistance because all the selection takes place during the exodemic. Plants are not assessed for their performance during the exodemic. This phenomenon is known as the vertifolia effect, after the potato cultivar in which it came to notice (van der Plank, 1963). This erosion of the horizontal resistance contributes to the 'bust' of the 'boom and bust' cycle because the low level of horizontal resistance present in the matched cultivar increases the effects of its susceptibility. The use of chemicals to protect crops from pests while breeding for yield and quality (the pesticide umbrella) has also contributed to the erosion of resistance.

5.3.2 The vertical pathosystem

Vertical resistance will not operate against a pathogen that has a virulent gene at the locus corresponding to the resistance gene of the host; a matching allo-infection will occur. A host plant that has only one gene for vertical resistance can be matched by a pathogen having a corresponding single virulence gene at the appropriate locus; however, a host plant having two resistant genes requires the pathogen to have two corresponding virulence genes at the appropriate loci. In such a situation a number of possibilities exist; an individual pathogen may have no corresponding virulence genes or virulence genes corresponding to one or both plant resistance genes. Thus there are four or 2^2 possible combinations of genes for virulence or avirulence and likewise, because of the gene-for-gene relationship, there are 2^2 possible combinations of genes for resistance and susceptibility. The same principle can be extended to n resistance genes. For n vertical resistance genes there are 2^n

possible genotypes for resistance and susceptibility.

In a subsystem where there is a total of 12 pairs of matching genes for resistance, each individual possesses only one gene; then the probability of matching occurring is 1:12 or 1/12. The situation becomes more complex when individuals possess more than one gene. If every individual possesses six genes, the probability of matching is equivalent to 1/924 or approximately 10^{-3} (according to the Pascal triangle, see Robinson, 1987), provided that equal frequencies of all six genotypes are maintained in both populations (Robinson, 1980b). The probability of a matching allo-infection decreases as the number of vertical genes in the subsystem is increased (Fig. 5.3) so that the probability of matching is as low as 0.1 in a subsystem having eight vertical genes. Robinson (1976) considered that in a natural vertical subsystem, there would be sufficient vertical genes so that the level of matching allo-infection could be reduced by at least a factor of ten.

For the vertical subsystem to be maintained, no individual must have a survival advantage or disadvantage over any other with respect to its vertical genes. Hence the probability of matching allo-infection must be constant, which is only possible if all vertical genotypes in both the host and the pathogen population have the same number of vertical genes, although the identity can be different (Robinson, 1980b). Vertical resistance will thus be most effective if each plant has a number of resistance genes and the population as a whole has a large number of different resistance genes. This will maximize the heterogeneity in the system over space. Any tendency for homogeneity will reduce the effectiveness of the vertical subsystem and any loss in effectiveness will mean the system will disappear, because matching would occur too often and too readily.

Another feature of the vertical subsystem is the requirement for sequential discontinuity of host tissue. This is necessary because once a matching has occurred

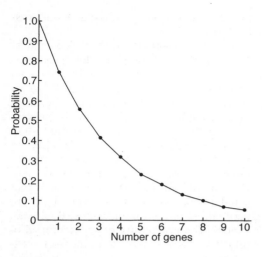

Fig. 5.3. The probability of a matching allo-infection as the number of vertical genes in the pathosystem is increased (after Robinson, 1976).

during the allo-infection, vertical resistance is inoperative and auto-infection of host tissue will then begin. If the vertical subsystem is to recover, then there must be a phase during which the esodemic is halted and the infection can start again. In the absence of the recovery phase, the vertical subsystem would disappear. The stability of the vertical subsystem is then dependent on two factors, sequential discontinuity and maximum heterogeneity of vertical resistance genotypes in space.

The need for a discontinuous pathosystem for the evolution of the vertical subsystem has already been mentioned; the other required characteristic is for a genetically mixed plant population (Fig. 5.4). In a wild pathosystem with natural levels of cross pollination, there will be a mixture of genetic lines within a population, even among clonal plant populations. Genetic mixtures have been utilized by the traditional farming methods of subsistence farmers who, even if growing clonal crops, still maintain different genetic lines

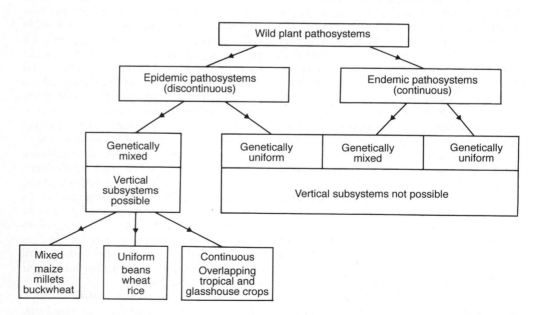

Fig. 5.4. Factors affecting the evolution of a vertical subsystem. A vertical subsystem can evolve only in a wild pathosystem that is both discontinuous and genetically mixed (after Robinson, 1987).

although they are only following ancient customs and cannot explain why they farm that way (Robinson, 1987). Crops produced from wild progenitors that evolved within a genetically mixed discontinuous pathosystem are those most likely to have a vertical subsystem. Plants that have evolved from progenitors within other types of pathosystem (Fig. 5.4) rely on the horizontal subsystem alone for their resistance to insect pests.

Case Study: The use of vertical resistance in wheat/rust pathosystems

Three species of rust are particularly damaging to wheat: the leaf brown rust, *Puccinia recondita*, the stem or black rust, *Puccinia graminis*, and the yellow (striped) rust, *Puccinia striiformis*. The brown rust is the most widely distributed of the three and occurs wherever wheat is grown, stem rust is common and is probably one of the most destructive diseases of wheat, while yellow rust is an important disease of bread wheat in temperate areas.

The breeding for resistance to these rusts has been a major preoccupation of wheat breeders since the beginning of this century and vertical resistance has represented the mainstay of their approach. Utilizing vertical resistance has advantages for the breeder, because the absence of infection can be readily detected which allows large numbers of plants to be screened, and because the resistance can easily be exploited due to its simple inheritance. However, breeding programmes have suffered from the continued need to identify and introduce new resistance genes due to the development of resistance breaking pathogen races. The varieties Newthatch, Stewart and Carlton were resistant to stem rust in USA and Canada from 1939 to 1950 but then a rust race (labelled 15B) became widespread and the resistance was broken to produce severe epidemics of stem rust disease in 1950, 1953 and 1954 (Russell, 1978). In Australia, a cultivar, Eureka, resistant to stem rust, was released in 1938 but a new pathogen race to which it was susceptible was identified in 1942. A cultivar, Fox, also resistant to stem rust, was released in the USA in 1970, but by 1972 the disease was prevalent and it never subsequently became an established cultivar (Johnson and Gilmore, 1980). During the 1950s in Canada, one of the most popular cultivars resistant to leaf rust was Lee; however, by 1960 it was susceptible to the disease (Anderson, 1961). In the UK, since the 1950s a succession of races of the yellow rust have been responsible for the breakdown of many new varieties (Russell, 1978). The length of time that a cultivar remains resistant varies considerably, as can be seen from the above examples, but a clear succession of resistance-breaking pathogen races has continually developed for each of these rusts, so that now there is a movement towards breeding for horizontal resistance, or 'slow-rusting' as it has been called (Johnson and Gilmore, 1980).

Host population; per cent (−) alleles												Mean
	100	90	80	70	60	50	40	30	20	10	0	
100	100	90	80	70	60	50	40	30	20	10	0	50
90	90	81	72	63	54	45	36	27	18	9	0	45
80	80	72	64	56	48	40	32	24	16	8	0	40
70	70	63	56	49	42	35	28	21	14	7	0	35
60	60	54	48	42	36	30	24	18	12	6	0	30
50	50	45	40	35	30	25	20	15	10	5	0	25
40	40	36	32	28	24	20	16	12	8	4	0	20
30	30	27	24	21	18	15	12	9	6	3	0	15
20	20	18	16	14	12	10	8	6	4	2	0	10
10	10	9	8	7	6	5	4	3	2	1	0	5
0	0	0	0	0	0	0	0	0	0	0	0	0
Mean	50	45	40	35	30	25	20	15	10	5	0	25

(left vertical axis label: Insect population; per cent (+) alleles)

Fig. 5.5. The horizontal pathosystem model (see text for further explanation) (after Robinson, 1979).

5.3.3 The horizontal pathosystem

The horizontal pathosystem is described by a model (Fig. 5.5) that assumes resistance and parasitic ability are controlled by many alleles which are either positive or negative. In the host each 1% of positive alleles increases resistance by 1% while each percentage point of negative alleles decreases resistance by a percentage point. Similarly, in the pathogen each 1% of positive alleles increase and 1% of negative alleles decrease parasitic ability by 1%. The form an interaction will take between a host and a pathogen parasite depends on the percentage of negative alleles in the host (% susceptibility) and the percentage of positive alleles in the pathogen (% para-sitic ability). Percentage parasitism is determined by the proportional suscepti-bility multiplied by the proportional para-sitic ability (i.e. % susceptibility × % parasitic ability /100). For instance, if a host had 80% negative alleles and a pathogen 50% positive alleles then the per-centage parasitism is 40%. Because there is no gene-for-gene relationship, horizontal resistance and virulence are independent and can thus increase and decrease regard-less of each other. Thus, the resistance genes show continuous variation from sus-ceptibility to resistance in the segregating populations of crosses between resistance and susceptible parents, i.e. the effects are quantitative (Gallun and Khush, 1980).

Case Study: Horizontal resistance to tropical rust in Africa

The introduction of the tropical rust *Puccinia polyspora* of maize into Africa is discussed at length by Robinson (1976, 1987), the subject is only summarized here to provide an illustration of the value of horizontal resistance.

Maize was introduced into Africa about 400 years ago from the Americas, but for one reason or another the tropical rust, normally a pest of this crop, was not evident in Africa until World War II. When it appeared it spread from west to east Africa within 8 years, causing devastating damage of subsistence maize crops. In an infested crop the majority of the plants were killed by the rust before they could even form flowers. This extreme susceptibility had been caused by the erosion of horizontal resistance by 400 years' absence of the rust. The frequency of the rust resistance polygenes, although still present, had declined due to this negative selection pressure to a neutral level maintained by the Hardy–Weinberg equilibrium. Under the extreme selection pressure now provided by the presence of the tropical rust the maize of few of the subsistence farmers survived, but that which did was harvested for its seed and resown during the next season and subsequent seasons. These surviving plants were resistant to the rust and under the high selection pressure, the resistance gradually accumulated over 10–15 host generations, which due to the two cropping seasons, meant that resistant plants were produced within 5–7 years. These land races grown by the subsistence farmers have now been resistant to the tropical rust for the last 30–40 years and are a testimony to the durability of horizontal resistance.

5.3.4 The gene-for-gene model and biotypes

The gene-for-gene relationship is most easily described by a lock and key analogy (Robinson, 1983) where the host plant is the lock and each resistance gene is the equivalent to a tumbler in that lock; the lock may have many tumblers. Each individual pathogen, a fungal spore for instance, represents a key and each gene for parasitism in the pathogen parasite is equivalent to a tooth on the key; the key can have many teeth. The key of an individual parasite either does or does not fit the lock of an individual host plant, i.e. a pathogen parasite either does or does not have the necessary parasitic genes to counteract the resistance genes of the host plant. In order to illustrate this, consider a situation where there are 1000 newly emerged host seedlings that are pathogen free and they can only be parasitized by immigration. Each seedling has a resistance lock different from every other (1000 different locks) and there are 1000 pathogen parasites each with a different key that will only open one type of lock. Each seedling receives only one immigrant. The chances of an immigrant's key fitting a plant lock is only 1/1000. On average therefore only one immigrant pathogen will open its seedling's lock and only one seedling will become infected. All the other pathogens will effectively be lost either because they germinate and then die because no gene-for-gene matching is achieved, or because conditions are such they remain dormant. The seedling that has been parasitized is then fully colonized by the pathogen (most often by asexual reproduction) and each descendant of this successful parent immigrant has the same key, either because the progeny are clones or because sexually produced offspring quickly become homozygous (Robinson, 1983). This means that the gene-for-gene relationship can only control the initial colonization, once the lock has been matched with a key; reinfection from genetically identical parasite progeny will not be controlled.

The whole basis of the vertical resistance mechanism expressed as a gene-for-gene effect is that it depends on heterogeneity in both the host and parasite populations for it to work. If there is uniformity in the gene population of the host then the vertical resistance mechanism breaks down. In the seedling analogy, if the 1000 seedlings all had the same lock and the 1000 pathogen parasites all had different keys, then the parasite which matched one of the seedlings would be able to reproduce and the genetically identical progeny would colonize all the other genetically similar seedlings. Once uniformity is introduced into the system then the system breaks down, in this case in favour of the parasite.

Given that both the host and parasite populations have equal frequencies of vertical genes then for a fixed population of the host there are two ways in which a parasite population can improve its chances of matching with a host. It can either reproduce at such a high rate that each host individual receives a number of parasite individuals, hence increasing the chances of a matching occurring, or the parasite could have a high dissemination efficiency increasing its chances of matching over space. Different parasites have different reproduction rates and dissemination efficiencies and the rate at which matching occurs within a reproductive cycle will be dependent on the combination of these factors. If a parasite has a high reproductive capacity and efficient dissemination capability then a large proportion of individuals within a population will be matched and the vertical subsystem will break down.

Many insect pest have a high dissemination efficiency, with migrations taking insects many hundreds or thousands of kilometres, e.g. *Spodoptera* spp. in Africa and planthoppers in South-East Asia. Although the ability of many insects to migrate long distances contributes to their dissemination efficiency it is their ability for controlled tactical mobility that has the greatest implications for the existence of vertical resistance. An insect that can readily detect and fly between potential host plants would, if it sampled sufficient plants, eventually find an individual plant with which it could match. If the hypothetical model of 1000 seedlings each with a different lock is considered again, and 1000 individual insects each having a different key are released, given an unlimited amount of time the insects could select and reselect plants until each located the plant with which it could match, 1000 seedlings would all be matched by 1000 insects. The limiting factor will always be the amount of time it takes to locate and sample a plant in relation to the period over which the insect retains its tactical mobility. Limitations on this are evident in some aphids which suffer wing muscle autolysis after a set time from adult moult, and variations in the pre-reproductive and reproductive flight ability of some insects. However, it is highly probable that insects with even a limited amount of tactical mobility would increase the probability of matching infection which would increase the tendency for homogeneity in the vertical subsystem. In the vertical subsystem any tendency for homogeneity would reduce the survival value of the system and hence the likelihood of its existence. If insect species were capable of always finding and matching with a host, the vertical subsystem would effectively disappear, or, conversely, would never develop. Thus, vertical resistance is likely to be inoperative against highly mobile insects that retain their mobility for a large proportion of their adult life and that are able to readily identify suitable hosts. These insects may also have an ability to migrate long distances but this is not necessarily an essential characteristic, since it is the ability to detect host plants that is important. Migrating long distances but having a poor host finding ability would not necessarily be advantageous. There are no recorded examples of vertical resistance to Lepidopteran pests which have a strong flight capability although there are examples cited of vertical resistance against aphid species (de Ponti, 1983) which have

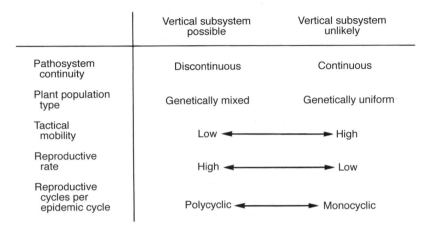

	Vertical subsystem possible	Vertical subsystem unlikely
Pathosystem continuity	Discontinuous	Continuous
Plant population type	Genetically mixed	Genetically uniform
Tactical mobility	Low ⟷ High	
Reproductive rate	High ⟷ Low	
Reproductive cycles per epidemic cycle	Polycyclic ⟷ Monocyclic	

Fig. 5.6. A summary of the factors influencing the likelihood of the existence of a vertical subsystem in insect host plant relations.

a relatively poor controlled tactical mobility, even though some of these can migrate over long distances, borne on weather fronts.

The number of generations of an insect life cycle that can take place during a plant's growth cycle will also influence the likelihood of whether vertical resistance is operative or not. Deciduous perennials and annual plants have distinct growth cycles which are discontinuous. Vertical resistance is impossible in continuous cycles and only horizontal resistance can exist in such circumstances. For plants with discontinuous cycles, the evolutionary value of vertical resistance will be increased where the pest insect has both a high number of cycles per plant season and a high reproductive rate. If the insect's reproductive cycle corresponds to the host's cycle then it is described as monocyclic, e.g. apple codling moth in temperate zones; if the insect has a few reproductive cycles in each cycle of its host it is described as polycyclic, e.g. aphids and whiteflies (Robinson, 1987). The more insect cycles there are, the greater the proportion of plants that will have become infested by the end of the epidemic. The reproductive rate will also influence the number of indi-

viduals available for infestation at the start of each insect cycle which will affect the number of insects that can potentially infest a single plant. Although the overall probability of matching infection always remains the same (see earlier), the higher the number of insects infesting an individual plant the greater chance there is of a matching infection. Thus, in pathosystems where the insect parasite has a relatively poor dissemination efficiency, vertical resistance is most likely to occur where the insect is polycyclic and has a high reproductive rate and much less likely if the insect is monocyclic and has a low reproductive rate. So the insect characteristics of tactical mobility, reproductive rate and the number of insect cycles per host cycle will influence the likelihood of a vertical subsystem existing within a given crop pathosystem (Fig. 5.6).

A biotype is a group of individual insects that are genetically identical with respect to their genes for virulence (Claridge and Den Hollander, 1983). Thus, when individual insects match with a plant, gene-for-gene, and produce offspring sharing the same virulence genes, after a number of generations a distinct biotype will emerge, especially where large areas

Table 5.1. Example of insect biotypes (from Panda and Khush, 1995).

Crop	Insect	Biotypes (no.)
Wheat	*Mayetiola destructor*	9 (field) 2 (laboratory)
	Schizaphis graminum	7
Raspberry	*Amphorophora idaei*	4
Alfalfa	*Acyrthosiphon pisum*	4
	Therioaphis maculata	6
Sorghum	*Schizaphis graminum*	5
Corn (maize)	*Rhopalosiphum maidis*	5
Rice	*Nilaparvata lugens*	4
	Nephotettix virescens	3
	Orseolia oryzae	4
Apple	*Eriosoma lanigerum*	3

are sown with a cultivar having a single resistance gene. The development of a biotype will be most pronounced among asexual polycyclic insects that can produce many genetically identical offspring (most biotypes recorded to date are aphids; Kogan, 1994). A biparental, sexually reproducing, monocyclic insect is less likely to develop insect biotypes because offspring will be genetically heterogeneous and only selection over a number of generations would produce a population genetically homogeneous for virulence. A low dissemination efficiency would reduce the rate of this development still further. There are few known examples of insect biotypes (Table 5.1) and hence too few detailed genetic analyses of their insect host plant interactions. Even among the insects for which claims of distinct biotypes have been made, there is still some dispute. The so-called biotypes of brown planthopper (*Nilaparvata lugens*) (Heinrichs, 1986; Khush, 1992) on rice have been shown to have a wide range of overlap in virulence both within and between cultivars and the resistance is in fact polygenically and non-monogenically inherited (Claridge and Den Hollander, 1980; Den Hollander and Pathak, 1981). Diehl and Bush (1984) also dispute the evidence of a gene-for-gene interaction in the hessian fly due to lack of hybridization experiments and in the raspberry aphid (*Amphorophora idaei*) due to small sample sizes used in the experiments producing inconsistent results. Hence, the

evidence for gene-for-gene interactions in insects is far from overwhelming and is certainly an area in which a great deal more research is required (Claridge and Den Hollander, 1980).

5.3.5 Incomplete, quantitative vertical resistance

One of the reasons why the pathosystem concept has had difficulty gaining credence among entomologists is the expression of incomplete or quantitative vertical resistance, a feature supposedly inconsistent with the pathosystem model. A prime example is the hessian fly, *Mayetiola destructor*, a stem boring insect pest of wheat. Quantitative resistance provides incomplete protection against non-matching insects and no protection against matching insects (Fig. 5.7). The cultivar in Fig. 5.7 possesses a single resistance gene R_1 that is matched by eight races (2, 6, 7, 8, 12, 13, 14 and 16) of a parasite (in this case, an insect pest) and resistance breaks down. The remaining races do not have matching interactions but the cultivar is still partially susceptible (compare with Figs 5.1a and 5.1b), i.e. its resistance is below the expected maximum level. A non-matching insect may infect a host plant with quantitative vertical resistance but its development and reproduction may be impeded. Quantitative vertical resistance has the biggest drawback associated with horizontal resistance, that of conferring on partial resistance, but it also has

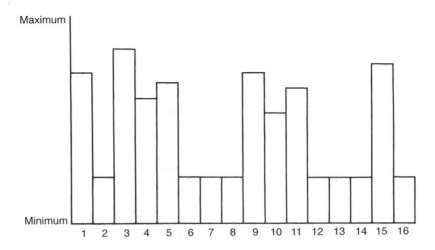

Fig. 5.7. The amounts of blight suffered by a potato cultivar with an R_1 resistance gene when exposed to 16 different races of *Phytophthora infestans,* in a situation where there is incomplete or quantitative vertical resistance (after Robinson, 1976).

the drawback of vertical resistance in that it is also liable to breakdown. It also poses a problem in resistance screening because it could be mistaken for horizontal resistance. The existence of quantitative vertical resistance represents an agricultural anomaly since there appears to be no evolutionary survival value associated with it. Quantitative vertical resistance does not represent a superior system to qualitative vertical resistance unless a survival advantage is obtained by slowing the epidemic by sacrificing an individual, e.g. the hessian fly larva bores into the stem of a wheat plant but the resistance is sufficient to prevent the larva maturing and reproducing; although the individual wheat stem is killed the epidemic is slowed down. This explanation would not, however, apply to polycyclic insects such as aphids and whitefly. An alternative explanation is that quantitative vertical resistance is an artefact of agriculture caused by a resistance mechanism, obtained from wild progenitors or from plants in other environments, and it fails to function properly in a new environment. Quantitative vertical resis-

tance is oligogenically inherited and has mostly been confined to pathogens of temperate cereals, but it has greatly contributed to the confusion surrounding the nature of pathogen and insect host plant relationships.

5.3.6 Durable major gene resistance and horizontal resistance

Another apparent anomaly that has reduced entomologists' interest in the adoption of the pathosystem concept has been the presence of durable resistance in crop cultivars where the resistance character has been controlled by major genes. If the gene-for-gene concept were applicable then a characteristic with a major gene inheritance would not be expected to be durable. One situation in which vertical resistance could be prolonged would be if the resistance were strong, i.e. the virulence gene in the insect were rare, and if an insect had a limited dissemination efficiency. If resistant cultivars were grown over large areas then it would be possible to envisage the evolution of resistance breaking insects but for most crops the use of insect resistant cultivars has been restricted, and hence few incidents of resistance failure have occurred. De Ponti (1983) described this by saying that 'resistance to insects is less operational than resistance to pathogens'.

If a gene-for-gene relationship is not evident, then by default the major gene resistance is horizontal resistance, which of course is durable. In cases where major gene inheritance is known to produce durable resistance, for example, jassid resistant cottons, cereals resistant to the stem sawfly and cereal leaf beetle, the characteristics that confer resistance are gross morphological features of the host plant (Russell, 1978). These morphological features, such as pubescence or solid stems, although monogenically inherited, are beyond the capacity of the insect for change, which obviously makes them durable. The search for such durable resistance based on gross morphological features has been promoted by the traditional entomological mechanistic approach to insect plant resistance, but it is basically horizontal in nature.

Horizontal resistance is, literally ubiquitous, but due to traditional breeding techniques it is often reduced to such a degree that it confers only a low level of protection against insect infestation. One of the reasons for this is that most breeding for plant yield and quality is carried out under the protection of insecticide applications. In the absence of selection pressure, i.e. presence of the pest, horizontal resistance is eroded. The use of insecticides to protect the screening populations reduces the insect selection pressure and so the level of horizontal resistance will diminish over successive generations of negative selection. Cultivars selected for high yield and quality, while under the protection of an insecticide umbrella, will thus be susceptible to insect attack. Such cultivars will then either require continued insecticide protection in the farmer's field, or it will be considered necessary to incorporate genes for resistance, which inevitably will involve either vertical genes or durable

major genes, and the cycle starts again. The need for screening plants under natural conditions of infestation is a cause supported by few breeders but, on its own, the use of insecticides in this way plays a major contribution to the problems of crop susceptibility and the lack of horizontal resistance.

Horizontal resistance to insects has many advantages: it is independent of the genetics of the pest, is permanent and readily accumulated over 10–12 host generations under high selection pressure, but it also has the disadvantage that it rarely offers complete protection. But then, in many situations this is not possible even with insecticides, so a certain minimum degree of damage is acceptable. Horizontal resistance is obviously not applicable in situations where a limited amount of damage is unacceptable, such as for instance, commercial tomato production.

5.4 Breeding Methods

The choice of breeding method depends on the objective of the breeding programme, the inheritance of the resistance characters and the reproductive system of the crop species, i.e. whether the species is self-pollinating or largely cross-pollinating (Kogan, 1994; Panda and Khush, 1995). The type of breeding populations used for self-pollinating crops is referred to as an inbred pure line whereas the cross-pollinating crops may be one of three types of breeding population, open-pollinated, hybrids or clones.

5.4.1 Plant breeding schemes
Plants can be divided into two pollinating groups, self- and cross-pollinating species, and asexually, vegetatively reproducing plants. Species that are self-pollinated tend to consist of mixtures of many closely related homozygous lines, which although they exist side by side remain more or less independent of each other in reproduction (Allard, 1960). Self-pollinators are referred to as inbreeders. Cross-pollinated species (outbreeders) tend to consist of highly

heterozygous populations in which enforced self-fertilizations can cause a loss of general vigour. The majority of plant species are outbreeders to a great extent (Lawrence, 1968) and there are more outbreeders proportionally among natural species than domesticated species (Allard, 1960). In general, most outbreeders are perennial while inbreeders tend to be annuals. Table 5.2 lists major crops according to pollination type.

Plant species within the self- and cross-pollinating groups have mechanisms that help to ensure that the required form of pollination predominates. Extreme forms of this in outbreeders are self-incompatibility and dioecy (the male and female reproductive organs exist on two separate plants). Most plants are, however, monoecious (having both male and female reproductive structures on a single plant), but a variety of mechanisms has evolved that either promote or restrict self-fertilization. Floral morphology can play an important part in the pollination mechanism and can restrict outbreeding, for instance, in some varieties of tomato, pollen is shed from the anthers onto the stigma as it grows up through the tube formed by the anthers, hence ensuring self-fertilization (Cobley and Steele, 1976). Cleistogamy, fertilization within an unopened flower, is another mechanism of ensuring self-fertilization, while cross-pollination is promoted if the maturation of the anthers and stigmas are separated in time (protandry and protogyny respectively). In some self-pollinating plants such as cotton and sorghum there exists a degree of cross-pollination, the extent of which may be greatly affected by environmental conditions. The amount of outcrossing may be as little as 1% in species such as rice, tomatoes and lettuce, but up to 50% in other species such as annual sweet clover. Where fully controlled crosses are required the degree of outcrossing in a particular plant species/cultivar needs to be known and precautionary measures should be taken to reduce unwanted contamination. Also, in situations where population enhancement

approaches are used outbreeding is a pre-requisite, hence the extent of outbreeding needs to be known. The degree of cross-pollination between different genotypes can be determined by interplanting strains having a recessive marker gene with strains carrying the dominant alternative allele. Seeds are harvested from the recessive type and the amount of natural crossing determined. Suitable markers are genes such as those for cotyledon colour in legumes or starchiness/glutinous endosperm in grasses.

The type of pollination control utilized will depend on whether or not prevention of contamination is required for the bulking of pure seed, or in breeding nurseries, or if control is required for specific matings between selected plants. The isolation of breeding material is the simplest method of control for the bulking of seed and is effective because intercrossing falls off with distance for both wind and insect pollinated plants. If the proportion of contamination (F) by outcrossing is small then it can be calculated, provided the amount of contamination at any two points is known.

$$F = \frac{y}{D_e^{KD}} \qquad (5.1)$$

where:
D = distance
y = contamination at zero distance
e = exponential constant
K = rate of decrease of contamination with distance (Lawrence, 1968).

For the prevention of contamination of specific plants, bags and cages can be placed over flowers to prevent cross-pollination by both insects and wind-borne pollen. Obviously the number of plants that it is practical to treat in this way is limited; this approach is only really applicable to plants required for specific matings. Bags also provide another means of ensuring monoecious plants self-pollinate.

Self-pollination can be prevented by emasculation, which involves removal of the anthers before pollen is shed. The method can be laborious and exacting,

Table 5.2. Crop plants categorized according to whether they are cross- or self-pollinated (after Allard, 1960).

Self-pollinated crop plants

Cereal grasses	Forage grasses	Fruit trees
Barley	Annual fescue	Apricot
Foxtail millet	Foxtail barley	Citrus
Oats	Mountain bromegrass	Nectarine
Rice	Slender wheatgrass	Peach
Sorghum*	Soft chess	
Wheat		

Legumes	Forage and green manure legumes	Other species
Broadbean*	Annual sweet clover	Cotton*
Chick pea	Bur clover	Eggplant
Common bean	Crotalaria juncea	Endive
Cowpea	Hop clover	Flax
Groundnut	Strawberry clover (common)	Lettuce
Lima bean	Subterranean clover	Okra
Mung bean	Velvet bean	Parsnip
Pea	Vetch (common, hairy and pannonica)	Pepper (Capsicum annum, C. frutescens)
Soybean		Tobacco
Sweet pea		Tomato
Urd bean		

Cross-pollinated crop plants

Forage grasses	Forage legumes	Fruits	Nuts
Annual ryegrass	Alfalfa	Apple	Almond
Buffalo grass	Alsike clover	Avocado	Chestnut
Meadow fescue	Birdsfoot trefoil	Banana	Filbert
Orchard grass	Crimson clover	Cherry	Pecan
Perennial ryegrass	Red clover	Date	Pistachio
Smooth bromegrass	Strawberry clover (Palestine)	Fig	Walnut
Tall fescue	Sweet clover	Grapes	
Timothy	White clover	American grapes	
		Mango	
		Olive	
		Papaya	
		Pear	
		Plum	

Cereal grasses	Legumes
Maize	Scarlet runner bean
Rye	

Other species

Artichoke	Cabbage	Chicory	Kohlrabi	Radish	Squash
Asparagus	Cauliflower	Chinese cabbage	Mangel	Raspberry	Strawberry
Beet	Carrot	Collard	Muskmelon	Rhubarb	Sunflower
Blackberry	Castorbean	Cucumber	Onion	Rutabaga	Sweet potato
Broccoli	Celery	Hemp	Parsley	Safflower	Turnip
Brussels sprouts	Chard	Kale	Pumpkin	Spinach	Watermelon

* Frequently more than 10% outcrossed.

especially if the flowers are small, but it does permit cross-fertilization to be controlled. Pollen from the required male parent can be introduced to the stigma of the female emasculated parent. However, emasculation is inappropriate for preventing self-pollination of large numbers of plants. In situations where outbreeding needs to be promoted then male gameticides need to be used to make inbreeders male-sterile. Other methods include genetic or cytoplasmic male sterility but gameticides have, to date, been mainly used to facilitate hybrid seed production; they have great potential for use in promoting outbreeding in population enhancement approaches.

Asexual reproduction is another means by which plants may be propagated and utilized in agriculture. The vegetative propagation of plants can take place by use of rhizomes, stolons, tubers, bulbs and corms or by budding and grafting techniques. Plants that are vegetatively propagated in agriculture include a diverse range of crops from potato, sweet potato, sugar cane, cassava, grapes, *Rubus* spp., strawberries to tree fruits and nut bearing trees. Where sexual reproduction does occur the plants are natural outbreeders and are highly heterozygous.

5.4.2 Inbred pure lines

The objective of an inbred scheme is to produce a highly uniform plant population having the desired characteristics that constitute an improved crop plant, for instance, high yield, drought tolerance or pest resistance. Pure lines are produced by continued selfing over a number of generations until they are genetically homozygous. Self-fertilization reduces heterozygosity according to classical Mendelian inheritance. A heterozygote Aa having a dominant 'A' gene and a recessive 'a' gene will if self-fertilized segregate in three genotypes, AA, Aa and aa in the ratio of 1:2:1 plants. If these genotypes are then self-fertilized, the AA and aa genotypes will breed true to type while the Aa will provide the same ratios as before, 1:2:1. In this way the heterozygosity is

reduced by half every generation and hence homozygosity is achieved within a few generations for a single gene pair and within a greater number of generations for an increasing number of gene pairs (Fig. 5.8). Once these pure lines have been achieved, then crosses are made between them to produce a new variable generation, but these plants are self-fertilized to produce nearly pure lines again, among which will be those lines having favourable characteristics as a result of the cross. These lines are then selected for further evaluation.

The three basic breeding schemes for inbred lines are pedigree breeding, bulk breeding and breeding through single seed descent. The choice of which scheme to use depends on the timing of intensive selection within the programme. Early generations are heterogeneous and hence only selection for highly heritable characters would be efficient. Later generations are more homogeneous and hence selection for plants having favourable characteristics is a choice between fixed lines (Table 5.3).

Pedigree breeding is a scheme that uses selection of single plants during early generations despite the high degree of heterozygosity amongst the plants at this stage

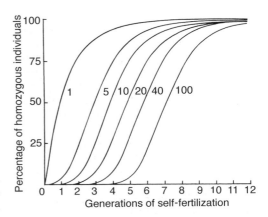

Fig. 5.8. The percentage of homozygous individuals after up to 12 generations of self-fertilization, when the number of independently inherited gene pairs is 1, 5, 10, 20, 40 or 100. The percentage of homozygosity in any selfed generation is given by the curve for one gene pair (after Allard, 1960).

Table 5.3. The different breeding schemes for inbred crops (after Simmonds, 1979).

Pedigree	Bulk	Single-seed descent
F_2 Maximal variability, high heterozygosity, heterosis persists		
Spaced planting: single parent selection (SPS)	F_2 bulk	Single seed taken from each plant of large sample
F_3–F_6 Declining heterozygosity and heterosis; lines assume individually		
Family-rows progressively replaced by plots and family selection; near-relatives eliminated; quality tests and yield trials started	Bulks, with or without some artificial selection	Single seed descent; two glasshouse generations per year
Large sample of near-homozygous lines isolated and selected		
F_5–F_8 Surviving lines effectively homozygous; variability greatly reduced		
Final family selection; detailed yield and quality evaluation; purification started	Selection and yield and quality evaluation completed; purification started	
F_7–F_{12} Only one–few genotypes survive, as pure lines		
Further trials in diverse locations and seasons; pure stocks multiplied; naming and release		

in the procedure. The plants are then self-fertilized and the seeds from each plant are used to produce family rows which in later generations are replaced by selection between family plots. In about the fourth generation the genetic differences between families are much greater than within families, hence it is at this time that the number of families can be reduced. The term pedigree breeding is derived from the use of records of lineage which permit selection among closely related families selected in the F_3 generation. Where there is little to choose between families, a single line is chosen to continue the lineage representative of a particular plant type. Numerous insect resistant varieties of various crops have been developed through the pedigree method (Panda and Khush, 1995). Johnson and Teetes (1980) describe the use of the pedigree method for breeding resistance to the sorghum midge (*Contarinia sorghicola*) in the following steps:

1. The highest level of resistance should

be transferred to agronomically acceptable types by hybridization and selection.

2. Agronomically acceptable lines with the least susceptibility to midge should be used as the non-resistant parent.

3. Grow a large F_2 population of at least 4000 plants. Selection can be accomplished up to this stage without midge being present. Selection for small-plumed types should increase the frequency of midge resistant types from the F_2 population.

4. Evaluate F_3 rows under midge infestation. To ensure a large midge infestation during the flowering stage, the F_3 should be planted at more than one date or location.

5. Evaluate F_4 selections in replicated progeny rows and backcross superior plants if necessary.

Pedigree breeding has also been used for resistance breeding against shootfly (*Atherigona soccata*) and greenbug (*Schizaphis graminum*) in sorghum (Johnson and Teetes, 1980; Teetes, 1980) and in rice against the green leafhopper (*Nephotettix virescens*) and the brown planthopper (*Nilaparvata lugens*; Khush, 1980). Modifications of the pedigree method and bulk breeding methods are used for breeding for resistance in cotton (Niles, 1980) and groundnut (Smith, 1980).

Bulk breeding involves leaving the selection for useful plants until the homozygosity is well advanced and if any selection is practised then it is only for highly heritable characteristics. The method is used for seed crops but is totally unsuited for fruit and most vegetable crops (Allard, 1960). The method has the advantage of reducing labour costs during the early stages of the scheme because it relies on natural selection for the elimination of undesirable genotypes and the changing of gene frequencies in the population. The bulking of seed increases the crop homozygosity before single plant selections are made and evaluated in the same way as for pedigree breeding. In general, most inbred line populations are handled by one of the bulk methods (Simmonds, 1979).

Single seed descent as a method of breeding that involves leaving the selection procedure as late as possible in the breeding programme. Single seeds are taken from each plant of a large sample. The seeds are sown in greenhouses with the object of getting through as many early generations (and hence the more heterogeneous generations) as quickly as possible. When homozygosity is achieved, plants are then selected for favourable characteristics and use in field trials. A single seed descent programme was used to develop inbreds with resistance to carrot fly (Ellis *et al.*, 1991).

5.4.3 Open-pollinated populations

Outbreeders naturally maintain a high degree of heterozygosity by cross-pollination and many suffer from loss of vigour if self-fertilization is enforced over a few generations. The breeding for improved open-pollinated crops is dependent on changing the gene frequencies of populations of plants in favour of useful characteristics, thus maintaining a high degree of heterozygosity, to prevent inbreeding depression. The gene frequency of a population is changed by selection of a large number of plants having the required characteristics over many generations so that the mean value of each characteristic shifts away from the original population (Fig. 5.9). Hence, in population breeding, crop characteristics are referred to in statistical terms, of means and variances, since the uniformity of inbred crops is

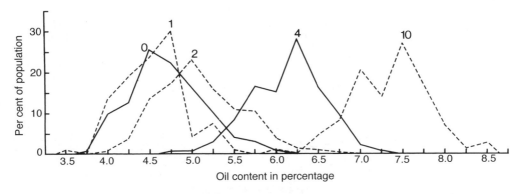

Fig. 5.9. The changing frequency distribution of percentage oil content in corn after selection over up to 10 generations. The numbers above each distribution denote the generation (after Allard, 1960 from Smith, 1980).

impossible for the heterogeneous out-breeders (Simmonds, 1979).

There are two types of open-pollinated populations used in plant breeding: the first is an isolated population indefinitely randomly mated within itself and the second is the synthetic population, which is an artificial population produced through combining randomly mated inbred lines (only appropriate for plants that can withstand a degree of inbreeding). The former, which is referred to as the population improvement technique, involves mass selection of superior plants, or some variation on this.

Mass selection is a simple selection procedure, requiring a low labour input and only one cycle of selection per generation. Desirable plants are selected from a large freely interbreeding source population. The seed is harvested, mixed and used as the next generation. No control is exerted over pollination; only the maternal parent is known, which means any heritability is halved. Mass selection is based on the choice of superior phenotypes or on char-acters that can be easily seen or measured and increasing the proportion of these within the population. Characters such as yield, oil and protein content in maize have been selected for (Lawrence, 1968; Woodworth et al., 1952; Fig. 5.10) as well as more easily identifiable characters such as plant height, maturity date and grain colour. Mass selection, more than any other plant breeding method, has been responsible for the improvement of open-pollinated crop varieties over many centuries. Farmers practised mass selection when choosing the plants for the seed that would be used for the next year's crop, and in doing so produced many locally adapted varieties. Mass selection has been used to improve varieties of sugar beet, maize and alfalfa and remains the most common form of population improvement.

The progeny testing and line breeding techniques are simple variations of the basic mass selection procedure. Progeny testing attempts to obtain a more accurate evaluation of the breeding potential of an individual by assessing some of it progeny.

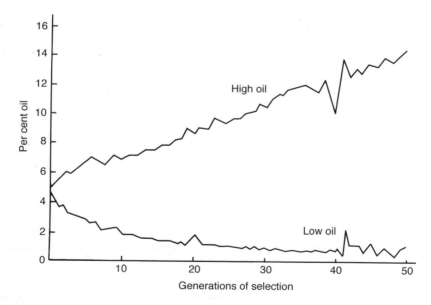

Fig. 5.10. The effect of 50 generations of selection for high or low oil content in corn (after Woodworth et al., 1952).

Progeny testing is important because it is not always possible to determine whether an individual's superiority is due to the environment or improved genotype. A small amount of seed is taken from the maternal parent and sown in rows or plots. The choice of parental plants that will be used to produce the next generations is then based on the performance of their progeny and not their own phenotypic appearance (Allard, 1960). However, the value of progeny testing is only to improve the selection of superior phenotypes; it has no intrinsic value since the seed from the progeny plots is not used in the breeding programme. Finally, with line breeding the progeny lines are combined to form a composite, and provided there is a large enough number of lines that are not too closely related, the inbreeding can be prevented. Line breeding is an important method used for maintaining a high level of resistance to pathogens (Lawrence, 1968).

The second type of population used in the breeding of open-pollinated populations is the synthetic population. The synthetic population consists of a number of plant genotypes that have been shown to combine favourably in all possible combinations with each other to produce agronomically valuable genotypes. The synthetic population involves testing the performance of progeny and it differs from line breeding because the latter is based only on composites of individually tested lines. Synthetic varieties have been produced in maize and forage crops.

Before leaving open-pollinated populations with their inherent need to maintain a high degree of heterozygosity in the final selection, there exists one other form of selection that should be considered, recurrent selection. Recurrent selection was developed by maize breeders with the ultimate objective of improving inbreds for hybrid production. The development of hybrids is considered in the next section but recurrent selection, which is essentially a method of population improvement, can be considered independently. Recurrent selection combines the advantages of line breeding and progeny testing with a limited amount of control over pollination. There are a number of different selection procedures, but the simplest involves selfing of superior plants and crossing their progenies in all combinations to provide seed for repeat cycles of selection and crossing. The more complex form of recurrent selection is the reciprocal recurrent selection procedure (Fig. 5.11) which is designed to improve two populations simultaneously with respect to their performance and to their mutual combining ability (Simmonds, 1979).

The recurrent selection techniques used in alfalfa breeding have been described by Nielson and Lehman (1980). Line selections are also used in breeding for alfalfa resistance but as a method it is less preferential to individual plant selections because there is less control of pollen sources. The individual plant selections involve an initial screening population of between 1000 and 10,000 plants, usually from a cross variety or plant introduction. Selection cycles are made where resistant material is selected, grown and seed resown each cycle. At each cycle the number of plants is gradually reduced until a desired number of superior plants remains. Progeny tests may then be used. For recurrent selection the number of selected plants or lines in each cycle should be at least 25 to 100 to ensure that the frequency of detrimental characters is maintained at a level no higher than in the original population. Selected plants/lines from one cycle are interpollinated and seed are produced for another cycle where selection is made for the same characters. Recurrent selection is used to breed alfalfa with multiple resistance (Hanson et al., 1972) and has been used to develop maize with resistance to *Spodoptera frugiperda* (Widstrom et al., 1992).

5.4.4 Hybrids

Hybrids are produced by crossing various inbred lines, usually those of outbreeders, and are considered important because some hybrids can produce outstanding

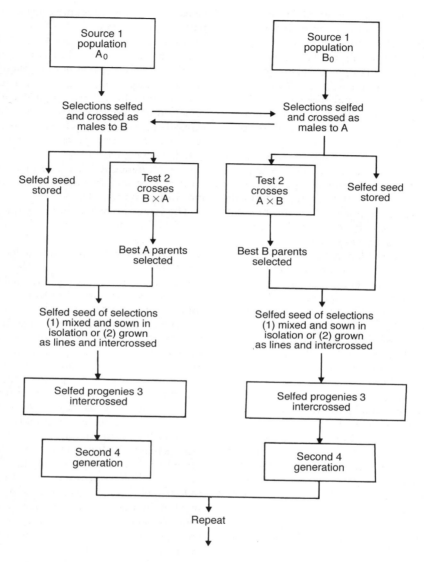

Fig. 5.11. A diagrammatic representation of a breeding scheme of reciprocal recurrent selection (after Simmonds, 1979).

yields, greatly exceeding those of the parents. The greatest drawback with developing hybrids has been the selection of the inbred lines that when crossed will produce a valuable hybrid. If two unrelated inbred lines are crossed, then, in general, the resulting hybrid will show some increased vigour relative to the parents, but out of the thousands tested too few hybrids have been produced that are economically valuable. It is the production of outstanding hybrids that is always sought but such crosses are relatively rare, to the extent that approximately 70% of US maize in 1979 was considered to be based on as few as six inbreds (Simmonds, 1979). Despite the great deal of work carried out in hybrid maize breeding the genetic mechanisms are not altogether clear. Inbreeding depression is usually attributed to the fixation of unfavourable recessives, and heterosis, essentially the opposite of this, has been ascribed to one of two mechanisms, either dominance or over-dominance of heterozygotes. The over-dominance hypothesis explains heterosis in terms of the superiority of heterozygous genotypes relative to homozygotes while the dominance hypothesis is based on the superiority of dominant alleles when the recessive alleles are deleterious (Strickberger, 1976). However, there is no clear cut evidence for one or other of these hypotheses and the true mechanisms await identification and clarification.

In maize, hybrid production is basically achieved in five steps:

1. Selection from an open-pollinated population of plants having favourable characteristics.
2. The selfing of these plants over a few generations to produce homozygous inbred lines.
3. Selection among the inbred lines.
4. A top cross testing for assessing general compatibility (normally referred to as crosses exhibiting a good general combining ability). A top cross is a cross between an inbred line and a common pollen parent variety, called the top cross parent.
5. A pair-wise testing for hybrid vigour.

Hybrids have also been developed among onions, sugar beet, Brussels sprouts, kale, castor, tomatoes and cotton, although dramatic increases in yields are not attributable to these hybrids. One of the driving forces behind the search for and development of hybrid crops lies in the economics of seed production and sales. Although hybrids may be expensive to develop they have the economic advantage that the grower cannot use harvested seed for the next season's crop, but must return to the wholesaler to purchase the F_1 hybrid seed. This commercial scheme seems to have worked well in developed countries where the necessary infrastructure exists to take advantage of it. Such an approach in developing countries may be less appropriate, especially since population improvement techniques have shown equal potential and are often less expensive.

5.4.5 Clones

Clones are groups of individuals that have descended from a common parent by mitotic division alone, meiosis has not been involved. Hence, clones are genetically identical plants that have been vegetatively propagated from a single parent. Vegetative propagation occurs naturally in the form of asexual reproductive structures such as stolons, rhizomes, tubers, bulbs and corms or by well established horticultural techniques such as budding, grafting, leaf cuttings and leafless stem cuttings. A wide variety of crops are produced as clones including perennial vegetables, potato and sweet potatoes, tree crops, apples, peaches and rubber trees, shrubby crops, bananas, sugar cane and pineapple, soft fruits, strawberry, blackberry and raspberry and certain species of grasses.

Breeding clones involves obtaining crosses between the flowering heterozygous clonal parents and then selecting among these F_1 individuals, and in subsequent vegetatively propagated clonal generations for individuals having favourable characteristics. The most valuable feature of clonal breeding is that favourable characteristics produced in the F_1 progeny of the sexual cross, are immediately fixed because subsequent multiplication is vegetative, producing only genetically identical individuals. However, one of the major problems with clonal breeding is that the plants used usually have reduced fertility thereby restricting the ease of the sexual cross.

There are basically two types of clonal crop: those that produce a vegetative product and those producing a fruit. Crops grown for their vegetative products can suffer from various degrees of reduced flowering and fertility making sexual reproduction and selected crosses difficult. Wild potato species flower readily while many potato cultivars either do not flower at all in the field, e.g. King Edwards, or their flower buds fall off prematurely. Pollen sterility is also a major drawback in hybridization preventing many desirable crosses (North, 1979). Although flowering is obviously not a problem among fruit producing clones, pollen sterility does provide a major drawback. Also a number of special reproductive problems exist, e.g. parthenocarpy in bananas, which have to be overcome before cross breeding is possible. Further problems for the breeder are encountered due to carry-over of virus diseases in clones, lengthy generation times of some perennial plants and a slow rate of multiplication of some cultivars (North, 1979). These problems are, however, being solved is some cases by the use of *in vitro* techniques.

In vitro methods involve the cultivation of plant tissues under sterile conditions in the laboratory using various nutrient media to promote growth and development. Meristem cultures utilize meristem tips of developing plants to rear virus free clonal stocks and for the rapid multiplication of some species having a slow rate of vegetative propagation. Embryo cultures, where developing embryos are removed from the young fruit and grown in artificial media, are used in situations where the embryo normally fails to develop, such as in the hybridization of related species. Cell and tissue culture, the most recent advances in *in vitro* methods, have provided some specific cases of potential uses in clonal multiplication, e.g. freesia, production of haploids from pollen cultures and hence homozygous diploids as an aid to inbreeding lines e.g. tobacco, and *in vitro* hybridization, e.g. soybean. The full potential of cell and tissue cultures is yet to be fully realized, but they will definitely play an important role in future developments in genetic manipulation.

5.4.6 Backcross breeding

Backcross breeding is used when breeding with inbred lines, open-pollinated populations and clones. It provides a method to incorporate a desirable trait into an otherwise acceptable variety of a crop plant (Mayo, 1987). Backcross breeding transfers a single or a few genes that have readily identifiable and desirable characters from a donor plant (often poor in general agronomic ability but having a useful trait such as pest resistance) to a generally superior variety lacking only the desired trait. The superior variety is referred to as the recurrent parent and the ultimate value of the new improved variety will be largely dependent on the quality of this parent. Hence, recurrent parents are usually established varieties that have a proven ability but lack a character that could potentially make them of more value. The agricultural value of the donor parent, for characters other than the one under selection, is unimportant but the trait to be transferred must be highly heritable and the intensity of its expression maintained throughout a series of backcrosses. Hence, traits that are used in backcrosses are usually controlled by major genes. Disease resistance, plant height and earliness are traits normally considered suitable for incorporation by backcross breeding (Mayo, 1987).

The general principle of backcrossing can be illustrated by considering two genotypes: aa the donor and AA the recurrent parent. If these are crossed, the resulting genotype is $\frac{1}{4}$AA, $\frac{1}{2}$Aa, $\frac{1}{4}$aa. If this F_1 generation is then backcrossed with the recurrent parent, the resulting genotype will be $\frac{3}{4}$AA, $\frac{1}{4}$Aa, i.e. the donor parents' contribution to the progeny will be halved in each successive backcross. Eventually, the genetic contribution of the donor parents becomes insignificant except for the desired character which is maintained during this reduction process by positive selection. The number of backcrosses that is required is dependent on the recovery of the essential characters of the recurrent parent, or alternatively the redirection of the characters, other than the selected character, associated with the donor parent. The recovery of the recurrent parent will be enhanced if in the early generations there is some selection for parental type. The number of generations required to achieve a suitable level of recovery can be determined in a similar way to the number of generations required to achieve homozygosity through selfing. Figure 5.9 can be used to find the percentage of plants homozygous for a given number of alleles entering the cross from the recurrent parent. If no selection is practised, then for parents differing in, for example, five gene pairs five backcrosses will produce a population in which approximately 85% of the individuals will be homozygous and identical with the recurrent parent at all loci (Allard, 1960).

Backcross breeding in open-pollinated crops differs only in the number of plants that must be used as recurrent parents. This number must be sufficient to ensure the recurrent parents represent the gene frequency characteristic of that particular variety (Lawrence, 1968). The situation is similar in clonal crops when dominant genes are being transferred; more than one recurrent parent is required to avoid inbreeding, and enable the progeny to be left heterozygous for the transferred gene.

5.4.7 Breeding for horizontal resistance

There are three possibilities to contend with in breeding for resistance to insects. The first is that the insect may exhibit gene-for-gene vertical resistance (although probably rare among insects) and the requirement for an appropriate breeding scheme to incorporate the genes into high yielding cultivars; backcrossing techniques should be appropriate. The second possibility concerns the identification of resistance characters controlled by major genes (preferably expressing characters that are beyond the insect's capacity for change) and the introduction of these genes into appropriate cultivars. The third possibility involves the selection and breeding for polygenically inherited resistance characters.

Utilization of characters that are beyond the insect's capacity for change will provide durable resistance in crops plants and although such characters are often controlled by major genes, the resistance is horizontal. Such major gene characters are amenable to traditional methods of breeding and can be readily incorporated into high yielding cultivars, but unlike vertical resistant characters, the resistance will be durable. In the short term the continued research effort for this durable major gene resistance is inevitable because at present it fits in with traditional methods of breeding.

The basic difference between breeding for horizontal resistance and the more traditional breeding techniques is that with the former outcrossing among individuals is actively encouraged while, in the more conventional approaches, inbreeding is considered the most useful form of fertilization. For horizontal resistance breeding, crops that are normally self-pollinating will have to be prevented from self-fertilization. The use of male gametecide sprays is one possibility, e.g. MS_3 for sorghum (Johnson and Teetes, 1980), although the frequent use of these will encourage resistant plant strains that do not respond to the spray treatment. The more conventional alternatives for preventing cross-pollination are plant nuclear and cytoplasmic sterility. Nuclear male sterility is controlled by male chromosomal genes that are usually recessive, e.g. *MsMS* and *Msms* are male fertile and *msms* are male sterile. Cytoplasmic sterility is maternally inherited, i.e. a sterile female crossed with a fertile male produces sterile progeny (Mayo, 1987). Male sterility facilitated recurrent selection has been successfully used to develop resistance to stem borers in rice (Chaudhary and Khush, 1990).

The general method for breeding for horizontal resistance is one of mass selection from a large random polycross. Plant populations must consist of thousands of individuals but need not cover an area greater than a few hectares if, for instance, cereals such as rice or wheat are being grown. Agronomically unsuitable and susceptible plants must be removed before flowering and selections should be based on assessments or measures of the relative amount of resistance or conversely the degree of susceptibility encountered. Horizontal resistance will have accumulated to useful levels after 10–15 generations which could mean only 5–7 year cycles with two cropping seasons per year.

Screening methods can then be devised to select plants according to their relative value measured in terms of numbers of insects present, their rate of development or growth, their fecundity and survival or the amount and type of damage they cause. Rarely are absolute measures of the above variables made; usually a scale or index scoring system is devised and a visual assessment of the variables translated into a relative estimate. Such visual assessments and scoring systems do not provide a very accurate evaluation of resistance but in most breeding situations large numbers of plants have to be screened quite rapidly, so other techniques are often too laborious and time consuming. Screening under controlled conditions in the laboratory may identify polygenically inherited resistance more readily than in the field. Among laboratory populations of organisms, selection occurs from among a relatively low number of individuals and because the distribution of phenotypes will be limited, the selection will tend to make use of existing common variation rather than drawing on novel variation from rare phenotypes (Roush and McKenzie, 1987). The application of insecticides to insects in the laboratory, where dosage and mortality can be precisely controlled, has resulted in the accumulation of polygenically inherited resistance to the insecticide by the small proportion of insects that survived each generation (Roush and McKenzie, 1987). Similar selection pressure to that exerted by the insecticide in the laboratory can be produced in host plant resistance studies with closely controlled infestations of insects for a given number of plants. The infestation has to be just great enough to permit sufficient plant survivors for the next generation. The work of Barnes *et al.* (1969) on laboratory screening of alfalfa for resistance to the alfalfa weevil (*Hypera postica*) represents the type of methodology that would

select for polygenically inherited resistance; the weevils were allowed to feed until 95% of the plants were destroyed and the 5% surviving plants were grown on for further resistance evaluation. The surviving plants in the above experiment were subjected to further tests and were shown to support significantly less adult feeding and produce smaller larvae than unselected plants of the same variety. The differences between the selected and unselected plants were small but this would be consistent with selection of polygenes which would produce a small effect each generation and would need to be accumulated over several generations to provide useful levels of resistance.

The majority of screening that is carried out, however, is not in the laboratory but in the field, where conditions are more variable and where screening and selection techniques need to be simple and economical of time and effort.

The selections must only take place during the esodemic if vertical resistance is suspected. Where vertical resistance is not present, the selection pressure (provided by the level of pest infestation) should, ideally, be both intense and uniform. Methods for increasing the level of infestation include planting susceptible varieties around and within the breeding plot (but not allowing them to flower) and the use of artificial infestations (inoculation from laboratory mass reared insects). Ensuring that regular severe infestations occur is important for two reasons: a patchy distribution of infestation will increase the likelihood of escapes, and horizontal resistance will be eroded in the absence of selection pressure. Plants that escape infestation are technically included as horizontally resistant but the greater the number of escapes, the slower the rate of accumulation of horizontal resistance since escapes that are selected may not contribute resistance genes to the gene pool. To prevent this, infestations need to be as uniform as possible over the breeding population.

In the same way that genes for resistance can be gradually accumulated in the presence of a positive selection pressure, resistance can also be lost or eroded in the absence of that pressure. The implication of this is that in situations where pest infestations are sporadic and highly variable between seasons, horizontal resistance will only accumulate at a slow or negligible rate unless the level of infestation is artificially maintained. The methods mentioned above are applicable. Susceptible varieties can be sown early around the block to encourage a quick establishment and build-up of the insect populations (Jackai, 1982). Infested plant material could then be distributed within the breeding block. As the infested plant sections wilt and die the insects will move on to adjacent test plants (Lowe, 1973).

Breeding for horizontal resistance will be easier in some crops than others. The difficulties with self-pollinating crops have been mentioned above but these can be overcome. The greatest problem with such crops comes from the reticence of the breeders to consider alternative breeding techniques rather than with technical difficulties related to the methodology. Long term perennial outbreeders are another matter. Many of these crops are difficult to breed because they occupy such large areas, some have a very low level of seed production, e.g. coconut, and long generation times, e.g. dates (7–10 years). Since there is little likelihood of perennial plants exhibiting vertical resistance to insects,

selection for horizontal resistance will be possible within existing crops. In forest trees, the selection for resistance may be hampered by the problems of screening such large plants. Where pests may be high in the crown developing appropriate sampling methodology will be difficult, although most trees can be vegetatively propagated by stem or root cuttings once resistant individuals have been identified. Recent advances in tissue culture methods with conifers and deciduous trees may stimulate vegetative production of insect resistant trees (Hanover, 1980).

Also little use has been made of mass selection for resistant phenotypes despite its successful use for other quantitatively inherited traits, including rust resistance. Mass selection has been used to improve crops of lobolly pine, Scots pine, black wattle and eastern cottonwood (Wright, 1976). The characteristics used were mostly related to growth and production such as growth rate, tree height, number of branches and diameter but work on slash pine (Goddard *et al.*, 1973) and eastern cottonwood (Jokela, 1966) has shown that mass selection techniques are also highly effective at selecting for horizontal resistance to rusts. Since the principles of breeding for horizontal resistance to pathogens apply equally well to insects, mass selection for resistance to insects in forest stands has a high potential for success.

Case Study: Directed mass selection for leafhopper, alfalfa aphid and diseases in alfalfa (Hanson *et al.*, 1972)

Alfalfa is a forage crop that is susceptible to leafhopper yellowing caused by *Empoasca fabae* and damage by the spotted alfalfa aphid, *Therioaphis maculata*, as well as a range of diseases including rust, common leafspot, bacterial wilt and anthracnose. Mass selection of two gene pools (A and B) was used to develop multiple resistance to these pests over 18 generations (Fig. 5.12). At the time the two pools were initiated, resistance to more than one disease or insect was infrequent in both exotic and domestic varieties of alfalfa. Pool A was initiated from 400 plants from four sources and pool B was established on 380 plants from nine sources (cultivars or synthetics). In all cycles, with the exception of the first cycle of selection for resistance to the spotted alfalfa aphid, no less than 80 plants and sometimes as many as 500 plants were selected to be intercrossed

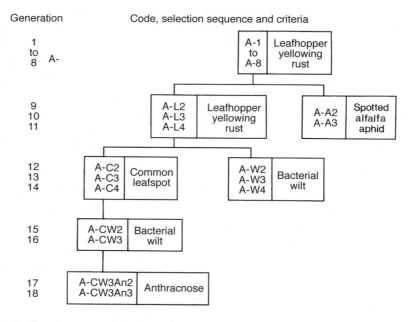

Fig. 5.12. Plan of recurrent phenotypic selection in germplasm pool A. A similar plan was used for pool B except that four cycles of selection for bacterial wilt resistance followed A-C4 instead of the two as in pool A (after Hanson *et al.*, 1972).

within each pool to produce seed for a new cycle. In contrast to this, the first cycle used for the selection of resistance to the spotted alfalfa aphid used 20 and eight plants respectively in pools A and B. During selection for resistance to this aphid a further exception was made. In each of the other field selections vigour and general plant appearance were always used as selection criteria for pest resistance, whereas with the spotted alfalfa aphid the reaction to the insect was the only criterion used. Selection in the field occurred in about half of the cycles of selection for each pest, the rest were based in the laboratory. The results of the selections for resistance to each pest were positive but only the results of the insect pests are discussed here.

The pools A and B responded similarly to selection for resistance to potato leafhopper yellowing (Fig. 5.13) but pool A responded more rapidly than pool B for resistance to spotted alfalfa aphid. Hence, the mass selection was an effective means of developing resistance to each of these pests, although yield decreased in the field during selection for spotted alfalfa aphid (Fig. 5.14). During selection for resistance to this aphid, a further exception was

vigour. Relaxation of the selection pressure for plant vigour was thought to be the likely cause of the much reduced yield, combined with the effects of using a narrower genetic base to initiate the cycle. The decrease in yield in the final generations of the cycle was also thought to be due to this relaxation of selection for vigour during these cycles which took place in the laboratory.

An unexpected bonus from the mass selection was that pool B was found to have a resistance to frost, probably as a result of selection in the field being made after the first frost of the season. In another gene pool a similar unexpected character, resistance to the pea aphid, was inadvertently selected for. The presence of these improved but unselected characters no doubt resulted from the selection of healthy, vigorous plants over the selection cycles and represents significant population improvement. The mass selection procedure has a great deal to recommend it as a means of improving resistance to specific pests and multiple pest resistances, as well as more complex agronomic characters.

5.4.8 Breeding trials design

The basics of trials design for field experiments have been dealt with in Chapter 3. Additional trial designs that are particularly relevant to plant breeding trials are included here. They are collectively referred to as incomplete block designs.

In plant breeding trials large numbers of breeding lines need to be screened and compared for yield, quality and resistance. Such large numbers of treatments with complete replication can extend over very large areas. One option in trials design would be to have controls randomly scattered throughout the experiment so that each of the treatments could be compared with the controls and hence with one another. With such a design the accuracy of comparison with the control is increased at the expense of other comparisons, hence use of random controls is usually considered inefficient (John and Quenouille, 1977). An alternative is the type of design in which comparisons between pairs of treatments are all made equally accurate. These designs are the incomplete block designs. This subject is dealt with only briefly here and readers requiring more information should consult Cochran and Cox (1957), John (1971) and John and Quenoulille (1977).

Incomplete block designs may require more planning than randomized blocks but are no more difficult as far as experimental

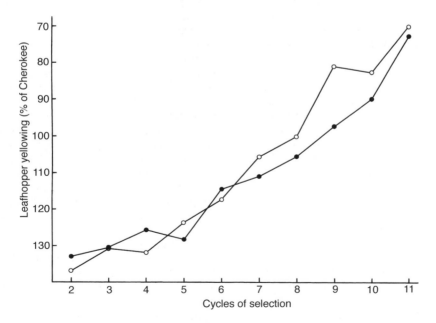

Fig. 5.13. The response to recurring selection for resistance to leafhopper yellowing (expressed as a percentage of yellowing on a check cultivar (Cherokee)), for germplasm pools A (●) and B (○) (after Hanson *et al.*, 1972).

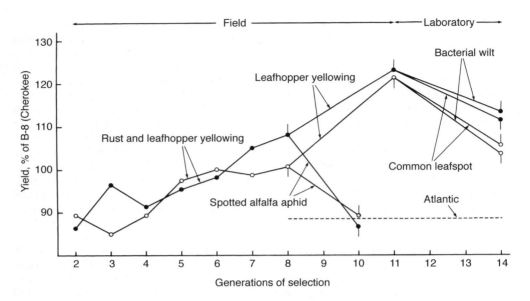

Fig. 5.14. Mean changes in alfalfa yield over 14 generations of selection in the field and the laboratory, for germplasm pools A (●) and B (○) (after Hanson *et al.*, 1972).

operations are concerned. Most of the incomplete block designs cover the range from 6 to 200 treatments and work has indicated that an average gain in accuracy of about 25% over randomized blocks can be obtained (Cochran and Cox, 1957). However, the number of treatments for which a substantial increase in accuracy is obtained needs to be determined by experience. It should be noted though that incomplete block designs are most useful when there are few or no missing data.

Three of the more simple incomplete block designs are depicted in Fig 5.15: the balanced, the incomplete, Latin Square and a partially balanced incomplete block design arranged in a Latin Square formation (Cochran and Cox, 1957). The balanced design shows seven treatments arranged in blocks of three units with every pair of treatments occurring once within some block. The incomplete Latin Square was designed for use in greenhouse experiments and is named after the man who developed it, Fouden, and hence the name Fouden squares. The example shown is a balanced design for seven treatments, so

that every treatment appears in each of the three rows and every pair of treatments appear together once in the same column. The example in Fig. 5.15b is of a partially balanced incomplete block design with six treatments in blocks of four. Each row forms a complete replication but some treatments occur less often than others in the same block, e.g. treatments 1 and 2 occur twice in the same block (1) and (4) whereas treatments 1 and 4 occur four times in the same block (1), (3), (4) and (6). This design permits greater flexibility over choice of replicates for particular treatments but their statistical analysis is more complicated and some pairs are more precisely compared than others. Visual assessment of plants using scales or indices are the commonest form of resistance evaluation. Bellotti and Kawano (1980) recommended the use of two selection scales, the first to evaluate a large number of varieties when the major objective is to reject susceptible material and the second allows a more accurate definition of the reaction of the selected plants (Table 5.4). The second scale (Scale B) in Table 5.4 uses three dis-

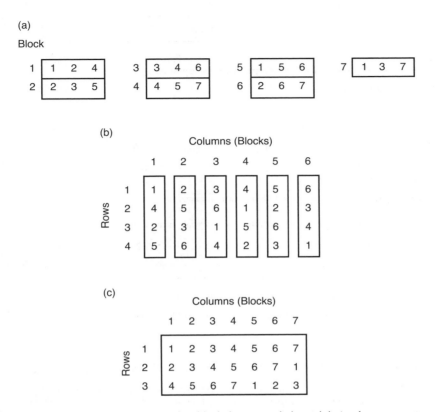

Fig. 5.15. Examples of three simple incomplete block designs: (a) balanced design for seven treatments in blocks of three units; (b) a partially balanced incomplete block design arranged in a Latin Square; (c) balanced design for seven treatments in an incomplete Latin Square (after Cochran and Cox, 1957).

tinct damage symptoms, leaf speckling, leaf deformation and bud reduction, which combined define the damage symptoms of *Mononychellus tanajoa* on cassava and permit small differences in damage to be detected. The development and use of scales that permit a comprehensive evaluation of damage or resistance should be used where possible. Although the need for subjective assessment in large scale screening is not denied, there is a need to ensure that the evaluation is based on actual discernible differences and scales that incorporate a number of important indicator characters that will provide a safety net during the selection process.

The value of the subjective assessment made by an assessor will depend on the person's perception of the crop and the pest infestation which, to a large extent, will be determined by an assessor's experience. The ability of assessors to discriminate between plant characters can also be improved through various physical means including the use of keys and regular inclusion of check plant material. Breeders must then be aware of two things in relation to the screening and selection techniques: firstly, the efficiency of the visual assessments for particular crop characters; and secondly, the variation in this efficiency between individual assessors. The easiest method of assessing these efficiencies is to compare a visual score assigned to crop characters with the actual quantitative estimate of the same characters. Alternatively,

Table 5.4. Plant damage scales that have been used to screen cassava plants for resistance to the cassava mite *Mononychellus tanajoa* (after Bellotti and Kawana, 1980).

Score	Description

A. Initial screening scale used to discard susceptible plants (up to 85%)

0	No mites or symptoms
1	Mites on bud leaves, some yellow to white speckling of leaves
2	Many mites on leaves, moderate speckling of bud leaves and adjacent leaves
3	Heavy speckling of terminal leaves, slight deformation of bud leaves
4	Severe deformation of bud leaves, reduction of bud, mites on nearly all leaves, with whitish appearance and some defoliation
5	Bud greatly reduced or dead, defoliation of upper leaves

B. The lines selected using scale A are re-evaluated using a scale of 0 to 10

0	No mites or symptoms
1	Plants with one or two bud leaves with light brown or whitish speckling located on a few lobes or dispersed over whole leaf. Average of less than 50 specks.
2	Light speckling distributed over all bud leaves, average 50–1000 specks
3	Moderate speckling of bud and terminal leaves. Attacked leaves fade in colour
4	Severe speckling of bud and adjacent leaves. One or two bud leaves show slight deformation. Leaves have whitish appearance
5	Severe speckling of apical and middle leaves. Light deformation on leaf margin, leaves turn whitish
6	Moderate deformation of leaf margin with indentation almost reaching mid-veins, and curling of apical leaves resulting in mosaic-like appearance. Basal leaves also show speckling. Slight bud reduction
7	Bud deformed and reduced, apical leaves with intense mottling
8	Total plants affected, severe reduction of bud and few new leaves developed, general yellow to white appearance with some apical leaf necrosis
9	Bud completely reduced, no new leaves developing, defoliation beginning with apical leaves
10	Bud dead and severe defoliation

visual scores can be compared with actual estimates between plants selected by an assessor and those randomly selected over the same area. Work with barley comparing scores of yield with actual yield has resulted in conflicting opinions. McKenzie and Lambert (1961) concluded that the use of visual scores for yield was unsatisfactory for evaluations, while Ismail and Valentine (1983) concluded that in the early generations visual selection using scores should be recognized as a basic tool of breeding. Briggs and Shebeski (1970) working with spring wheat and testing the visual selection efficiency of 14 assessors, using comparisons between random and assessor selections, found that assessors had a general ability to improve yield by selection, but that individual selectors demonstrated a rather limited ability to identify the actual highest yielding plots in the experiment. The variation between assessors in their ability to select visually for a specific character such as yield (the same applies equally to characters for resistance) is only a major concern if they are unaware of their inefficiency. When an assessor's selection efficiency is known, even if it is low, the number of selections that are made can be chosen on the basis of this information.

Perception of favourable characteristics in crop plants may be improved if an assessor can make reference to a standard plant or plant character, or if comparisons are made on a scale that reduces environmental variation between plants: hence the use of check varieties in trials plots, the use of keys to aid visual assessment and the development of Gardner's grid system and nearest neighbour techniques (Gardner, 1961; Townley-Smith and Hurd, 1973).

The general practice in plant breeding is to intersperse numerous check varieties within plots and to select other plants on the basis of their characteristics relative to the standard check. In resistance studies the check variety is usually a universally accepted highly susceptible variety. The more numerous the check variety the easier it is to compare with other plants, since the efficiency of selection is related to the distance from the nearest standard (Briggs and Shebeski, 1968). In practice the breeder has to balance the number of checks and selection efficiency with the area available for trials and the number of plants that need to be screened.

The use of keys has been shown to produce slight increases in selection efficiency (Ismail and Valentine, 1983) but to be useful the key must be carefully designed and the assessors must be familiar with its use. One of the biggest drawbacks with the key developed by Ismail and Valentine (1983) for yield components in barley was that for only a small increase in selection efficiency, the key took an unacceptable increased length of time to use, but a greater familiarity with the key might have reduced this time. The use of keys is probably more important in situations where checks cannot be used or only used sparingly, such as in the large outcrossing populations needed in horizontal resistance breeding where the susceptible checks must not cross-pollinate with the resistant breeding population. The Gardner's grid system (Gardner, 1961) and the moving means of nearest neighbours (Townley-Smith and Hurd, 1973) are also appropriate to situations where checks provide an additional complication. The moving mean method of selection is dependent on the means of nearest neighbours as a comparative standard, while the grid system reduces the area over which comparisons between plants are made and selections chosen. Both these methods are thought to reduce the environmental variation among selected plants but neither has been used extensively, despite even the grid system's simplicity and ease of use.

Case Study: The efficiency of visual assessment of grain yield and its components in spring barley rows (Ismail and Valentine, 1983)

The principles involved in evaluating the efficiency of visual assessments made by breeders during plants selections are the same whether the characters selected for are those of agronomic quality or for resistance to insect pests. The efficiency of the visual assessment of yield is used here as an example because yield is a particularly complex variable to assess. It has a number of components and assessment involves the mental integration of a number of different variables.

The data for assessments of five assessors are considered; 99 spring barley lines were sown in single rows in a randomized block design with three replicates. Visual assessments were made on a 0–9 scale where 9 = high expression of the component, and the separate components assessed were: yield/row, tillers/row, grains/ear and 1000 grain weight. Characters were assessed individually, the rows harvested and the actual values for each of the characters were determined.

Table 5.5. The r^2 values (expressed as a percentage) from the regressions of visual assessment of yield against actual yield estimates (from Ismail and Valentine, 1983).

Assessor	Yield/row	Tillers/row	Grains/ear	1000/grain weight
B	43.2	55.6	21.7	–
C	36.6	52.4	14.7	37.9
E	25.9	39.8	30.4	43.7
A	43.1	–	–	–
D	13.3	–	–	–
Mean	32.4	49.3	22.3	40.8

The multiple regression of the actual measures of the three yield components with yield per row accounted for 82.5% of the total variation. The associations between the visual assessments (dependent variable) and the actual measurements were determined using linear regression analysis. The r^2 values (the proportion of the variation accounted for by the regression relationship) were used as a measure of the efficiency of the visual assessment (Table 5.5). The number of lines to save, in order to retain a given proportion of the best lines, was also used as a measure of efficiency (Table 5.6). The proportion of the variation accounted for by the relationship between visual and actual measures of yield components was generally poor, with only two assessors having r^2 values greater than 50%. This suggests there are large differences between what the assessors perceive as a high yielding component and the actual yield estimates. The second method, based on the number of lines to save was considered by Ismail and Valentine to be more relevant to the breeder than the r^2 values. Breeders have to be prepared to lose some of the best lines during the selection process simply because the alternative would be to reduce the number of genotypes (and hence the potential variability of material) that could be screened. The results in Table 5.6 indicate that although visual selection may be inefficient it is still better than random selection (50 lines would have to be selected to save at least 50% of the 10 best lines from the complete 99 lines if random selection were used). However, considerable variation does exist between assessors. The two types of analysis point to the same component, tillers/row, as being the variable most accurately assessed and selected.

Table 5.6. The number of lines that would need to be saved in order to ensure at least 50% of the best 10 lines were saved by the individual assessors when evaluating yield on the basis of four different criteria (yield/row, tillers/row, grains/ear and 1000/grain weight). The total number of lines is 99 (from Ismail and Valentine, 1983).

Assessor	Yield/row	Tillers/row	Grains/ear	1000/grain weight
B	25	18	12	–
C	25	19	28	14
E	16	15	20	27
A	10	–	–	–
D	26	–	–	–
Mean	20.4	17.3	20.0	20.5

Case Study: Gardner's grid system and plant selection efficiency in cotton (Verhalen *et al.*, 1975)

Gardner's grid system (Gardner, 1961) was developed for maize breeding but its effectiveness as a method of improving plant selection efficiency is independent of crop type, quantitative trait or breeding method. The method appears to be particularly useful for discriminating between small inconsistent difference in plant character, which makes it highly relevant to breeding for horizontal resistance.

In 1967, a 60 × 60 m block of Westburn cotton was arbitrarily subdivided into three 20 × 60 m grids. One hundred individual plant selections per grid were made on the basis of boll type and apparent yield. The seed cotton from each selected plant was sawginned and the fibre length determined. On the basis of this measure the upper and lower 10% of the plants were selected over the whole block and within each grid. There was a total of 85 selections overall and seed from these plants was sown in 1968 and 1969 and samples sawginned and measured as before.

Thirty five of the plants selected over the block were also selected within the grids; 25 selections were not. If the two methods of selection (block vs. grids) were identical then theoretically the same plants would be selected in each case. The observed difference in the plants selected would suggest that one technique was superior to the other. One of the main reasons for the difference between grids and the block could be due to environmentally induced phenotypic effects. If this were the case then smaller differences between selections would be expected within the grids than over the whole block. An analysis of variance of within grid and between grid (block effect) fibre lengths showed that variation within grids was less than between them (Table 5.7). The estimates of the component variances indicate that the between grid component of the phenotypic variation accounted for 22% of the variance for the whole block (0.000321/(0.001131 + 0.000321)). Then the between grids component is mainly environmental and since phenotypic variance was 22% less within grids, selections within grids should be more efficient than those over the whole block.

Table 5.7. Partition of total variance into its between and within grid components (0.000321 is derived from 0.033257 − 0.001131)/100) (after Verhalen *et al.*, 1975).

Source	df	Mean square		Estimate of variance components
		Calculated	Expected	
Between grids	2	0.033257	$0_w^2 + n0_B^2$	0.001131 + 100(0.000321)
Within grids	297	0.001131	0_w^2	0.001131

Selection efficiency can be assessed in terms of selection differentials, selection response and heritabilities. The selection differential is calculated by subtracting the mean of the lower 10% from that of the higher 10% selections for each grid and the block. Since the variation within grids was less than that over the whole block the selection differential would be expected to be lower within each grid than for the block. Five of the six differentials at 10 and 5% selection intensity were less than for the block differential and the sixth grid differential equalled that of the block (Table 5.8).

Table 5.8. Estimates of selection differentials, selection responses and heritabilities for two selection methods (within grids or over the block), at 10 and 5% levels of selection. Locations one and two were sampled on years 1968 and 1969 respectively (after Verhalen *et al.*, 1975).

Selection method	Selection differential	Response		Realized heritability	
		1	2	1	2
10% level of selection intensity					
Within grids					
Grid 1	0.112	0.054	0.055	0.482	0.491
Grid 2	0.104	0.061	0.045	0.587	0.433
Grid 3	0.130	0.043	0.062	0.331	0.477
Block	0.130	0.041	0.040	0.315	0.308
5% level of selection intensity					
Within grids					
Grid 1	0.129	0.068	0.052	0.527	0.403
Grid 2	0.121	0.062	0.096	0.512	0.793
Grid 3	0.140	0.076	0.052	0.543	0.371
Block	0.152	0.057	0.056	0.375	0.368

The selection responses were also calculated as the difference between the two directions of selection but in this case the progeny of the initial selections have been used (Table 5.8). Ten out of the 12 responses are greater within grids than over the block. The realized heritabilities are calculated by dividing the selection response by the selection differential. In each case the heritability is greater within grids than over the block (Table 5.8).

The selections made with the grid method have been shown to be superior to those made over the block in terms of reduced phenotypic variation (22%), a higher differential (11–14%), a higher selection response (20–35%) and a higher estimated heritability (41–52%).

5.5 Evaluating Resistance

The traditional approach to insect host plant resistance has followed the ideas espoused by Painter (1951) which emphasize the mechanisms employed by the plant to reduce insect infestations and the morphological and biochemical bases involved in this. The emphasis on mechanisms of resistance combined with the need to incorporate resistance genes and breeding schemes into established breeding programmes have dominated insect host plant resistance studies.

5.5.1 Mechanisms of resistance

Painter (1951) divided insect resistance mechanisms into three categories: non-preference, antibiosis and tolerance. The term non-preference has subsequently been replaced by antixenosis (Kogan and Ortman, 1978), because non-preference refers to the insect and this is incongruous with the notion of resistance being a property of the plant.

Antixenosis is the resistance mechanism employed by the plant to deter colonization by an insect. Insects may orientate towards plants for food, oviposition sites or shelter but certain plant characteristics may be a biochemical or morphological factor, or a combination of both. Plants that exhibit antixenotic resistance would be expected to have reduced initial infestation and/or a higher emigration rate of the pest than susceptible plants.

Antibiosis in contrast to antixenosis is the mechanism by which a colonized plant is resistant because it has an adverse effect on an insect's development, reproduction and survival. These antibiotic effects may result in a decline in insect size or weight, an increased restlessness, poor accumulation of food reserves affecting the survival of hibernating or aestivating stages, or have an indirect effect by increasing the exposure of the insect to its natural enemies (Singh, 1986).

Plant tolerance can be described as the extent to which a plant can support an insect infestation without loss of vigour and reduction of crop yield. Beck (1965) does not consider that tolerance falls within the definition of resistance. Plant tolerance is usually taken to mean that when two cultivars are equally infested the less tolerant one has a smaller yield. At the plant physiological level the loss of tolerance is due to an abnormally heightened response to infestation, at the epidemiological level tolerance is considered a component of resistance (Robinson, 1976).

The three mechanisms of resistance will influence the population dynamics of a pest insect throughout a season (Fig. 5.16) by their action on the life history parameters: initial colony size, duration of larval period, fecundity of adults, mortality of larvae and adults. The last four of these parameters are used to determine the intrinsic rate of increase, r (Section 2.3). Thomas and Waage (1996) investigated the effect of each of the four parameters on the rate of increase assuming that host plant resistance seeks to reduce 'r' below zero, the point at which the pest population will decline over time (Fig. 5.17). The analysis revealed that not all the components of antibiotic resistance have equivalent effects, the effects of the individual mechanisms depend on the life history of the pest. With a slow growing population, increased development time or increased juvenile mortality have roughly the same effects on reducing 'r'. As the population growth rate increases however, the value of increasing development time markedly increases (Thomas and Waage, 1996). The relative advantage of manipulating adult mortality increases only as fecundity declines, as the product of juvenile mortal-

ity and development time gets larger and as population growth rate increases. In addition the relative value of a decline in fecundity is dependent on juvenile mortality and development time. The impact of antixenosis on the population dynamics is no less complex, with some of the effects paralleling those of antibiosis. For instance, reduced oviposition through non-preference is equivalent to reduced fecundity, and it can also increase larval movement thereby slowing development time or increasing juvenile mortality. Increased emigration from the crop due to antixenosis has the equivalent effect of increased adult mortality (Thomas and Waage, 1996). Overall, therefore, not only different mechanisms but also different strengths of resistance may be required to effect equivalent levels of population suppression of pests with different life histories. Clearly an evaluation of host plant resistance should look further than just assessing damage or numbers of insects to gain a full appreciation of the ways in which resistance will have an impact on pest populations.

5.5.2 Evaluating antixenosis and antibiosis

During a breeding programme plants may be screened for resistance to insect pests but most experimental work on resistance mechanisms takes place either with released cultivars or with resistant cultivars prior to release. Experiments to evaluate antixenosis and antibiosis are usually labour intensive and can only be carried out on relatively few cultivars and hence are rarely included in routine screening programmes (although there are exceptions, e.g. screening in alfalfa and rice).

Antixenotic resistance can be assessed in preference tests either in a choice or no-choice situation (Ng et al., 1990; Jackai, 1991) or by comparing the behaviour of the insect on plants having a range of susceptibilities. For example, experiments can be designed which compare the number of adults alighting on plants (e.g. Singh et al., 1994) or their oviposition response

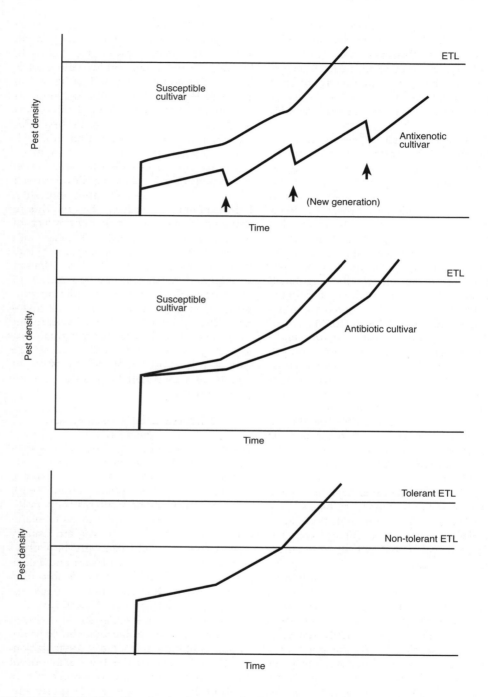

Fig. 5.16. Effects of three functional resistance mechanisms on pest population dynamics where the top graph shows the effects of antixenosis; in the middle, antibiosis; and at the bottom, tolerance (modified from Kennedy *et al.*, 1987).

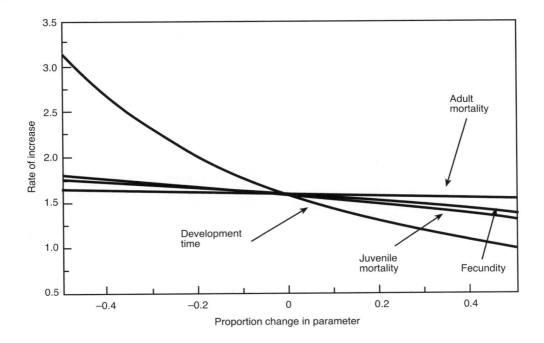

Fig. 5.17. The effect of changes in life history parameter values on the intrinsic growth rate of a pest population. The zero value for each parameter relates to that given in the text: development time (2 weeks), fecundity (140 eggs per week), juvenile mortality (0.5 per week) and adult mortality (0.5 per week). The population growth rate with these parameter values is $r = 1.6$ (after Thomas and Waage, 1996).

on a range of cultivars, either in a choice or no-choice situation (e.g. Jager *et al.*, 1995). The disadvantage with no-choice situations is that spurious results may be obtained if the response of the insects to a particular cultivar is conditioned by the presence of other cultivars. Tests such as these have been carried out in both the laboratory and the field. The importance of choice in influencing insect response to resistant and susceptible plants in the field has been demonstrated in the work of Cantelo and Sanford (1984). Intermixed and isolated pure stands of resistant and susceptible varieties of potato, cabbage and lima beans were sown to test for resistance to the potato leafhopper, *Empoasca fabae*, the imported cabbage worm, *Pieris rapae*, and the Mexican bean beetle, *Epilachna varievestis*, respectively. The

response of the insects to the mixed and isolated pure stands was different in each case. The plant array did not have a significant effect on the size of the Mexican bean beetle population in the resistant and susceptible cultivars of lima bean, but antixenosis was more apparent against the potato leafhopper in isolated pure stands of susceptible and resistant potato lines than in mixed resistant and susceptible lines. The opposite was true with the imported cabbage worm on cabbage. When planted in a mixed stand the preference for egg deposition on susceptible cultivars was greater than in pure isolated stands of either resistant or susceptible types. This example illustrates the differences between insect responses to their host plants and the need to identify the effects of choice on the preference before designing experiments to evaluate the potential of antixenosis as a resistance mechanism.

Antixenosis has also been measured in terms of the number of insects leaving cultivars, both in the laboratory and the field. The rationale behind this approach is that insects that have located a susceptible plant will be less inclined to leave it than an insect on a resistant plant, hence the numbers leaving susceptible and resistant plants should differ. However, differences in both the field and the laboratory are not necessarily large. Müller (1958) observed the arrival and departure of the black bean aphid, *Aphis fabae*, on two bean cultivars in the field, a resistant Rastatter and a susceptible Schlandstedter. He observed that while equal numbers of alatae land on both varieties only 1% remained to reproduce on the resistant variety while 10% remained on the susceptible variety. The mean staying time for those adults that landed and then took off again was higher on the susceptible beans ($6\frac{1}{2}$ min) than on the resistant beans ($3\frac{1}{2}$ min) and these differences were significant. So even though large numbers landed on both cultivars the difference in the numbers taking off was quite small, although sufficient to account for differences in population sizes on the two cultivars. A laboratory study of the reproduction and flight of alatae of the grass aphid, *Metopolophium festucae cerealium*, from a number of grasses and cereals also indicated that adults even reselect after settling on susceptible host plants (Dent, 1986). Most alatae flew from the grasses *Festuca rubra* (94%) and *Festuca arundinacea* (84%) after producing only a few nymphs. Whereas between 38% and 64% of alatae flew from *Lolium multiflorum*, oats, *Lolium perenne* and wheat, these alatae deposited more nymphs on the hosts before flight. This study indicated the importance of measuring both the flight response and reproduction, since an evaluation of flight response alone might provide misleading information about antixenotic resistance. Also the ranking of host according to alatae nymph production

differed from that of apterous virginoparae on the same host plants (Dent and Wratten, 1986) emphasizing the dangers inherent in assessing host resistance only in terms of the antibiotic effects on the non-selective life stage of an insect, e.g. apterous aphids or lepidopterous larvae.

Antixenosis, and to a lesser extent antibiosis, is often evaluated by studying the behaviour of insects on potentially resistant and susceptible cultivars. An introductory guide to measuring behaviour has been written by Martin and Bateson (1993) and further more detailed studies relevant to resistance studies are given in Wyatt (1997) and Eigenbrode and Bernays (1997). A few elementary points are mentioned here in relation to evaluating resistance mechanisms.

The initial temptation to devise behavioural experiments testing for resistance must be avoided until the observer has had time to observe the insect and its interactions with the plants in question. This period of preliminary observation enables the observer to become acquainted with the insect's behaviour and enable better formulation of appropriate questions and ensure the correct choice of measures and recording methods.

The first step after this observation involves describing the insect's behaviour. Martin and Bateson (1993) describe behaviour in terms of structure, consequences and relations. The 'structure' describes the behaviour in terms of the subject's posture and movements, e.g. flying, walking, feeding. The consequences are the effects of the insect's behaviour on the plant (and vice versa), e.g. insect impaled, insect takes off, whereas the relations describe where, when and with whom an event is occurring, e.g. the insect on the adaxial leaf surface. Behaviour is, of course, continuous but it must be broken down and divided into discrete units to allow it to be measured. There must be enough measurement categories included to describe the behaviour adequately, and each should be precisely defined to allow any other experimenter to make the same observation. There are four types of behavioural measure: latency, duration, frequency and intensity. Latency is the time from some specified event to the onset of the first occurrence of a behaviour. Givovich *et al.* (1988) measure the behaviour of 50 alatae of *Aphis craccivora*, on each of three cowpea lines and found that once transferred to the plant the aphids took longer to decide whether to feed on the two resistant than on the susceptible line. The measure of latency was also combined with a measure of duration to assess the response of the aphid to the cowpea lines. Duration is the length of time for which a single occurrence of the behaviour pattern lasts, and in the above experiment duration, total probing time was longer on the two resistant than the susceptible lines. The frequency is

the number of occurrences of the behaviour per unit of time. The most common frequency measure used in behavioural studies of homopterans is the number of probes made per unit time. Intensity, in contrast to the other behavioural measures, has no universal definition and is a subjective assessment, e.g. the extent of restlessness. This type of measure should be avoided if possible unless it can be scored in terms of the number of movements of a particular kind per unit time. For example, Bernays *et al.* (1983) observed the climbing speed of *Chilo partellus* larvae at different temperatures on two different sorghum cultivars (Fig. 5.18). Behavioural studies are often aided through the use of physical models (e.g. Harris *et al.*, 1993; Vaughn and Hoy, 1993), electroantennagrams (EAG) or single

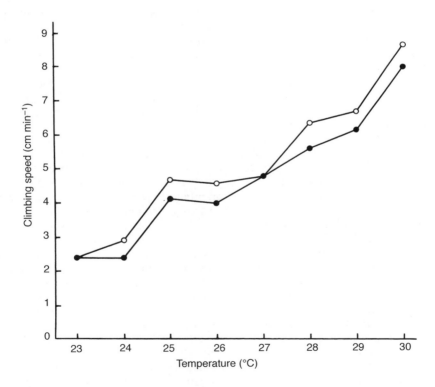

Fig. 5.18. The climbing speed of newly hatched larvae of *Chilo partellus* on two sorghum cultivars (CV.IS 1151, ○; CV.IS 2205, ●) over a range of temperatures (after Bernays *et al.*, 1983).

cell recording (SCR) (Blight *et al.*, 1995; Pickett and Woodcock, 1993) and video recording (see Wyatt, 1997 and Eigenbrode and Bernays, 1997).

Tests for antibiosis mechanisms of resistance are usually carried out under no-choice conditions, with the insects confined on plants or plant materials inside a cage. Most cages consist of a fine mesh material that can be used to cover the plant or plant part as a sleeve, or a cage either to cover the whole plant or to isolate specific areas such as a section of leaf.

Tests for antibiosis among plant cultivars usually assess the performance of pest individuals to obtain a mean estimate of development, reproduction and survival. Insect development can either be measured as a rate or expressed in terms of insect size or weight. The development rate is usually considered in terms of the length of time taken between one stage and another on resistant and susceptible cultivars. For instance, the larval and pupal periods were shorter and adult longevity longer for individuals of *Chilo partellus* reared on susceptible maize than on resistant maize (Sekhon and Sajjan, 1987). The more resistant varieties also reduced larval weight by 51 to 60 mg per larva and pupal weight by 49 to 52 mg per pupa. The lengths of insects have also been used as a measure of size in resistance studies, both in the bean flower thrips, *Megalurothrips sjostedi* (Salifu *et al.*, 1988) and in the leafhopper, *Amrasca devastans*, a pest of okra (Uthamasamy, 1986).

Differences in antibiotic resistance between cultivars can also be assessed by measuring the fecundity and fertility of insects. This can be extended to population effects by taking mortality into account and obtaining r_m values (birth rate – mortality rate; Birch and Wratten, 1984; Holt and Wratten, 1986). Life table analyses have also been used to evaluate cultivars for antibiotic resistance (Easwaramoorthy and Nandagopal, 1986). Tests to determine antibiotic resistance may be undertaken utilizing artificial diets, leaf discs, excised leaflets,

lyopholized resistant plant materials or membrane filters with incorporated leaf extracts (e.g. Wiseman, 1989; Allsopp *et al.*, 1991; Allsopp, 1994; Hammond *et al.*, 1995). In addition a variety of innovative techniques have been developed measuring electrical signals to study insect probing behaviour (e.g. Kimmins, 1989), the activity of enzymes (e.g. Wu *et al.*, 1993) and honeydew excretion (e.g. Pathak *et al.*, 1982; Cohen *et al.*, 1997).

5.5.3 Morphological and biochemical bases of resistance

Plant resistance to insects is rarely totally dependent on a single mechanism, there are often overlaps between the morphological and biochemical bases of resistance, e.g. trichomes that exude substances that are toxic to insects as in some species of *Nicotiana* which exude alkaloids toxic to aphids (Thurston *et al.*, 1966). The mechanisms of resistance are divided into their morphological and biochemical bases simply for convenience of discussion. Morphological bases of resistance can be classified as remote factors, plant architecture/anatomical features, surface factors, and subsurface factors according to the level at which they function against the insect.

Remote factors influence the orientation of the insect towards the plant, and hence can have an antixenotic effect. Although plant shape may have some effect on insect orientation the most important remote factor is plant colour. The attraction of aphids to yellow reflecting surfaces (Kennedy *et al.*, 1961) is now well known and has been utilized in the construction of yellow sticky traps for monitoring aphid numbers. Yellow is thought to be attractive because it is the colour, or the wavelength of the colour, associated with the senescing tissue favoured by aphids. Adult *Pieris rapae* prefer green and blue green surfaces for pre-ovipositional displays (Ilse, 1937), the cabbage aphid *Brevicoryne brassicae* is less attracted to red Cruciferae (Singh and Ellis, 1993; Ellis *et al.*, 1996) and the boll weevil is attracted less to red cotton plants than green (Stephens, 1957). The work of Prokopy *et al.* (1983) has shown the importance of the leaf colour for the visual selection of plants by *Delia radicum*. More females landed on mimics coloured as radish leaves than on green or red mimics of cabbage. Although these differences in preference occur, it is debatable whether such characters can be used as a mechanism of resistance since it is unlikely that the resistant effect of colour will persist in the absence of hosts having a preferred colour. Also little can be done by way of genetic manipulation to affect plant colour without affecting some fundamental physiological plant process (Norris and Kogan, 1980).

Anatomical features of plants can have a major influence on plant resistance and often they do not conflict with other essential crop characters. For instance, genotypes of spring wheat that possess awns were more resistant to grain aphid *Sitobion avenae*, than awnless plants (Acreman and Dixon, 1986), the difference was due to lower fecundity on and increased likelihood of dislodgement from the awned wheat. The combination of these factors reduced the growth of the aphid populations in the field by two-thirds on the awned plants. A resistance base in rice grain is due to the physical state of the rice husk (Cogburn and Bollich, 1980). Breese (1960) reported rice varieties having intact husks prevented penetration by *Sitophilus sasakii* and *Ryzopertha dominica*. Plant anatomy can also confer resistance indirectly by its influence on natural enemies. Some anatomical features allow natural enemies greater access to their prey or host than others (e.g. Heinz and Zalom, 1995). Open panicled sorghum supports fewer *Helicoverpa* larvae than closed panicle plants, probably because natural enemies are prevented from locating and attacking the larvae in the tight confines of the closed panicle.

Surface factors that provide mechanisms of resistance include the presence or absence of pubescence and the presence and type of cuticle waxes. The surface

waxes of leaves provide a means of controlling water loss and act as a natural barrier to insect attack. *Myzus persicae* prefers to settle on non-waxy leaves of Brussels sprouts (Way and Murdie, 1965) whereas *Brevicoryne brassicae* prefers the waxy leaves of both Brussels sprouts and forage kale (Thompson, 1963; Way and Murdie, 1965) but not cabbage (Singh and Ellis, 1993). The response of natural enemies to the presence of waxy leaves may also contribute indirectly to the resistance of the plant. In the above situation, the natural enemies of *B. brassicae* such as Coccinellidae and Anthocoridae prefer to oviposit on non-waxy leaves (Russell, 1978) hence the pest mortality due to natural enemies would be reduced on waxy leaved plants.

Pubescence on plant surfaces is made up of individual trichomes or hairs. The trichomes may be unicellular or multicellular, simple or branched, scale like or glandular and they may cover both surfaces of leaves, stems, shoots and roots. The absence as well as the presence of trichomes may provide a mechanism for resistance. Glabrous strains of cotton are more resistant to *Helicoverpa zea* and *Heliothis virescens* than pubescent cultivars (Table 5.9). Oviposition of the female moths was reduced on the glabrous cultivars (Lukefahr *et al.*, 1971; Robinson *et al.*, 1980). When pubescence is present the mechanism of resistance may depend on one or more of four characteristics of the trichomes, their density, erectness, length and shape. In addition, some trichomes also possess glands, the exudates of which confer resistance against some insects.

Pubescence can affect the oviposition, locomotion, feeding and ingestion of the plant's insect pests. The oviposition of the cereal leaf beetle, *Oulema melanopus*, was reduced by high trichome density and trichome length in four wheat lines (Webster *et al.*, 1975) while the Mexican bean beetle, *Epilachna varievestis*, was observed to fall from cultivars having leaves with long trichomes (van Duyn *et al.*, 1972). Leaf

Table 5.9. The number of eggs oviposited by *Helicoverpa* spp. on pubescent (P) and glabrous (G) cotton strains (SG = semi-glabrous) and the number of trichomes on the upper leaf surfaces (after Lukefahr *et al.*, 1971).

Year	Plant character	Seasonal totals of eggs per acre	Number of trichomes in a square inch on upper leaf surface
1965–8	P	16,948	3,833
	P	17,424	2,821
	G	9,082	402
	G	8,606	386
1968–9	P	16,724	2,640
	SG	10,230	1,007
	G	6,521	589
1969–70	P	21,058	2,430
	P	17,005	1,550
	P	30,157	1,772
	G	2,426	40
	G	5,936	0
	G	4,766	0

trichome density on cotton affects the size of whitefly (*Bemisia tabaci*) populations, such that higher numbers are supported on highly pubescent cultivars than on glabrous types (Ozgur and Sekeroglu, 1986). This may be because eggs laid at the base of trichomes are less susceptible to parasitoid mortality since searching efficiency of the parasitoid decreases with increasing trichome density (Hulspas-Jordan and van Lenteren, 1989; Heinz and Zalom, 1995). The locomotion of aphids *M. persicae* and *Macrosiphum euphorbiae* was severely impeded by the presence of glandular trichomes on three wild potato species, *Solanum polyadenium*, *S. tarijense* and *S. berthaultii*. The glandular trichomes exuded a substance that hardened on the aphids' limbs and if it accumulated it would eventually stick the aphid to the plant where they would remain immobilized until they died (Gibson, 1971). The effectiveness of this resistance mechanism was linked to the density of the glandular trichomes since the contact frequency increased with higher trichome density. There are few detailed analyses of the influence of trichomes on the feeding behaviour of insects as a mechanism of resistance (Norris and Kogan, 1980),

although it is generally assumed that the feeding of sap sucking insects may be prevented on pubescent plants because the proboscis cannot reach the mesophyll of vascular bundles. The trichomes of soybean influenced the ability of the leafhopper *Empoasca fabae* to get to the leaf surface in order to feed on it (Singh *et al.*, 1971) but in general it is not known to what extent reductions in population size on pubescent plants are due to interference with feeding (Norris and Kogan, 1980). The larval weight of the cereal leaf beetle was found to be negatively correlated with increased pubescence and high mortality was explained by larvae ingesting unusually large amounts of cellulose and lignin as they consumed the trichomes to reach the epidermis (Schillinger and Gallun, 1968). Surfaces other than leaves are also attacked by insects, and their structure can greatly influence the insects' response to the plant. The bark beetle, *Ips paraconfusus*, preferred fissured bark to smooth bark in which to tunnel on both a host tree, *Pinus ponderosa*, and a non-host tree, *Abies concolor* (Elkinton and Wood, 1980).

The subsurface morphological factors that affect the resistance of plants to insects are usually concerned with tissue toughness. The feeding rate and larval growth of the mustard beetle, *Phaedon cochleariae*, were related to the leaf toughness of turnip, kale and Brussels sprout leaves (Tanton, 1962). For the internode borer *Proceras indicus*, which attacks sugar cane in India, a correlation was found between the percentage of damaged internodes and percentage fibre and rind hardness (Agarwal, 1969). The observed resistance was thought to be due to one or more characters that contribute to rind hardness, the number of vascular bundles per unit areas of rind and the lignification of cell walls of the vascular bundle sheaths, the parenchyma of the subepidermal layers and the intravascular parenchyma cells (Agarwal, 1969).

The biochemical bases of resistance can be divided into two broad categories, those influencing behavioural responses and those influencing the physiological responses of insects. Insect behaviour modifying chemicals are further divided into attractants, arrestants, stimulants, repellents and deterrents, while plant chemicals affecting the physiological processes of insects may be classified as nutrients, physiological inhibitors and toxicants (Hsiao, 1969). The range of responses elicited by chemicals and their effects on insects are quite diverse and complex. Even simple chemicals consisting of a few molecules may have multiple sites of action and influence the insect in a number of ways.

The types of chemical responsible for insect resistance are numerous but the major classes include the terpenoids, flavonoids, quinones, alkaloids and the glucosinolates. These are all secondary metabolites, chemicals that are not required for the general growth and maintenance of the plant but which serve as plant defence products. The organic isothiocyanates (mustard oils) are the main biologically active catabolites from the glucosinolate components of crucifers, which like other glucosinolates defend plants against generalized insects including aphids and grasshoppers (Panda and Khush, 1995) but

can also act as phagostimulants and as kairomones, crucifer specific pests (e.g. Dawson *et al.*, 1993). Some primary and intermediate metabolites such as citric acid and cysteine can also act in plant defence chemistry (e.g. Jager *et al.*, 1996).

There are a number of examples of plant chemicals that have been used in promoting resistance to insects. Oxygenated tetracyclic triterpenes, commonly called cucurbitacins, have been shown to provide antixenotic resistance against Luperini beetles in cucurbits. The adult beetles have a great affinity for the cucurbitacins which act as strong arrestants and feeding excitants for the beetles. Cucurbit varieties which are low in cucurbitacins are 'resistant' to the beetles (Metcalf *et al.*, 1982; Metcalf, 1986; Metcalf and Metcalf, 1992).

The European corn borer, *Ostrinia nubialis*, has two generations per season throughout most of the USA corn growing belt. The first brood damages the leaves from the early to mid whorl stage and reduces kernel filling (Ortega *et al.*, 1980) and it was in this brood that a resistance mechanism was discovered. Resistance was found to be due to a chemical, 2,4-dihydroxy-7-methoxy-1,4-benzoxazin-3-one or DIMBOA. The levels of DIMBOA were found to be high in the seedling stage but decreased as the plant matured, hence DIMBOA levels are low at the time of the second brood infestation. Screening populations for this chemical could provide a simple method for evaluating resistant plants; in practice, however, the analytical method proved too slow to be of value in corn-breeding projects (Guthrie, 1980).

Gossypol is a polyphenolic yellow pigment of cotton plants that has been shown to confer antibiotic resistance to *H. zea* and *H. virescens* (Kumar, 1984). Experiment has shown that the gossypol (11,6,6,7,7-hexahydroxy-5,5-diisopropyl(-3,3-dimethyl-2,2-binaphthalene)-8,8-dicarboxaldehyde) content of cotton buds can be increased genetically from a normal 0.5% to 1.5%, and a larval mortality of 50% can be expected when cotton square gossypol content is increased above 1.2% (Schuster, 1980). The combination of a high bud gossypol with glabrous cotton strains can result in as much as a 60–80% reduction in *Helicoverpa zea* and *Heliothis virescens* larval populations (Lukefahr *et al.*, 1975; Niles, 1980).

These examples illustrate both the potential uses of biochemical bases for resistance, in that they can be highly effective and could potentially provide a fast means of screening segregating plant populations, and the reality of the practical situation where reliable and fast analytical methods are required if such an approach is to be incorporated into a breeding programme.

5.5.4 Diagnostic characters and genetic markers

The identification of a clearly recognizable characteristic of resistance is of great value in a breeding programme. The ability for instance, to be able to screen plants for an obvious morphological trait that confers resistance, such as presence or absence of pubescence, can markedly improve the efficiency of evaluation and hence progress towards developing resistant varieties. The presence of biochemical bases for resistance provide the opportunity for analytical methods which will speed up the process of screening segregating plant populations. It should be easier to screen for, for instance, hydroxamic acid in plant lines than to evaluate the numbers of insects present in field plots of each line. This is why so much research effort concentrates on linking insect performance on cultivars to chemical content and concentration (e.g. Givovich *et al.*, 1992; Nicol and Wratten, 1997).

The availability of molecular markers is likely to have a significant impact in the future on the ease and precision with which screening for a resistance character can be achieved. Genes that are close together on a chromosome tend to be inherited together (i.e. they are not separated during recombination). Hence a gene that can be used as a marker that lies close to a resistance gene may be used to track the inheritance of that resistance gene. The molecular markers currently available are isoenzymes (enzymes associated with a particular gene, identified using gel electrophoresis), restricted fragment length polymorphism (RFLPs: presence or absence of a restriction enzyme site near a particular length of DNA identified by a radioactive probe; Beckman and Soller, 1983) and more recently marker systems based on the polymerase chain reaction (PCR) (Saiki *et al.*, 1988) (which amplifies specific lengths of DNA). The identification and use of markers allows the tracking of different resistance genes and the construction of complex resistance lines, the latter enabling combinations of minor gene assemblies and pyramiding. The present state of marker assisted breeding for insect resistance must be seen only as an intermediary stage restricted at present by the labour required to identify suitable markers (Chapman, 1996). With increased efforts over the coming decade to map plant genomes, new classes of markers will inevitably be identified and more rapid procedures and automation will reduce the burden of work necessary to extend their application into routine plant breeding.

5.6 Genetic Manipulation

The traditional plant breeding methods used to develop insect resistant cultivars involves a lengthy process by which appropriate characteristics are selected over a number of generations. This process has only been of limited success and for a number of key insect pests it has proven very difficult to identify and select for suitable resistance characters (e.g. yellow stem borer resistance in rice; Bottrell *et al.*, 1992). Genetic manipulation of crop plants and the introduction of novel genes for resistances (e.g. from bacteria, viruses or unrelated plants) could markedly improve levels of resistance obtained and in some cases reduce the time and cost of more conventional methods.

5.6.1 Techniques in genetic manipulation

Genetic manipulation is the process by which potentially useful characteristics of one organism are transferred in the form of recombinant DNA to another organism in order that the new organism has enhanced value such as insect resistance. Transferring genes from one organism to another requires the availability of: (i) a DNA vector, which can replicate in living cells after foreign DNA has been inserted into it; (ii) a DNA donor molecule to be transferred; (iii) a method of joining the vector and donor DNA; (iv) a means of introducing the joined DNA molecule into the recipient organism in which it will replicate; and (v) a means of screening for recombinant lines that have replicated the desired recombinant molecule (Lindquist and Busch-Petersen, 1987).

The two basic methods for plant transformation currently in use are: (i) *Agrobacterium* mediated transfer; and (ii) direct gene transfer (the DGT technique).

The *Agrobacterium* technique makes use of the characteristic of *Agrobacterium* to deliver a segment of extra chromosomal plasmid (the Ti plasmid) called the T-DNA genes which cause host plant tissue to proliferate to form a tumour (galls) and to synthesize novel metabolites which are used by the pathogen as carbon and nitrogen sources (Lazzeri, 1998). This naturally occurring transformation system has been modified to enable almost any gene to be transferred into plant cells. Horsch *et al.* (1985) developed a simple and general procedure for transferring genes into plants with *Agrobacterium* Ti vectors. Until recently, however, *Agrobacterium* transfer techniques were confined to the natural hosts of *Agrobacterium* which restricted its value. This led to the development of DGT methods which include: microinjection of DNA into cells; electroporation of cells or protoplasts; puncturing cells with microscopic silicon carbide fibres; and particle bombardment (Barcelo and Lazzeri, 1998). Each of these techniques have been used to manipulate crop species but because most of them require complex cell culture and

regenerative procedures to function efficiently, their broad range application has been limited (Lazzeri, 1998).

5.6.2 Genetic manipulation for insect resistance

Two classes of genes have received most attention for enhancing crop plant resistance to insects: (i) genes encoding for plant genes; and (ii) those encoding for *Bt* endotoxins or other non-plant 'toxins'.

The main plant proteins that have been investigated for development of resistance in crop plants are inhibitors of digestive enzymes, in particular protease inhibitors, lectins and enzymes (Gatehouse *et al.*, 1998). In addition there is some potential in the manipulation of plant secondary metabolites to produce crops resistant to insects (Hallahan *et al.*, 1992).

The first example of a foreign plant gene conferring resistance to insects was the transfer of a trypsin inhibitor (CpTi) gene from cowpea in tobacco *Nicotiana tabacum* (Hilder *et al.*, 1987). The trypsin inhibitors affect insect digestive enzymes and the cowpea trypsin inhibitors provide protection against species of Lepidoptera, Orthoptera and Coleoptera (Gatehouse *et al.*, 1992) including the vine weevil *Otiorhynchus sulcatus* in strawberry (Graham *et al.*, 1996). The tomato proteinase inhibitors expressed in transgenic tobacco have been shown to confer resistance against tobacco hornworms *Manduca sexta* (Johnson *et al.*, 1989) and the potato proteinase inhibitor PPi-11 is effective against *Chrysodexis criosoma* (McManus *et al.*, 1994). The α-amylase inhibitor (from the common bean *Phaseolus vulgaris*) in seeds from the transgenic pea plants and Adzuki bean have been shown to confer bruchid resistance (Schroeder *et al.*, 1995; Ishimoto *et al.*, 1996).

Lectins are carbohydrate-binding proteins found in many plant tissues and are abundant in the seeds of some plant species (Panda and Khush, 1995). The first lectin to be expressed in transgenic plants was the pea lectin in tobacco which resulted in enhanced resistance against

Heliothis virescens (Boulter *et al.*, 1990). Since then the snowdrop lectin (GNA) has been transferred to potato and tomato where it has reduced damage by Lepidoptera (Gatehouse *et al.*, 1997) and aphids (Down *et al.*, 1996). Despite these successes and the wider potential for use of lectins (e.g. Allsopp and McGhie, 1996) the use needs to be tempered by the possibility of toxicity to consumers and hence the need for vigorous testing for potential human toxicity.

The potential of plant derived genes encoding insecticidal proteins for resistance against insects has as yet only been demonstrated in the laboratory and glasshouse trials. They are yet to undergo the large scale trials and commercialization process of *Bt* transgenic crops (Gatehouse *et al.*, 1998).

The bacterium *Bacillus thuringiensis* (*Bt*) produces a number of insect toxins which are protein crystals formed during sporulation (Section 6.8). Preparations of *Bt* spores and crystals have been used as commercial biopesticides for over 20 years. The first isolation and cloning of a *Bt* gene was achieved in 1981, the transformation

of the endotoxin into tobacco occurred in 1987 (Vaeck *et al.*, 1987) and the first field trials with transgenic crops began in 1993 in the USA and Chile (Merritt 1998). *Bt* transformed agricultural crops now include tomato (Delannay *et al.*, 1989), cotton (Perlack *et al.*, 1990), maize (Koziel *et al.*, 1993) and potato (Merritt 1998), while in forestry, transformed *Populus* has also been shown to be effective (McCown *et al.*, 1991).

Transgenic cotton containing *Bt* genes has been field tested in replicated field crops in 1987, and in 1995 bulking of seed for commercial sale was undertaken (Harris, 1997a). Transgenic *Bt* cotton strains produced enough toxin to provide excellent control of *Heliothis zea*, *Trichophlesia ni* and *Spodoptera* species. The *Bt* was first commercialized in the USA in 1986 on 729,000 ha, then increased to just over 1 million ha in 1997 and 2 million ha in 1998 (Table 5.10) (Merrit, 1998). The growth in *Bt* maize is no less spectacular with five companies supplying seed for the 1998 season for a total area of 14 million acres in the USA (Anon., 1997). The *Bt* maize is aimed at control of the European

Table 5.10. Approximate total areas of commercial plantings (thousands of hectares) of insect protected crops containing *Bt* genes from Monsanto (from Merrit, 1998).

Country	Year	Cotton*	Corn	Potato
USA	1996	729		
	1997	1000+	1300	10
	1998	2025+	4000	20
Canada	1997		4	2
	1998		122	4
China	1998	53		
Mexico	1997	15		
	1998	40.5		
Argentina	1998	8	1.6	
Australia	1997	60		
	1998	81		
South Africa	1998	12		

*Figures for cotton in the USA in 1997 and 1998 include a proportion of herbicide tolerant and combined insect protected and herbicide tolerant cotton.

corn borer which annually costs between $10 and $30 million to control (Ostlie, 1997). The use of *Bt* maize should reduce the number of chemical applications and provide additional control from preserved natural enemies.

Insect protected crops using *Bt* genes have shown rapid uptake by growers which will encourage further development and investment into this technology by the commercial companies. *Bt* crops and those modified to express other proteins, etc. are going to have a significant impact on the pest management in the next 20 years.

5.7 Durable Crop Resistance to Insects

The primary goal of conventional (Mendelian) breeding is to produce a higher yielding, better quality crop and once that is achieved then resistance characters can be incorporated, provided the method for introducing the resistance can itself be readily integrated into the breeding programme. In this way, the breeding of pest resistant plant material is considered as an adjacent to the main breeding programme. Resistance, once it has been identified, is incorporated into plant material already having high yield and other favourable agronomic characteristics (Fig. 5.19 provides a generalized scheme). For this to be possible, a good source of resistance is required, and it must be controlled by simple inheritance so that it can be easily incorporated using backcross breeding methods. The good source of resistance may come from plants within a screening population or from pre-adapted plants from wild relatives. Provided the resistance is simply inherited, i.e. the resistance is controlled by major genes, then both sources are readily used.

The need for a good source of useful characteristics led to the establishment in 1973 of the International Board for Plant Genetic Resources (IBPGR) by the Consultative Group for International Agricultural Research (CGIAR) (Harlan and Starks, 1980). The role of IBPGR is to identify crops and regions where the need for conservation is greatest, to coordinate efforts for collecting germplasm and to promote the cooperation and free exchange of germplasm between nations and organizations. Seven of the CGIAR's International Agricultural Research Centres (IARCs) have germplasm collections of their designate crops, hence there are germplasm collections for all the major food crops (Sattaur, 1989). The collections are maintained as either seed, seedling lines or clones (Simmonds, 1979) (Fig. 5.20).

In contrast to the traditional breeding approaches utilizing existing related crop species as a source of useful genes, transgenic techniques make it possible to introduce non-plant genes or unrelated plant genes into the required crop cultivars. Both approaches, however, are largely dependent on the identification and use of single genes, an approach that makes both types of crop susceptible to resistance breakdown due to adaptation of the insect pest. A number of tactics have been proposed to manage and prolong the use of particularly useful characters including provision of refugia, pyramiding genes, multilines and sequential release of cultivars incorporating different resistance genes.

Refugia work on the basis that the individuals that survive 'exposure' to the transgenic resistant crop are double recessive individuals and these are likely to mate with susceptible individuals from the refuge areas, the cross of which maintains a low number of double recessives in the next generation thereby preventing the build up of resistant populations. In the case of *Bt* cotton, a system of 'refugia' has been introduced whereby growers are obliged for every 100 acres of *Bt* varieties, either to plant 25 acres (i.e. 20% of the total acreage) to varieties which do not contain the *Bt* gene, and on which they are able to use standard insect control measures or alternatively to plant 4 acres with non-*Bt* varieties and leave these untreated with any control measure (Merritt, 1998). For less crop specific pest species, refugia may be maintained in surrounding alternative hosts, e.g. with *Bt* potatoes and Colorado beetles, hence crop refuges may not be necessary.

Pyramiding genes for resistance involves combining two or more resistant major genes in a single variety and thus reducing the possibility of a pest insect

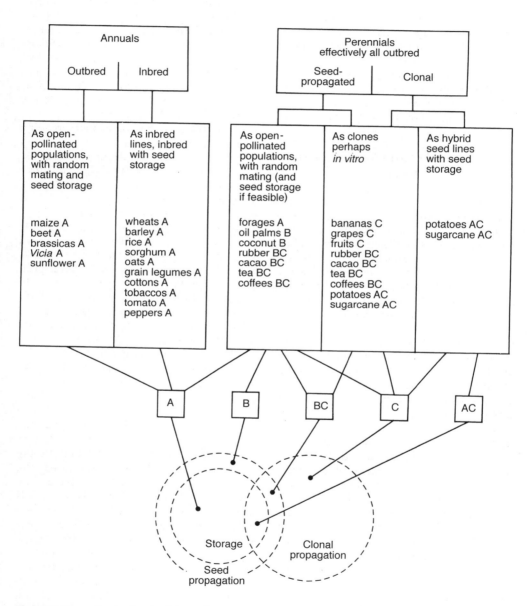

Fig. 5.19. The methods of maintaining collections in relation to the crop biology. A = seedlines with storage, B = seedlines without storage and C = clonal plants. These three methods generate five groups (after Simmonds, 1979).

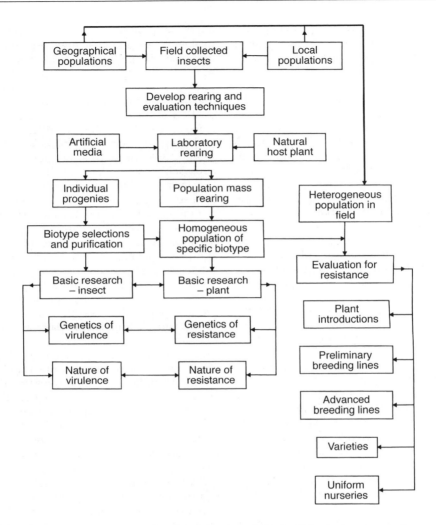

Fig. 5.20. Generalized diagram depicting the entomological phases of developing plants resistant to insects and mites (after Gallun, 1980).

assembling the right combination of virulence genes to be able to attack the variety (Barrett, 1983; Gallun and Khush, 1980), thus prolonging its life. The disadvantage of using this approach is that it could cause the simultaneous loss of several resistance genes if a virulent biotype happened to develop. Also, when a number of resistance genes are introduced into a single variety, there will be no corresponding virulence genes that can be used to provide

evidence of their presence, and in the transfer process some of the genes can be lost (Singh, 1986). Pyramiding genes has been tried in conventional plant breeding with rice against the brown planthopper (Khush, 1980) and has been proposed for genetically manipulated crops.

The multiline strategy uses the same principles as variety mixtures for reducing the level of initial infestation and the rate of spread of the insect, but differs in that multilines are composed of phenotypically similar component lines, derived from a common breeding programme. The component lines differ only in their genes for resistance, with as many as 6–15 lines being combined to produce a single multiline. The multiline can be produced using lines resistant to all prevalent races of the pest (a 'clean crop' approach) or by including lines that are not resistant to all the races (a 'dirty crop' approach). The former approach aims to keep the crops as pest free as possible while the latter provides greater flexibility to the breeder for selecting other characters. It also frees the breeder from continually isolating and evaluating new sources of resistance (Singh, 1986) because the life of strong resistance genes will be extended (van der Plank, 1963). However, while the use of multilines has proved successful with the control of plant pathogens, the application of multilines to insect pest problems awaits further research.

The sequential release of resistant genes is another strategy that could potentially be used to prolong the life of a particularly useful cultivar. Resistance genes are held in reserve and as soon as resistance breakdown becomes inevitable, a new resistance gene is incorporated into the cultivar. The drawback of this approach is that new genes for resistance must continually be selected and evaluated before the old lines become susceptible, a continual race that the plant breeder can never be guaranteed to win. An alternative approach would be to recycle or rotate the use of resistant genes or old resistant varieties. Varieties can, after widespread use and breakdown of resistance, be withdrawn and reintroduced at a later period when the original virulent pest races are known to have become scarce. The success of this approach will mainly depend on whether or not the old varieties are equal in yield, quality and agronomic attributes to the more recent varieties. If they are not, then recycling of the resistant genes alone could provide another option.

All of the above approaches or tech-niques for prolonging the usefulness of vertical genes are aimed at simulating the permanence of horizontal resistance, with some also attempting to recreate the heterogeneity of the wild vertical subsystem, but they all inevitably create a cycle of introduction and breakdown of vertical resistance genes. Were the identification and evaluation of vertical resistance genes a simple and cheap process then such an approach could be readily justified. However, this is not the case, since it requires many scientist years and resources to produce a resistant cultivar and in some cases breakdown occurs before the cultivar has even been released. The identification and development of cultivars having vertical resistance, whether used sequentially, in pyramids or as multilines, requires an enormous investment for the purpose of prolonging the life of an impermanent form of resistance, a high cost to pay for a defective strategy.

The mechanistic approach of Painter (1951) set entomologists on the search for readily identifiable, simple inherited characteristics that confer resistance to plants against insects. This approach, rather than the epidemiological and genetic one taken by pathologists, is responsible for the entomologists' emphasis on the durable resistance obtained from major genes. The characters identified are usually morphological and confer durable resistance because the characters are beyond the capability of the insect for microevolutionary adaptation. Superficially, the identification and use of this form of resistance would seem sensible, having the advantage of simple inheritance, so the characters are readily transferable and it also provides durable resistance. In practice, the drawback of this approach is that selection and incorporation of a resistance character for a particular pest may confer durable resistance to the one pest but it can also make the plant more susceptible to other pests, normally of secondary importance. Glossy non-waxy Brussels sprouts are resistant to cabbage aphids (*Brevicoryne brassicae*) but are susceptible to several other pests including *Myzus persicae*, a major virus vector (Russell, 1978). Hairy cultivars of cotton are more susceptible to some Lepidoptera than others, e.g. glabrous cotton cultivars resistant to *Helicoverpa* spp. are susceptible to *Spodoptera littoralis* (Norris and Kogan, 1980). Glandless cottons are more susceptible to *Helicoverpa* spp. and blister beetles *Epicauta* spp. but are resistant to the boll weevil *Anthonomus grandis*. These resistant/susceptible relations are a result of an imbalance in the plant pathosystem because of an emphasis on one particular pest. One problem is simply being replaced by another: an undesirable situation but one exacerbated by the lack of a holistic approach to breeding resistant plants.

Durable resistance to insects is possible, provided breeders select for horizontal rather than vertical resistance and preferably for horizontal resistance that is not selected for, purely on the basis of a single mechanism of resistance (durable major gene, horizontal resistance). Horizontal resistance is particularly applicable to insects because they have a high dissemination efficiency and tactical mobility, which suggests the general lack of a vertical subsystem. This makes breeding for horizontal resistance relatively simple but it still requires a major shift in approach by many breeders. The problem lies in the reluctance of breeders to consider a new line of reasoning and a consequent change in breeding methodology. With the advent of genetic manipulation, which represents an even more extreme single gene based approach, it is unlikely in the near future that horizontal resistance breeding will gain much attention.

In the same way that genetic manipulation is not a panacea for insect pest control, there are of course situations in which breeding for horizontal host plant resistance would be totally inappropriate. For instance, crops with a high commercial value that have special qualities, e.g. wine grapes, date palm and pineapple, the essential characteristics of which could be lost during the breeding process. Also in highly commercial crops the costs of controlling the pest insect may be small relative to the crop value, so there is little incentive to use horizontal resistance, even if it were possible. Robinson (1987) argues that the agricultural value of horizontal resistance tends to be inversely proportional to the commercial value of the crop (Fig. 5.21). Therefore breeding for horizontal resistance will be most appropriate to subsistence farming and pasture crops where profit margins will not permit expenditure on insecticides. The need for developing higher yielding, high quality food crop cultivars resistant to major pests for the benefit of farmers in developing countries has long been recognized. The use of resistant cultivars provides the most appropriate means of control for subsistence farmers and is one form of new technology that does not require the farmers to make fundamental changes in their way of life or farming methods.

5.8 Discussion

Host plant resistance to insects has, in theory, always held an important position as a potential control option to be utilized in insect pest management. In practice, however, it has rarely lived up to its potential. The absence of complete resistance to insects (vertical resistance), its breakdown where it has occurred and the general acceptance that partial, incomplete resistance is all that can be expected, have fuelled the belief that the potential of host plant resistance will never be fulfilled and that any resistance obtained will not provide long lasting control (Buddenhagen, 1983). This view is understandable given the relative lack of success of scientists in resistance breeding but it is certainly not an acceptable situation.

Despite the emphasis on the identification of mechanisms of resistance to insect pests, relatively few cultivars have been produced that are resistant to insects. This is partly due to the nature of resistance to insects and to the relatively late interest

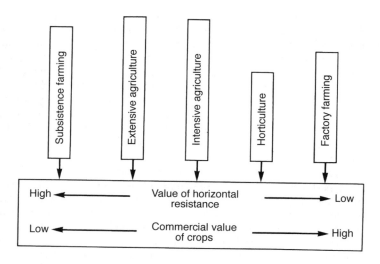

Fig. 5.21. The value of horizontal resistance to different types of agriculture (after Robinson, 1987).

shown in developing insect resistant culti-vars. Resistance to insects is often only par-tial (de Ponti, 1983) and since this can produce problems for the breeder, the development of resistance to insects has taken second place to that of resistance to pathogens where large discrete differences have been found. Secondly, the progress in plant breeding technology and procedures has mainly centred around the temperate zones where most of the major crop prob-lems are associated with pathogens. Also, for the years in which plant breeding gained in importance, insect pests were being controlled effectively and cheaply by insecticides, so that insect resistant crop plants were considered unnecessary, or at least of only secondary importance. The role of resistance to either pathogens or insect pests though, has always been sec-ondary to that of improving crop yield and quality, and it is this obsession with maximizing yield that has fashioned the approach of breeders to that currently followed today.

Where resistance has been incorporated into crops either by vertical resistance from conventional breeding or genetic manipula-tion, then it will always be susceptible to resistance breakdown. This could involve farmers in the 'boom and bust cycle' evi-denced within pathogen resistance – it can be very costly! The costs to the farmer and to the region or state can be enormous if a resistant cultivar breaks down, especially if the crop is a staple food. An epidemic can result in widespread damage and some-times total crop loss, so that individual farmers can lose their income, source of sta-ple food and seed for the following year's crop. Governments may find that they have to spend more money on insecticides, on importing alternative foodstuffs or request-ing aid from other countries. As far as the farmer who utilized the resistant crop culti-var is concerned, the costs of a resistance breakdown can vary from a mild nuisance to economic and social disaster. The real-ization of the potential for the latter places a grave responsibility on the breeders and gene manipulators of resistant cultivars. The adoption of horizontal resistance as a long term durable method of control is unlikely until the evidence for its value is overwhelming. However, in the present cli-mate of enthusiasm of genetically manipulated crops, horizontal resistance research will not receive the funding it requires to prove its case. The consequence will inevitably be a benefit to the commercial seed producers but not to sustainable resistance in crop plants and farmers.

6

Biological Control

6.1 Introduction

The interactions between insects and their natural enemies are essential ecological processes that contribute to the regulation of insect populations. In situations where this interaction is disrupted potential pest populations may develop unconstrained, and excessive population growth, which constitutes a pest outbreak, may occur. Pest outbreaks can occur when alien insects are introduced into new geographic areas without their natural enemies or when insecticides destroy natural enemy populations (Price, 1987). Insects can also become pests when dissociated from their natural enemies due to a habitat modification that differentially favours the pest, e.g. habitat simplification with a monoculture. The use of natural enemies in pest management is mainly concerned with redressing the imbalance that has occurred through this dissociation, whether by reintroducing natural enemies into the system or by trying to recreate conditions where an association can occur.

6.2 Objectives and Strategies

The five methods of biological control are introductions (often referred to as classical biological control), inundation, augmentation, inoculation and natural enemy conservation. Historically most emphasis has been placed on classical biological control although more recently (the last 15–20 years) a great deal more effort has been directed at inundative and augmentative control. Classical biological control has been most successful with pests of fruit, forest and range crops (Greathead and Waage, 1983; Huffaker, 1985) where the perennial nature of the crop permits a continuous interaction between the natural enemy and the pest host without the ecological upheavals that are associated with the management of annual crops (Hassell, 1978). In annual field crops, the instability of the system produces a separation in both space and time between the natural enemy and the host insect pest population. The extent of this separation can sometimes be reduced by the use of management techniques that conserve natural enemies and enhance the stability of the cropping system. This approach of natural enemy conservation can also be used to improve the establishment of introduced natural enemies in classical biological control. In some situations, conservation techniques alone may not allow sufficient build up of natural enemies to control the pest, in which case natural enemy numbers may be augmented with releases of laboratory reared organisms. This augmentative release method of control can serve to maintain the pest population below a threshold density during a critical growth stage or outbreak period, but the control is only temporary as a natural balance has not been established.

Inoculative release of natural enemies may be used in situations where a native pest has extended its range and has been separated from its natural enemies or when an introduced species of natural enemy is unable to survive indefinitely. The release is made at the start of a season and will last for the duration of the crop or season (Greathead and Waage, 1983). A further form of insect pest control with natural enemies has made use of entomopathogens which has often been referred to as an inundative release and allows entomopathogens to be used to suppress pest outbreaks in much the same way as chemicals are used. However, these biopesticides have added advantages of being most host specific, safer to use, more environmentally friendly and preserve the integrity of other biological control agents.

The science of biological control is the science of insect population ecology, since biological control is essentially an interaction between predator and prey populations. There can be little doubt that classical biological control has contributed to the development of insect population ecology but there is less evidence that biological control has benefited from the theory of population ecology (Waage, 1989). Although the theory has provided a scientific framework for the practice of biological control, improved our understanding of predator–prey interactions and helped to identify suitable attributes for potential biological control agents. In practice, the assessment of natural enemies for such characters is rarely carried out prior to release due to constraints in time and resources or because of poor scientific technique. Also the value of theoretical and analytical models to biological control is quite limited since they tend to operate within quite narrow criteria. Simulation modelling on the other hand is increasingly used because when combined with appropriate experiments it can identify the constraints on pest systems, gaps in experi-

mentation, as well as potential candidate species for control (Wratten, 1987).

The use of natural enemies has long been recognized as a fundamental aspect of insect pest management but too few systems have been studied sufficiently for the pest–natural-enemy interactions to be adequately understood. Ecological theory undoubtedly has a role to play in the future but hopefully in more diverse ways than at present. The approach presented here tries to provide a balance between the theoretical and the practical, the potential and the realistic, since it is only with a more pragmatic approach that the use of natural enemies in insect pest management will continue to gain credence.

6.3 Micro- and Macro-biological Control Agents

The types of natural enemy used in biological control of insects includes pathogens, and a surprising range of invertebrates acting as predators, true parasites and parasitoids. Insect parasites usually kill their insect hosts and hence differ from true parasites that do not. To recognize this fundamental difference insect parasites have been referred to as parasitoids (Reuter, 1913).

6.3.1 Pathogens
The entomopathogens that have been used in biological control include representatives of bacteria, fungi, viruses, nematodes and protozoa. There tends to be an inverse relationship between the attributes of these pathogens that can successfully be employed as biopesticides and those that are appropriate for introductions, augmentation, inoculation and conservation. The pathogens effective in these longer term, limited release strategies generally have lower virulence than their insecticide counterparts but can survive for longer periods in the host population, or in the environment, for example through the production of resistant stages (Payne, 1988). For instance, the European corn borer

(*Ostrinia nubilalis*) is host to *Nosema pyrausta* which has a low pathogenicity, causing some larval mortality especially under environmental stress but most host individuals survive to adults but have a reduced longevity and fecundity (Canning, 1982). The parasite is transmitted transovarially and is highly prevalent in the field. Other examples include the protozoan *Nosema locustae*, the bacterium *Bacillus popilliae*, the *Oryctes* (rhinoceros beetle) baculovirus, the nuclear polyhedrosis virus for the sawfly *Gilpinia hercyniae*, and the fungus *Beauveria brongniartii* used in the control of the cockchafer *Melontha melontha*. However, examples of sustained natural insect population regulation in crop systems by entomopathogens are rare. This is because regulation demands stable ecosystems, such as forests and rangelands, and a capacity for the pathogen to spread (Payne, 1988). Hence for the most part, entomopathogens have been considered only as biological alternatives to chemical insecticides (Waage, 1997).

Primary among the pathogens used as environmentally friendly alternatives to chemicals and commercialized for use in this way have been bacteria, particularly *Bacillus thuringiensis* (*Bt*). *Bt* was first recorded in 1901 in Japan but was named by Matte in 1927 from a strain isolated in Germany. By the mid-1940s, the first commercial preparation named 'Sporeine' was available in France (Deacon, 1983). Currently, *Bt* products are being used on several million hectares annually to control lepidopteran pests of agriculture, forestry and stored products (Smits, 1997) and represent 39% of the 185 biopesticide products on the market (Copping, 1999). The *Bt* bacterium (a Gram-positive spore forming rod) is characterized by an intracellular protein crystal which contains a toxin that acts as a poison for lepidopteran larvae. When a mixture of the spores and crystals are ingested by the insect feeding on treated vegetation, the protein crystal is solubilized in the alkaline mid-gut (pH 10.2–10.5) and the toxin is released; the toxin causes gut paralysis and the bac-

terium is then able to invade the weakened host and cause a lethal septicaemia. There are thousands of *Bt* isolates grouped into serotypes or subspecies that are characterized by flagellar antigens (Smits, 1997). Well known subspecies are *Bt kurstaki* and *Bt azaiwai*, which are active against lepidopteran larvae, *Bt israeliensis* which is active against dipteran larvae and *Bt tenebrionis* which is active against coleopteran larvae. Another species of *Bacillus*, *B. popilliae* is also active against beetles, causing what is known as 'milky disease' of beetle larvae. A further species, *Bacillus sphaericus*, is also of some practical importance because of its use against mosquito larvae. However, it is *Bt* that has dominated biopesticide development and has found most commercial success to date, although viruses, fungi and entomopathogenic nematodes continue to grow in importance.

There are six main groups of insect viruses but only three are sufficiently different from human viruses to be considered safe and these are: the nuclear polyhedrosis virus (NPV), the granulosis virus (GV) and the cytoplasmic polyhedrosis virus (CPV). All three are occluded viruses, i.e. the virus particles are enclosed in a proteinaceous shell which has a paracrystalline structure called an inclusion body (Payne, 1982). Around 125 types of NPV have been described, isolated from Lepidoptera (butterflies and moths), Hymenoptera, Diptera (cranefly and mosquitoes) and the Orthoptera (grasshoppers and locusts). NPVs tend to be family specific with little or no cross infection between insect families. There are only about 50 GVs recorded, mainly form Lepidoptera, and over 200 CPVs but the CPVs are not very host specific and hence have less potential as biopesticides. All three types of virus need to be ingested, after which the inclusion body dissolves in the insect's mid-gut, releasing the virions which penetrate the epithelial lining and start to replicate. The time taken to kill varies according to dose, insect development stage and environmental factors but generally ranges from 6 to 24 days. The CPVs can take longer to kill and can be relatively unstable (Deacon, 1983). The number of commercially successful products is limited (there are 24; Copping, 1999) in comparison with the number of baculoviruses studied (Smits, 1997); however, there are a number that proved economically viable, e.g. *Heliothis zea*, NPV for cotton, vegetables and tomato, *Spodoptera exigua* NPV in vegetables, cotton and grapes (Georgis, 1997).

More than 750 species of fungi have been recognized as entomopathogens (McCoy *et al.*, 1986) with the most potential as biopesticides from the Deuteromycetes ('imperfect' fungi), namely species of *Beauveria*, *Metarhizium*, *Verticillium*, *Nomuraea* and *Hirsutella*. Individual species of fungi such as *Metarhizium anisopliae* have wide host ranges including species of Coleoptera, Lepidoptera, Orthoptera, Hemiptera and Diptera (Hall and Papierok, 1982). In addition to their wide host range, a further attribute that makes these fungi attractive as biocontrol agents is their route of infection. Entomopathogenic fungi do not need to be consumed by the insect, they can penetrate the cuticle, which means that they can be used against insects with sucking mouthparts. Up until 10 years ago, the potential of fungi seemed limited by their need for high humidities in order to germinate. However, more recent developments in formulation technology have removed this constraint so that it is now possible to use fungi for control of pests in semi-arid environments (Neethling and Dent, 1998). Worldwide, there are currently 47 fungal based products used as biopesticides (Copping, 1999) and it is likely that many more will be developed in the near future.

The same is true of entomopathogenic nematodes of which there are currently 40 products available worldwide (Copping, 1999). The prospects for nematode products have been aided by the lack of registration requirements for these biocontrol agents. The main interest in nematodes has been concerned with those species that kill their hosts in a relatively short time. There are three families of nematodes to which this can be considered to apply: the Steinernematidae, the Heterorhabditidae and the Mermithidae. The first two families are terrestrial nematodes that are associated with symbiotic gut bacteria that kill the host by septicaemia, and the Mermithids are aquatic nematodes that kill their host upon exit through the cuticle. In agricultural systems, the searching ability of nematodes (although limited) makes them potentially ideal candidates for use in situations where chemical insecticides and microbial formulations cannot be targeted effectively, for instance the cryptic habitats of pod borers, tree boring insects or for root attacking insects in soil (Poinar, 1983). The nematode infective (which is the third stage larva ensheathed in the second stage cuticle and referred to as a dauer larva) can detect its host by responding to chemical and physical cues. The Steinernematidae enter the host via the mouth, anus or spiracles while the Heterorhabditidae also have the ability to penetrate the host cuticle (Kaya, 1987). Once inside the host, the dauers exsheath and mechanically penetrate through the haemocoel where symbiotic bacteria are released to kill the host by septicaemia within 24–48 hours (Kaya, 1985). The nematodes develop and reproduce, the Steinernematidae need both male and females per host for reproduction while Heterorhabditidae are hermaphrodites. As the resources are depleted, the nematodes leave the cadaver in search of new hosts.

Steinernema carpocapsae products have been used in a wide range of systems including home and garden, berries, turf grass, ornamentals, citrus and mushrooms (Georgis, 1997) against Lepidoptera and Coleoptera (Curculionidae and Chrysomelidae). *Steinernema feltiae* has been targeted successfully at dipteran pests, particularly Sciaridae, while *Heterorhabditis* species have been used against lepidopteran and coleopteran pests. Provided limitations of production and formulation can be overcome, the potential use of entomopathogenic nematodes look promising.

The situation, however, is less promising for protozoa as biocontrol agents. Their low levels of pathogenicity causing chronic rather than acute infections, and difficulty of large scale production makes these pathogens unattractive prospects as biopesticides. This is reflected in the fact that there are only two protozoan based biopesticides available worldwide (Copping, 1999). Despite this, further research could exploit protozoan potential for use as inoculative augmentations or introductions in stable habitats such as forests and pastures.

6.3.2 Predators

Predators of insects may be entomophagous insect or vertebrates. Vertebrate natural enemies of insects pests can be found in each of the five animal groups, although only birds and mammals have attracted any serious attention. Reptiles and amphibians have such low consumption rates (Buckner, 1966) that they have little potential as biocontrol agents and although fish have been shown to be important in the control of mosquitoes (Hoy *et al.*, 1972; Legner, 1986; Bence, 1988) and rice stemborers (Durno, 1989), more work needs to be done to evaluate the potential of this major group. Birds have long been recognized as predators of insect pests but it is only occasionally that such effects are quantified (Dempster, 1967; Stower and Greathead, 1969; Atlegrim, 1989). Buckner (1966, 1967) considered the role of both birds and mammals in the control of pest insects in forest ecosystems. The two groups prey on insects in different parts of the forest system, with the birds feeding on free flying adult insects and on larvae inhabiting the trees while the mammals were mainly restricted to feeding on ground inhabiting stages of the insect's life cycle, mainly the pupae. The importance of these natural enemies is potentially quite high because they are feeding on the later stages of the insect's life cycle, i.e. pupae and adults, rather than the more abundant early stages. The animals may concentrate

their feeding on pest prey when they are readily abundant but are also likely to have alternative sources of prey that they can switch to as the pest numbers diminish. Such generalist feeders are thought to be of little value as biological control agents and numbers of mammals and birds can not be easily manipulated at man's convenience (Harris, 1990). Hence, the potential for such animals as biological control agents is low except perhaps in very special circumstances, such as the masked shrew (*Sorex cinereus cinereus*) which was introduced into Newfoundland from the mainland USA to control the larch sawfly, *Pristiphora erichsonii* (Buckner, 1966).

Predators feed on all stages of the host, eggs – larvae (nymphs), pupae and adults – and each predator requires a number of prey individuals to enable it to reach maturity, unlike parasitoids which require only a single individual. The immature stages as well as adult predators then have to search, find, subdue and consume prey.

Predators can be crudely divided into those with chewing mouthparts and those with sucking mouthparts but generally they lack the highly specialized adaptations associated with parasitism. Predators are found among the Coleoptera, Neuroptera, Hymenoptera, Diptera, Hemiptera and the Odonata, but more than half of all predators are coleopterans (DeBach, 1974). The most important families within the Coleoptera for biological control have been the Coccinellidae and the Carabidae. Other arthropod natural enemies include predatory mites and spiders. Predatory mites have played an important role in biological control both in orchards and glasshouse systems by feeding on phytophagous mites (Gerson and Smiley, 1990), but too little attention has been paid to spiders although they are abundant in perennial systems such as orchards. A number of studies have identified numerous species of spider as important predators of arthropod pests (Riechert and Lockley, 1984; Riechert and Bishop, 1990) and capture efficiencies have been determined for some species (e.g.

Sunderland *et al.*, 1986). Individual species of spider appear incapable of tracking population changes in specific prey species, either through increased rates of attack or through changes in effective population densities in local areas (Riechert and Lockley, 1984). However, spiders are for the most part polyphagous and as such are not considered ideal as candidates for control agents.

6.3.3 Parasitoids

A parasitoid is only parasitic during its immature stages when the larvae develop, either within (endoparasite) or on (ectoparasite) their host, from eggs that were either oviposited inside or near the host. The developing larva(e) usually consume(s) all or most of the host and pupate either within or near it. The adult parasitoid is free-living and usually feeds on substances such as pollen, nectar, honeydew or sometimes on the body fluids of its host.

Parasitoids exhibit a number of different life habits and are themselves parasitized by secondary or hyperparasites. Endo- and ectoparasites can be solitary or gregarious; when a single larva is produced per host individual then the parasitoid is solitary but if a number of larvae develop within a single host and all the larvae survive to maturity, then the parasitoid is described as gregarious. However, if more than one species of parasitoid parasitizes a host then it is known as multiple parasitism. In such circumstances it is rare for all species to complete their development. Where a number of individuals of a single parasitoid species occur in one host but only a few survive to maturity, it is referred to as superparasitism. The host can sometimes die prematurely with the result that all the developing parasitoids are lost (van den Bosch and Messenger, 1973), or if the host does not die and a number survive then the adults may be smaller than normal (Samways, 1981). Different species of parasitoid attack different life stages of the pest. Thus, *Trichogramma* spp. which attack the egg stage of insects are known as egg parasitoids, Braconidae such as *Cotesia glomer-*

ata which attack larvae are larval parasitoids and so on for adult and nymphal parasitoids (van Driesche and Bellows, 1996). Further details of parasitoids can be found in Askew (1971) and Waage and Greathead (1986).

The types of natural enemy used to control pest species through introductions, augmentations, inoculations and conservation has largely been dominated by entomophagous insects, and within these groups, primarily by host specific parasitoids in preference to more generalist species. The use of birds and mammals has largely been restricted to a few specific situations and is now generally considered inappropriate by responsible biocontrol practitioners. The opportunities for the use of pathogens for inundative release is now becoming increasingly important as new developments improve the performance and attributes of these agents for use as biopesticides.

6.4 Agent Selection

The choice between the various forms of biological control – classical, augmentation, inoculation or conservation – for the control of a particular pest will depend on a number of characteristics of the insect pest combined with those of the cropping system in which it is a problem (Greathead, 1984). The following account aims to highlight some of the more important attributes of pests, cropping systems and the biological control agent that should influence the choice of control option.

6.4.1 Pests and cropping system

Insect pests cause different amounts of damage and contribute to yield loss to varying extents according to where and how they feed and the amount they consume (Chapter 3). Insects that feed on leaves or stems are less likely to be as much of a problem as those pests that feed directly on the harvestable plant product. Solomon (1989) categorizes pest insects of top fruits according to the thresholds at which they cause damage (Fig. 6.1). High threshold pests are those that have a high economic injury level, generally feed on the foliage but not the fruit and at low population levels cause no damage at all. Low threshold pests are at the other extreme of the continuum, feeding directly on the fruit and causing damage even at very low pest densities.

The use of natural enemies for the control of low threshold pests is generally not feasible, which is why biological control is rarely used for control of disease vectors or of pests of stored products. The tolerance levels in stored products are usually very low so that even a minor infestation may prove to be unacceptable. However, some parasitoids can exert control to levels of 98% (e.g. *Choetospila elegans* against *Rhyzopotha dominica*: Flinn *et al.*, 1996) and in rural storage bins in developing countries subsistence farmers may have a higher tolerance for damaged products (de Lima, 1979). This increase in tolerance has the effect of increasing the threshold of the pest, with the result that biological control then becomes a possible option for pest control. In the case of stored products, however, there are still few insect parasitoids that appear to exert sufficient control to be of practical use (de Lima, 1979; Parker and Nilakhe, 1990) although some

species of predator hold some promise (Arbogast, 1979).

Other pest characteristics apart from the type of damage they cause that can influence the choice of control method are reproductive rate and dissemination efficiency. Pests having a high reproductive rate and good dissemination ability will make poor candidates for control by classical introduction (Southwood, 1977a) but such pests may be controlled by augmentation. With a classical introduction, a high reproductive rate may allow a pest to outstrip the population growth of the natural enemy whereas an augmentation may provide sufficient numbers of natural enemies at an appropriate time to suppress the pest. Mobile pests may simply disperse before natural enemy populations build up, although augmentation might prove sufficiently quick acting to provide control. As a general rule, the characteristics of the pest will influence the type of biological control option that is appropriate, so that pests amenable to control by one option such as augmentation may not be suitable candidates for control by another, such as introductions. The same general rule applies to the type of cropping system in which the pest control is sought.

The characteristics of cropping systems that may influence the likelihood of suc-

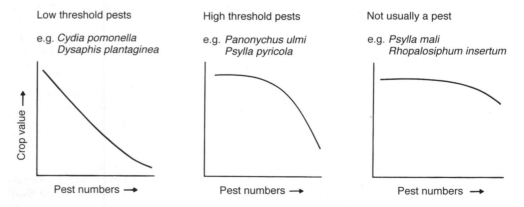

Fig. 6.1. A classification of fruit pests according to their economic impact on the crop (after Solomon, 1989).

Table 6.1. Crop suitability assessment for 'classical' biological control. A range of scores is used to account for the variability in cropping conditions. A low range of scores indicates that augmentation and microbial pesticides are most likely to succeed while a high range of scores means that classical biological control through introductions may succeed as well (after Greathead, 1984).

Crop type	Score	Classical biological control		
		Attempts	Successes	%
Cereal	1–6	21	8	38
Root	3–10	3	2	67
Vegetable	0–7	88	27	31
Oil seed	3–8	2	0	–
Plantation	7–12	152	50	33
Fruit	4–12	185	107	58
Other cash	2–8	26	3	12
Forage	4–12	23	9	39
Total		506	208	41

cess of introduction, augmentation, inoculations or the use of biopesticides can be considered under climate, crop duration, scale of planting and cultural/agronomic practices. Greathead and Waage (1983) produced an assessment of crop suitability for biological control based on a score for each of the above cropping system attributes (Table 6.1). The scores for climate, crop duration, scale of planting and cultural/agronomic practices were each graded from 0 to 3, with the higher score indicating an increasing chance of achieving long-term biological control. Hence, augmentative control would be favoured in an extreme climate having an erratic rainfall, wide ranging temperatures, a short growth season and scattered small areas of crops being intensively cultivated. An introduction, however, would be more likely to succeed in a uniform climate, long duration crops (e.g. more than six years) on large plantations, estates or community farming schemes that maintain a low level of cultural/agronomic practices such as weeding. The different crops thus vary in their suitability for long-term prospects for biological control, e.g. millet and sorghum have low scores and poor prospects for long-term control whereas the opposite is true of sugarcane (Table 6.1). The influence and importance of such characters has long been recognized but the above scoring sys-

tem provides a synthesis of the different cropping system characters into a simple easily interpreted summary. Such scores could also allow for the inclusion and evaluation of opposing characters, e.g. large farms (which should favour introductions) and intensive cultivation (which is less favourable for introductions).

Cropping systems have usually been considered in terms of their stability, since it is normally assumed that habitat stability makes the establishment of natural enemies more likely due to the presence of alternative hosts or prey, nectar, pollen or overwintering sites (Hoy, 1985). The successful establishment of natural enemies introduced against exotic pests is definitely affected by stability of the cropping system (Southwood, 1977a). Hall and Ehler (1979) considered habitats in terms of their stability; forests and range lands are more stable than orchards which are themselves more stable than annual horticultural or agricultural field crops. Natural enemies released into forests and rangeland habitats had the highest rate of establishment (36%) followed by orchards (32%) and then the annual crops (28%). These results support those of Beirne (1975) who found that the establishment of natural enemies released in orchards and ornamental shrubs was higher (43%) than those released in annual crops (16%). Monocultures may also make establishment difficult because the habitat lacks diversity, especially if intensively managed. Diversification may be enhanced by leaving areas unweeded or maintaining hedgerows or ground cover, although such vegetation can provide refuges for pests as well as natural enemies (Solomon, 1981).

The requirement for habitat diversification, a stable climate and long duration crops planted over large areas, for long-term biological control to succeed stems from the need to maintain predator–prey interactions. If a pest–natural enemy relationship is disrupted over space or time then the success of the control may be put in jeopardy. The conditions necessary for successful short-term control may not be as critical, since the pest–natural enemy interaction does not have to be maintained indefinitely. If cropping systems can be evaluated for the likelihood of successful control and this is combined with the relevant information about pest damage and ecology then the choice of the type of biological control that is appropriate should be self-evident.

6.4.2 Biocontrol agent selection criteria

Biological control has a long history and in the past the introduction of biological control agents to control exotic pests was as much an art as science. However, because of this long history, it is now possible to look back over time to try to determine which attributes of natural enemies contributed to the success or otherwise of past introductions. In general, data indicate that there has been only one successful introduction in every seven attempts (Clausen, 1978; Hokkanen and Pimentel, 1984), which would suggest that either the natural enemies selected were inappropriate or that procedures for introduction were inadequate. Hokkanen and Pimentel (1984) claim that the choices of natural enemies for biological control are based on the wrong criteria. Historically, programmes in classical biological control have sought potential agents for introduction from the area of natural origin of the pest (an old association between pest and natural

enemy), but it has been suggested that potentially more successful agents for control are to be found among natural enemies that are not normally associated with the target pest, either because they do not come from the native area of the pest or because they come from a related pest species (new associations; Hokkanen and Pimentel, 1984). It is claimed by Hokkanen and Pimentel (1984) that the use of new associations would make it possible to increase the chances of successful biological control from about 7:1 to about 4:1, which would represent a significant improvement. This result was obtained from a database analysis and was used as evidence in support of the new association hypothesis that had been a subject of debate during the 1960s. The analysis of the database has refuelled the debate (Greathead, 1986; Waage, 1989; Waage and Greathead, 1988). The contention of Pimentel (1963) has been that the co-evolution of the parasitoid/predator and its host leads to an increased resistance in the host with a concomitant decrease in the effectiveness of the natural enemy. Hence, the longer the association between a pest insect and its natural enemy, the less effective the enemy will be. If this hypothesis is correct, then it would imply that natural enemies introduced as control agents have only a limited time over which they will be effective and following on from this, natural enemies introduced as new associations will be more effective and exert control for longer than an old association. These two implications have been questioned since there are many examples of long-standing and effective control by introduced natural enemies such as the vedalia beetle for cottony cushion scale in California (80 years), *Crytorhinus* for sugarcane leafhopper in Hawaii (50 years), *Eretmocerus* for citrus blackfly in Cuba (40 years) and *Cryptognatha* for coconut scale in Fiji (40 years; Huffaker *et al.*, 1971), and it seems that the evidence for the effectiveness of new associations depends on the procedure used in the analysis and the database used (Waage, 1989).

In an analysis of a different database (BIOCAT, database of the International Institute of Biological Control), Waage and Greathead (1988) found no significant difference between old and new associations for the number of introductions that provided complete, partial or no successes. In another analysis using the same database, Waage (1989) found that for all realistic methods of analysis, old associations were associated with a significantly higher probability of establishment and success: the complete opposite of Hokkanen and Pimentel (1984). However, given the unreliability of the records that constitute these databases (including misidentification of both pests and natural enemies, errors in dates and places and the arbitrary interpretation of successes) such database analyses are not really considered a sound enough basis on which to identify universal selection criteria (Waage, 1990; Waage and Mills, 1992). In practice, studies on the target pest in its region of origin represents the most promising approach to identifying suitable biocontrol agents, but the potential usefulness of new associations should not be disregarded (Waage and Mills, 1992).

Decisions on the relative merits of types of natural enemy tend to be based on reductionist or holistic criteria (Waage, 1990). The reductionist approach involves selecting agents on the basis of particular biological attributes, e.g. searching efficiency, aggregative response. The reductionist criteria are mostly derived from the parameters of the analytical parasitoid-host or predator–prey population models (see below). In particular, the parameters that are important for lowering of host or prey equilibria and/or which promote population stability (stability of the natural enemy pest interaction reduces the risk that the pest population will be driven to extinction by the control agents which themselves would otherwise become extinct) (Kidd and Jervis, 1996). This reliance on theory to identify selection criteria has been questioned (Gutierrez et al., 1994) since theory seems to have contributed little to improving the success of introduc-

tions or understanding the reasons for failure. In addition, the difficulty of estimating parameter values makes practical use of these theoretical models very difficult (Godfray and Waage, 1991).

A more holistic approach to the selection of appropriate biocontrol agents reduces the emphasis on the attributes of one particular agent and instead considers the interaction between different agents and mortalities acting on the pest in its area of introduction (Kidd and Jervis, 1996). The approach considers the selection of agents that follow rather than precede major density dependent mortalities in the pest life cycle (e.g. van Hamburg and Hassel, 1984), that are constrained in their area of origin (e.g. by hyperparasitism; Myers et al., 1988) and that involve the use of communities of natural enemies and their complementary action (May and Hassell, 1981).

6.5 Predator–Prey Theory and Analytical Models

The parasitoid/predator–prey interactions involved in biological control through introductions are largely very simple ones between a single species of parasitoid or predator and a pest species. Such simple interactions rarely occur in the natural world but their use in biological control has provided an applied context for models of two species interactions. The majority of the theoretical work carried out has been based on these relatively simple parasitoid–prey models. These theoretical models have, to a limited extent, been useful to biological control for the identification of important aspects of interactions and potentially (and apparently) useful attributes of potential control agents.

There are two basic types of a mathematical model relevant to insect parasitoid/predator–prey interactions: they are deterministic and stochastic models. The majority of models developed in this field have been of the deterministic kind despite a number of important drawbacks associated with their use. Firstly, to provide an adequate picture of the behaviour of populations, a deterministic model has to assume that populations of predator and prey are of infinite size (Maynard-Smith, 1974). They also assume that any reproductive individual within the population gives rise to a set number of offspring within a defined period of time. In contrast, a stochastic model assumes a probability distribution of offspring over time providing a greater degree of realism as well as population means and variances. Deterministic models ignore environmental randomness, that is, the effects of random fluctuations in environmental conditions such as the weather, although the mathematical difficulties of stochastic models which allow for such randomness are formidable (Maynard-Smith, 1974). This mathematical inconvenience has been one of the main reasons for the use of deterministic in preference to stochastic models. Further, only certain types of deterministic models have been used. Deterministic predator–prey models have traditionally been written as either differential or difference equations (Hassell, 1978). Differential equations are used where an insect has overlapping generations and birth and death is a continuous process, for instance with tropical or glasshouse insect pests. Most predator–prey interactions have been modelled with difference equations which are mostly only applicable to temperate insect pests having discrete single generation cycles.

It is obvious from the above remarks that the types of model used to study predator–prey interactions have been largely to deterministic difference (equation) models which apply to insect pests in the temperate areas. The situation is further limited in that the models that have been used have concentrated on the simple dynamics of the host-specific parasitoids simply because such systems may be collapsed into single age-class models and also because the number of parasitized hosts closely defines the number of subsequent parasitoid progeny. This contrasts with predator models where age-class is an

essential feature of predator dynamics and the number of prey consumed does not readily relate to the production of progeny. These points have been mentioned because this restricted approach to modelling has had some influence in the type of agent selected for biological control programmes.

The following sections consider the traditional approach to the theory and use of deterministic models and the factors thought to influence the establishment and maintenance of low, stable equilibrium levels before turning to more recent events and the challenges posed to this traditional approach by the use of alternative approaches.

6.5.1 The general model

The basis of the theoretical approach to parasitoid–prey interactions has largely revolved around the generalized model:

$$N_{t+1} = Fg(N_t)N_t f(N_t, P_t) \qquad (6.1)$$

$$P_{t+1} = cs\,N_t\left[1 - f(N_t, P_t)\right] \qquad (6.2)$$

where:

N_t and P_t are the sizes of the initial prey and parasitoid populations and N_{t+1} and P_{t+1} are the number in the next generation; $Fg(N_t)$ is the per capita net rate of increase of the host population and $g(N_t)$ is a density dependent function ranging from zero to one;

c is the average number of progeny produced per prey attacked and s is the proportion of progeny that are female.

The function (N_t, P_t) defines prey survival and includes all assumptions about parasitoid searching efficacy (Hassell and Waage, 1984; May and Hassell, 1988). (Some of these functions are given in Table 6.2.)

The progress in the development of these models has led to a step-wise identification and exploration of the important factors influencing parasitoid–prey dynamics, always with the intention of explaining their role in the regulation of the two populations. The factors have been evaluated on the basis of the magnitude of their effects on the prey equilibrium levels and the

Table 6.2. A number of examples of specific expressions that have been used for $f(N_t, P_t)$ in equations 6.1 and 6.2. a, a' and Q are searching efficiencies, Th is the handling time as a fraction of total time, m is an interference constant, α and β are the fraction of hosts and parasitoids respectively in the ith of n patches, k is the clumping parameter from the negative binomial distribution (after Hassell and Waage, 1984).

Form of f	Reference
$f = \exp\left(-aP_t\right)$	Nicholson and Bailey (1935)
$f = \exp\left(-\dfrac{a'P_t}{1 + a'ThN_t}\right)$	Rogers (1972)
$f = \exp\left(-QP_t^{1-m}\right)$	Hassell and Varley (1969)
$f = \sum\limits_{i=1}^{n}\left[\alpha_1 \exp\left(-a\beta_1 P_t\right)\right]$	Hassell and May (1973)
$f = \left(1 + \dfrac{aP_t}{k}\right)^{-k}$	May (1978)

stability of this equilibrium population. Both these criteria have obvious importance for classical biological control because a successful introduction should reduce the prey population to a new low level and maintain it there indefinitely. The effects of different components of the models (6.1 and 6.2) on equilibrium levels and their stability are considered separately in the following sections. Although treated in this way, it should be realized that all factors affect the potential equilibrium levels to some degree.

6.5.2 Equilibrium levels
When a natural enemy is introduced in classical biological control, it should, if it establishes itself, reduce the abundance of the pest to a level below the pre-introduction population size. With a successful introduction, this new population level will be well below the economic damage threshold. The formal relation between the initial population (equilibrium) level (K) of the pest and its new equilibrium level (N^*) in the presence of the natural enemy is described by the ratio:

$$\frac{N^*}{K} = q \qquad (6.3)$$

Typical values for q obtained from field studies range from between 0.002 and

0.025 (Beddington et al., 1978) which provides evidence of the success of some natural enemies since these figures indicate a severe depression of pest populations. By returning to the models (6.1 and 6.2), the factors that could be responsible for such marked depressions can be considered.

One of the major factors that could influence the equilibrium level would be the rate of increase (F) of the prey population. This rate of prey increase will be influenced by factors other than the mortality caused by the introduced parasitoid such as fecundity, sex ratio, immigration and emigration, the availability and quality of the host plant as well as the presence of other natural enemies. Hence, although the rate of increase (F) is affected by many factors, it is that part of the host's potential that the parasitoid must act on to prevent further host increases (Waage and Hassell, 1982). A high rate of increase would raise the equilibrium level, so it is important for biological control that a knowledge of the factors affecting pest survivorship be known and taken into account. This is particularly important for biological control when density dependent mortalities ($g(N_t)$) occur since their existence and timing could affect the success of an introduction (Hassell and Waage, 1984). For instance, where a pest rate of increase is influenced by density-dependent mortality during its larval stages, then any mortality caused by an egg parasite may be compensated for during the larval stage. The introduction of an egg parasite would mean that very high parasitism would be needed to achieve any significant increase in total mortality and the introduction would be unsuccessful.

A second factor that could influence the pest equilibrium level is parasitoid mortality during its development in the host. However, one of the simplistic assumptions that tends to be made in parasitoid–host models is that the parasitoids suffer no developmental mortality. It is generally assumed that a host parasitized in one generation will produce a parasitoid in the next, or at least a fixed number of parasitoids in the next generation (Waage and Hassell, 1982). Mortality may occur in the developing parasitoid because of the immunological defences of the host or through external factors such as superparasitism, multiparasitism, hyperparasitism or detrimental climatic changes. These mortality factors can be important, especially if any are density dependent, because any reduction in survival will affect the value of c and a reduced value of c increases the host's equilibrium level. The value of s, the proportion of female parasitoid progeny produced (equation 6.2), can also influence the host equilibrium level. Most parasitoid–host models assume that all adults that emerge are female which, for the majority of parasitoid species, is an oversimplification. Many species of parasitoid can control the sex of their progeny and hence vary the proportions of males and females produced. While experimental evidence for density-dependent sex-ratios in parasitoids is limited, theory predicts that it will be widespread in parasitoids that exploit and mate on patchily distributed hosts (Hassell et al., 1983). Although there are a variety of possible mechanisms for density-dependent sex-ratios, the possibilities when sex-ratio is a function of adult female population size or where it depends on the ratio of females to hosts,

have shown that the host equilibrium level increases with increasing sex-ratio $(1-s)$ (Hassell et al., 1983), i.e. the greater the proportion of males, the lower the level of parasitism and hence the higher the equilibrium levels. An increase in the level of parasitism would be achieved with a greater proportion of female parasitoids, alternatively it could also be achieved if a parasitoid had a high searching efficiency.

The biological characteristics that affect searching efficiency include everything that alters the amount of time available for parasitoid searching and everything that alters the proportion of hosts parasitized in a given amount of time. The amount of time a female parasitoid has available for searching for prey will depend on the amount of time that has to be spent on other activities such as mating, feeding, oviposition and resting. The greater the allocation of time spent on these activities, the less time is available for searching for prey. The proportion of prey parasitized in a given amount of searching time will depend on the fecundity of the parasitoid, the density of the prey and the prey distribution, as well as factors such as the density of parasitoids, the presence of alternative host prey and of competing natural enemies (Hassell, 1978). Not all of these factors have a significant effect on equilibrium levels but on population stability; the latter will be considered in Section 6.5.3.

In order to maximize its opportunities for encountering prey an insect parasitoid needs to minimize the amount of time spent on activities other than those related to searching. One of the important factors that affects this, and ultimately in some species longevity and fecundity, is the amount of time spent feeding and the type of food utilized. Adult parasitoids can be either pro-ovigenic or synovigenic. Pro-ovigenic parasitoids emerge with a full complement of eggs and the necessary resources for their maturation whereas synovigenic species emerge with only a fraction of their total egg complement and need to feed to maximize egg production.

These adults may feed on host haemolymph or on food that is quite independent of their prey such as pollen, nectar and aphid honeydew (Hassell, 1978). Kidd and Jervis (1989) have calculated that at least one-third of the world's parasitoids are species that feed on their hosts. They have considered the effects of such feeding on host equilibrium levels. There are a number of possible combinations of feeding and oviposition, non-concurrent oviposition with non-destructive feeding, concurrent oviposition with non-destructive feeding, and lastly, non-concurrent oviposition with destructive feeding. For non-concurrent oviposition and non-destructive feeding, the parasitoid either oviposits or feeds on encountering a host; feeding encounters are equivalent to a failed oviposition encounter, so if the probability of feeding or oviposition is constant, the net effect will be a reduction in the searching efficiency, which would increase the host equilibrium level. With concurrent oviposition and non-destructive feeding, provided the feeding does not affect the survival of the host, the dynamics would be identical to those of a parasitoid which does not feed. If the feeding does damage the host then the effect would be to decrease c (parasitoid reproduction rate) with a resultant increase in the host's equilibrium level. With non-concurrent oviposition and destructive feeding the situation becomes more complex. Changes in equilibrium levels will be dependent on energy obtained by parasitoids' feeding on the host, the maintenance and search costs of parasitoids and the costs of egg production. Equilibrium levels of the host will be lowered by high energy extraction during feeding and low maintenance, search and egg production costs (Kidd and Jervis, 1989).

One of the other factors limiting searching time which can also affect host equilibrium levels is the amount of time it can take to locate, subdue and oviposit within a prey individual, collectively known as prey handling time. A reduction in the maximum attack rate due to an increased handling time would reduce the per capita parasitism, thus raising the equilibrium level (Hassell and Waage, 1984). There will always be a definite minimum time below which it would be impossible for parasitism to occur and this will be determined as much by the number of eggs available and the time taken to mobilize each egg after oviposition as by the time taken to quell and parasitize the prey. As prey densities increase, the handling time will become an increasing limitation on the number of prey attacked. The relation that typifies the situation is given by the type II functional response (Fig. 6.2).

This concludes a brief introduction to the factors that are thought to influence prey equilibrium levels in parasitoid–prey interactions described by models 6.1 and 6.2. Next, it is important to consider the role of factors that contribute to population stability and persistence at low equilibrium levels.

6.5.3 Stability

The stability of a host–parasitoid interaction refers to the persistence of both the prey and parasitoid populations at a low average level of abundance. Prey numbers will fluctuate around this average value but will be restrained by the parasitoid population from increasing to higher levels. A problem that intrigued population ecologists and biological control practitioners alike has been to determine what factors are responsible for maintaining stability at the low equilibrium levels. Knowledge of the type and the way in which stabilizing factors operate would be useful for biological control, since selection of agents having suitable stabilizing attributes could improve the success rate of introductions. A number of factors have been considered as possible stabilizing mechanisms. These include host density dependence, sigmoid functional responses, mutual interference, host susceptibility, variable parasitoid sex ratios, synovigeny and non-random search in a patchy environment. A short account of the potential role of each of these factors in producing stability is given below.

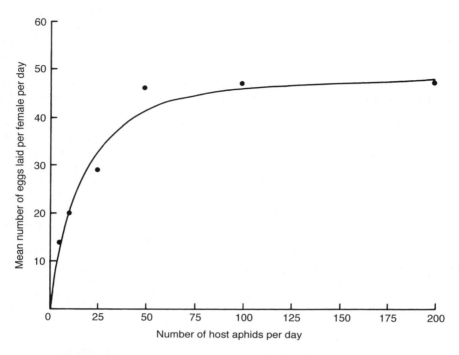

Fig. 6.2. A type II functional response curve for the aphid parasitoid *Aphidius sonchi* (after Shu-Sheng, 1985).

Density dependence

In the absence of natural enemies, the size of a prey population will be limited by the carrying capacity (K) of the environment, which in effect will be the first resource used by the prey that becomes limited as the population size increases. The availability of that resource will affect the prey density that is sustainable, i.e. a density-dependent resource. In order to study the effects of density dependence on the host population, a density dependence has to be incorporated into the basic model. Beddington *et al.* (1975) incorporate:

$$g(N_t) = \exp\left[r\left(1 - \frac{N_t}{K}\right)\right] \quad (6.4)$$

where: $r = \log_e F$ and $1 - N_t/K$ defines the change in prey numbers as a resource becomes limiting (as the numbers approach the carrying capacity K, $1 - N_t/K$ will decline to zero making $r = 0$ and hence no

further increase), into the Nicholson-Bailey model (see Table 6.2) to give:

$$N_{t+1} = N_t \exp\left[r\left(1 - \frac{N_t}{K}\right) - aP_t\right] \quad (6.5)$$

$$P_{t+1} = N_t\left[1 - \exp\left(-aP_t\right)\right] \quad (6.6)$$

This density dependence has been shown to be stabilizing at high prey densities when resources become limiting. However, if a parasitoid is introduced so that the equilibrium level is lowered to a point well below that at which resources are limiting, then the density-dependent mechanism will not come into play and the prey population is no longer self-regulating. The model 6.6 will not permit more than a 40% depression ($q = 0.4$) of the host equilibrium level and still remain within the bounds of stability (Fig. 6.3). This value of q is well above those cited earlier from actual field

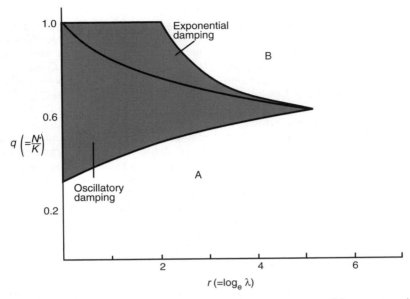

Fig. 6.3. Stability boundaries for the model 6.5. The parameter q is a measure of the extent to which the prey equilibrium N^* is depressed by predation below its carrying capacity K. The equilibrium point is stable within the shaded area only. In the region there will be population fluctuations within the bounds found in natural populations. The fluctuations in region would be so great that the predator would effectively become extinct (after Beddington *et al.*, 1975).

examples of depressed equilibrium levels (such as $q = 0.002$ to 0.025) which suggests that stability obtained in the field has not been achieved through this type of density-dependent relation of the prey population (Beddington *et al.*, 1975). The stability at the depressed equilibrium levels must be due to some property of the parasitoid–prey interaction. One possible contender that has been considered for this role is the sigmoid functional response, a relation that will influence the function (N_t, P_t) of the models 6.1 and 6.2.

Functional response
Parasitoids can exhibit a type III functional response, a sigmoid response that can, over part of its range, exert density-dependent mortality (Fig. 6.4). For the response to be sigmoidal, the parasitoid must search more actively as prey density increases. This would mean that one or more of the components of searching activity must be dependent on prey density, which would suggest

parasitoids either have an innate limited ability to find prey up to a certain density or there is a tendency to switch between prey types according to their relative densities. Under certain conditions, this density dependence will operate up to a threshold host density and may lead to stable equilibria (Hassell, 1984). However, in the case of specific parasitoids having discrete synchronized generations, the time delays between changes in population density and the level of prey mortality in the subsequent generation make it unlikely that a sigmoid functional response could stabilize the populations (Beddington *et al.*, 1975).

Mutual interference
At high parasitoid densities, individuals will come into frequent contact and in doing so could mutually interfere with each other's ability to parasitize hosts. This would cause a decline in the per capita searching efficiency with increasing parasitoid density (Fig. 6.5) due to a reduction

in searching time during and following encounters with other parasitoids or with already parasitized prey. This density-dependent reaction could provide a powerful stabilizing mechanism (affecting $f(N_t, P_t)$) but it is difficult to see how such interference could have a significant stabilizing effect at the extremely low population densities of parasitoids and hosts evident in successful biological control programmes (Hassell, 1984).

Host susceptibility

Any heterogeneity in population distribution is likely to accentuate the impact of density-dependent processes. One such form of heterogeneity could be created by variation in the susceptibility of individuals in the prey population to parasitism. The probability of prey survival or of achieving reproductive potential may not be the same for all individuals within a population (Hassell and Anderson, 1984). For instance, it has been shown that larvae of the larch sawfly (*Pristiphora erischsonii*) vary in their ability to encapsulate the eggs

of the parasitoid *Mesoleieus tenthredinis* (Turnock, 1972). If the prey population exhibits a range of such susceptibility then this will create an uneven exploitation by the parasitoid population. Some prey individuals will be more at risk that others and this will help promote stability. The density dependence necessary for the stabilizing reaction will come about because of the changes in searching efficiency as the parasitoid density increases (affecting (N_t, P_t)). Because the exploitation of the more susceptible hosts is high at increasing parasitoid densities, the searching efficiency per individual over all hosts is lower than at reduced parasitoid densities (Hassell and Anderson, 1984).

Variable parasitoid sex-ratios

Hassell *et al.* (1983) considered the effects on stability of the changes in sex-ratios with the density of adult females, changes in the ratio of parasitoids to hosts and in aphelinid wasps, the situation where females develop as primary parasitoids while males develop as hyperparasitoids.

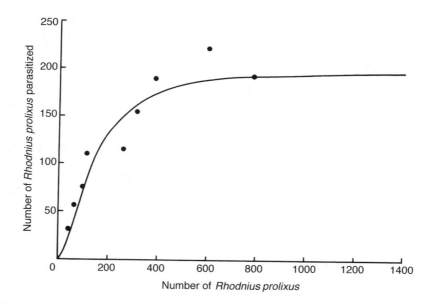

Fig. 6.4. A type III functional response for the parasitoid *Ooencyrtus trinidadensis* that attacks *Rhodnius prolixus* (after Hassell and Rogers, 1972).

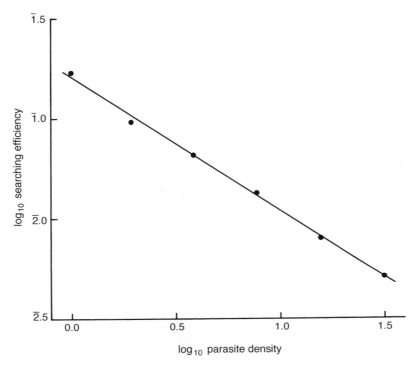

Fig. 6.5. The relationship between searching efficiency of the parasitoid *Hemeritis canescens* and parasite density (after Hassell and Rogers, 1972).

This treatment was expanded to consider how sex-ratios vary with host density and with the density of male parasitoids (Comins and Wellings, 1985). The proportion of females produced tends to decrease with increasing density of female parasitoids (*P*) (for example Fig. 6.6). Such density-dependent relationships would be expected to contribute to model stability. In general, Hassell *et al.* (1983) found that sex-ratio shifts, dependent on the ratio of females to hosts (*P/N*), were less stabilizing that those dependent on *P* alone, but that this latter stabilizing effect was itself dependent on a very narrow range of conditions. It seems likely that this kind of density dependence will only be stabilizing in concert with other factors. In contrast, the aphelinid models showed a high degree of stability. This group of heteronomus hyperparasitoids exerts little control over their sex-ratio but has contributed to

some of the most stable and hence most successful host-parasitoid interactions in biological control, which may bear some relation to the striking effect on stability of this method of sex allocation.

The approach used by Comins and Wellings (1985) involved the introduction of density dependence of the host population growth rate into their model. This provided a zone of stability which then gave a more sensitive indication of the stabilizing or destabilizing effects of a density-dependent sex-ratio. Under these conditions, the models considered were shown to be locally stable over a broad range of host depressions.

Synovigeny
Kidd and Jervis (1989) in modelling the effects of synovigeny, concluded that egg limitation or resorption, in the way portrayed in their particular models,

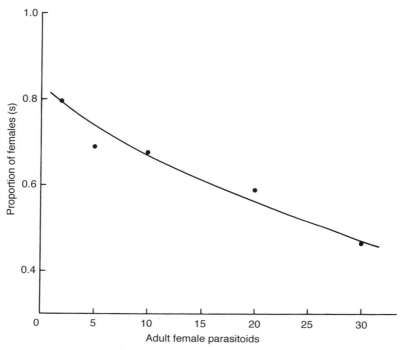

Fig. 6.6. The proportion of female progeny as a decreasing function of the density of adult female parasitoids *Hasonia vitripennis* parasitizing pupae of *Musca domestica* (after Hassell *et al.*, 1983).

constituted potentially important destabilizing influences on parasitoid–prey interactions. This could have important implications for selection of biological control agents because many parasitoids used are synovigenic.

Non-random search in a patchy environment
Spatial heterogeneity in the distribution of prey and parasitism provides the most likely key to stability at low equilibrium levels in biological control. In the discussion above on the sigmoid functional response and mutual interference, it was assumed that prey were distributed uniformly and that parasitoids searched randomly for prey. If searching by parasitoids is non-random and prey are distributed heterogeneously, then a situation arises where stability is more plausible. In the real world, prey populations will tend to have clumped distributions and parasitoids and predators may then spend

more of their time in areas where prey are plentiful, leading to an aggregation of natural enemies. By tending to aggregate in certain patches, particularly those of high prey density, the parasitoids are leaving other patches as partial refuges from parasitism (Waage and Hassell, 1982). This tendency for aggregation and thereby the creation of prey refuges provides a powerful stabilizing mechanism even at low host population levels (Beddington *et al.*, 1978). The extent of this stability hinges on at least three things: firstly that it is enhanced by increasing amounts of parasitoid aggregation provided the host distribution is sufficiently uneven; secondly that it is enhanced by clumping in the host distribution (k in Table. 6.2) for a given amount of parasitoid aggregation – if there is too little contrast, no amount of aggregation will provide stability; and lastly that any tendency towards stability is counteracted by an increased host rate of increase

(F) (Hassell, 1984). Overall, the most important point about this mechanism for stability is that it can stabilize populations at different equilibrium levels, which makes it the factor most likely to be responsible for stabilizing low equilibria in successful biological control (Beddington *et al.*, 1978).

The prey–parasitoid models have thus been used to identify characteristics of the species interactions that most contribute to low equilibrium levels and stability. They have also helped identify specific attributes of parasitoids which may be indicative of a potentially successful agent for introduction. The following criteria were identified by Beddington *et al.* (1978): high effective attack rates, high egg complement, synchronous emergences of parasitoid with appropriate life stage of prey and lack of polyphagy and low handling time. In general, a high searching efficiency relative to the host's rate of increase would achieve sufficient depression and a marked aggregation in patches of high prey density would be required to provide a stable equilibrium level (Hassell, 1984).

6.5.4 Metapopulation dynamics and models
The effects of spatial heterogenicity essentially have been viewed in two different ways. The first considered above emphasizes patchiness and movement within populations; these ideas have largely dominated predator/parasitoid–prey research (e.g. Hassell, 1978). The other concept, of metapopulations, although not new (Nicholson and Bailey, 1935; Murdoch and Oaten, 1975; Murdoch, 1979) received relatively little attention until the 1980s (Murdoch *et al.*, 1985; Reeve, 1988, 1990; Taylor, 1988, 1990, 1991). The metapopulation concept revolves around the idea of the 'population' or 'cell', the unit within which will occur various types of interaction – reproduction, population regulation, predation – and within which most movement is confined (Taylor, 1990). A metapopulation is then a collection or ensemble of such local populations and 'dispersal' refers to movement, often once

in a lifetime from one population to another. Although in abstract terms this distinction between populations and metapopulations is quite logical, in practice defining a metapopulation on the basis of whether movement is within or between populations is very difficult (Taylor, 1991). However, it is central to the metapopulation concept. The proposition that population subdivision and dispersal provide an alternative stabilizing mechanism of 'classical' predator–prey theory requires that distinctions can be made between versus within population movement. If all local populations are considered as discrete entities that are essentially equivalent and equally affected by dispersal then Taylor (1990) proposes three hypotheses that may be tested to determine if dispersal is important to persistence of metapopulations.

1. Local extinctions and recolonizations occur frequently.
2. Isolated local populations frequently would go extinct but migration (usually) prevents this.
3. Isolated local populations usually would persist but fluctuate and migration reduces the magnitude of the fluctuations.

All of these hypotheses have been explored using theoretical models (reviewed by Taylor, 1988; Reeve, 1990) but only the first has been studied in the laboratory and the field.

The modelling approach has consistently shown that persistence can be enhanced by dispersal among local populations within a metapopulation provided three conditions are met.

1. Asynchrony of local population fluctuations.
2. Dispersal rates must be such that if local extinction is inevitable then predator invasion must not be too rapid relative to prey colonization.
3. Some local density dependence is present (Taylor, 1990).

These conclusions are consistent with observation in several laboratory studies of population structure, dispersal and persistence in predator–prey systems (Huffaker, 1958; Pimentel, 1963). However, while such studies agree with the theoretical models there are sufficient methodological reasons, e.g. extremely artificial scale and structure, to question their value in representing real field situations.

Murdoch *et al.* (1985) reviewed a number of examples of successful classical biological control and concluded that extinction of local populations may well occur in the systems studied and that migration might be essential to their regional persistence. However, extinction in the examples cited was certainly not proven or occurred at a scale more coincident with local population processes (Taylor, 1990). One of the problems with field observational approaches is that while local extinction and recolonization may be observed (and is thus adequate proof of both the absence of local stability and metapopulation structure and dispersal), it is difficult to prove that such extinction has actually occurred in any but the simplest situations. Evidence for the importance of metapopulation processes in the field is thus lacking, chiefly because few studies have been conducted to specifically obtain such evidence (Kidd and Jervis, 1996). In the absence of such experimentation the real importance of metapopulations to biological control is yet to be determined.

6.5.5 Multiple species models

Not all of the deterministic models that have been developed have been describing two species interactions. There has been a limited interest shown in multiple species interactions mainly because the value of multiple introductions has always been a question of debate among biological control practitioners.

Multiple species models may involve one or more of the following interactions:

1. One parasitoid species attacking two or more (probably competing) prey species.
2. Two or more parasitoid species attacking a single prey species.
3. Prey–parasitoid–obligate hyperparasitoid interactions.

4. Prey–parasitoid–facultative hyperparasitoid interaction.

The first possibility is of little interest here since it is unlikely that a parasitoid would knowingly be released against more than a single pest species. The third possibility is of interest only from the point of view that it has been shown that obligate hyperparasitoids will have little effect other than to improve stability, but their presence can mean an increase in the equilibrium level of the prey–parasitoid interaction. This is one of the reasons why parasitoids are screened for hyperparasitoids prior to release in any biological control programme.

Interactions 2 and 4 are of interest because many biological control programmes have led to the establishment of more than one parasitoid and/or because indigenous natural enemies have had an influence (May and Hassell, 1988). There has, however, been a long-standing debate over the merit of such multiple introductions. It has been argued that inter-specific competition between the two parasitoids may occur which may produce an increase in the pest equilibrium level relative to control by a single parasitoid system (Turnbull and Chant, 1961; Watt, 1965; Kakehashi *et al.*, 1984). Models of multiple species interactions have tended to predict in favour of multiple introductions (May and Hassell, 1981, 1988; Waage and Hassell, 1982). The model of the two species of parasitoid interaction with a single prey species is an extension of 6.1 and 6.2 and takes the form:

$$N_{t+1} = F.g(N_t)N_t.f_1(N_t, P_t).f_2(N_t, Q_t) \quad (6.7)$$

$$P_{t+1} = c_1 s_1 N_t [1 - f_1(N_t, P_t)] \quad (6.8)$$

$$Q_{t+1} = c_2 s_2 N_t.f_1(N_t, P_t)[1 - f_2(N_t, Q_t)] \quad (6.9)$$

where:
functions f_1 and f_2 are the negative binomial function (May, 1978) in Table 6.2 and P and Q are the population sizes of each of the parasitoid species (May and Hassell, 1981).

This model is appropriate for two different types of interaction, both frequently found in real systems, where a parasitoid species P attacks the prey first followed by parasitoid species Q attacking the surviving prey (usually a different developmental stage), and where P and Q act together on the same developmental stage but the larvae of P are superior larval competitors and always eliminate Q in situations where multi-parasitism occurs (Waage and Hassell, 1982). The dynamics of these two situations are the same. The following points emerge from the models of the interactions:

1. The two parasitoids are most likely to coexist if each contributes to the stability of the interactions, i.e. if k_1 and k_2 (Table 6.2) are small and correspond to a marked ability for non-random search by both species.
2. Coexistence is more likely if species Q has the higher searching efficiency, although too high an efficiency will cause a replacement of P by Q, too low an efficiency and Q will fail to establish at all.
3. In general, the presence of Q will further reduce the pest equilibrium level but there is a slight chance that the level of depression achieved will be no greater than that with Q alone.

However, given the difficulties of pre-release evaluation, the differences are only slight and in general the use of deterministic models predicts that the strategy for multiple introductions is a sound one (May and Hassell, 1988).

6.5.6 Life table analysis

In 1970, Varley stated that 'it would be pleasant if we could leave behind all the fog generated by these theories, built with deductive logic on sandy foundations of guesswork and start a new era in which theories are built inductively on the firm basis of accurate measurements' (Varley, 1970). Varley was referring to the use of analytical techniques to help understand the processes involved in regulating pest populations. The simplest of these has been the measure of percentage parasitism while the more complex and labour intensive analytical methods involve construction of life tables for pests and parasitoids.

Percentage parasitism is commonly used as a simple estimate of the mortality attributable to a parasitoid species. At different intervals of time, samples of the pest may be collected and reared through in the laboratory. The percentage parasitism is calculated on the basis of the number of insects in the sample from which parasites emerge. However, such methods only provide a snapshot of parasitism and it is well known that such estimates vary widely with time and location (Mills, 1997). In the first instance, the number and timing of samples taken are usually inadequate for the task (van Driesche, 1983) and, secondly, the impact of parasitism can be overestimated if single samples are taken from a host population because parasitized hosts persist for longer at a particular stage (Russell, 1987). This problem is greatest with pests having short development times at a particular stage which itself leaves few traces, such as insect eggs or homopteran nymphs. The problem is least important when traces of some or all stages in a generation are preserved with evidence of their various mortalities (leaf miners, gall forming insects, bark beetles) (Hassell and Waage, 1984). To assess a parasitoid's contribution to host population mortality, it is the percentage attacked for the generation which must be determined and this may be best done within the context of a life table study (Kidd and Jervis, 1996).

Age-specific life tables are more commonly used in entomology than time-specific life tables. The former are based on the fate of a real cohort throughout a generation while the latter are based on the fate of an imaginary cohort, identified by determining the age structure of a sample of individuals, from what is assumed to be a stationary population with overlapping generations. The stationary population refers to a constant recruitment rate and mortality rate during the life of the individuals in the table (Southwood, 1978).

Age-specific life tables are only applicable to pest insects having discrete generations which largely limits their use to temperate situations. Both time- and age-specific life tables have been dealt with by Begon and Mortimer (1981), Horn (1988) and Dent (1997a). The concern here is more with the role of age-specific life tables in key factor analysis, since this technique provides a means of identifying the potential role of parasitoids and predators in the regulation of pest populations.

The data required to construct a life table for key factor analysis are a series of successive samples taken from each life stage of a generation, over eight to 15 generations (Southwood and Jepson, 1962; Bellows et al., 1989; Manly, 1990). The numbers are most usefully expressed as numbers per metre squared or some other unit measure, since this allows direct comparisons to be made between samples taken at different times. For each sample, identifiable causes of mortality such as parasitism will be recorded. Varley et al. (1973) defined a set of rules that should be adhered to when mortality data are recorded during the construction of life tables:

1. Where mortality events from definable causes are well separated in time, they are treated as if they are entirely separated with no overlap, while events which seriously overlap in time may be more conveniently considered as acting contemporaneously.
2. Every insect must be considered either as alive and healthy or as certain to die or dead, e.g. a parasitized insect although alive is considered as certain to die.
3. No individual can be killed more than once. If the host is attacked by two parasitoids, then the death of the host must be credited to the first parasitoid species. If the second parasitoid is, in fact, the eventual victor, it must be credited with the death of the first parasitoid. The second attack is entered in the life table of the parasitoid, but not that of the host (Varley et al., 1973).

The first step in constructing a life table is to obtain an estimate of the potential natality. This is calculated from an estimate of the mean fecundity per female which is multiplied by the number of females of reproductive age. The values for each generation are converted into logarithms. The difference between actual and potential natality (k_o) can then be calculated by subtracting the log of the actual number of eggs produced from the log potential natality. The k_o value in Table 6.3 has been calculated on the basis of each adult. When negative k_o values are obtained (as in 1970 generation I, Table 6.3), it suggests either an immigration of adult females or a greater than assumed maximum potential fecundity (Benson, 1973).

The population density sampled at each stage is similarly converted into logarithms and the mortality ($k_1 - k_n$ values) calculated as the difference between two stages (Table 6.4). The total mortality per generation (K)

Table 6.3. The mean fecundity of adult cabbage root fly in Canada, based on the assumption of 80 eggs per female and equal numbers of males and females, the observed fecundity and the subsequent mortality factor k_o (after Benson, 1973).

	Generations								
	1967			1968		1969		1970	
	I	II	III	I	II	I	II	I	II
Log potential eggs per adult	1.6	1.6	1.6	1.6	1.6	1.6	1.6	1.6	1.6
Log observed eggs per adult	–	1.02	1.47	–	0.88	1.53	0.65	1.71	1.26
k_o	–	0.58	0.13	–	0.72	0.07	0.95	−0.11	0.34

Table 6.4. A life table for the carrot fly *Psila rosae* over a 2-year period (after Burn, 1984).

Age class (mortality factor)	Population density (per 50 roots)		Log$_{10}$ population density		k value for mortality	
	1975	1976	1975	1976	1975	1976
Potential fecundity (eggs) (k_o loss in natality)	340	33	2.531	1.518		
					0.199	−0.746
Egg input (k_1 egg mortality)	215	184	2.332	2.264		
					0.107	0.209
Larvae I and II (k_2 early larval mortality)	168	113	2.225	2.053		
					0.322	0.366
Larvae III (k_3 late larval mortality	80	49	1.903	1.690		
					0.040	0.027
Pupae (k_4 pupal mortality)	73	46	1.863	1.662		
					0.064	0.201
Adults plus parasites (k_5 parasitism)	63	29	1.799	1.462		
					0.109	0.015
Adults (k)	49	28	1.690	1.447		
					0.778	0.072

is the sum of all the k_n mortalities occurring at each stage. The contribution of each mortality k value to total mortality can be assessed graphically (Varley and Gradwell, 1960; DeGrooyer *et al.*, 1995) or quantitatively (Podoler and Rogers, 1975). The graphical method simply involves a visual assessment of the correlation between K and $k_o...k_n$ when the K and k values are plotted against their corresponding generations (Fig. 6.7). The k mortality that most closely correlates with the overall K is referred to as the key mortality factor. A more quantitative approach (Podoler and Rogers, 1975) involves a regression analysis of the k values against total mortality K. The mortalities (k values) that contribute little to changes in K have very low regression coefficients and those that play a major role have high values. However, this is not a particularly precise test because the sampling errors incorporated in k will also appear in K but it may prove useful where the identification of the key factor is in doubt (Southwood, 1978). A more sophisticated approach was developed by

Manly (1977) involving partition of variances in a multiple regression analysis.

Where life tables and key factor analysis are carried out to determine the role of natural enemies in pest population changes, the relationship between mortality and pest density will need to be known. Only when the mortality is density dependent will the natural enemies have a potential role as control agents. The various k mortalities can be tested for density dependence if each value of k is plotted against N_t, the logarithm of the number of pest individuals entering the stage on which it acts. These data should be analysed using a regression technique and provided that the results are significant, the regression coefficient will provide a measure of how the mortality factor will act. A regression coefficient of exactly 1.0 means that the mortality factor is perfectly density-dependent and will compensate completely for any changes in density. A slope of greater than 1.0 will mean over-compensating density-dependence and of under 1.0 under-compensating density dependence. The plot of data

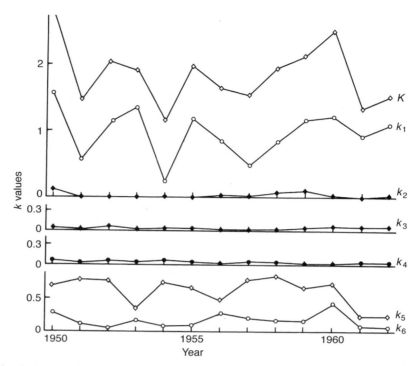

Fig. 6.7. k values plotted against generation for the winter moth (*Operophtera brumata*) to determine the key mortality factor (k_1 winter disappearance, k_2 parasitism by *Cysenis*, k_3 parasitism by other insects, k_4 microsporidian disease of larvae, k_5 pupal predation and k_6 pupal parasite). The key factor is k_1 identified because it made the biggest contribution to change in the generation mortality ($K = k_1+k_2+k_3+k_4+k_5+k_6$) (after Varley *et al.*, 1973).

could also reveal an inverse density-dependent relationship, with the mortality factor having greatest effect at low densities. A non-significant relationship could imply density independence or delayed density dependence. If the values of k and N_t are plotted and the points joined in a time sequence, then the presence of delayed density dependence can be identified if the points form a spiral (Fig. 6.8). A delayed density-dependent relationship will occur when the effects of mortality are not immediately effective but influence the size of the subsequent generation.

Life tables, although most often used for insect pests, can also be constructed for parasitoids. Data from such life tables can be used to calculate the area of discovery 'a' of the parasitoid that can then be used in the type of population models consid-

ered above (Varley *et al.*, 1973). The role of hyperparasitoids in the population dynamics of parasitoids can also be investigated using key factor analysis.

The construction of life tables is an important component in the understanding of the population dynamics of a species (Southwood, 1978) but it does take considerable time and manpower to obtain realistic results (Pschorn-Walcher, 1977). They also have the drawback that they are based on correlation rather than demonstration of cause and effect (DeBach, 1974; Huffaker and Kennet, 1969) and are very sensitive to an underestimation of mortality unless taken frequently (Wilson, 1985). However, if carried out correctly life tables still remain the most important analytical technique available for identifying key mortality components in an insect pest's life cycle.

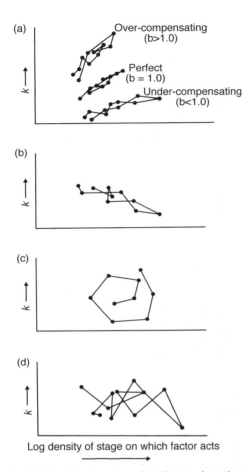

Fig. 6.8. The time sequence plots showing how the different density relationships may be recognized. (a) Direct density dependence (b positive); (b) inverse density dependence (b negative); (c) delayed density dependence (b = 0); (d) density independence (b = 0). b refers to the slope of the relationship (after Southwood, 1978).

6.6 Practical Approaches to the Evaluation of Natural Enemies

The experimental methods used to evaluate the potential role of natural enemies in pest regulation are very diverse; in the field, they range from intensive sampling and data collection to observations of individual behaviour, and in the laboratory from studies of maximum prey consumption to the use of serological techniques to establish the presence of prey in the guts of predators. All of the techniques have a number of advantages and disadvantages, or are appropriate only to specific situations. Despite this, it is possible to generalize to some degree and suggest a certain logical sequence of experiments that should enable natural enemies to be ranked according to their effectiveness in reducing prey numbers (Fig. 6.9). This sequence of experiments will be considered in more detail below.

6.6.1 Field survey, collection and observation

Inevitably, the first step in any investigation of the role of natural enemies in pest control involves a field survey to determine which species are present and how their numbers vary in relation to those of the pest insect (e.g. Sivasubramaniam *et al.*, 1997). This usually means that intensive field sampling is carried out using a variety of techniques over a number of seasons in order to gain some insight into the magnitude of seasonal fluctuations and some indication of the relative importance of the different natural enemies. The researcher would be looking for apparent associations and correlations between insect pest numbers and the numbers of a particular group or species of natural enemy (Figs 6.10 and 6.11) (e.g. Heong *et al.*, 1991), although the existence of such correlations is not proof of a causal relationship between pest and natural enemy numbers. One way round this problem is to make a step-by-step analysis of the role of each of the other possible causes of pest suppression such as emigration and weather. However, such studies may produce only nagging doubts which will then suggest the need for other qualifying information on the dynamics of predation processes from the field (Wratten, 1987). Provided these field surveys are seen as an initial first step in the study of pest/natural enemy associations, and too much emphasis is not placed on their independent value, then they represent an important springboard for further research. The problem comes when decisions about the potential value of natural

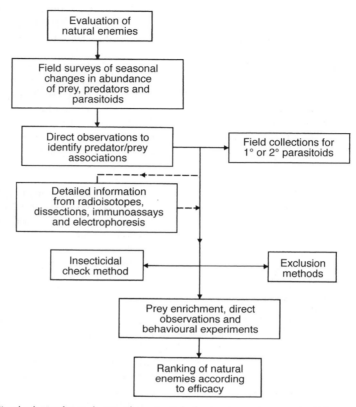

Fig. 6.9. A generalized scheme for evaluating the potential of natural enemies as potential control agents.

enemies as control agents are based solely on such associations.

Supplementary information about pest and natural enemy associations can be obtained by direct observation of predation events or parasitoid oviposition. Parasitoid/prey associations can be determined from field collections of potential prey which are reared through in the laboratory and emerging adult parasitoids identified. Direct observation can be used in the laboratory or the field to establish the diet breadth of a predator and with detailed observation, the rate of predation can be determined (Mills, 1997; Edgar, 1970; Nyffeler *et al.*, 1987). The use of sentinel prey (non-mobile stages such as eggs or pupae) placed in the field can be used to monitor losses by predation (e.g. Andow, 1992; Cook *et al.*, 1994, 1995; Berry *et al.*, 1995). Such approaches can provide

useful information on predation as long as the prey are no more or less susceptible to predation than the wild population.

Where more detailed information about predator–prey relationships is required at this early stage in the study then the use of dissection, serological tests, electrophoresis and radio-isotope techniques may be considered appropriate.

6.6.2 Dissection and biochemical techniques
Predators and parasitoids may either chew their prey or feed on their body fluids. For predators that chew their prey, it may be possible to identify their presence from unconsumed prey remains (e.g. Andow, 1992; Heinz and Parrella, 1994) or dissect the predator's gut to identify the consumed remains of the prey (e.g. Sunderland, 1975). Although Sunderland (1975) was

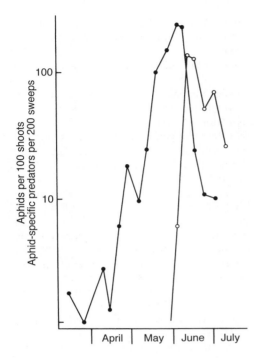

able to identify fragments of prey cuticle from earthworms, spiders, aphids, beetles and diptera from the guts of carabid beetles, such an approach has the obvious drawback that soft bodied prey may not provide detectable fragments and fluid feeding predators do not ingest recognizable fragments of their prey (Mills, 1997). There now exists a range of biochemical techniques that can be used to indicate the presence or absence of a particular prey species in an individual predator and more recently techniques have been developed that permit quantitative estimates of prey remains in predators at the time of death (Sopp *et al.*, 1992). The use of biochemical techniques for studies of predation and parasitism have recently been reviewed by Symondson and Hemingway (1997), Powell *et al.* (1996), Greenstone (1996) and Powell and Walton (1995).

One of the simplest techniques for identifying prey in the guts of natural enemies are serological tests, mainly because it is possible (depending on the type of test) for someone with little training to examine large numbers of insects with very simple apparatus (Boreham and Ohiagu, 1978;

Fig. 6.10. The changes in numbers of aphids (●) and the active predatory stages of aphid-specific predators (○) in a cereal field in the UK in 1980 (after Chambers *et al.*, 1983).

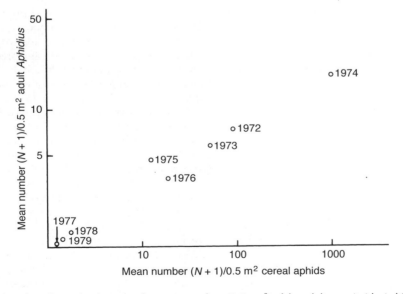

Fig. 6.11. The relationship between the mean numbers (0.5 m^{-2}) of the adult parasitoids *Aphidius* spp. and cereal aphids in UK grass fields in May 1972–1979 (after Vickerman, 1982).

Crook and Sunderland, 1984). The theory behind the serological tests is simple. Each prey insect is composed of many chemical substances, some of which are unique to particular species. A chemical extract of a prey species (the antigen) is introduced into the blood serum of an animal (usually a rabbit) so that antibodies are produced, then, when the serum (containing the antibodies) is mixed with the antigen, a precipitate is produced. Hence, if a predator has consumed the prey insect, this will be indicated by the presence of a precipitate when an extract from the predator's gut is mixed with the appropriate antibody serum. Simple serological tests can be used to visualize this reaction of the gut contents in predatory insects (Boreham and Ohiagu, 1978; Southwood, 1978). However, these tests are not much used today (Symondson and Hemingway, 1997). The assay of choice is now the enzyme-linked immunosorbent assay (ELISA) due to its extreme sensitivity and suitability for quantitative analyses. Symondson and Hemingway (1997) list 21 such studies carried out in the last 15 years, investigating diets of insects as widely different as Carabidae, Culicidae, Salticidae and Pentatomidae. Greenstone (1996) highlights the advantages of the ELISA technique which is routinely capable of detecting <10 ng of prey antigen.

A further refinement for use with the ELISA assay involves generation of monoclonal antibodies (Symondson and Liddell, 1996) which, because they can bind to a single epitope (representing 8–10 amino acids) on a target prey protein, means they can be used to distinguish between very closely related prey (e.g. different species of whitefly, Hagler et al., 1993) and particular insect stages (e.g. Greenstone and Morgan, 1989).

Another biochemical technique that has found an application in the analysis of predator diets is that of electrophoresis. Electrophoresis can be defined as the movement of ions, or charged molecules, in a fluid under the influence of an electric field, with positively charged particles moving to the cathode and negative particles to the anode (Symondson and Hemingway, 1997).

Thus the aim of the electrophoretic process is to separate biological materials (usually proteins) into clearly resolved bands that can be visualized by staining on a gel. Electrophoresis has been used to identify to species the early stages of parasitoids (Wool et al., 1978; Walton et al., 1990a,b) and for detection of predation, especially where it is difficult to gather sufficient prey material for antibody production (Murray and Solomon, 1978; Fitzgerald et al., 1986; Solomon et al., 1996).

One other technique that is used to identify the prey consumed by predators is that of marking prey with radio-isotopes. Potential prey are fed with radioactive tracers (usually ^{32}[P]; Southwood, 1978) and then exposed to natural predation (e.g. McCarthy et al., 1980). James (1966) tagged mosquito larvae with radioactive ^{32}[P] and released them in woodland pools to identify their predators. Predators are collected and assayed for the presence of radioactivity using a scintillation or Geiger counter, or with autoradiography. Fourteen of 38 species of aquatic insects collected by James (1966) ingested the tagged larvae. The value of this technique lies in its potential to measure the number of prey consumed, because if individual prey carry a similar burden of radio-isotope, the level of radioactivity detected will identify the number of prey consumed. In practice, however, it is extremely difficult to make this technique quantitative because there is usually considerable variation in the level of radioactivity of individual prey (Kiritani and Dempster, 1973). Added to this, many predators only consume part of their prey and the rate of excretion of the radio-isotope by a predator can depend upon the amount of food it subsequently eats.

The real significance of these biochemical assays is their potential for quantifying predation rates. If a predator gives a positive reaction, then provided a number of other conditions can be met predation rates can be estimated. The other information required includes the density of the predator and prey population, the length of time taken for a predator to digest a given meal

Table 6.5. Predation rate equations (from Sopp *et al.*, 1992).

Equation	Reference
1 $r = \dfrac{pd}{t_{DP}}$	Dempster (1960, 1967)
2 $r = pr_1d$	Rothschild (1966)
3 $r = \dfrac{pr_1d}{t_{DP}}$	Kuperstein (1974, 1979)
4 $r = \dfrac{\left[\log e(1-p)\right]d}{t_{DP}}$	Nakamura and Nakamura (1977)
5 $r = \dfrac{Q_0 d}{f t_{DP}}$	Sopp *et al.* (1992)

r = Predation rate (biomass or numbers of prey).
r_1 = Predation rate measure in an insectary.
p = Proportion of predators containing prey remains.
d = Predator density.
t_{DP} = Detection period.
Q_0 = Quantity of prey recovered.

(calculated from laboratory experiments) and the proportion of predators sampled that produced a positive reaction. If all this is known then the minimum predation rate can be obtained; Table 6.5 lists a number of commonly cited predation rate equations (Sopp *et al.*, 1992).

The first three equations assume that a positive result from a predator gut analysis represents a fixed number of prey eaten (which is normally unrealistic). The equation used by Nakamura and Nakamura (1977) for predation of the chestnut gall wasp (*Drycosmus kuriphilus*) by spiders, assumed their predation was random and thus it followed a Poisson distribution (Equation 4, Table 6.5) to get around the problem of a fixed number of prey. Sopp *et al.* (1992) took the whole process one stage further through use of ELISA (but also applicable to electrophoresis) which permits the quantity of prey biomass to be calculated. The value 'p', the proportion of predators, is replaced by Q_0/f where Q_0 is the quantity of prey mass recorded in the predators and f the average proportion of the meal remaining. The value of 'f' is calculated from a digestion decay function (Sopp and Sunderland, 1989; Symmondson and Liddell, 1995). Although the Sopp *et*

al. (1992) method represents a significant advance on the other models it still involves several assumptions (see Kidd and Jervis, 1996) which could cause serious errors if not satisfied.

6.6.3 Exclusion/inclusion methods

The majority of experimental methods used in this context are of the exclusion type rather than inclusion methods. Exclusion usually involves the elimination and subsequent exclusion of established natural enemies from plots which are then compared with another comparable set of plots having natural enemy populations that have been undisturbed (DeBach and Huffaker, 1971). The natural enemies can be eliminated from a plot by chemical or mechanical means and then excluded by mechanical means, usually cages.

Cages can cover part of a plant (e.g. Chandler *et al.*, 1988), a whole plant or group of plants (e.g. Rice and Wilde, 1988) and may take the form of muslin sleeves or large wooden or metal framed mesh cages. Cages must be used for the controls as well as the treatment plots but the control cages should allow access by natural enemies. The mesh size of the cages may be varied to allow access of some flying natural ene-

Table 6.6. Use of cages to study effects of predators on the damson-hop aphid.

Dates	Exclusion cages		Open-ended cages		Uncaged strings	
	Adults	Nymphs	Adults	Nymphs	Adults	Nymphs
12 August	41	85	23	78	15	52
19 August	87	608	0	3	0	6
26 August	181	1652	0	8	0	1
2 September	870	6193	0	27	0	12

mies but not others, or the sides left partially open to allow entry (King *et al.*, 1985). The use of control cages reduces the problem of differences between controls and treatments due to variables other than natural enemies. However, the effect of cages on the interior micro-environment and predator–prey behaviour remains one of the primary concerns with this technique. Cages screen the plants, providing shading and reducing wind speed while increasing the temperature and humidity relative to uncaged areas. Net cages used to cover bushes of *Euonymus europaeus* decreased light intensity by 18% and wind speed by 24% (Way and Banks, 1968). Temperatures within closed cages were within one degree of field temperatures in a study conducted by Frazer *et al.* (1981). The differences between cages and natural conditions could affect the physiology of the plants on which the insect pests are feeding, the growth rate and activity of the insect pest and hence their exposure to predation (Luck *et al.*, 1988). The cages impede emigration of both the pest and natural enemy and may increase predation rates because the predator will repeatedly search the same area beyond the point at which they would normally have emigrated (Luck *et al.*, 1988). A further drawback with cages is that predator free controls are rarely free of predators.

Nevertheless, exclusion remains the most appropriate experiment for testing whether natural enemies have potential for control or not (Luck *et al.*, 1988). Exclusion cages have proved effective in identifying the role of predators in control of a number of different pests, particularly aphids (Chambers *et al.*, 1983; Dennis and Wratten, 1991). The pea aphid (*Acyrthosiphon pisum*) in alfalfa

was shown to be controlled by a complex of predators each responding to changes in aphid density (Frazer *et al.*, 1981). Use of cages to study the damson-hop aphid (*Phorodon humuli*) and its predators on hops provided conclusive evidence of the controlling influence of predators (Table 6.6), while cage studies carried out by Way and Banks (1968) provided evidence that natural enemies which attack *Aphis fabae* on *Euonymus europaeus* and on its summer hosts, were responsible for the common 2-year cycle of aphid abundance. Likewise, when excluded from plots by cages, the ground living predators of *Pieris rapae* on cabbages were shown to account for about 23% of larval mortality (Ashby, 1974).

Exclusion can also be obtained by hand removal of predators, a technique that circumvents objections about manipulations affecting pest and predator population dynamics, since it should have no effect on the environment. The technique is, however, extremely time consuming and labour intensive. A group of workers would be needed on a continuous basis, night and day to hand remove any natural enemies which settle on the leaves, twigs, branches, etc. of the protected area. This must continue for the duration of the test and long enough to produce a large differential (van den Bosch and Messenger, 1973). Such labour intensive studies are really only applicable to inactive predators, at reasonable densities and those that are diurnally active or undisturbed by night light (Luck *et al.*, 1988). Pollard (1971) used the technique to remove syrphid eggs and larvae from Brussels sprout plants but found no differences between the size of aphid infestations on plants where syrphids had been removed and the controls, at least for two out of three sites, and concluded that the searching was too inefficient for the technique to be of value. Luck *et al.* (1988) contends that the technique of hand removal deserves more attention especially as a method of checking for bias in other exclusion methods.

Ingress and egress barriers are used in field crop situations to allow ground dwelling predators either into plots but not out, or out but not in (ingress and egress barriers respectively). The ingress barrier should thus increase the number of predators in a plot, whereas an egress barrier should decrease the number. Such barriers usually consist of a small trench with an overhang into this created by a board or rigid felt material, which will allow insects into the trench from both sides but out of the trench only on the opposite side to the overhang.

In studies carried out in New Zealand using ingress and egress barriers to assess predator effects on populations of *Myzus persicae* on sugar beet, the peak number of aphids in the egress plots was 150 times greater than that in the control (Wratten and Pearson, 1982). These results suggested that relatively low numbers of predators in New Zealand sugar beet could have a substantial effect on keeping aphid numbers low. Wratten and Pearson (1982) concluded that the biggest drawback with this type of work was the lack of information concerning where the predation occurs and by which species, and although pest numbers can be correlated with predator numbers if pitfall traps are used, immunological or electrophoresis studies of predator gut contents are required to establish a causal link.

Other types of barrier included grease bands around tree trunks preventing entry or exit for climbing predators and trenches treated with insecticide. Straw treated with DDT or aldrin placed in trenches has been used as a means of preventing movements of predators in and out of plots (Wright *et al.*, 1960; Coaker, 1965) and methods using other insecticides could work equally well.

Case Study: The effects of predator exclusion and caging on cereal aphids in winter wheat (Chambers *et al.*, 1983)

Field cages were used to exclude aphid specific predators from small areas of wheat during 1976, 1977 and 1979. The changes in numbers of aphids (*Sitobion avenae*) inside and outside the cages were determined and related to changes in the density of the predators and other factors.

The cages used to exclude the aphid specific predators were 1.8 m high, rectangular (3 × 1.5 m) cages consisting of a tubular metal frame covered with polyester netting (mesh size 1 × 0.3 mm, thread thickness 0.1 mm). The netting extended below the soil surface to a depth of about 30 cm. Access to the cage was possible via a vent held closed by 'Velcro' strip.

Table 6.7. The mean number ($0.5\ m^{-2}$) of predators (for the period in which they were active in the field) and the totals for each year. The ratio peak number/shoot is the ratio of the peak number of aphids per shoot in the cages relative to the open plots (after Chambers *et al.*, 1983).

	Coccinellid adults and larvae	Syrphid larvae	Chrysopid larvae	Total	Ratio peak no. shoot
1976	20.22	8.92	3.02	32.16	6.41
1977	3.75	9.67	2.00	15.42	3.68
1979	0.45	0.64	0.27	1.36	2.82[*]

*Corrected for different starting numbers in open and caged plots.

Different numbers of cages and plots were used during the three years and artificial infestations of the aphids were necessary to supplement natural infestations during 1977 and 1979. The aphids were counted weekly in 1977 and 1979 and every 10 days in 1976. Each plot was divided into four subplots and aphids counted on a row of shoots each with a minimum of 40 shoots sampled per plot. The subplots were examined by four different members of the sampling team to minimize the variability due to counting error. Alate aphids were removed by hand from the interior walls and roof of each cage in 1979 when they were not removed in order to try and determine whether the population size was increased by their subsequently returning to the plants.

The aphid specific predators were sampled each year at the same intervals as the aphids but with a Dietrick vacuum suction sampler. The plant growth stage was assessed weekly in open and caged plots.

Three phases of aphid population development were distinguished in this experiment: an initial rapid growth phase, a divergence phase and a decline phase (Fig. 6.12). During the growth phase, the aphid population both inside and outside the cages increased at the same rate but during the divergence phase, the aphid populations inside the cages continued to increase to a peak, while populations outside increased at a slower rate or decreased. The rate of decline of populations outside the cages was faster than that inside the cages.

For each of the three years the divergence phase coincided with an increase in the numbers of aphid specific predators and a negative association between predator numbers and aphid abundance was apparent (Table 6.7). Up to a six

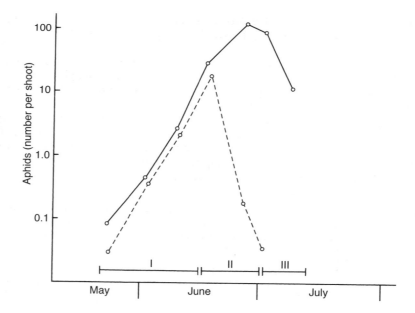

Fig. 6.12. The numbers of aphids on wheat in cages (———) and in open plots (– – –) for the three phases of population development: (I) the rapid growth phase; (II) a divergence phase; and (III) a decline phase (after Chambers *et al.*, 1983).

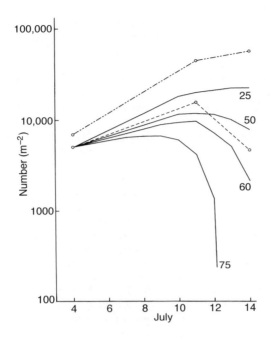

Fig. 6.13. Aphid populations during phase II in cages (o —·—·— o) and open plots (o – – – o) and the calculated population trends (——) based on different values for per capita daily consumption rates of the predators (after Chambers *et al.*, 1983).

times difference in aphid numbers occurred between predator–excluded plots and uncaged plots. These observed relations between aphid numbers and predator numbers (Table 6.7) are not proof of causal relations and it is not known whether sufficient predators were actually present to account for the differences between cages and plots. The effectiveness of the predator is partially dependent on its per capita consumption rate of aphids. If the aphid population trends coincided with expected population sizes based on a realistic consumption rate then this would add force to the argument that the predators were responsible for the observed differences between cages and open plots. Figure 6.13 shows that consumption rates of between 50 and 60 aphids per predator per day would have been required to bring about the observed differences between cages and plots. This falls within the ranges of published values and was found to do so for the other two years as well. It was concluded that predation was likely to be the major cause of the differences between cages and open plots at the time of the aphid population peak.

There was no difference in plant growth stages inside and outside the cages during the three years and the divergence phase occurred at different growth stages in each of the three years. This does not, however, rule out possible subtle effects of the cages on the physiology of the wheat, or the aphid, acting only during the divergence phase (Chambers *et al.*, 1983).

6.6.4 Insecticidal check method

The insecticidal check method makes use of the physiological or ecological selectivity of insecticide application to kill a large proportion of the natural enemies while leaving the pest population unaffected. The development of pest populations is then compared in treated and untreated plots and the difference provides an indication of natural enemy effectiveness (DeBach and Huffaker, 1971). This technique has been successfully applied to studies in citrus of the effects of natural enemies on the long tailed mealybug, the cottony cushion scale, the California red scale, the yellow scale, the six-spotted mite (van den Bosch and Messenger, 1973), leafhoppers and planthoppers (Kenmore et al., 1985; Ooi, 1986), aphids (Milne and Bishop, 1987) and thrips (Tanigoshi et al., 1985). Three species of beetle, Stethorus spp., were found to be important predators of Tetranychus urticae in orchards in Australia by the use of an insecticide to which the mite was known to be resistant (Readshaw, 1973), a small positive advantage associated with the development of insecticide resistance.

The insecticidal check method has the advantage that it does not markedly alter the environment or prey movements; it does, however, have a number of disadvantages. The use of insecticides may appear simple but in practice a great deal of information is required before an insecticide can be chosen and applied selectively to affect natural enemies alone. If physiologically selective insecticides are known there is little problem with this approach, provided that sufficient is also known about appropriate field doses and application rates. The problem with the ecologically selective use of insecticides is the prior knowledge that is required about the pest and its natural enemies. Other problems with this method include the stimulatory effects of pesticides on the reproduction of the prey, pesticide induced biases in sex-ratio and pesticide induced physiological effects on the plant (Luck et al., 1988). Because the insecticides potentially stimu-

late arthropod reproduction, affect plant physiology, and are difficult to apply selectively, this method should only be used to estimate crude predation rates.

6.6.5 Behavioural studies and prey enrichment studies

Research programmes aimed at evaluating the effectiveness of natural enemies should ideally start with behavioural studies to immediately identify important predators/parasitoids and their consumption/parasitism rates (Luck, 1990). However, this is rarely done and if carried out at all is left until the programme is well established.

Behavioural studies can be carried out in semi-natural conditions in the laboratory and can provide important information about predator behaviour (Marks, 1977b; Gardner and Dixon, 1985) but it is impossible to entirely recreate field conditions in the laboratory and this may lead to important mechanisms being overlooked. Field studies provide the opportunity to identify some of the underlying behavioural and ecological mechanisms involved in the process of predation and parasitism; as an approach, however, it does offer some formidable experimental problems.

Field studies of insect behaviour are highly labour intensive and data only accrue slowly because of the difficulties involved in observing small invertebrates at low densities and often also at night. High densities of the predator have been used in field arenas to overcome the problems of trying to observe the beetle Agonum dorsale at night (Griffiths, 1982; Griffiths et al., 1985). Red torch light was also used to aid observations after laboratory experiments had established that it did not affect the behaviour of the beetle. The problems associated with observations of small highly mobile Hymenoptera are even more horrendous, although Waage (1983) at least showed that it was possible using artificially infested plants and two observers with binoculars. Waage (1983) collected and dissected the larvae of Plutella xylostella observed during the experiment to determine how many had been parasitized. Similar collections and dissections of artificially infested insects have been used to determine levels of parasitism (Way and Banks, 1968) and are generally referred to as prey enrichment studies. They usually work best with non-mobile stages such as eggs but care must be taken to arrange them in realistic positions on host plants (Luck *et al.*, 1988).

6.6.6 Ranking of natural enemies

The ultimate aim of using experimental methods to evaluate natural enemy effectiveness is to identify those species that have the most potential for use as biological control agents, whether through introduction, augmentation or conservation. The approach taken rarely follows the protocol suggested in Fig. 6.9, rather, a piecemeal approach develops with different researchers in different institutes working independently on a common problem, usually with a method appropriate to the expertise and facilities available at any given place. Apart from the fragmentation of research among different workers one of the reasons for this piecemeal approach has been a preoccupation with, and the need to develop, better techniques. While the use of cages and barriers has long been established the biochemical analyses investigating gut contents of predators and parasitoids are now making more detailed and precise evaluation possible. Thus, a variety of accurate and practical methods now exist that permit a logical progression of experimental procedures that will inevitably provide the information necessary to reliably rank natural enemies according to their potential effectiveness as biological agents.

6.7 Classical Biological Control

International agricultural trade is increasing rapidly which is leading to an increase in the rate at which new alien pests are introduced (Waage, 1984; van Lenteren, 1995). In the USA, new species are added to the fauna at a rate of between 11 and 13 per year (Hoy, 1985; Sailer, 1983) and with new trade routes being opened such as the North–South trade in horticultural products, then pest-like aphids, whiteflies, leaf miners, thrips and mealybugs are likely to spread into new areas at an alarming rate. Classical biological control, which started in the 1800s, remains a very effective option today for controlling alien species. As an approach classical biocontrol offers permanent control that has few risks associated with it and above all it provides a highly cost-effective solution. A study of seven biological control programmes in a 40-year period provided estimated benefits to costs of 4.5 to 1 (Simmonds, 1968). A further seven programmes implemented during the 1970s provided enormous benefits obtained through savings on both chemical control and crop yield gains (Greathead and Waage, 1983). Tisdell (1990) reports an average benefit to cost ratio of 10.6:1 for biological control efforts in Australia, with maximum benefits exceeding 100:1. The biological control of the cassava mealybug *Phemacoccus manihoti* in Africa by introduction of a parasitoid from South America generated an estimated benefit to cost ratio of nearly 200:1 (Herren and Neuenschwander, 1991). Thus, classical biological control offers an economically attractive means of control of exotic pests which will ensure its continued use as a primary control tactic. It has also been suggested that it might be worth trying to reattempt introductions against past failures (Hoy, 1985) especially as new information on geographic ranges, biosystematics, biology and host preferences has been accumulated, continually improving the chances of selecting a more appropriate control agent. Pimentel (1963) argues that introductions should be used against native pests as well which would further increase the scope of biological control, for instance control of *Helicoverpa* in Australia (Michael, 1989) and the USA (Powell, 1989).

The process of introducing an exotic natural enemy can take between two and

Table 6.8. A summary of steps normally part of programmes for introduction of natural enemies (from van Driesche and Bellows, 1996).

Step	Objectives
1. Target selection and assessment	Identify target pest, define biological, economic and social attributes which relate to biological control; establish objectives for introduction programme; resolve any conflict of interest
2. Preliminary taxonomic and survey work	Determine current state of taxonomic knowledge of pest and natural enemies; conduct literature review on natural enemies of target species and relatives; survey in target area for any existing natural enemies
3. Selecting areas for exploration	Define native home of target pest and other possible areas of search for natural enemies
4. Selecting natural enemies for collection	Choose which of various candidate natural enemies encountered may be appropriate to collect for further study in quarantine
5. Exploration, collection and shipment of candidate natural enemies	Obtain and introduce into quarantine candidate natural enemies
6. Quarantine and exclusion	Process shipped material to destroy any undesirable organisms
7. Testing and selecting of natural enemies for additional work	Conduct research as necessary in quarantine on natural enemies
8. Field colonization and evaluation of effectiveness	Release natural enemies in field and monitor for establishment and efficacy
9. Agent efficacy and programme evaluation	Evaluate degree of achievement of overall programme goals and objectives

three years (Greathead and Waage, 1983) especially in situations where there is little recorded information available about the pest and its native complex of natural enemies. An ideal sequence of events involved in the introduction of such natural enemies is outlined in Table 6.8.

6.7.1 Target pest identification and area of origin

Lindroth (1957) described five criteria that could be used to identify a pest as alien: insect history, geography, ecology, biology and taxonomy. Pschorn-Walcher (1977) added a sixth: parasitology. In brief, the first of these criteria applies to the rare situation where the history of an introduction has been well documented; geographical criteria may be inferred when an insect has a very patchy distribution over the range of its host, perhaps because insufficient time has elapsed to allow colonization. Insects

that have a strong association with the human environment such as cockroaches would be included as alien under the ecology criterion. Host specificity would be a character associated with biology while the taxonomic criteria would apply to species holding an isolated taxonomic position with no close relatives among the native fauna. The parasitological criterion would include pests that had no obvious natural enemy complex associated with them, suggesting recent colonization (in evolutionary terms).

Once a pest has been identified a literature review would be undertaken to find out as much as possible about the species, its origins and its native natural enemies. The basis on which a pest is identified as alien should provide some information about its centre of origin. The ever increasing knowledge of insect taxonomy and distribution makes it increasingly easy to identify the native area of an invading species (van den Bosch and Messenger, 1973). However, failure to identify correctly the centre of origin could mean the importation and introduction of less effective natural enemies than could potentially be available, since a mature, well-balanced complex of natural enemies would only exist in the native home area (Pschorn-Walcher, 1977).

It is not necessarily a simple matter to identify the centre of origin, often because the insect(s) concerned have such vast ranges; for instance, the alfalfa weevil (*Hypera postica*) has a range that extends across Europe, Africa, the Middle East and Asia, while the codling moth (*Cydia pomonella*) covers the Palaearctic region (van den Bosch and Messenger, 1973). The distribution of the target host plant or areas having appropriate climates may provide useful indicators. Information on the biology of the rice stem borers *Chilo suppressalis* and *Tryporyza incertulas* provided some circumstantial evidence for the area of origin of these insects (Yasumatsu, 1976). *C. suppressalis* is distributed northward from the tropics into the temperate zone where it enters diapause during the winter. It seems to prefer well drained paddies and its larvae remain high in the stalk at harvest. By way of contrast, *T. incertulas* is distributed from the tropics north to the southern regions of the temperate zone. It appears to favour poorly drained situations and its larvae bore towards the roots even in deep water. Yasumatsu (1976) concluded that these differences suggested that *C. suppressalis* originated in the dryer temperate area and *T. incertulas* in a damp tropical situation.

6.7.2 Foreign exploration, selection and field evaluation

A great deal of information about pests and their natural enemy complexes can be obtained from the literature and correspondence but generally foreign exploration is necessary and desirable in order to achieve a thorough check of the natural enemy complex (DeBach, 1974) and to ensure that

a proper inventory of such complexes can be prepared. First it is necessary to find the pest species in the countries in which it is endemic and verify it taxonomically to avoid the type of confusion that occurred with the misidentification of the cassava mealybug, *Phenacoccus manihoti* (Bennett and Yaseen, 1980; Cox and Williams, 1981; Neuenschwander and Herren, 1988). Then an investigation into the natural enemy complex of the pest can begin. Field studies are important in this context because it is only under natural conditions that the interactions between natural enemies and their host can be truly evaluated. When competition between guilds of parasitoids or predators exists through an overlap in resource use, e.g. larval parasitoids emerging from a host pupa may interfere with a pupal parasite, then it is only through field studies that a hierarchy between and within guilds can be established (Pschorn-Walcher, 1977). A number of factors must always be taken into account, such as the degree of synchronization between the natural enemy and the host, host specificity (although this has been questioned), physiological tolerance, as well as genetic variability of natural enemy species. The value of collecting natural enemies from different areas is that individuals with a wider range of physiological tolerances or adaptations can be obtained, although there is always the risk that these may turn out to be biogeographical or host races. There is a great need for genetic studies in biological control to determine whether individuals selected are genetically similar and whether those imported remain so throughout culturing and release (DeBach, 1974; Caltagirone, 1985; Hoy, 1985).

6.7.3 Quarantine and mass production

The mass production of alien species is a crucial aspect of every classical control programme because sufficient numbers of individuals of a suitable quality are needed to make release and establishment possible (Hopper *et al.*, 1993). Culturing of natural enemies is, however, a complex business and each species will need to be reared for

at least one generation to ensure hyperparasitoids are screened for as a quarantine
measure. Quarantine presents a problem in
relation to the number of insects that can
be handled, since it is impossible to scruti-
nize and maintain very large numbers, the
process tends to restrict sample size and
hence genetic variability. Sample sizes and
genetic variability may also be restricted in
other ways during the introduction
process. In some cases the entire importation consists of a single small sample containing a few individuals and at times it
has consisted of only a single female
(Messenger and van den Bosch, 1971).
Under such genetic bottle-necks, it might
be expected that the loss of adaptability
would be phenomenal and the success of
establishment threatened. However, theory
suggests that such loss is not great even
with severe bottle-necks (Nei *et al.*, 1975)
and experimental evidence seems to support this (Bryant *et al.*, 1986 a,b). Although
in cultures of aphid parasitoids, inadvertent selection resulted in low founder numbers, genetic drift and bottle-necks, factors
that can strongly influence biological and
behavioural experiments (Powell and
Wright, 1988), such factors have important
implications for pre-introduction evaluation, as well as for the chances of successful establishment. The latter may also be
threatened by the conditions under which
the insects are maintained since most
insect rearing rooms are kept at relatively
high (within the physiological limits of the
insects) constant temperatures – conditions
that are far from natural.

Many natural enemies have complex
life cycles making rearing difficult and on
occasion laboratory colonization can fail.
The ease of culturing tends to be one of
the factors that influences the selection of
potential agents. Culturing requires the
simultaneous rearing of the host insect
pest and its associated plant host as well
as the natural enemy itself. A break in the
sequence of either of the first rearing procedures and the natural enemy species
could be lost. For many species though,
the problems have been overcome and

thousands of individuals can be produced
for release.

6.7.4 Release and establishment

The first step, prior to release, is to ensure
that the target pest is present in sufficient
numbers at an appropriate life stage at the
different release sites. Other factors that
need to be taken into consideration
include:

1. Ensuring that conditions at each
release site are optimal for establishment.
2. Use of an adequate number of natural
enemies in a given release.
3. An appropriate number of such
releases.
4. The use of a number of release sites
over the geographical and ecological ranges
of the pest (van den Bosch and Messenger,
1973).

If pesticides are used at the release site
or in the surrounding areas, then the success of establishment may be affected. If
insecticides have to be used in the vicinity,
then pre-introduction studies should
include work on insecticide selectivity and
their application postponed in the immediate area and period after release. The successful establishment might also depend
on other characteristics of the site such as
adult food availability and the presence of
secondary hosts if a multivoltine natural
enemy is released against a univoltine pest.
The presence of indigenous natural enemies, especially hyperparasitoids, may
cause a problem by attacking the introduced natural enemy. For example, in
south-western Nigeria, hyperparasitoids
destroyed up to 50% of all the cassava
mealybug mummies parasitized by
Epidinocarsis lopezi during the first season
of its release (Neuenschwander and
Herren, 1988).

The number of individuals released on a
single occasion may also influence the
chance of establishment. In Canada, an
analysis by Beirne (1975) of successes and
failures of introductions indicated that successful establishment may be favoured by
release of large numbers of individuals.

Only 10% of introductions were successful when a total of 5000 individuals or less were released, 40% were successful when introductions involved totals of between 5000 and 31,200, whereas 78% of introductions were successful where releases involved over 31,200 individuals. A sequence of releases is also important. Again, in the analysis by Beirne (1975), 70% of species that were released more than 20 times achieved successful establishment compared with 10% when release occurred less than 10 times. However, Beirne (1975) warns that the decision whether or not to continue with a sequence of releases may be influenced by the success or failure of the earliest releases and that this could have affected the values cited.

Releases are usually made at a number of different sites selected not only for conditions appropriate to establishment but also to reflect the range of environments occupied by the host pest insect. The climatic and ecological conditions at some part of the site should be suitable to allow establishment of the introduced natural enemy. Once established, selection will occur and those best adapted will survive. Further genetic changes will occur if a natural enemy is able to extend its range into all the environments occupied by its host.

6.7.5 Post-establishment evaluation

There are few examples in classical biological control programmes where extensive evaluation studies have been made after the initial release and establishment of natural enemies. The successful introduction of the natural enemy and its subsequent

control (or lack of it) of a pest may appear self-evident and it is perhaps due to this that funds are rarely made available for properly conducted post-establishment evaluation. After all, from a funding agency's point of view, provided that an introduction produces an adequate level of control then the problem has been solved, or if it has not, then it would appear unwise to fund a project further, especially if this will only tell you why it went wrong. From a biological control practitioner's point of view, however, the need for proper post-establishment evaluation of a release programme is paramount. Such studies are required to allow practitioners to evaluate the true level of success that has been obtained and to allow them to gain an understanding of the processes involved in the insect pest/natural enemy interaction. This is especially important when programmes fail at the release and establishment phase, when the reasons for failure need to be known for future attempts and for generally increasing our understanding in order to improve the likelihood of developing future successful biological control programmes.

One of the problems with advocating such post-establishment evaluations is associated with the type of work and the resources required to carry out the job properly. The most appropriate type of evaluation involves obtaining data suitable for constructing life tables, both before and after establishment over a number of years and at different sites. However, sufficient funds are rarely made available for such studies despite their obvious importance.

Case Study: Control of the cassava mealybug (*Phenacoccus manihoti*) in Africa through the introduction of the exotic parasitoid *Epidinocarsis lopezi* (Neuenschwander and Herren, 1988)

Cassava (*Manihot esculenta*) is the staple food of over 200 million Africans living in the humid and subhumid tropics as well as an emergency food reserve in many arid zones. Since its introduction into Africa over 500 years ago, cassava had remained relatively pest free until the new pest *Phenacoccus manihoti* appeared in Zaire in the early 1970s. This parthenogenetic pest spread rapidly through Senegal and The Gambia in 1976, into Nigeria and Benin Republic in 1979 and by 1985 it had spread into Sierra Leone and Malawi and from there onwards at a rate of 300 km a year. By the end of 1986, it had reached about 25 countries and covered about 70% of the African cassava belt. This major pest was causing damage by stunting the growth points and defoliating the cassava plants. Because of the extensive distribution of the pest, in largely mixed crop-ping subsistence farming communities, the most appropriate means of control appeared to be the introduction of an exotic natural enemy.

Cassava had been introduced into Africa from South America, hence searches for natural enemies of the mealybug started there. The search began in 1977 and in 1981, *Phenacoccus manhoti* was finally discovered in Paraguay. Among the parasitoids isolated from the mealybug was *Epidinocarsis lopezi*, an endophagous, solitary encyrtid. This parasitoid has four larval instars, passes the nymphal stage inside the mummified host mealybug and can develop twice as fast as its host at 27°C. The adult females are attracted to mealybug infested cassava leaves and are thought to home in on the odours (synomones) of the attacked cassava to locate the mealybug (Nadel and van Alphen, 1986). Thus the parasitoid has a very good host locating capability. After oviposition, a female will often feed on the host so although its reproduc-tive capacity is limited, the parasitoid also causes mortality by host feeding and mutilation of its hosts.

The first releases of the parasitoid were made in Nigeria in 1981 and they were found the following year to have become established. By March 1983, *Epidinocarsis lopezi* was recovered from all sampled fields within 100 km of the release point. *Epidinocarsis lopezi* was exported to 50 other sites in countries in Africa. It is presently established in 16 countries with an estimated distribution of 750,000 km², which exceeds that of any other agent introduced into Africa for the biological control of insect pests.

Post-establishment studies have included physical and chemical exclusion experiments (Neuenschwander *et al.*, 1986). A sleeve cage experiment on cassava demonstrated that two months after artificial infestation, the mealybug popula-tions were 7.0 and 2.3 times lower on the growing tips covered with open cages than on tips in closed cages that excluded most parasitoids. On similarly infested but uncovered tips the mealybug populations were 24.3 and 37.5 times lower and parasitism rates were higher. In the chemical exclusion experiment, an artifi-cially infested field treated with weekly applications of carbaryl had a peak mealybug population of 200 insects per tip. The adult *Epidinocarsis lopezi* had been killed by the insecticide whereas in the untreated plots, parasitism was higher and mealybug infestation lower than ten individuals per tip.

The control of cassava mealybug with the parasitoid *Epidinocarsis lopezi* rep-resents one of the largest and most successful biological control programmes of all time.

6.8 Inundation and Biopesticides

Biological control through inundation involves the release of the agent in massive numbers within a very short space of time. The agent is usually not persistent, is expected to kill the pest relatively quickly and is usually only relevant to use of pathogens, formulated as biopesticides, which can be utilized as alternatives to chemical insecticides. Viruses, bacteria, fungi and entomopathogenic nematodes have all been developed as biopesticides and used for inundative control (Section 6.3.1). Such research and development has been largely undertaken in the public sector (Waage, 1997; Dent, 1997b) and to a large extent has treated the biological agent much like a chemical active ingredient with the subsequent expectation that it will perform to the same standards. In almost all cases this has not been possible and hence all the shortcomings of biopesticides relative to chemicals emerge, and few of the benefits (Waage, 1997). It is not surprising that living organisms do not make as good chemical pesticides as chemicals do. However, some pathogens have desirable properties that chemicals do not and can-

not possess. For instance, nematodes have a capacity to find their pest host, to kill them and then reproduce in them. All pathogens have the capacity to reproduce and hence produce another generation able to attack and kill other individuals in a pest population. This numerical response, of which chemicals are incapable, has rarely been exploited with biopesticides. Opportunities exist to utilize these additional properties of host finding and reproduction which make biopesticide superior to chemicals (Waage, 1997).

Case Study: Green Muscle – a mycoinsecticide for locust and grasshopper control (Neethling and Dent, 1998)

Locusts are major pests of agriculture worldwide against which large volumes of chemicals are applied during outbreak years. An international research programme called LUBILOSA has developed and commercialized a mycoinsecticide trademarked Green Muscle for use against the desert locust (*Schistocerca gregaria*), the brown locust (*Locustana pardalina*), the Moroccan locust (*Dosiostaurus maroccanus*) and a number of Sahelian grasshoppers in Africa.

Green Muscle is based on the fungal pathogen *Metarhizium anisopliae* var. *acridum* (previously known as *Metarhizium flavoviride*). It is applied using formulations of conidia in an oil suspension, allowing the product to be used under dry and arid conditions. This ability to utilize a mycoinsecticide in arid conditions represents a significant advance in biopesticide technology since application of mycoinsecticides has traditionally only been considered possible in more humid environments.

Green Muscle achieves 90% kill in 7–21 days. In field trials in West Africa it has been demonstrated that in comparison with an ultra low volume formulation of fenitrothion, the advantage of an initial high kill by the organophosphate has been countered by the more sustained killing power of the mycoinsecticide. The product also has a good toxicological and ecotoxicological profile. The acute oral toxicity LD_{50} in rats is >2000 mg kg^{-1} and the acute pulmonary LC_{50} in rats is >4850 mg m^{-3}. The fungal isolate is non-irritant and non-infective to mammals. An extensive range of ecotoxicological tests have indicated only low or moderate infectivity to some species of Isoptera, Coleoptera and some Hymenoptera. Hemiptera, Dictyoptera and other Coleoptera and Hymenoptera are not infected (Prior, 1997).

The mycoinsecticide Green Muscle provides an effective, environmentally friendly alternative to chemical control for the management of locusts and grasshoppers in Africa.

6.8.1 Biopesticide development

Commercial companies have rarely invested large sums of capital into the early stages of development of biopesticide products (Jenkins *et al.*, 1998). More often fundamental research and development up to the stage of small scale field testing is carried out by public funded research laboratories and universities (Bowers, 1982). Hence, the general knowledge base for biopesticides is built up in a haphazard way, through the uncoordinated efforts of many scientists all pursuing their own individual research objectives and interests. This contrasts markedly with the more focused factory-like screening and development process which characterizes agrochemical R&D that produces new chemical insecticides (Dent, 1997c). Private sector R&D tends to be carried out within a single organization involving multidisciplinary teams organized to address problems in a step-wise process of development. Public sector R&D tends to involve a single institute and individual scientists or small teams of insect pathologists working on related problems. Rarely will a team include experts from all the different disciplines required to develop a biopesticide, including exploration, identification and screening of pathogen isolates, ecology, production, storage, formulation and application. The majority of research organization involved in biopesticide development have in-house expertise in exploration, identification and screening of pathogens but tend to lack skills and the necessary facilities to undertake more complex storage, formulation and application studies (Harris and Dent, 1999). Investigations into host–pathogen interactions by ecologists are also rare.

The longest period of time in the life of a product elapses during storage (Jones and Burgess, 1998). During storage, biopesticides must remain viable, with minimum loss of potency, there should be no loss or breakdown of the desired formulation properties, e.g. clumping or caking of powders, flocculation and settling of suspensions, or breakdown of additives that

protect the pathogen against the environment. Production methods and packaging can affect storage and the active ingredient may need to be stabilized and purified to reduce contaminants. The product needs to be evaluated for its storage capability as shelf lives usually need to be in the order of one year (minimum). There is also a need to be able to predict product performance under the conditions that will be experienced in actual use. The use of models may assist in this process (Hong *et al.*, 1998).

Selection of the appropriate formulation for a biopesticide can improve product stability, viability and may reduce inconsistency in field performance (Burgess, 1998 for a review). Formulations are essential to protect against environmental extremes of moisture and temperature, as well as provide protection from UV light and desiccation (Boyetchko *et al.*, 1999). Sunlight blockers and UV-absorbing compounds can be added to formulations or starch encapsulation to increase survival and shelf-life (Shapiro and Argauer, 1995; Shapiro and Argauer, 1997). The formulation should also be appropriate for the application process – for instance ensuring that the particulate matter (e.g. fungal conidia) does not clump and block the spray nozzle.

Application technology for biopesticides is relatively under-developed (Dent, 1997c). The key requirement is to ensure that the biopesticide is sprayed in such a way that the droplet size is optimized for pick-up by the insect and that the amount of biological agent within each droplet is sufficient to ensure infection and subsequent mortality. However, in order to achieve this, application technologists consider formulation properties in the spray tank, atomization, transport to and impaction on the target surfaces, distribution of the deposit and subsequent environmental degradation, to biological effects (Chapple and Bateman, 1997; Bateman, 1999).

Studies of the ecology of the insect–pathogen interactions can provide valuable information that can be used to define a 'good use strategy' for the biopesticide. Information about insect feeding behaviour (Thomas *et al.*, 1997, 1998) and the thermoregulatory behaviour of the insect after infection (Blanford *et al.*, 1998; Blanford *et al.*, in press) and secondary cycling of the biocontrol agent (Thomas *et al.*, 1997) forms the basis for development of an appropriate use strategy for the biopesticide. The secondary cycling aspects of pathogen biological control agents set biopesticides apart from chemical insecticides and offer many opportunities for longer term benefits from biological control.

6.8.2 Production of pathogens
The main reason for *Bt* predominance in biocontrol is that it is easy to produce by simple fermentation techniques in relatively inexpensive media for controlling a broad range of hosts (Guillon, 1997). The scale-up from the laboratory to the industrial scale of pathogens remains one of the biggest constraints to the future success of biopesticides. Industrial scale processes exist for *Bt*, entomopathogenic nematodes, the fungi *Beauveria* spp., *Metarhizium* spp., *Paecilomyces* spp. and *Verticillium* spp. and for a range of nucleo-polyhedrosis and granulosis viruses (e.g. Guillon, 1997). The manufacture of a biopesticide involves producing sufficient quantities of the active ingredient at an economic cost, of consistent quality and compatible with the selected formulation and application equipment (Jenkins *et al.*, 1998). Production systems can be developed for research use (mainly laboratory studies and small scale field trials) but to scale up for operational trials or actual use, the production parameters need to be defined and optimized. This necessitates an understanding of the temperature, airflow, humidity and nutrient requirements for optimal production, drying and extraction processes. Commercially *B. thuringiensis* is produced using batch deep liquid fermentation techniques and it is regarded as an easy organism to produce. *B. sphaericus* has similar media requirements but pilot batches of a commercial product have

given variable results (Burgess, 1982). The ease with which the insecticide can be produced will obviously affect its commercial viability. One of the drawbacks of the production of *B. popilliae* insecticide has been the difficulties of large scale production since this bacterium can only be produced *in vivo*. The problem is exacerbated further as the beetle larvae are difficult to rear in the laboratory.

The problem with viruses is that they have to be produced *in vivo* which tends to be labour intensive and hence costly, although some cell culture work shows promise. Large scale rearing of Lepidoptera larvae is possible using artificial diet but artificial diets are not available for all insect pests which therefore have to be reared on host plant material. The high costs of these production techniques are partly offset by the high yield of inclusion bodies from each larva; approximately 25 larvae provided sufficient inclusion bodies to spray and protect 0.4 ha of cotton infested with *Helicoverpa zea* (Ignoffo and Couch, 1981). Guillon (1997) provides production costs and methods which demonstrate that such industrial systems can produce granulosis virus at an economically viable cost.

6.8.3 Use of biopesticides
There is now a considerable body of research results which testify to the potential of biopesticides, especially for bacteria, viruses, entomopathogenic nematodes and fungi. However, converting these results into viable products has proved a long and arduous task. Despite this, research has now reached a stage where the prospect for new products looks very promising. A number of technical breakthroughs such as ULV formulations and improved storage capability for fungi, are opening up new market opportunities (see Case Study), viruses are increasingly being evaluated and used on a wider scale (e.g. Bell and Hayes, 1994; Hayes and Bell, 1994), important niche markets are being created for nematodes (e.g. Gouge and Hague, 1993; Renn, 1995; Gouge *et al.*, 1997) and fungal

products are now more increasingly available for pests ranging from the European corn borer and whitefly (Guillon, 1997) to locusts and grasshoppers (Neethling and Dent, 1998). Overall, the market for biopesticides has grown c. 80% in four years (1991–1995; Georgis, 1997) and although a slower growth is anticipated by 2000 the market for biopesticides is expected to exceed £141 million (Wood MacKenzie Consultants Ltd, 1995).

6.9 Augmentation and Inoculation with Natural Enemies

In situations where natural enemies are absent or population levels are too low to be effective, numbers may be augmented by the release of laboratory reared insects. Where augmentative control is used, there is no interest in long-term stability but merely in the suppression of pest numbers below the economic threshold (van Lenteren, 1986) and multiple releases may be needed because the control is only temporary. With inoculative releases control will be seasonal or for the duration of the crop. The view that augmentation of natural enemies should be given the lowest priority in biological control research (DeBach, 1974) and that it should only be introduced where the level of control desired is so low that it may be impractical by any other method (DeBach and Hagen, 1964) is now totally invalid. A great deal of R&D that has been carried out over the last 20 years is currently yielding benefits with an increase in both the natural enemies commercially available for augmentative control and the number of commercial companies producing natural enemies for sale. In 1968, there were two commercial biological control companies in Europe whereas today there are 26 in Europe and a total of 64 companies worldwide (van Lenteren et al., 1997). At this time, there are about 100 biological control agents available with about 30 beneficial species making up 90% of the total sales (van Lenteren, 1997).

Augmentative control has been used successfully in over 14,000 ha of glasshouse crops for a number of insect pests (van Lenteren and Woets, 1988; van Lenteren, 1995). This successful use of augmentative releases in glasshouses has been due to a number of factors. Glasshouses represent closed or nearly closed systems in which one or only a few major pests exist, the glasshouse environment can be controlled and the crops grown are of high value and intensively produced. Resistance to some chemical insecticides has also promoted interest in the technique.

Natural enemies suitable for release in an augmentative release programme in glasshouses and other systems should possess a number of attributes. The natural enemy development should be synchronous with the pest, so that the adult natural enemy is available when suitable host stages are present (although multiple releases can alleviate this problem). The natural enemy must reproduce, develop and migrate under the conditions of release (van Lenteren, 1986). Further desirable attributes include ease of culture, preference for pest species if a non-pest alternative is present, a potential maximum reproductive rate equal to or greater than that of the pest, and a good response to pest density. The natural enemy should also not attack other beneficials.

The two most celebrated examples of successful augmentative release of natural enemies in glasshouses are those of Encarsia formosa against the whitefly Trialeurodes vaporariorum and the use of the phytoseiid mite Phytoseiulus persimilis to control the two spotted spider mite Tetranychus urticae. P. persimilis is the most widely available species of all marketed natural enemies in a survey carried out by van Lenteren et al. (1997). It is produced by nine companies in even larger quantities than E. formosa. P. persimilis is widely used in control of T. urticae attacking tomato, cucumber, egg plants (aubergines), sweet peppers and gerbera crops (van Lenteren, 1995).

In field crops, the species of natural enemy that has been used more than any other are the egg parasitoids of the genus *Trichogramma*. Today, mass release of *Trichogramma* has achieved significant results especially in cotton in the USA, Mexico and the former USSR and in India for the control of cotton bollworms and sugarcane borers (King *et al.*, 1985; Manjunath, 1998). It is the costs of producing the natural enemies and releasing them at the required intervals that limits the usefulness of the augmentative approach. The situation with inoculative releases is not as marked because fewer releases are necessary. Also the rearing of insects for augmentative releases needs to be a commercially viable proposition, whereas with inoculative releases, particularly those aimed at controlling the spread of a pest species (e.g. *Dendoctronus micans* in Europe), the funding is provided by governmental organizations.

6.9.1 Mass rearing of natural enemies for release

The augmentation approach is restricted to natural enemies that can be mass reared and for which suitable storage and packaging methods exist or can be readily developed. Most natural enemies are reared on their natural hosts which means the development of a triple phase programme where (i) the host plant is produced for (ii) the pest insects, and (iii) the natural enemy is then propagated onto the pest insect. Such programmes can be quite involved and pose a great challenge to the insectary specialist (van den Bosch and Messenger, 1973). For several natural enemies, mass rearing on the natural host is either too expensive or impossible because of the risk of contamination with pest organisms or with other pests or diseases. In the circumstances, artificial host media or unnatural hosts may provide the only alternative (van Lenteren, 1986). Artificial diets have the advantage that they probably reduce costs of production, whereas an unnatural host is not normally attacked by the natural enemy because of some isolating mechanism, but it can serve as a host in the insectary (Morrison and King, 1977). The

disadvantage of using unnatural hosts (and artificial media) is the possibility that the natural enemy's host preference will be altered, rendering it useless for augmentation purposes, although experiments to test this phenomenon have failed so far to show any change in effectiveness when released against the natural host.

When a mass rearing unit is established the initial stock should not be less than 1000 individuals and should consist of genetically diverse material (van Lenteren and Woets, 1988). This high level of initial stock material is important to reduce the 'founder effect' where the genetic makeup of the isolated population represents only a small fraction of the original genetic variability present in the parent population (Mackauer, 1972). The larger the founder number and the range of environments/areas from which the insects were collected, the greater the reduction in the founder effect and the more genetically diverse and representative the insectary population. Once a colony has been established, it is very important to ensure that inbreeding is kept to a minimum and where possible to maintain random mating between family lines. The selection for insects suited for insectary condition can be prevented, or at least reduced, if the insectary environment is similar to natural conditions. However, there is a tendency for insectaries to be maintained at constant temperature and humidities with a fixed photoperiod and light intensity. Not surprisingly, these conditions are incompatible with rearing insects representative of wild populations. Other conflicts between rearing and release requirements in insects for glasshouses are listed in Table 6.9. The role of insect behaviour and its effects under selection in the insectary is also a major factor influencing the quality of the reared insect that is not normally taken into account (Boller, 1972). Quality in insectaries is normally divided into four components: adaptability; host selection; sexual activity; mobility. Most routine measurements of traits are production orientated, e.g. percentage parasitism,

Table 6.9. Comparison of conditions for mass rearing of parasitoids and those of the release environment in the glasshouse (van Lenteren and Woets, 1988).

Aspect	Mass rearing	Greenhouse
Host plant	Usually different	
Host	Usually the same	
Host density	High	Low
Dispersal	Not appreciated	Essential
Ability to disperse	Small	Large
Population size	Minimal size sufficiently large	
Temperature	Temperature higher in mass rearing	
Humidity	Usually the same	
Pesticides	Not applied	Applied

emergence success, egg production, while traits more directly related to behavioural performance such a diurnal rhythmicity, flight propensity and genetic variation are rarely measured (King *et al.*, 1985). There is perhaps a need to move away from production orientated evaluation of efficiency, in terms of number of insects produced per monetary unit, to a production/quality approach where the success of the project is evaluated in terms of the relations of responses between natural insects and reared insects to a given objective (Boller, 1972).

One of the other key factors in the success of mass rearing of natural enemies for augmentation is the ability to store the insects for relatively long periods; this is required to allow some kind of buffer for changes in demand. If storage were possible, continual rather than seasonal labour could be used and reserve quantities of insects used to counteract periods of low production or high demand (Morrison and King, 1977). Low temperatures that reduce development rates have been the most common method used to facilitate storage but often not without some loss of insect vigour or viability (Deng *et al.*, 1987; Gilkeson, 1990). Continued research is needed in this area.

The packaging and shipping of natural enemies it also not without its problems

(e.g. Morewood, 1992). With augmentative releases in glasshouses, the frequency of dead or injured natural enemies is a serious problem and various types of packaging have been used, from cut leaves in boxes to insect coated sticky pads, or loose insects in sealed plastic containers. The least vulnerable stages such as eggs and pupae are normally most appropriate but even these are subject to large losses and packaging and distribution problems remain one of the main restrictions to the application of this method of biological control.

For further information on mass rearing techniques for natural enemies in augmentative release programmes readers should refer to Morrison and King (1977), Laing and Eden (1990) and Gilkeson (1992).

6.9.2 Costs and benefits

The cost of reared natural enemies must be judged in terms of the value of the crop protected by using the agent and in comparison to the cost of competing pest control options such as chemicals (van Driesche and Bellows, 1996). Use of *T. cryptophebiae* against citrus pests in South Africa is economically viable even though the level of pest control is less than 60% (Newton and Odendaal, 1990). This is because the chemicals used provide only similar levels of control and are at least

Table 6.10. Average price (and range; in US$) for the most commonly sold organisms used to control pests in Europe (from van Lenteren *et al.*, 1997)

Biological control organisms	No. of companies	Species life stage shipped	Average price per unit (range)
Amblyseius californicus	3	Mixed life stages	$0.031*
Amblyseius cucumeris	5	Mixed life stages	$0.0094 ($0.00032–$0.0019)
Amblyseius degenerans	5	Mixed life stages	$0.1 ($0.065–$0.147)
Aphelinus abdominalis	3	Adult	$0.98 ($0.092–$0.104)
Aphidius colemani	5	Mummy	$0.035 ($0.024–$0.049)
Aphidius ervi	3	Mummy	$0.13*
Aphidoletes aphidimyza	4	Pupa	$0.0318 ($0.0243–$0.0485)
Chrysoperla carnea	2	Egg	$0.0209*
Cryptoleamus montrouzieri	5	Adult	$0.472 ($0.362–$0.588)
Dacnusa sibirica	3	Adult	$0.026*
Delphastus pusillus	3	Adult	$0.706*
Diglyphus isaea	5	Adult	$0.204 ($0.065–$0.349)
Encarsia formosa	7	Pupa	$0.0081 ($0.0028–$0.016)
Eretmocerus californicus	3	Pupa	$0.010 ($0.0065–$0.013)
Harmonia axyridis	2	Adult	Price not known
Hypoaspis miles	4	Mixed life stages	$0.00097 ($0.00058–$0.0014)
Leptomastidea abnormis	3	Adult	$0.263 ($0.141–$0.436)
Leptomastix dactylopii	5	Adult	$0.246 ($0.141–$0.436)
Leptomastix epona	4	Adult	$0.343 ($0.165–$0.441)
Macrolophus caliginosus	5	Adult	$0.179 ($0.129–$0.226)
Orius insidiosus	4	Adult	$0.0404*
Orius laevigatus	5	Adult	$0.0725 ($0.0404–$0.105)
Orius majusculus	5	Adult	$0.0865 ($0.0404–$0.114)
Phytoseiulus persimilis	9	Mixed life stages	$0.1013 ($0.00647–$0.17647)
Trichogramma brassicae	2	Parasitized host egg	$0.0003* (per unit) $102.79* (per ha)
Trichogramma evanescens	2	Parasitized host egg	$60.35* (per ha)

* Price given by one producer.

twice the cost. In Europe, the costs of biological control agents for use mainly in protected crops and horticulture have proved economic (Table 6.10; van Lenteren *et al.*, 1997) and compare favourably to costs of chemical insecticides (van Lenteren, 1989). Augmentative control seems to have proved viable in high value horticultural or agricultural crops and especially where there is a premium for chemical free produce.

Case Study: Augmentative releases of *Encarsia formosa* to control the greenhouse whitefly *Trialeurodes vaporariorum*

Encarsia formosa is an aphelinid parasitoid of the greenhouse whitefly *Trialeurodes vaporariorum* which is used extensively in greenhouses for augmentative control in tomato, sweet pepper and cucumber in both the UK and Netherlands (King *et al.*, 1985). *E. formosa* is a parthenogenic species and each female can oviposit an average of 50 eggs (Parr *et al.*, 1976). These are oviposited within the third and fourth instar *T. vaporariorum* and the pre-pupa (Vet *et al.*, 1980). The adult feeds on the host's body fluids as well as honeydew.

A prerequisite for any augmentative release programme is the ability to produce the natural enemy in sufficient numbers at economic costs. The economic
feasibility of mass rearing *E. formosa* to control the whitefly in cucumber crops
in the UK was calculated first by Scopes (1969). The methods used were further
expounded by Scopes and Biggerstaff (1971) (Fig. 6.14) and have since been
developed commercially for 20 years (van Lenteren, 1992a). *Nicotiana tabacum*
is used exclusively for the production of both the pest and parasitoid with a production cycle of between 36 and 40 days for multiplication and distribution to
nurseries of the adult parasitoid. During the summer when the adult whiteflies
are most abundant, the production of honeydew and subsequent development of
sooty mould can be a problem. Spraying plants with water twice weekly removes
much of the honeydew.

The damage that the whitefly cause to the crop plants can be direct physical
damage or indirect damage through the accumulation of honeydew, resulting in
the development of sooty moulds (*Cladosporium sphaerospermum*) which
reduce photosynthesis and respiration (Vet *et al.*, 1980). The economic damage
threshold is represented by the number of whitefly that produce the quantity of
honeydew that is insufficient to produce damaging levels of sooty mould
(Hussey and Bravenboer, 1971). The release of *E. formosa* to control *T. vaporariorum* is based on one of two approaches, either the 'classical method' or the 'dribble method'.

The classical method involves the introduction of the whitefly into the crop to
ensure an even distribution and level of infestation and then subsequently *E. formosa* is introduced at rates totalling 120,000 parasitoids per hectare at the most
appropriate time, dependent on the rate of scale development and temperature
(Parr *et al.*, 1976). The dribble method involves the introduction of *E. formosa*
after whitefly have been observed in the crop and multiple introductions at set
intervals after this. The classical method was found to give more predictable

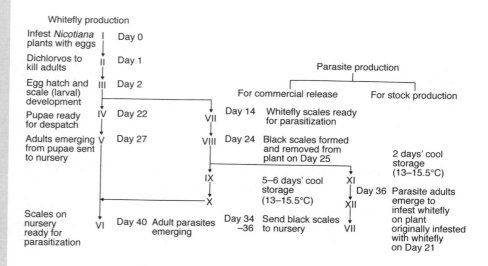

Fig. 6.14. A system for producing both whitefly and its parasite (after Scopes and Biggerstaff, 1971).

results in both cucumbers and tomato crops (Gould *et al.*, 1975; Parr *et al.*, 1976) but the establishment of whitefly in the crop prior to parasitoid release was not acceptable to growers. Hence, the dribble method has been the most widely adopted technique. The use of *E. formosa* to control the greenhouse whitefly has proved to be an outstanding success so that now a number of companies exist to supply commercial growers throughout the season with the required numbers.

Case Study: The use of inoculative releases of *Rhizophagus grandis* to control the greater European spruce beetle (*Dendroctonus micans*) (Grégoire *et al.*, 1990)

Dendroctonus micans inhabits the Eurasian conifer forests from eastern Siberia to central France and the UK. The pest is still extending its distribution with an estimated 200,000 ha of forest suffering from outbreaks in invaded areas in the UK, France, USSR and Turkey. *D. micans* invaded the French Massif Central during the 1970s and is still spreading from the north to the south-west. The inoculations reported here were made in this area. *Rhizophagus grandis* was considered a suitable candidate for inoculative releases to control *D. micans* for a number of reasons. Firstly, *R. grandis* is very common and abundant in the inner part of the bark beetles' range and was considered to be responsible for maintaining low population levels. Secondly, field and laboratory work had shown that *R. grandis* had characters which indicated its potential value as a control agent, an extraordinary capacity to locate its prey, a flexible phenology, a high fecundity, matching that of *D. micans* larva during their development, and pairs of *R. grandis* and their broods can reduce a brood of *D. micans* by at least two-thirds.

The rearing method for *R. grandis* was refined from the use of *D. micans* infested logs to the use of cylindrical clear polystyrene boxes containing *D. micans* and rehydrated spruce bark powder. Under room temperatures, 1.5–3.5 young adult *R. grandis* could be produced per *D. micans* larva with a generation time of 60–80 days; 30–70 young adults of *R. grandis* were produced from each adult female. Some 84,000 predators have been produced since 1983.

The criteria for release of the predators were 50 pairs of adult beetles at the base of each attacked tree if the site contained less than ten attacked trees per hectare, or at least 500–1000 pairs per site if there were more than ten trees attacked per hectare. A total of 2350 *R. grandis* was released in 1983, 8500 in 1984, 16,350 in 1985 and 41,800 in 1986 in about 50 sites over an area of 5000 km². Six release sites have been surveyed to assess the establishment of the predator and *R. grandis* has been found in all of them. The lowest colonization after one year was 14.4% of brood systems but generally there was an increase in colonization each year. In one area, the colonization rate progressed from 17% in 1983 to 72% of the brood systems colonized in 1986. The maximum colonization rate over all sites was 75%, which provides a substantial level of control. *D. micans* is still progressing south-westwards and so further mass rearing and release of *R. grandis* are required, but the inoculative method has proved successful in keeping control of the pest within acceptable limits.

6.10 Conservation Biological Control

Conservation biological control attempts to make use of indigenous natural enemies by manipulating the environment in such a way that their activity and effectiveness as biocontrol agents is enhanced. The premise on which conservation biological control is based is that the intensification of cropping systems has reduced the effectiveness of indigenous natural enemy populations and that this can be compensated for by appropriate manipulation of the habitat. The process of intensification in Western Europe, for instance, led to the removal of hedges to produce larger fields (Davies and Dunford, 1962; Edwards, 1970), a process that continues today (Greaves and Marshall, 1987). In addition, the use of herbicides to control weeds in the remaining hedgerows has reduced their 'quality' as field boundaries in terms of natural enemy conservation. Conservation biological control, to a large extent, is involved with the management of the field habitat to recreate the essential aspects required to maintain the effectiveness of natural enemies without disrupting the efficiency or profitability of farm operations. Habitat manipulation techniques involve creating habitat refuges such as headlands, hedgerows or grass-banks (e.g. Thomas *et al.*, 1991 – see Case Study) which serve as recolonization foci or providing key resources such as food or nectar which attract natural enemies to their site of action (Lövei *et al.*, 1993). Such relatively small discrete features such as strips of food plants and overwintering shelter may be thought of as micro-ecotones, zones where, through the provision of the 'right kinds' of physical and floral diversity, pest suppression can be achieved (Gurr *et al.*, 1998).

6.10.1 Habitat refuges

Invertebrate predators and parasitoids in arable land occupy an unstable, fragmented and often hostile environment. Without an ability to leave an arable field in the autumn after harvest for example, many species will be exposed to the extremes of burning, ploughing, harrowing, drilling and autumn applied pesticides in a habitat low in food resources and in its range of available microclimates. Hence, insect natural enemies require both refuges from such perturbations and reservoirs from which decimated populations can be replenished. Alternative ecotones, which can substitute for lack of hedgerows or poor quality hedgerows that remain, revolve around management of headlands, existing hedgerows or provision of beetle banks. Studies have shown that higher winter densities of predators are found in field boundaries with deeper soil, greater vegetation height, having an east–west orientation, lower soil moisture and warmer mean daytime temperatures (Dennis *et al.*, 1994). Higher overwintering survival of natural enemies provides a greater number of individuals that can colonize the crop in the spring and raises the natural enemy : pest ratio in favour of the natural enemies. The creation of linear island habitats in the centre of cereal fields, and thus reduction of the field size (Thomas *et al.*, 1991), has been shown to produce in two years densities of polyphagous predators up to 1500 m^{-2} which greatly exceeds densities previously recorded in field margins around the UK (Thomas *et al.*, 1992). In the USA, uncultivated 'corridors' of grasses and other herbaceous plants within soybean fields reduced populations of *Empoasca fabae* (Rodenhouse *et al.*, 1992). Predator populations were higher in the corridors than in the crop itself. Hence the effectiveness of predators such as carabids depends not only on their ability to penetrate fields from overwinter sites in field boundaries but also on movement on a larger spatial scale, from one field to another (Frampton *et al.*, 1995). The rates of movement of predators between fields will in some cases be impeded by field boundaries but the degree to which they act in this negative way is not clear and it is likely that throughout the year the relative importance of their roles may change.

Case Study: Beetle banks – as overwintering habitats (Wratten *et al.*, 1990; Thomas *et al.*, 1991, 1992)

Beetle banks are grass-sown (e.g. *Dactylus glomerata*) earth ridges at the centre of cereal fields which recreate those aspects of existing field boundaries that favour high overwintering densities of polyphagous predators. Destructure sampling of these strips of grass during their second and third winters following establishment revealed high predator densities. Two highly ranked predators of cereal aphids were particularly prevalent, *Dometrias atricapillus* and *Tachyporus hypnorum*. Further studies indicated food supply to be important for both predator species during the winter period and that the grass banks provided temperature stabilizing effects throughout winter. The raised earth banks which can be 0.4 m high, 1.5 m wide and 290 m long were first shown to be economic in 1988 (Wratten, 1988). Figures updated in 1994 would also indicate that the costs of establishment (including grass and labour costs) and loss of cereal crop due to the beetle banks are outweighed by the savings on pesticides if the enhanced natural enemies prevent a cereal aphid outbreak. The combination of labour costs for bank establishment (1–2 days) with yield loss due to land taken out of cereal production (*c.* £30 or US$50 assuming the average yield of 6 tonnes ha^{-1} @ £110 or $180 tonne^{-1}), together with the cost of grass seed (£5 or $8), would amount to approximately £85 or $140 in the first year for a 20 ha field of winter wheat. There would be no establishment costs in subsequent years only gross yield lost due to the area occupied by the bank (£30 or $50 per year). An aphid outbreak kept below a spray threshold by the action of the natural enemies from the beetle bank could save £300 or $495 per annum in labour and pesticide costs for a 20 ha field. Alternatively, prevention of an aphid induced yield loss of 5% could save £660 ($1090) for a field the same size (Wratten and van Emden, 1995).

6.10.2 Food sources

Several species of parasitoid wasps and hoverflies require nectar and/or pollen as food to provide energy and protein for egg maturation (Schneider, 1948; Jervis and Kidd, 1986). On the basis of this, it is reasonable to assume that an increased availability of suitable flowers in or near crops can increase natural enemy effectiveness (Powell, 1986). Hence, a number of studies have tried to identify suitable flowering plants to include in field boundaries to attract parasitoids and predators (Zandstra and Motooka, 1978; Lövei *et al.*, 1993; Hickman *et al.*, 1995; Cowgill *et al.*, 1993). A number have been able to establish that the presence of suitable flowering plants has increased the incidence of natural enemies (Molthan and Rupert, 1988; von Klinger, 1987). Few, however, demonstrate subsequent reduction in pest populations. Numbers of the bean aphid *Aphis fabae* in

sugar beet plots drilled with tansy leaf (*Phacelia tamacetifolia*) between rows or at plot edges were reduced by syrphids compared with populations in control plots (Sengonca and Frings, 1989).

Manipulation of habitats can improve the availability of alternative hosts or prey for important natural enemies. Natural enemies that are not host or prey-specific will search for alternative food or hosts in the absence of the pest prey or host species. If these alternative sources of prey are not found in the vicinity of the cropping system then there will be a tendency for the natural enemies to disperse away from the crop. This may mean that when the pest population in the crop starts to increase, the natural enemies that could potentially reduce the rate of growth are not within the crop vicinity and a delayed return, perhaps later in the season, may be too late for them to significantly affect the pest population. The presence of a secondary prey source has been shown to be important in a number of cases. Anthrocorid bugs (*Orius majusculus*) which are used for the control of western flower thrips (*Frankliniella occidentalis*) have been shown to benefit from the presence of alternative prey species in situations where thrips are scarce (Brodsgaard and Enkegaard, 1997). In rice systems, the egg parasitoids *Anagrus* spp. and *Oligosita* spp., which are important control agents for rice planthoppers, use some delphacids as alternative hosts. The delphacids survive in non-rice habitats such as field bunds, irrigation canals and roadsides (Yu *et al.*, 1996). Also in rice systems, Settle *et al.* (1996) hypothesized that the abundance of alterna-

tive detritivore prey for generalist predators give these natural enemies a 'head start' on later developing pest populations leading to a stability in rice ecosystems by decoupling predator populations from a strict dependence on herbivore populations. The early appearance of the parasitoid *Anagrus epos* of the grape feeding leafhoppers *Erythromeura* spp. was explained by companion planted prune trees being a refuge for an alternative host for the parasite, the prune leafhopper *Edwardsiana prunicola* (Pickett *et al.*, 1990). There are probably many such examples of such relationships; the difficulty is identifying them and then being able to manipulate the habitat in such a way as to make effective their use.

Case Study: Impact of wild flowers in orchards (Leius, 1967a,b)

Direct observation of the parasitoid *Scambus buoliariae* in paired flower tests showed a preference for wild parsnip *Pastinaca sativa* when offered along with wild carrot *Daucus carota*, buttercup *Ranunculus acris* and water hemlock *Cicuta maculata* (Fig. 6.15) – plants all found in areas where the parasitoids are present (Leius, 1967b). In general, it would seem that the flowers of the Umbelliferae are better sources of nectar for adult parasitic Hymenoptera than

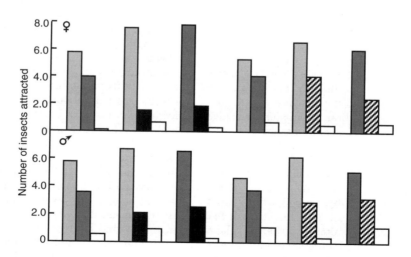

Fig. 6.15. The number of *Scambus buoliana* alighting and feeding on four different kinds of flowers in paired comparisons (▨ wild parsnip, ▰ wild carrot, ■ buttercup, ▨ water hemlock, ▢ no response) (after Leius, 1967b).

most other plants because the structure of the individual flowers is amenable to feeding with biting mouthparts. Proportionally about 18 times as many test caterpillar pupae were parasitized in orchards with rich undergrowths of wild flowers as in orchards with poor floral undergrowths (Table 6.11; Leius, 1967a).

Table 6.11. Percentage parasitism of tent caterpillar pupae and codling moth in apple orchards having rich, average and poor floral undergrowths (Leius, 1967a).

Type of floral undergrowth	Tent caterpillar larvae		Codling moth	
	% parasitism	ratio	% parasitism	ratio
Rich	65.2	18	33.6	5
Average	19.7	5	18.2	3
Poor	3.7	1	7.1	1

6.11 Discussion

The control that can be exerted over pest insect species by their natural enemies needs to be harnessed and used to its maximum potential in any insect pest management programme. The way in which the natural enemies are used, i.e. introductions, inundation, augmentation, inoculation or conservation, will depend on the characteristics of the pest and the cropping system in which it causes damage or yield loss. As has been described, more stable systems lend themselves best to use of natural enemies, but augmentations can be used in less stable systems and even in highly intensive systems such as glasshouses, provided that the approach is economically feasible. Techniques that conserve natural enemies are already part of traditional mixed subsistence farming methods but more needs to be done to promote natural enemy conservation in more intensive cropping systems. This would require a greater interest by scientists and investment of funds for research in this subject than is currently apparent. Research into techniques for conservation of natural enemies has its dedicated advocates but as a subject it lacks the glamour and prestige that might be associated with control through introductions and augmentations. The problems associated with implementation and adoption of conservation techniques can also be formidable, hardly an incentive to budding applied scientists. However, a greater emphasis needs to be placed on the conservation of natural enemies and work on this given a higher priority, especially because of its relevance to insect pest management in developing countries.

Biological control was superseded earlier this century by the development of and increased use of chemical insecticides so that research into biological control decreased markedly. Although biological control has once more gained its due recognition as an extremely valuable form of control, especially within the context of the developing philosophy of IPM, on the whole chemical and biological control are incompatible techniques, mainly because so many insecticides are detrimental to natural enemies. There have been moves to evaluate the effects of insecticides on beneficial insects and a general commitment to recommend use of insecticides less toxic to natural enemies or to apply insecticides at times when natural enemies are less at risk. However, in practice, natural enemy populations are consistently being annihilated by insecticide use.

An increasingly appropriate alternative to chemicals is now being provided by a whole gamut of biological control agents formulated as biopesticides. A greater emphasis on rational, more selective chemical insecticide use, combined with biopesticides, should provide an effective means of preserving natural enemy popula- tions and gaining maximum impact from biocontrol agents in integrated programmes of pest management.

7

Cultural and Interference Methods

7.1 Introduction

Cultural control of insect pests is affected by the manipulation of the environment in such a way as to render it unfavourable for the pest. Many of the methods interfere with the pests' ability to colonize a crop, promoting dispersal, reducing reproduction or survival. This may be achieved through techniques such as crop rotation, hermetic storage systems, intercropping, manipulation of planting dates and management of field margins. Interference methods include the use of semiochemicals to disrupt insect communication, preventing colonization or mating, and the insect sterile technique which reduces the number of females that can successfully reproduce through interference with male fertility.

Like host plant resistance, cultural control is a prophylactic method of control. It can rarely be used as a tactical means of control except perhaps in stored product systems where spot check 'fumigation' can be achieved by modification of the storage environment. Cultural control should be considered as the first-ditch defence around which to build other control options (Coaker, 1987). On its own, it is unlikely to reduce pest infestation to below economic damage thresholds. Since the levels of control obtained with cultural control are less dramatic than those achieved from the use of other techniques,

such as chemical insecticides, and as a great deal of pest control research has been aimed at developing products rather than techniques, there has been a tendency among research workers to neglect cultural control. For instance, van Lenteren (1987) noted that in the study of environmental manipulation for the purpose of improving natural enemy efficacy, the only area that had received attention appeared to be behaviour modifying chemicals; a typical example of the emphasis placed on the development of products (Dent, 1993). Of all the control techniques, cultural control is rarely dependent on a particular product and as a consequence of this the approach offers few opportunities for capital return and few potential commercial markets. Hence cultural control has attracted little interest and funding for research. Cultural control is largely dependent on techniques and practices and since it is difficult to sell someone an idea or an approach, it is left to the universities and government research institutes rather than industry to carry out the necessary research. Here though, cultural control has suffered in that the techniques on their own are rarely capable of exerting the required levels of control and there is no glamour or prestige associated with control options that only partially solve problems. The panacea mentality has not really been subsumed by the concept of IPM and cultural control techniques remain one of the casualties of this approach (Section 10.3). Cultural control represents an important component in the development of a coherent, holistic approach to pest management and warrants a great deal more attention than it is currently receiving.

Perhaps in contrast to this, the sterile insect technique and the application of semiochemicals have always promised more than they have actually been able to deliver as control measures. A great deal of research has been conducted to identify potentially useful semiochemicals, a large proportion of which has involved the study of lepidopteran sex pheromones. From this have developed a range of monitoring devices that detect pest presence (but rarely provide more quantitative monitoring or forecast information) and a number of area wide mating disruption programmes. The sterile insect technique was developed as an area wide approach but has lacked a low cost and simple means of achieving insect sterility and delivery systems for it to be truly widely applicable. However, both these interference methods have over the last 10 years moved in a direction that can address previous failings. For insect sterilization, the development of chemo-autosterilization techniques (e.g. Wall and Howard, 1994) paves the way to a cheap and effective technique with widespread applicability, and for semiochemicals, the prospect for more integrated use in IPM is improved by new strategies defining a more coherent framework for their development and use (see below) (Smart *et al.*, 1997).

7.2 Approaches and Objectives

Many of the problems involved in understanding the underlying processes and mechanisms of cultural control are fundamental to the ecology of insect/host interactions and thus are of immense interest to some ecologists: for instance, the role of insect dispersal in the colonization and exploitation of crop plants. It is this interest in ecological principles that has produced some of the more illuminating work on the processes involved in cultural control. Knowledge and understanding of these processes will enable entomologists to predict more accurately the potential value of similar control techniques in other unstudied systems. For the large part, however, research into cultural control revolves around the evaluation of various combinations of practices and techniques for specific crop systems. These studies make little or no attempt to identify the processes that account for the differences between treatments. While there is a need for both approaches, it would be difficult, and often impossible, to predict the wider value of new techniques and practices using the latter approach. There has been a tendency for a greater emphasis to be placed on limited evaluations of treatments rather than on experiments to evaluate processes and as a result, cultural control is often considered to include a mismatch of techniques that can have no cohesive relations. However, over recent years more has been done to develop an integrated philosophy based on a greater knowledge of the host, crop/pest biology and their underlying ecological principles, providing a framework for a more integrated approach in the future.

The approaches that are currently influencing the development of semiochemical use are referred to as 'push-pull' or 'stimulodeterrent diversionary' strategies (SDDS) (Smart *et al.*, 1997; Miller and Cowles, 1990). These are perhaps best explained through example. Semiochemicals can be used in a variety of different ways to control pests: the harvestable crop can be protected by means of repellents, antifeedants or egg-laying deterrents ('push'), and parasitoids and predators are attracted in to mop up pests undeterred by the protectants ('pull'). At the same time, aggregative semiochemicals, including host plant attractants and pheromones, stimulate pests to enter traps or to colonize a trap crop ('pull') where selective chemicals or biopesticides are applied (Smart *et al.*, 1997). Such strategies provide a framework for a systems approach to semiochemical R&D that promises much more than that based on an isolated independent development of semiochemical products. The opposite is almost true for the sterile insect technique SIT, however, where having identified an approach (chemo-autosterilization) that has general applicability, target/sterilization systems need to be developed on a case by case basis.

7.3 Condition of the Host

The ability of a host to withstand pest infestation may be due as much to its general state or condition as to any inherent resistance. A healthy and fit animal or plant can have a greater tolerance of pest attack than an unhealthy one. The immunological response of humans and animals can be impaired by any number of factors that can indirectly affect their susceptibility to pest attack. The immunological response of cattle to the ticks *Boophilus microplus* is affected among other things by stress, photoperiod and infestation with other parasites, while their resistance to the cattle grub *Hypoderma lineatum* can be impaired by a vitamin A deficiency (Drummond *et al.*, 1988). In human, animal and plant systems such factors may be influenced by the use of cultural control techniques. These are techniques that can be used to improve the condition of the host and thereby make it more tolerant of pest attack. Of course, the converse is also possible, since crop plants provided with adequate nutrition and water availability may prove to be more attractive to pests, the populations of which may increase to even greater levels on healthy crops. In these circumstances, the benefits of increased yields from, for instance, the use of fertilizers and improved water availability, must be balanced against the increased size of the pest population (van Emden *et al.*, 1969). For example, inorganic fertilizer applied to tomatoes was related to higher incidence of aphids compared with tomatoes treated with organic fertilizer

(Edwards *et al.*, 1996) and high densities of the cereal aphids *M. dirhodum* and *S. avenae* were associated with plots that received the higher spring application of fertilizer (Gash *et al.*, 1996). However, there are no hard and fast rules. Coaker (1987) states that irrigation, mulching, manuring and fertilization are practices that in general promote rapid growth and shorten the time the susceptible plant stage is available for attack, providing the crop with greater tolerance and the opportunity to compensate for insect damage. However, such practices can also change the physiology of the plant as food for insect pest species, making it more lush and thereby enhancing pest survival (Coaker, 1987). The value of these and other cultural practices must at present be assessed on an individual crop/pest basis since there is little current information on which general conclusions can be based.

Pest outbreaks may occur because crop hosts suffer environmental stress (Brodbeck and Strong, 1987; Mattson and Haack, 1987), the most devastating of which can be drought stress caused by high temperatures and decreased water availability. Under these conditions virtually every plant process is affected. Water stress induces changes in trichome size and density, their thickness and waxiness (as a means of increasing reflectance and decreasing transpirational water loss), the plants are generally warmer due to less transpirational cooling (Mattson and Haack, 1987) and have increased levels of soluble carbohydrates and amino acids present in the leaves (Wheatley *et al.*, 1989). Depending on the pest, these changes may or many not influence their ability to survive, increase reproduction and development rates. In water stressed groundnut, it has been shown that the leaf miner *Aproaerema modicella* was most abundant on the plants suffering the greatest stress and where leaf surface temperatures were highest, whereas the cicadellid *Empoasca kerrii* tended to concentrate where there was no drought stress and where leaf temperatures were lowest (Wheatley *et al.*,

1989). Hence, in situations where irrigation was introduced to reduce the effects of drought, it would be expected to have differential effects on the two pest species.

The physiological condition of plants can also be influenced by low temperatures. For instance, strawberry cultivars are sensitive to environmental factors and their physiological responses are determined in large part by vernalization or chilling (Bringhurst and Galleta, 1990). The chilling history of plants has been shown to affect their subsequent growth and pattern of resource partitioning (Gutridge, 1958) and pest abundance (Walsh *et al.*, 1997). Chilled strawberry plants show increased vegetative vigour and reduced susceptibility to two spotted spider mites *Tetranychus urticae* (Walsh *et al.*, 1997). Chilling cabbage has also been shown to affect the abundance and distribution of *Bemisia argentifolii* (Zalom *et al.*, 1996). More detailed studies are required on the effects of such factors as irrigation, fertilization and vernalization which can have a major influence on the condition of crop hosts and thus on the levels of pest infestation likely to occur. Often the introduction of these practices is studied purely from the point of view of agronomy and the effects they have on crop yield. Perhaps the research should be extended to also evaluate their concomitant effects on pest infestation, just to ensure that the problem is being fully solved rather than another being created.

7.4 Modifying the Physical Environment

The physical environment can have a marked effect on the development of insect pest populations and in some circumstances it may be possible to manipulate or modify it to obtain a required level of control. The term physical environment refers to any abiotic variable that may affect the insect in any way and includes temperature (e.g. Yokoyama and Miller, 1996), humidity (e.g. Santoso *et al.*, 1996), light

intensity, atmospheric conditions (e.g. Hodges and Surendro, 1996), and soil composition and structure (e.g. Blackshaw and Thompson, 1993; Leather, 1993). The range of conditions and situations in which the physical environment plays a role in diminishing or increasing pest numbers is wide but the number of situations in which the environment is under the control of man and hence can be actively used as a means of cultural control is more limited. Irrigation of field crops is one obvious example of how a crop environment may be physically changed according to man's wishes and there are situations where irrigation has been shown to influence pest population development (Tabashnik and Mau, 1986; Wheatley *et al.*, 1989). Another obvious example, this time from medical entomology, is the modification of mosquito breeding grounds by drainage, as a control for malaria. It is stored product systems, however, in which the modification of the physical environment has had its most important role in the control of pests.

Stored product systems are effectively simple man-made habitats in which temperature and other environmental variables may be ideal for pests but in which, in many situations, the conditions can also be controlled by man. Each system is easy to understand and can often be changed according to need, especially with regard to physical factors. There are, however, almost as many different systems as there are stored and processed materials which makes it difficult to develop wide ranging coherent IPM strategies (Benz, 1987). Strategies have to be developed according to each individual circumstance (a common feature among cultural control techniques) although the physical factors that are used or modified to control pests tend to be restricted to temperature, humidity or effects on the interstitial atmosphere.

There are about 20 economically important insect pests of stored products and each species has a characteristic range of limiting and optimal temperatures and humidities (Evans, 1987). Generally, insect development is possible in storage systems at temperatures within the range of 10–40°C but tends to be optimal at 25–30°C, while access to free water is generally not important (Evans, 1987). If temperature and humidity are to be modified so that conditions are unfavourable for the pest, it is first necessary to determine what range of temperatures and humidities are detrimental to each species. The range of temperatures and humidities that favour development of the grain mite *Acarus siro* are shown in Fig. 7.1. Irrespective of temperature, the mite cannot develop when the relative humidity falls below 60% and although some adults may survive for limited periods, mortality is usually rapid, few eggs are laid and none subsequently hatch (Cunnington, 1976). *A. siro* does, however, show a high tolerance to low and high temperatures provided humidity is kept high (Fig. 7.1). To ensure control of the pest, temperatures/humidities must be increased or decreased either above or below the tolerance limits of the insect or to levels that, if they do not kill the pest, will reduce its rate of development to an acceptable level. For insect control, the aim is usually to reduce temperatures to 15–17°C although

stored product pests from tropical climates may be inactivated even at relatively high temperatures. Beetles such as *Tribolium castaneum* and the cowpea beetle *Callosobruchus maculatus* all require temperatures above 20°C for normal development and reproduction (Benz, 1987), thus any reduction in temperature below this will have a significant effect on population development. Cooling large masses of product by refrigeration may be uneconomical but cooling by aeration during a cold season in a temperate climate is feasible, although the level of disinfection obtained will depend on the intrinsic cold tolerance of the species, the development stages present, the rate at which cooling is achieved, the temperature attained and the time for which it is maintained (Evans, 1987). With the grain weevil (*Sitophilus granarius*) −20°C for 8 hours, −16°C for 12 hours and −12°C for 24 hours are enough to kill all the stages of the beetle (Benz, 1987). High temperatures in the range 35–40°C are also sufficient to control most stored product pests.

The control of pests through changes in the interstitial atmosphere is a technique

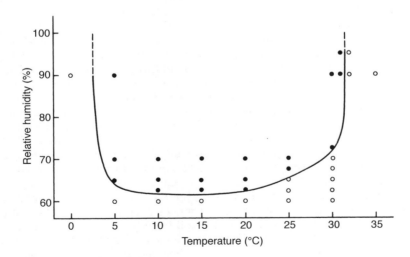

Fig. 7.1. The effect of physical conditions (relative humidity and temperature) on the development of the grain mite, *Acarus siro*, showing conditions under which the life cycle was completed (○) and those under which eggs failed to hatch or development was incomplete (●) (after Cunnington, 1976).

that has ancient origins in hermetic storage systems. Hermetic storage involves the modification of the atmosphere in sealed underground storage pits, where the respiration of the stored product increases the CO_2 level and decreases the level of oxygen to an extent that kills any insect pests contained within the store. Hermetic storage is still used in some developing countries in northern Africa. The more modern equivalent of hermetic storage systems and their effect on the storage atmosphere is the use of gas tight storage facilities in which oxygen deficient, carbon dioxide or nitrogen enriched atmospheres are used to control insect pests (e.g. Soderstrom and Brandl, 1982). These atmospheres may affect various pest species and their developmental stages differentially. Grain weevil, *S. granarius*, adults were the most susceptible stage to atmospheres of CO_2 and N_2 followed by larvae, pupae and eggs whereas in the rice weevil, *S. oryzae*, the pupae were the least susceptible stage followed by eggs, larvae and the adults (Lindgren and Vincent, 1970). The management of storage pests can be achieved by appropriate timing of aeration (Hagstrum and Flinn, 1990; Flinn *et al.*, 1997), by compression of air (e.g. Yokoyama *et al.*, 1993b) or by a combination of fumigant and aeration (and/or compression; e.g. Yokoyama *et al.*, 1993a). Controlled atmospheres have been shown to be economically competitive with chemical treatments such as phosphine (Soderstrom *et al.*, 1984) but if they are used as longer-term protectants, they have the disadvantage that tight sealing of the system needs to be ensured and access for inspection and handling is restricted. However, controlled atmospheres remain less hazardous than chemical fumigants for use by control operatives and are less likely to leave residues in the stored product.

Case Study: Modification of habitat to control the lone star tick (*Amblyomma americanum*), a pest of cattle (Meyer *et al.*, 1982)

The physical environment plays an important role in the distribution and abundance of the lone star tick, *Amblyomma americanum*. Thus techniques that modify the habitat in such a way as to render the physical environment unsuitable for the tick, could be used as a means of cultural control to reduce the tick infestations of cattle.

The habitat modification evaluated in this study consisted of a total mechanical clearing of all accessible land on the chosen field, followed by new pasture establishment. After the application of fertilizer, fescue, annual ryegrass and white clover were sown. Data on the size of the tick populations and on the physical parameters (percentage soil moisture, soil temperature, soil surface temperature, air temperature, percentage relative humidity and light intensity) were collected in 1978 prior to habitat modification. These same parameters were then measured and compared after habitat modification to evaluate their effect.

The changes in the physical environmental characteristics of the field and the mean number of nymph and adult ticks sampled in the field in 1978, 1979 and 1980 are shown in Table 7.1. The mean number of adults and nymphs is markedly reduced in 1979 and 1980 from the values in 1978 prior to treatment. Regression analyses of the physical parameters against the numbers of nymphal and adult lone star ticks showed that soil temperature (depth not given) and soil surface temperature were significantly associated with reductions in nymphal populations, whereas only soil surface temperature was associated with adult population reductions. From this it might be postulated that nymphs are more dependent on shelter in close association with the soil and are more sensitive to temperature changes in the shelter than the adults. The removal of leaf litter in

Table 7.1. A comparison of the mean values of a number of physical parameters and of the mean number of adult and nymphal lone star ticks sampled in 1978 prior to habitat modification, with conditions and mean insect counts in the same field after habitat treatment in 1979 and 1980 (Meyer *et al.*, 1982).

	Year		
	1978	1979	1980
% Soil moisture	61.6	72.7	30.9
Soil temperature (°C)	21.9	27.3	30.1
Soil surface temperature (°C)	27.1	32.1	37.7
Air temperature (°C)	25.8	29.5	33.5
% RH	57.2	46.1	39.8
Light intensity	4.9	7.9	8.4
Nymphs	10.8	0.7	0.2
Adults	4.8	0.5	0.2

the clearing process during habitat modification eliminated one of the major sources of shelter for the ticks and thus probably made them more susceptible to the effects of extreme temperatures.

An economic evaluation of the costs of the habitat modification technique was made and compared with the costs of the alternative treatments, insecticide use and animal control through fencing (restricting animals to areas unlikely to be infested with ticks). In terms of total annual costs, the habitat modification was not the cheapest treatment but when total costs were evaluated in terms of the cost of production of a kilogram of plant dry matter per hectare, the habitat modification proved most profitable, costing $0.14 compared with $0.74 and $0.33 for fencing and insecticide use, respectively.

7.4.1 Physical barriers and mulches

The environment of a pest can be disrupted using physical barriers or mulches. Barriers may be quite subtle in form, such as those composed of soil particles that are too large for termites to displace with mandibles, yet too small for termites to pass between (Su and Scheffrahn, 1998) or obtrusive, such as stainless steel mesh barriers used for preconstruction installation in houses to prevent termite foraging (Lenz and Runko, 1994; Grace *et al.*, 1996). Other types of barriers include reflective plastic mulches and netting tunnels which can repel insects or reduce the transmission of viruses (Stapleton *et al.*, 1994). Nets have been shown to offer protection to cauliflowers against *Delia radicum*, *Brevicoryne brassica* and Lepidoptera, and to Japanese radishes against *Delia radicum* (Ester *et al.*,

1994). Reflective mulches have proved particularly effective in preventing insect colonization. Flying aphids are not attracted to plants growing close to white or reflective surfaces. Reflective surfaces such as aluminium foil, reflective polyethylene films and reflective powders have been used to repel aphids and reduce incidence of virus diseases (Henshaw *et al.*, 1991; Stapleton *et al.*, 1993). More recently, biodegradable spray mulches have been successfully used (Stapleton *et al.*, 1994; Summers *et al.*, 1995).

7.5 Agronomic Practices

Crop rotation, tillage and planting date are three agronomic practices that can directly affect crop yield. They may also have a secondary effect in that they can sometimes also influence the level of insect pest infestation that may occur within a crop. Provided this additional role of pest control does not conflict with the primary objective of directly improving crop yield, then agronomic practices may be recommended as a means of cultural control for use in IPM.

7.5.1 Crop rotation

The rotation of crop provides a means of maintaining soil fertility so that an appropriate rotation can produce better average yields than continuous cultivation of the same crop (Table 7.2; Webster and Wilson, 1980) without the need for additional fertilizers. The type of crops used in the rotation will often be very important for maximizing yield. There may be some advantage in using crops that have different rooting habits and hence vary demand between different soil layers, or using leguminous plants that can fix soil nitrogen. Another factor that could also influence the choice of crop types in a rotation is that combination which best reduces pest damage. Part of the value of crop rotations is their ability to prevent the build up of insect pests, as well as other pests such as weeds and pathogens.

Generally, crop rotation is most effective against pest species that have a narrow host range and limited range of dispersal. A new generation of an insect pest that may have overwintered in the vicinity of its host crop will be faced with a different, non-host crop plant in a subsequent season. The insect pest will be obliged to disperse and for insects with poor dispersive powers this could reduce the likelihood of finding a host. The result will be that subsequent colonization of some fields of the host crop species may be retarded. The extent to which colonization is reduced will determine the value of the technique. In the event that absolute numbers locating the crop are not diminished, satisfactory control may still be achieved if the arrival of the colonizers is delayed to a less critical plant growth stage.

Rotations have been shown to be effective against the Colorado potato beetle (*Leptinotarsa decemlineata*) by creating a delay in the time of crop infestation (Lashomb and Ng, 1984; Wright, 1984). In a rotation of potato and wheat, the oviposition and first appearance of the beetle were delayed when compared with an unrotated potato field. This delay was attributed to physical and environmental barriers that slow emigration from the wheat by the overwintering adults (Lashomb and Ng, 1984). Since rotations typically involve planting the potato crop in an adjacent field (often separated from the field in which the potatoes were grown the previous year by only a drainage ditch, canal, field road or hedgerow) and the dispersal

Table 7.2. The influence of cropping systems on the yield (kg ha^{-1}) of sorghum, groundnut, cassava and cotton at two different sites in Nigeria. Crops are used in rotation or as a continuous cultivation of the same crop (monocropping) (from Webster and Wilson, 1980).

Site	First cycle (1950–52)			Second cycle (1953–55)		
	Sorghum	Groundnut	Cassava	Sorghum	Groundnut	Cassava
Bida						
Monocropping	353	305	5094	397	143	3128
Rotation	709**	388	6059**	749**	183	5296**
Samaru	Sorghum	Groundnut	Cotton	Sorghum	Groundnut	Cotton
Monocropping	1053	804	471	1062	781	402
Rotation	1341**	1160**	498	1248**	1027*	489

*Significant increase, $P = 0.05$; ** $P = 0.01$.

by Colorado potato beetle from overwintering sites is mainly by walking (French *et al.*, 1993), it should be possible to determine the optimum distance for spacing rotations. This has been done using mark-recapture experiments which have indicated that an effective field rotation to reduce beetle population densities from year to year will require a distance of ≥ 0.5 km (Follett *et al.*, 1996).

Crop rotations have become less popular as more intensive farming methods have been utilized but there is little doubt that they can provide an effective means of pest suppression. The choice of crops used in the rotation must be appropriate and should not introduce other pests into the system. It will be more difficult to devise appropriate rotations against polyphagous and/or mobile pests. In Europe, typical rotations involve grasses, legume and root crops which have been used to control wireworms (*Agriotes* spp.), chafers (*Melolontha melolontha* and *Amphimallon solstitalis*) and leatherjackets (*Tipula* spp.) (Coaker, 1987). These and other insects that have soil inhabiting stages are also susceptible to control through different types of tillage practices.

7.5.2 Tillage practices

The type of tillage practice used can have an impact on labour costs, machinery wear, soil erosion and water quality. There has been a general change away from more traditional mould-board ploughing (which turns over the soil integrating all crop residue) to various forms of conservation tillage (Hammond, 1997). The total number of hectares where conservation tillage is used currently accounts for >8 million ha in the mid-west of the USA alone (CTIC, 1995). Conservation tillage may be defined as those practices leaving >30% residue cover on the soil surface after planting and includes no-till, ridge till and mulch till (CTIC, 1995).

The type of cultivation can also markedly influence the soil environment and affect insect survival either indirectly by creating inhospitable conditions and by exposing the insects to their natural enemies or directly by physical damage inflicted during the actual tillage process (Stinner and House, 1990). Often the actual means by which insect survival is reduced remains unknown and studies are conducted simply to quantify the extent of the differences in insect infestations between tillage treatments. However, in two grasshopper pests *Kraussaria angulifera* and *Oedaleus senegalensis*, cultivation is known to affect the numbers of eggs and nymphs present by exposing the egg pods to desiccation, by reducing the level of food, shelter and vegetation available and by making the soil rough and unsuitable for egg laying (Amatobi *et al.*, 1988). Hence, the longer a field was left fallow influenced the build up of the grasshoppers in these fields.

Comparisons of the results obtained from different tillage practices may shed some light on the way in which infestation is reduced. Emergence of the sunflower seed weevil (*Smicronyx fulvus*) was reduced by 29–56% with the use of a mould-board plough which turns over the soil, effectively burying the late larval and pupal stages. However, this covering effect was thought to be only partially responsible for the increased mortality as a chisel plough resulted in a reduced emergence of between 36 and 39% without moving the larvae substantially deeper in the soil profile. Thus, it was concluded that factors such as aeration, soil temperature and drying and physical damage resulting from tillage were also important contributory factors (Gednalske and Walgenbach, 1984). A disc-plough treatment caused up to 73.5% mortality of the overwintering larvae of *Dectes texanus*, a pest of sunflower, compared with 39.7% mortality using a sweep plough. The higher mortality in the disc treatment was probably due to greater root destruction by discing, exposing larvae to the soil environment. The sweep blades tend to sever the tap roots while leaving the upper part of the roots and stubble intact, permitting larval survival (Rogers, 1985). Such observations can lead us to some tentative conclusions about the likely means

Table 7.3. The effect of no tillage and of strip rotary tillage on peppermint fresh hay weights at harvest, oil yields and *Fumibotys fumalis* cumulative season totals (from Pike and Glazer, 1982).

Treatment	Fresh hay weight (kg ± SE 10 m^{-2})	Oil yield (ml ± SE 3 kg^{-1} fresh hay)	*Fumibotys fumalis* cumulative season Total ± SE 3 m^{-2}
1980			
Strip rotary tillage	13.0 ± 0.2	11.0 ± 0.8	6.0 ± 1.9
No tillage	13.0 ± 0.3	11.9 ± 0.7	28.0 ± 4.3
1981			
Strip rotary tillage	10.5 ± 0.3	9.6 ± 0.3	5.6 ± 0.7
No tillage	10.8 ± 0.4	10.4 ± 0.6	32.5 ± 4.8

by which tillage practices reduce infestations of pests, but unless studies evaluate the actual mechanisms involved then there can be little hope of extrapolating the results of treatments to other cropping systems. The same is generally true of studies of the effects of the timing of tillage practice on insect pest infestations.

Some studies will speculate on the reasons for reductions in pest numbers (e.g. Kay *et al.*, 1977) while others attempt to identify the reasons for difference in pest numbers with cultivation treatments (e.g. Hammond, 1995). Too few studies also evaluate the economic impact of cultivation practices that lead to reductions in pest numbers. Pike and Glazer (1982) considered the effects of tillage both on the level of pest infestation of the rhizome borer *Fumibotys fumalis* and of yield of the peppermint crop.

F. fumalis overwinters in the soil of peppermint fields, and the crop is grown in pure stands for 4–5 years until damage from *F. fumalis* or other pests renders it unprofitable. The prepupae of *F. fumalis* overwinter near the soil surface and it was this characteristic that made cultivation a possible option for control. The tillage method was selected on the basis that it was already in use by some growers as the means of regenerating the peppermint crop. Strips of peppermint were tilled to a depth of 14 cm and the number of adults emerging from the stripped plots was reduced by 79 and 83% compared with solid stand peppermint plots. Also, there was no significant difference between strip plots and no tillage plots for fresh hay weights or oil yields (Table 7.3). Such estimates of both yield and infestation levels should be taken into account in all experiments where tillage practices are being evaluated, otherwise the implications of the technique cannot be fully appreciated and farmers are unlikely to adopt the idea.

Case Study: The effects of crop rotation and tillage on the size of infestations of the black cutworm (*Agrotis ipsilon*) in corn (Johnson *et al.*, 1984)

The black cutworm (*Agrotis ipsilon*) is a sporadic pest of corn in the USA during May and June. The level of infestation has been shown to be influenced by the presence of large amounts of crop debris and weeds and, hence, the use of a crop rotation and an appropriate tillage practice could provide a possible means of cultural control.

The rotations evaluated in this study were continuous corn, continuous soybean, a corn followed by soybean rotation, and a soybean, wheat, corn rotation with each crop present in each year of the study. Each rotation treatment

Table 7.4. The effect of crop rotation and tillage system on black cutworm (*Agrotis ipsilon*) infestations (from Johnson *et al.*, 1984).

Treatment	Crop rotation					
	Corn/corn		Soybean/corn		Soybean/wheat/corn	
	1981	1982	1981	1982	1981	1982
No till	4.0	0.5	6.2	0.9	–	1.9
Chisel plough	1.4	0.5	3.4	0.7	–	0.8
Mould-board plough	1.5	0.6	0.8	0.6	–	0.6

included three tillage systems: autumn mould-board plough, autumn chisel plough/spring disc plough, and no tillage. The rotation/tillage combinations were replicated four times in a randomized complete block design. The extent of the black cutworm damage in each treatment was evaluated by stand counts and the number of plants cut off near the soil surface, including plants exhibiting wilting ('deadheart') from subsurface or internal tunnelling.

The results of the experiment showed that corn following soybean or wheat in a rotation, combined with reduced (no) tillage produced higher black cutworm damage in the plots (Table 7.4). The black cutworm moths appear to be less attracted to corn residue than to soybean or wheat debris for oviposition. Damage was consistently lower in the mould-board plough treatments. The complete soil turnover by mould-board ploughing should negate any influence of crop residue on black cutworm oviposition. Hence, infestations of the black cutworm are least likely to occur if corn is grown continuously but the mould-board plough is used to reduce the amount of crop residue remaining for cutworm oviposition.

7.5.3 Planting date

The time or date at which crop plants are sown can have a significant impact on subsequent yield. In general in the tropics, early planting at the start of the rainy season is essential if the best yields are to be obtained. As a rule, yields are markedly and progressively reduced the longer planting is delayed after the onset of the rains (Webster and Wilson, 1980). Early sown crops benefit from a full season's rainfall, suffer less weed competition and benefit from the initial high soil nitrate levels available at the beginning of the rains. Crops may also benefit from a reduced insect pest infestation (A'Brook, 1964; Gebre-Amlak *et al.*, 1989; Emehute and Egwuatu, 1990) although early planting is not always the most appropriate strategy to ensure reduced levels of attack (Starks *et al.*, 1982; Sloderbeck and Yeargan, 1983;

Bergman and Turpin, 1984; Rogers, 1985; Zeiss and Pedigo, 1996).

Varying the planting time of crops works on a means of cultural control by creating asynchrony between the crop phenology and the population dynamics of the pest species. Asynchrony, particularly with an oligophagous pest, can retard the rate of colonization, reproduction and survival of the pest and reduce damage with respect to the susceptible crop growth stage (Ferro, 1987; Teetes, 1991; Metcalf and Metcalf, 1993). For such methods to have an impact though, planting times need to be synchronized between farms within a region (particularly for highly mobile pests) to reduce the variation in availability of susceptible crop stages.

The adoption by farmers of different planting times as a means of control will largely depend on what other work has to be completed during the period recommended from planting. Damage by pests to direct drilled, autumn sown grass was greatest in the autumn compared with spring (Bentley and Clements, 1989) so sowing in spring would seem to be a good cultural practice if pest damage is to be minimized. However, farmers would often rather sow in autumn when the land has the lowest potential for production than in spring when the disruptive effects of reseeding are greatest (Bentley and Clements, 1989). Thus, the recommendation of a particular planting date to control a specific pest in a single crop must not be made in isolation of other farm activities. The control of insect pests must work within the context of farming systems and the needs of the farmer should represent a key factor in the choice of dates for evaluation in field trials.

Harvest dates, as well as planting dates, can have an effect on pest incidence or the amount of damage caused (e.g. Berry *et al.*, 1997). In crops where insect pests reach outbreak proportions during a season and produce a resting stage at the end of a season it is possible that the area will act as a major source of pest inoculum during the subsequent season. This can sometimes be avoided if the crop is harvested before the insect develops its resting stage (Ferro, 1987). In other situations, such as forage crops which are regularly harvested, timely cutting can be used to control pest numbers (Penman *et al.*, 1979; Hagen, 1982; Onstad *et al.*, 1984). The early cutting of a second harvest of alfalfa interrupts the development of the potato leafhopper offspring, *Empoasca fabae*, before any become nymphs or adults. This is important because it is only these two stages that can survive the harvest (Onstad *et al.*, 1984).

Case Study: The effect of sowing date on infestation and damage caused by the maize stalk borer (*Busseola fusca*) on maize in Ethiopa (Gebre-Amlak *et al.*, 1989)

The maize stalk borer is an economically important pest of maize and sorghum in Ethiopia but only conflicting data are available on the effects of sowing time on the size of borer infestation and the damage.

Successive sowings of maize were made at 10 day intervals between 10 April and 7 July in 1985 and 1986. There were ten different planting dates with three replications, using a randomized complete block design. There were two sets of experiments, in one the plots were left untreated and in the other they were treated with cypermethrin at the rate of 0.3 kg a.i. ha^{-1}. At the end of the season, yields of maize in both sets were compared and the difference in yield between the two was considered to be the crop loss due to *B. fusca* infestation at that sowing date, although the degree of control of the respective infestations is not given.

The level of infestation of *B. fusca* was estimated every two weeks as the percentage of infested plants and deadhearts. There were two peaks of infestation representing two generations of the insect. Plants sown in April and early May had significantly lower infestations of first generation larvae while levels of infestation by second generation larvae were significantly higher on later sowing dates. The infestation by the second generation larvae had a significant effect on the grain yield of the late sown maize (Table 7.5). Even with insecticide use, the yield of the late sown crop was markedly reduced. Hence, to obtain better yields without the application of insecticide the planting of maize in this region of Ethiopia should not be later than April.

Table 7.5. Yield (quintal ha^{-1}) of maize with and without insecticide use to control *Busseola fusca* and yield loss due to this pest following different dates of sowing (Gebre-Amlak *et al.*, 1989).

Sowing date	Yield (q ha^{-1}) *Busseola fusca* controlled	Yield (q ha^{-1}) *Busseola fusca* not controlled	Yield loss due to *Busseola fusca* infestation (q ha^{-1})	(%)
10 April	74.9	67.6	7.3	9.8
20 April	80.4	77.3	3.1	3.9
30 April	75.7	79.7	–	–
10 May	76.6	64.7	14.9	18.7
20 May	65.4	58.0	7.4	11.3
30 May	49.5	18.2	31.3	63.2
9 June	40.0	3.2	36.8	92.0
19 June	33.0	1.7	31.3	94.9
29 June	24.2	0.0	24.2	100.0
9 July	14.8	4.9	9.9	66.9

7.5.4 Sowing/planting density

The spacing of plants will have a significant effect on their yield, with the maximum weight of the plant material from a unit area of land only being obtained when the individual plants are competing with each other (Finch *et al.*, 1976). Obviously there are limits where plant density will severely influence plant yield and/or quality but within these limits, there may be plant densities which also reduce pest insect abundance and can potentially be used as a form of cultural control. Ultimately, however, yield advantages due to plant spacing will tend to over-ride the effects of plant density on pest incidence or damage. The density of sorghum seedlings affects the number of eggs that are laid by the sorghum shootfly (*Atherigona soccata*) (Fig. 7.2) but at the higher densities where the pest control effect is greatest, the stem height, width of last expanded leaves and leaf stage of the sorghum are reduced to a level that would not give a satisfactory yield (Delobel, 1981).

In general, an increase in plant density, i.e. a reduction in plant spacing, seems to reduce pest numbers (A'Brook, 1964, 1968; Farrell, 1976; Tukahirwa and Coaker, 1982) but not in all cases (Mayse, 1978; Troxclair and Boethel, 1984). One of the main reasons for the response of insects to varying plant density has been the contrast between plants and their soil background and the effect of this on the optomotor landing response of the flying insects (A'Brook, 1964, 1968; Smith, 1969). *Aphis craccivora*, *Aphis gossypii* and *Longiunguis sacchari* were trapped at approximately 1 m above ground more often over widely spaced than over close-spaced groundnuts. At crop height *A. craccivora* and *A. gossypii* showed a similar but even greater response to plant spacing (A'Brook, 1968; Table 7.6). This effect was attributed to an optomotor response to the contrast between bare earth and the plants, and hence, the stimulus was greater over the wider spaced plants. Such responses may be explained if the insects concerned have evolved to find host plants at low densities, perhaps as early colonizers of disturbed ground or as small patches isolated from adjacent host plants (Coaker, 1987).

Other reasons that have been given to

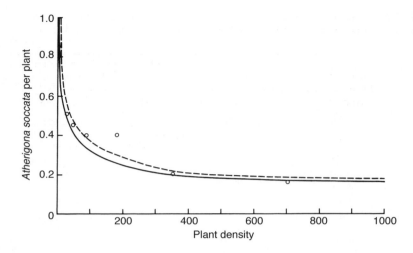

Fig. 7.2. The relationship between the host plant density and the number of *Atherigona soccata* eggs (– – –) and surviving first star larvae (——) per plant (after Delobel, 1981).

explain lower insect numbers in dense plantings have included host plant condition (Farrell, 1976), the presence of excess vegetation acting as a deterrent (Delobel, 1981), changes in the microenvironment favoured by the pest and its natural enemies and the crop's attractiveness (Coaker, 1987).

Table 7.6. Percentage total trap catch at different plant spacings of groundnut and bare earth (from A'Brook, 1968).

	Vertical trap at 0.91 m			Horizontal trap at 0.30 m		
	Close spacing	Wide spacing	Bare earth	Close spacing	Wide spacing	Bare earth
Aphis craccivora	15.1	37.4	47.5	2.6	27.4	70.1
Aphis gossypii	22.2	30.6	47.2	3.7	40.9	55.5
Longiunguis sacchari	13.5	38.5	47.9	6.1	19.3	74.6

Case Study: The effect of plant density on populations of the cabbage root fly on four cruciferous crops (Finch *et al.*, 1976)

The cabbage root fly (*Delia radicum*) is a pest of the cruciferous crops swede, cauliflower, cabbage and Brussels sprouts in the UK. To study the effects of planting density on infestation by the root fly, plots of each crop were sown as concentric circles (24 per plot) of plants at spacings ranging from 10 to 90 cm between individual plants, which provided plant densities between 1.5 and 68.3 plants m^{-2}. The number of cabbage root fly pupae was estimated from 15 cm diameter soil cores each containing the root system of one plant. The pupae were collected and weighed.

The number of pupae increased in each crop with increasing plant density (Fig. 7.3). The slopes of the regression for Brussels sprouts, cabbage and cauli-

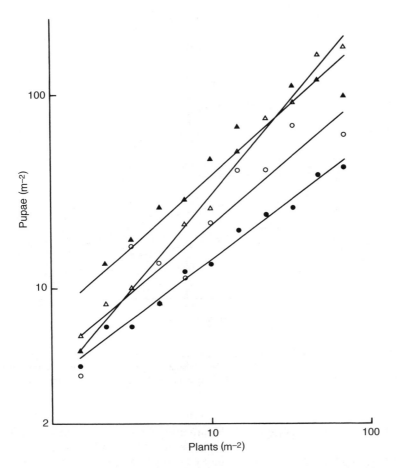

Fig. 7.3. The relationship between plant density and the number of pupae of the cabbage root fly produced per m² on four types of cultivated plants: swede (△), cauliflower (▲), cabbage (○) and Brussels sprouts (●) (after Finch *et al.*, 1976).

flower were not significantly different but the intercepts were significantly different. The regression slope of the relationship between the number of pupae and plant density for swede differed from the combined value for the slope of the other three crops. Although the number of pupae increased with plant density, the mean weight decreased in every crop except cauliflower (Fig. 7.4).

These results suggest that for a given cabbage root fly population, the size of the subsequent population is determined both by plant density and the type of host crop present. The numbers of pupae were related to plant density by the power of 0.7 (the value of the combined slopes, Fig. 7.3) indicating that a tenfold increase in plant density in summer planted brassica crops should result in an approximate fivefold increase (antilog 0.7) in the cabbage root fly population available for infestation the following spring, provided no density dependent mortalities occur during the winter period.

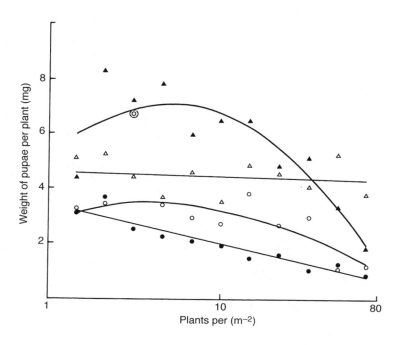

Fig. 7.4. The relationship between plant density and the weight of pupae of the cabbage root fly produced per plant for swede (△), cauliflower (▲), cabbage (○) and Brussels sprouts (●) crops grown during the second and third generations of flies in 1974. ◎ point omitted from cabbage regression (after Finch *et al.*, 1976).

7.6 Mixed and Intercropping

One of the features of modern agriculture is an ever increasing tendency towards the use of monocultures (Theunissen and Den Ouden, 1980). These have allowed agriculture to become intensified and more easily manageable with a high input mechanized approach to farming. In tropical countries, the use of monocultures is practised only on larger farms or estates, while the more traditional approaches to farming utilize polycultures. The polycultures include mixed intercropping (no distinct row arrangement), row intercropping (one or more crops planted in rows), strip intercropping (crops grown in different strips, wide enough to permit independent cultivation), relay intercropping (two or more grown simultaneously for part of the life cycle of each, a second crop being planted

before the harvest of the first) (Andrews and Kassam, 1976) and alley intercropping (annual crops are grown in strips between trees) (ICRISAT, 1989). Most of the food consumed in tropical Asia, Latin America and Africa is produced in such systems which often more readily meet the needs of the smaller scale farmer than does the monocrop (Perrin and Phillips, 1978). A mixed or intercropping regime can provide a greater total land productivity as well as insurance against the failure or unstable market value of any single crop. In addition crops in intercropping systems may improve soil fertility and the availability of alternative sources of nutritious products (Risch *et al.*, 1983) as well as reducing the incidence of insect pest attack (Tingey and Lamont, 1988) and thereby maintaining lower pest control costs. Intercropping has been studied sufficiently for there now to be a considerable body of evidence to show that it can be used to reduce the incidence of pest insects (e.g. Theunissen and Den Ouden, 1980; Tukahirwa and Coaker, 1982; Uvah and Coaker, 1984; Tingey and Lamont, 1988; Edwards et al., 1992; Khan et al., 1997). For this reason, policy makers in tropical countries continue to give attention to improving production in traditional intercrops, rather than replacing them with capital and energy intensive technology (Perrin and Phillips, 1978). In more intensive systems, monocrops are maintained because with high inputs, they are economic despite susceptibility to pests. For these reasons, it is unlikely that intercrops will play any major role within the framework of intensive crop production, except perhaps in smaller scale farming enterprises and horticulture, and as a result most intercrop studies will remain pertinent to situations in tropical countries (Coaker, 1990).

One of the major problems has been predicting which cropping systems will reduce pest abundance, since not all combinations of crops will produce the desired effect and blind adherence to the principle that a more diversified system will reduce pest infestation is clearly inadequate and often totally wrong (Gurr et al., 1998). In an examination of 150 studies involving a total of 198 plant damaging insect species, 53% were found to be less abundant in a more diversified system, 18% were more abundant, 9% showed no difference and 20% showed a variable response (Risch et al., 1983). Clearly the majority of the species were less abundant in more diversified systems but 38% were either more abundant or produced a variable response. This indicates the need for caution and a greater understanding of the mechanisms involved to explain how, where and when such exceptions are likely to occur. It will only be through detailed ecological studies that such an understanding can be gained and an appropriate predictive theory developed. This means a greater emphasis has to be placed on detailed ecological experiments rather than on purely descriptive

studies of relative insect abundance under different cropping systems. Since there are numerous possible intercrops and a multitude of conditions under which such systems could be utilized, there is little likelihood that each of these could be investigated (Herzog and Funderburk, 1986). Hence, it is necessary to understand the underlying mechanisms to allow extrapolation to various other cropping systems and situations (Gurr et al., 1998).

7.6.1 Ecological theory and experimentation

The ecological theory relating to the benefits or otherwise of mixed versus simple cropping systems revolves around the two possible explanations for experimental evidence that insect pest populations obtain higher levels in less diverse cropping systems compared with diverse ones. The two hypotheses proposed by Root (1973) are:

1. The natural enemy hypothesis which argues that pest numbers are reduced in more diverse systems because the activity of natural enemies is enhanced.

2. The resource concentration hypothesis argues that the presence of a more diverse flora has direct negative effects on the ability of the insect pest to find and utilize its host plant.

The resource concentration hypothesis (alternatively called the disruptive crop hypothesis; Vandermeer, 1989) predicts lower pest abundance in diverse communities because a specialist feeder is less likely to find its host plant due to the presence of confusing masking chemical stimuli (e.g. Uvah and Coaker, 1984), physical barriers to movement (Perrin and Phillips, 1978) or other environmental effects such as shading (Risch, 1981); it will tend to remain in the intercrop for a shorter period of time simply because the probability of landing on a non-host plant is increased; it may have a lower survivorship and/or fecundity (Bach, 1980).

The extent to which these factors operate will depend on the number of host plant species present and the relative preference of the pest for each, the absolute density and spatial arrangement of each host species and the interference effects obtained from non-host plants (Risch, 1981). If the density of a host species is low and it is well distributed among non-host plants, then an insect approaching the habitat will have greater difficulty in locating its host than if the host density is high relative to non-hosts and if its distribution is clumped. Also, the concentration of the host resource may influence the probability of the insect staying in a habitat once it has arrived. For instance, an insect pest may tend to fly earlier, further or straighter after landing on a non-host plant resulting in a more rapid movement from habitats with a low resource concentration (Risch, 1981).

The natural enemies hypothesis attributes lower pest abundance in intercropped or more diverse systems to a higher density of predators and parasitoids (Bach, 1980). The greater density of the natural enemies is caused by an improvement in conditions for their survival and reproduction, such as a greater temporal and spatial distribution of nectar and pollen sources, which can increase parasitoid reproductive potential and an abundance of alternative hosts/prey when the pest species are scarce or at an inappropriate stage (Risch, 1981). These factors can in theory combine to provide more favourable conditions for natural enemies and thereby enhance their numbers and effectiveness as control agents.

The question then arises, which of the two hypotheses is the most important for influencing the relative abundance of pest insects in diverse systems. The question has been approached in two ways: (i) reviews of the literature relating to crop diversity and pest abundance; and (ii) by experimentation. The first of these studies by Risch *et al.* (1983) has been referred to above, and this concluded that the resource concentration hypothesis was the most likely explanation for reductions in pest abundance in diverse systems, but it pointed out that the mechanisms to support this had rarely been studied. Nineteen studies that tested the natural enemy hypothesis were reviewed by Russell

(1989). Of these 19 studies, mortality rates from predators and parasitoids in diverse systems were higher in nine, lower in two, unchanged in three and variable in five. Russell (1989) concluded that when evaluated, the natural enemy hypothesis was confirmed and considered the two hypotheses complementary. When studies on intercrops were compared with crop/weed systems, Baliddawa (1985) found that 56% of pest reductions in crop/weed studies were caused by natural enemies compared with 25% in the intercrop studies, although intercrops probably slowed pest colonization.

The most comprehensive of the studies (Andow, 1991) classified pests as monophagous or polyphagous and reviewed 254 herbivore species of which 56% were lower in diverse systems: 66% for monophagous herbivores and 27% for polyphagous ones. Predator and parasitoid numbers were higher in diverse habitats in 48% and 81% of studies respectively. Andow (1991) concluded that of the two hypotheses, resource concentration was probably of more importance. When annual and perennial crops are considered separately, the resource concentration hypothesis seems most likely for annual crops whereas the natural enemy hypothesis is most likely for perennial and some annual crops (Cromatie, 1991). Overall, the review studies conducted would indicate that the two hypotheses are probably complementary in many systems (Wratten and van Emden, 1995) although there have been too few studies that clearly elucidate the mechanisms involved.

Experiments to determine the impact of the two mechanisms are possible because they each make different assumptions about how diversification will affect monophagous and polyphagous insects (Risch *et al.*, 1983). The natural enemies hypothesis predicts that the numbers of monophagous and polyphagous pests should be equally reduced in a monocrop and an intercrop because natural enemies do not prey differentially on these groups. However, if the resource concentration hypothesis is more important, then only the monophagous insect species should be less common in the intercrop than the monocrop. The polyphagous pest species that has an equal preference for all the crops in an intercrop should be equally abundant in both cropping systems. Risch (1981) carried out an experiment to test which hypothesis influenced pest abundance when the pests were six species of chrysomelid beetle and the polycrop and monocrop were composed of corn (*Zea mays*), beans (*Phaseolus vulgaris*) and squash (*Cucurbita maxima*; Fig. 7.5). There were both monophagous and polyphagous species among the six pests studied and measures of rates of parasitism, beetle counts and beetle movement all suggested the resource concentration hypothesis accounted for the differences in beetle

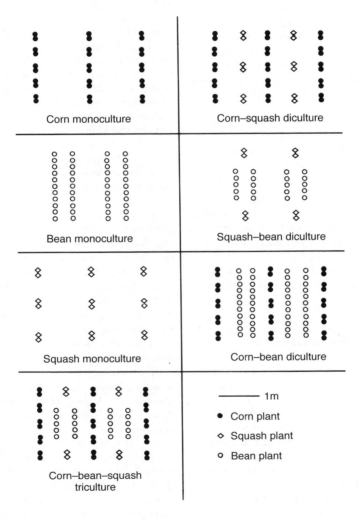

Fig. 7.5. The spatial arrangement of corn, bean and squash plants within monocultures, dicultures and tricultures, in an experiment to determine the relative effect of resource concentration or natural enemies on the abundance of six species of chrysomelid beetles (after Risch, 1981).

abundance between the monocrops and polycrops. The evidence supporting this was summarized by Risch (1981) as follows:

1. There were no measurable differences in predation or parasitism of beetles between treatments.
2. The number of beetles per host was lower in the polycrop than the monocrop only when there was a non-host plant present in the polycrop. When two host plants were present in the polycrop, then the numbers of beetles per host plant were in fact higher in the polycrop than in the monocrop. This pattern of abundance would be predicted only if beetle movements and not mortality due to natural enemies was responsible for differences in beetle abundance.
3. Direct measurements of beetle movements in the field showed that beetles tended to emigrate more from polycrops that included a non-host plant than from host monocrops.

Such experiments are very involved and complicated, especially the work needed to assess insect movement (Wetzler and Risch, 1984) but more such work and experiments are required if it is ever going to be possible to predict reliably pest abundance in yet untried intercrop or polycrop systems.

7.7 Semiochemicals

Semiochemicals are chemicals that mediate interactions between organisms (Nordlund, 1981). They are divided into two major groups: *pheromones*, which mediate in intraspecific interactions and *allelochemicals* for interspecific interactions. Each of these groups can be subdivided further: allelochemicals as allomones, kairomones, synomones and apneumones (Table 7.7) and pheromones as sex pheromones, alarm pheromones and epideitic or aggregation pheromones. While the majority of research and development has concentrated on pheromones, particularly those of

lepidopteran pests, there is now an increasing diversity of interest in the isolation and use of semiochemicals (behaviour modifying chemicals) that act as repellants, antifeedants, attractants and stimulants.

7.7.1 Types of behaviour modifying chemicals

Sex pheromones are used by males to locate females for mating. Typically, female Lepidoptera emit pheromone for a specific period during the day in order to attract males. The pheromone gland in the final segments of the abdomen is extruded so that the pheromone is released to move downwind as a plume of odour. A resting or flying male that detects the pheromone will fly into the wind and follow the plume by upwind anemotaxis (orientation with respect to the direction of the wind). On arriving near the female, the high concentration of odour perceived by the male causes a reduction in forward flight speed and further orientation over the last few centimetres will be both by chemical sensing and visual means (Shorey, 1977). Copulation may then take place. A large number of the sex pheromones of Lepidoptera have now been identified but the behavioural sequences involved in the detection and tracking leading to successful mating has rarely been studied. If insect behaviour is to be manipulated successfully, a greater emphasis will need to be placed on understanding mate-finding behaviour. In this way, the weakest link in a response sequence can be identified and more effectively exploited.

In some insect–plant interactions, there is an optimum density for the survival of the insects. Among several insect species, there exist epideitic pheromones that enable insects to aggregate to an extent necessary to efficiently exploit a host or habitat. Such aggregation pheromones can also prevent the accumulation of insects at a given site extending beyond the optimum range (Jackson and Lewis, 1981). These aggregation pheromones are found mostly in beetle species of which some of the best studied examples are of forest insect pests such as *Dendroctonus* spp., *Trypondendron*

Table 7.7. The different types of semiochemical involved in insect communication (after Nordlund, 1981).

Pheromone

A substance that is secreted by an organism to the outside and causes a specific reaction in a receiving organism of the same species

Sex pheromone: a substance generally produced by the female to attract males for the purposes of mating

Aggregation pheromone: a substance produced by one or both sexes, and bringing both sexes together for feeding and reproduction

Alarm pheromone: a substance produced by an insect to repel and disperse other insects in the area. It is usually released by an individual when it is attacked

Allelochemical

A substance that is significant to organisms of a species different from its source, for reasons other than food

Allomone: a substance produced or acquired by an organism that, when it contacts an individual of another species in the natural context, evokes in the receiver a behavioural or physiological reaction that is adaptively favourable to the emitter but not the receiver, e.g. repellents, antifeedants, locomotor excitants

Kairomone: a substance produced or acquired by an organism that, when it contacts an individual of another species in the natural context, evokes in the receiver a behavioural or physiological reaction that is adaptively favourable to the receiver but not the emitter, e.g. feeding or oviposition stimulants

Synomone: a substance produced or acquired by an organism that, when it contacts an individual of another species in the natural context, evokes in the receiver a behavioural or physiological reaction that is adaptively favourable to both emitter and receiver

Apneumone: a substance emitted by a non-living material that evokes a behavioural or physiological reaction that is adaptively favourable to a receiving organism but detrimental to an organism of another species that may be found in or on the non-living material

---

lineatum and *Gnathotricus sulcatus* (Borden, 1990).

Alarm pheromones, epitomized by (E)-β-farnesene in aphids, are released when insects are under threat. The aphid alarm pheromone is released when individuals are attacked by predators or parasitoids eliciting increased activity amongst nearby individuals. Alarm pheromones basically elicit an escape response increasing movement and thus preventing or reducing their chances of attack. They have had only limited use in applications for pest control although they have potential for use in novel ways. For example, (E)-β-farnesene has been used to increase the mobility of aphids so that contact pesticides such as pyrethroids are more readily picked up (Pickett, 1988), while simple components of the honey bee sting and mandibular gland can be used to repel beneficial foraging bees from oilseed rape during insecticide applications (Free *et al.*, 1985).

Among the allelochemicals, there are probably fewer practical examples of the use of apneumones (an attractant produced by a non-living substance such as food or oviposition substrate) than with pheromones, allomones and kairomones. Allomones are substances that when produced by plants can reduce insect feeding and/or fecundity by their toxicity, or act as antifeedants and repellents (Panda and Khush, 1995). The identification of the phenylpropanoid, 4-allylanisole, a compound produced by many conifers including loblolly pine (a preferred host of the southern pine beetle *Dendroctonus frontalis*) as a repellent, enabled its use to protect pines in urban environments (Hayes *et al.*, 1994; Hayes and Strom, 1994; Hayes *et al.*, 1996). Methyl salicylate and (−)-(1R, 5S)-myrtenal are both plant derived repellents for the black bean aphid *Aphis fabae* (Hardie *et al.*, 1994). Methyl salicylate is associated with secondary metabolite based defence in plants,

and the monoterpenoid $(-)$-(1R, 5S)-myrte-
nal is metabolically related to $(-)$-(1S, 5S)-α-
pinene, a component of defensive resins
produced by gymnosperms. Hardie *et al.*
(1994) argue that these two compounds are
employed by *A. fabae* as indicators of nutri-
tionally unsuitable or non-host plants.
Methyl salicylate has also been demonstrated to act as a repellent in other aphid
species, *Rhopalosiphum padi* (Wiktelius and
Pettersson, 1985), *Sitobion avenae* and
Metapolophium dirhodum (Pettersson *et al.*,
1994).

Antifeedants have potential in affecting
levels of pest colonization when applied to
crops thus reducing feeding damage and
virus transmission (Pickett *et al.*, 1987). A
number of compounds have been impli-
cated for aphids including azadirachtin,
polygodia and hop β-acids (Powell *et al.*,
1997) but to date the most successfully
deployed antifeedant in the field has been
drimone $(-)$-polygodial, extracted from the
water-pepper *Polygonum hydropiper*
which, when applied to barley against
aphids, decreased the transmission of barley yellow dwarf virus (Dawson *et al.*,
1986; Pickett *et al.*, 1987).

Kairomones are used by insect pests to
locate host plants and by natural enemies to
locate their prey. Volatile compounds
released by a plant or from prey gland
secretions, frass, honeydew or cuticular
secretions, may be detected from minute
traces in the air and are used to assist host
location. Aphids have been shown to utilize kairomonal olfactory cues in their environment to locate host plants (Nottingham
et al., 1991; Lösel *et al.*, 1996) while contact
with aphid, scale and mealybug honeydew
arrest both parasitoids, lacewings and
cocinellid predators (Budenberg, 1990;
Hågvar and Höfsvang, 1991; Heidari and
Copland, 1993; McEwen *et al.*, 1993; Merlin
et al., 1996; Lilley *et al.*, 1997). Kairomones
are also sequestered and utilized by insects
as pheromones. The female western pine
beetle *Dendroctonus brevicornis* which
attacks ponderose pine in North America is
attracted to its host by oleoresin. The
female attracts males by a pheromone, one
of whose chemical constituents have been
sequestered from the tree's oleoresin
(Edwards and Wratten, 1980).

When herbivore pests damage crop
plants, they may release synomones that
attract parasitoids. The parasitoid
Epidinocarsis lopezi is attracted to cassava
plants that release a synomone when fed
upon by mealybugs *Phenacoccus manihoti*
(Nadel and van Alphen, 1987). The parasitoids of the brassica pod midge
Dasineura brassicae are attracted to isothiocyanates released by brassicas. It has
been possible to attract these natural enemies into a trap using 2-phenylethyl isothiocyanate (Murchie *et al.*, 1995; Pickett *et
al.*, 1995).

7.7.2 Mass trapping

Mass trapping, as its name implies, is the use
of large numbers of pheromone traps to catch
a large proportion of the pest population.
This technique has proved unsuccessful for a
whole range of lepidopteran and coleopteran
pests (Campion, 1989; Jones and Langley,
1998). Lepidopteran sex pheromones only
attract the males, hence to be successful,
highly efficient traps are needed in order to
catch a high enough proportion of the male
population to prevent mating with females.
Mass trapping is more appropriate with
aggregation pheromones, since these attract
both males and females. As part of an IPM
programme, mass trapping has proved useful
in the control of the cotton boll weevil
(*Anthonomus grandis*; Dickerson, 1986) and
the spruce bark beetle *Ips typographus* (Raty
et al., 1995; Grégoire *et al.*, 1997). In a large
mass trapping programme aimed at the
control of *I. typographus* in Norway
(1979–1980), the aggregation pheromone
component 2-methyl-3-buten-1-ol was dis-
pensed in over 600,000 traps and captured
7.4 billion beetles in 2 years (Bakke *et al.*,
1983). Damage caused by the beetle was
significantly reduced in the forest around the
traps.

7.7.3 Mating disruption

The use of pheromones for mating disruption is potentially a very powerful tool in

insect pest management. The technique is based on the premise that male insects would be unable to locate females if the environment around the female is permeated with sex pheromone (Jackson and Lewis, 1981). However, the precise mode of action remains speculative (Cardé, 1990). Three factors were proposed by Shorey (1977) that may act alone or in combination to produce the mating disruption effect, sensory adaptation, habituation and direct competition. Sensory adaptation may occur because after prolonged exposure, the olfactory sensory neurones no longer detect the pheromone. Habituation is the situation in which the insects stop responding to a stimulus if earlier responses did not lead to a proper result, while direct competition may occur if males are flying to pheromone sources instead of females, which reduces the chances of successful mating. Cardé (1990) proposed a further three: camouflage of the natural pheromone plumes by a high concentration of synthetic pheromone, an imbalance in the sensory input where the synthetic pheromone has unnatural ratios of components, and pheromone antagonists that reduce the attractiveness of pheromones. Cardé (1990) concluded that for most mating disruption systems, several mechanisms were operative and that

the biggest outstanding problem was knowing the behaviours to be modified and the sensory inputs that modulate these reactions. Only with this knowledge would mating disruption be fully understood.

The technique of mating disruption has been successfully used to control the pink bollworm *Pectinophora gossypiella* (Campion, 1989), the lesser peach tree borer *Synanthedon pictipes* (Pfeiffer *et al.*, 1991), the European grape berry moth *Eupaecilia ambiguella* (Neuman, 1990), the artichoke plume moth *Platyptila carduidactyla* (Campion, 1989), codling moth *Cydia pomonella* (Charmillot, 1990; Barnes *et al.*, 1992; Howell *et al.*, 1992; Pfeiffer *et al.*, 1993) and gypsy moth *Lymantria dispar* (Kolodny-Hirsch and Schwalbe, 1990). The mating disruption technique has proved very successful on an area-wide basis and probably has yet more to offer in terms of species and cropping systems where it may be suitably operated. Pheromone formulations appear to be completely non-toxic to man and other higher animals and natural enemy numbers are increased in pheromone treated areas (see Case Study). Close cooperation among growers, government research bodies and the agribusiness is needed to yield further exciting possibilities in this field.

Case Study: Mating disruption for the control of the pink bollworm (*Pectinophora gossypiella*) in cotton (El-Adl *et al.*, 1988)

Pectinophora gossypiella is a pest of cotton that is difficult to control by conventional insecticides because the timing of application is critical to ensure that the newly hatched larvae are killed before they penetrate the flowers and bolls. The control of this pest by sole use of the mating disruption technique was first achieved in Egypt in 1981 using a micro-encapsulated formulation of the bollworm pheromone (Critchley *et al.*, 1983). The use of a micro-encapsulated formulation produced results equal to those of insecticide use. The following account is of work conducted to compare the efficacy of three pheromone formulations, hollow plastic fibres, plastic laminate flakes and the micro-encapsulated formulation with the standard insecticide application in larger scale field trials. Each treatment was applied to a 100 ha block of cotton and there were replicate blocks only for the hollow plastic fibres. The insecticide treatment was carried out on a site of similar size in the same locality.

The micro-encapsulated formula was applied at 10 g a.i. ha^{-1} from the air at a height of 3 m using conventional boom and nozzle equipment. The hollow fibre formulation was also applied by air at a rate of 3 g a.i. ha^{-1} from a height of 15 m using specially designed applicators. For logistic reasons, the first two applications of laminate flakes were made by hand, later aerial applications at a rate of 7 g a.i. ha^{-1} from a height of 15 m were made, also using a specially designed applicator. The pheromone formulations were each applied four times while six applications were made in the insecticide treatment. There was some partial insecticide use in some of the pheromone treated areas to control *Spodoptera littoralis*.

The following criteria were used to assess the efficacy of the pheromone and insecticide treatments:

1.   The daily catches of three pheromone traps in each treatment area.
2.   Weekly samples of 500 bolls from each treatment to evaluate boll infestation.
3.   Yield loss estimates were made at the end of the season by determining the proportion of incorrectly opened bolls in a sample of 100 taken from each treatment.
4.   The average yield from each treatment replicate was estimated by weighing the yield from all plants within a 10 m^2 area in the middle of the plot.
5.   Weekly sweep net catches of natural enemies were made in a 1 ha site in each treatment.
6.   The relative costs of the pheromone and insecticide applications.

The results of this extensive experiment can be summarized as follows:

1.   In all the pheromone treated areas moth catches were greatly reduced following treatment compared with those in the insecticide treated area.
2.   In none of the areas did boll infestations reach 10% damage levels.
3.   The estimated yield losses due to *P. gossypiella* were all low, with a maximum in the insecticide treatment of just under 5%.
4.   The same trend was reflected in the average yields with all the treatments having a high yield but the lowest yield occurring in the insecticide treatment (Fig. 7.6).
5.   Beneficial insects were found in greater numbers in the pheromone treated areas compared with those treated with insecticides. The seasonal totals were 6620 in the laminate flake treatment, 14,463 in the micro-encapsulated treatment, 10,747 and 18,743 in the two replicates of the hollow-fibre treated area compared with a total of 4094 in the insecticide treated area.
6.   The cost of the seasonal application of the pheromones (1988 prices) was lower (£78.90) than that of an insecticide (£107.79) mainly because of the lower fixed cost of the pheromones. The cost of aerial application was the same for the different products at £14.55 for four applications.

The three pheromone formulations offer different advantages and disadvantages. The glue based hollow fibre and laminate flake formulations can be used to treat plots faster (Table 7.8) and with greater pilot safety in the presence of obstacles on the ground such as trees and telegraph poles as they give a flying altitude of 14 m. Spraying is also possible throughout the day because the glue based formulations are unaffected by high temperatures. The micro-encapsulated formulation (unlike the other two pheromone formulations) requires no special equipment for application or pilot training since it uses conventional spraying equipment. If required, application costs may be reduced by mixing additional substances such as fertilizers or trace elements to the spray mix. In rural areas, and where aerial application is not possible, application of the glue based formulations would have to be made by hand whereas a knapsack sprayer could be used for the micro-encapsulated formulation.

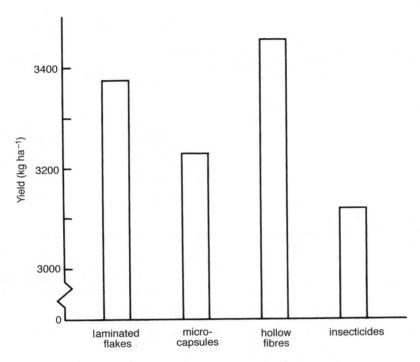

Fig. 7.6. The estimated averages of cotton yield weights in pheromone and insecticide treated fields (after El-Adl *et al.*, 1988).

Table 7.8. A comparison of the area treated in one flight, the application and reloading times for the hollow fibre, laminate flake and micro-encapsulated *P. gossypiella* pheromone formulation, when compared with spraying conventional insecticides, all using an Air Tractor aircraft (after El-Adl *et al.*, 1988).

	Micro-encapsulated pheromone formulation	Laminate flake pheromone formulation	Hollow fibre pheromone formulation	Insecticide
Number of hectares treated in one flight	50	200	130	50
Application time in minutes	30	75	60	30
Time for aircraft reloading in minutes	15	5	5	15

7.7.4 Lure and kill target systems

The lure and kill approach utilizes semio-chemical attractants (pheromones, kairo-mones, apneumones or empirically derived chemical attractants) to lure the insect to the attractant's source which has been treated with chemical insecticides or biopesticides. The insect picks up a lethal dose of the insecticide (referred to in this case as an affector; Lanier, 1990) and subsequently dies. The treated source may be a part of a crop, a tree or a specifically designed target. When applied to crops and trees, sprayable formulations of the insecticide and attractants are used. Unlike with conventional insecticide application which has to be applied to the whole crop to provide protection, sprayable lure and kill formulations can be applied to only part of the crop, thereby reducing the amount of active ingredient applied and levels of environmental contamination. An aerial application of pheromone/insecticide mixture for control of the olive fly *Bactrocera oleae* could be delivered in a 20 m swath every 100 m of olive grove so that only 20% of the crop is treated (Howse *et al.*, 1998). The use of such approaches have been shown to be practical and economically viable (Montiel, 1992). The majority of lure and kill systems have however, not been applied to crops but to target devices, which differ from mass trapping only in that the insect is not entrapped at the source of the attractant but is subjected to a dose of insecticide. The first successful attempt at using a target lure and kill strategy was made by Steiner *et al.* (1965) on the island of Rota in the Marianas to eradicate *Bactrocera dorsalis*. The potent parapheromone methyl eugenol was mixed with an insecticide and applied to 5 cm fibreboard squares which were thrown from aircraft. Eradication was achieved within 6 months (Jones and Langley, 1998). The same approach has been successfully utilized for control of *B. dorsalis* in Northeastern Australia (Broughton *et al.*, 1998) and on the island of Mauritius (Seeworthrum *et al.*, 1998) and in the Japanese Ogasawara islands (Koyama *et al.*, 1984).

Target devices treated with food attractants and insecticides to control *Bactocera oleae* in olive groves in Greece (Haniotakis *et al.*, 1991) and tsetse flies *Glossini* spp. in Zimbabwe baited with acetone and octenol (two components from the odour of host cattle) (Vale *et al.*, 1985) have also been shown to be successful. Targets that utilize insect growth regulators and chitin synthesis inhibitors have also been developed (e.g. Howard and Wall, 1996a; Langley, 1998) (these are considered in Section 7.8.1 dealing with autosterilization).

7.7.5 Manipulating natural enemies

Carbohydrate and protein food sources are important for natural enemies for energy and reproduction. In nature, carbohydrates may be obtained from prey or host fluids, homopteran honeydew and plant nectar (van Driesche and Bellows, 1996). Protein is obtained from pollen of wild flowers and weeds. Under circumstances where it is thought these resources may be limiting, they can be provided as an artificial application, e.g. as sugar or molasses. In addition, since these resources can act as attractants to natural enemies, applications can be used to manipulate populations.

Chrysoperla carnea is an important generalist egg predator that has been the subject of a number of attempts at manipulation using semiochemicals. Field applications of artificial honeydew (usually a mixture of sucrose, yeast and water) have been used to enhance the impact of chrysopid populations in various cropping systems including potato (Ben Saad and Bishop, 1976), cotton (Hagen *et al.*, 1971), apples (Hagley and Simpson, 1981) and olives (McEwen *et al.*, 1993, 1994; Liber and Niccoli, 1998). L-Tryptophan, a component of some artificial honeydews, has been shown to be effective at attracting and/or arresting chrysopid adults in treated olive trees (McEwen *et al.*, 1994) (Table 7.9). However, it has consistently proved more difficult to demonstrate reductions in pest numbers or reduced damage as a result of such applications (McEwen *et al.*, 1993). The use of semiochemicals for

Table 7.9. Numbers of green lacewings trapped 1, 2 and 6 days after application on 23.6.92 of a tryptophan spray that delivered 2 g of L-tryptophan per tree. There were 64 non-coloured sticky traps in each treatment.

Days after spraying	Numbers of lacewings		t-test P value	χ² P value
	Sprayed trees	Control trees		
1	30	7	0.0004	$P < 0.001$
2	39	15	0.0014	$P < 0.010$
6	82	36	0.0018	$P < 0.001$

manipulating natural enemies provides some promise but it is clear that there is an area where more research is required before semiochemicals are in practical use as a component of an IPM system.

7.8 Sterile Insect Technique and Autosterilization

The sterile insect technique (SIT) is most well known for its use to control the screw worm of cattle *Cochliomya hominivorax*, in the USA (Davidson, 1974) and more recently in North Africa (Lindquist *et al.*, 1992). Female flies oviposit on the wounds and cuts of living animals where the larvae feed and cause harm to the animal, damage the hide and carcass and, in severe cases, death of the animal (Drummond *et al.*, 1988). The first large-scale trial of SIT was made on the island of Curaçao in the Lesser Antilles. Flies, sterilized by gamma-rays, released by air at the rate of three flies per hectare per week, eradicated the screw worm in three to four generations (Davidson, 1974).

After this initial success another programme was initiated in Florida in 1958, where four flies per hectare per week were released. In June 1959, the last screw worm infestation was recorded and the cost of the programme had only been (approximately) US$7 million, estimated to be about a third of the potential annual losses in livestock to screw worms in the area. This success provided the incentive for the next even more ambitious programme, to eradicate the screw worm from 12.9 million hectares

in Texas and other Western States of the USA and to establish a 160 km wide barrier zone along the Mexico–US border to prevent reinfestation from Mexico.

In 1962, a rearing facility produced 100 million flies a week that were being used for release in Texas. These releases caused a dramatic decline in the number of reported screw worm cases between 1962 and 1964, and by 1967, screw worms had been officially declared as 'eradicated' from the USA, although invasions from Mexico did still occur. To reduce the number of invasions, an agreement was reached to extend the SIT programme in Mexico. This programme also proved effective and by 1983, there were no screw worm cases reported in the USA (Drummond *et al.*, 1988).

Since this major success, a number of other attempts were made to control a range of pests by SIT including mosquitoes (Morlan *et al.*, 1962; Weidhaas *et al.*, 1962), horn flies *Haematobia irritans* (Kunz *et al.*, 1974), heel fly *Hypoderma lineatum* (Kunz *et al.*, 1984), the Mediterranean fruit fly *Ceratitis capitata* (Mellado, 1971) and the blowfly *Lucillia sericata* (MacLeod and Donnelly, 1961). However, while in many of these and other cases experimental work has suggested a good likelihood of success, in practice the technique has proved too expensive to be used on a large scale or little more than temporary local population suppression was achieved (Wall, 1995). In general, SIT has found very limited practical use and tends to be feasible only when dealing with isolated populations of a particularly virulent pest and a high value

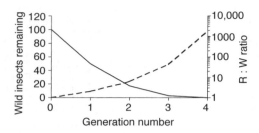

Fig. 7.7. The number of wild insects of either sex remaining in a population subjected to the release of a constant ratio of released males, equivalent to the initial number of fertile females (solid line). Also the change in ratio of released males (R) to wild (W) insects of either sex (dashed line) (after Wall, 1995).

product. The extensive rearing facilities required to breed and release large numbers of insect makes application expensive and the additional public relations problem of releasing billions of insect pests (often noxious blood feeding adult insects) makes effective control problematic (Wall and Shearer, 1997).

The use of SIT involves the mass rearing of the pest species, its sterilization and its release. Control is theoretically obtained through the mixing of the sterile insects, usually male, with the wild population, leading to a reduction in the fertility and, hence, population suppression. The degree of suppression obtained will be dependent on the competitiveness of the released males compared with their wild counterparts and the ratio of sterile to wild males (Fig. 7.7). Sterility may be induced through cytoplasmic incompatibility, hybrid sterility or through dominant lethal mutations. Sterility through cytoplasmic incompatibility occurs when a cross between two apparently conspecific populations results in only a partial embryonation in some ova, i.e. no fusion occurs between the spermatozoa and the ova (Boller, 1987). Hybrid sterility, which is available at present only for the suppression of *Heliothis virescens* (Lindquist and Busch-Peterson, 1987) involves the release of backcross insects derived from a cross between females of

Heliothis subflexa and *H. virescens* males. The progeny from this consist of fertile females and infertile males. When the fertile females mate with the males, the progeny again consists of the fertile females and infertile males, and this can continue indefinitely, or at least until there is a dilution of males due to wild populations entering the area. This technique is thought to have some potential, but little work is presently being done to locate closely related species or strains which result in hybrid sterility (Lindquist and Busch-Peterson, 1987).

The most commonly used SIT involves dominant lethal mutations that are induced either by irradiation or chemosterilization. Irradiation with gamma- or X-rays used to be considered preferable to chemosterilants because available chemosterilants posed a risk to humans and the environment (Campion, 1972). However, the availability of safer chemical sterilants (IGRs and chitin synthesis inhibitors) have now made it possible to develop alternative, cheaper systems for achieving insect sterilization. These techniques combine target systems with chemical sterilants and are referred to as autosterilizing systems (Jones and Langley, 1998).

7.8.1 Autosterilization systems

Autosterilization systems overcome many of the drawbacks of SIT; there is no need to rear and release large number of insects or to sterilize them by irradiation. The autosterilization systems utilize target systems that attract insects to a surface treated with IGRs or chitin synthesis inhibitors (Wall and Howard, 1994; Howard and Wall, 1996a; Jones and Langley, 1998), which then affects subsequent reproductive success. Although in principle, the use of autosterilizing systems is the same as target systems (Section 7.7.4), the potential for effective efficient pest suppression by autosterilization is in some respects much greater (Howard and Wall, 1996a). This is because the sterilized insects cannot reproduce which is equivalent to killing and secondly these individuals continue to compete for mates with remaining normal individuals which will result in non-viable progeny (Wall and Howard, 1994; Wall, 1995). The affect is most pronounced if both male and females of the target species can be attracted and sterilized (Langley and Weidhaas, 1986).

Examples of autosterilization systems include use of yellow cloth or plastic sheet treated with pyriproxifen to control green-house whitefly, black and white targets treated with triflumuron to control the lesser housefly *Fannia canicularis* in commercial rabbit houses, and control of midges in sewage beds (Langley, 1998). Populations of housefly *Musca domestica* have also been suppressed using autosterilization in commercial poultry houses in India (Howard and Wall, 1996a).

Case Study: Control of *Musca domestica* in poultry units using autosterilization (Howard and Wall, 1996a,b)

The housefly *Musca domestica* is amongst the most well known and familiar nuisance pests of human and livestock habitations (West, 1951). The abundance of organic waste material in animal houses combined with shelter and elevated temperature provide an ideal habitat for houseflies. Poultry houses are particularly prone to severe infestations with losses due to *M. domestica* estimated at US$60 million in 1976 in the USA (Anon, 1976; Axtell and Arends, 1990).

The use of autosterilization of *M. domestica* dates back to the 1960s when alkylating agents such as tepa, metepa, hempa and apholate were used as chemosterilants (La Brecque and Meifert, 1966; La Brecque *et al.*, 1962, 1963; Gouck *et al.*, 1963). More recently, the chitin synthesis inhibitor, triflumuron has been shown to prevent egg hatch and larval development (e.g. Table 7.10) in the laboratory populations.

Field trials carried out at a poultry house in Bhubaneswar in North-East India indicated the potential of autosterilization systems. Targets consisted of 20×50 cm rectangles of white polyester cloth treated with a triflumuron suspension and 50% w/v sucrose solution and stapled to a wire frame. Control targets were treated with 50% w/v sucrose solution alone. Fifty treated targets were placed in each of two treatment houses, suspended in five rows of ten. Similarly, 50 untreated (sucrose only) targets were suspended in the control house. Sampling had been initiated 2 weeks prior to introduction of the targets and this was continued at every 48–72 hour intervals for a further 6 weeks.

Table 7.10. The detransformed mean percentage egg hatch (±SE) in 5 day periods following 24 h exposure of adult houseflies, *Musca domestica*, to sucrose-baited targets dosed with 3% triflumuron suspension concentrate. In treatment 1, both sexes were exposed to the triflumuron treated targets; in treatment 2, only females were exposed; in treatment 3, only males were exposed; in the control, neither sex was exposed to triflumuron.

Days after exposure	Detransformed mean percentage egg hatch (±SE)			
	Treatment 1	Treatment 2	Treatment 3	Treatment 4
≤5	3.4 (±2.8)	7.8 (±4.6)	42.5 (±10.3)	91.9 (±2.4)
6–10	42.0 (±4.3)	53.8 (±8.1)	65.8 (±4.6)	91.2 (±1.5)
11–15	62.9 (±4.8)	77.7 (±2.8)	74.0 (±3.4)	92.6 (±1.3)
≥16	76.8 (±3.6)	83.5 (±2.1)	80.2 (±5.3)	88.2 (±1.5)

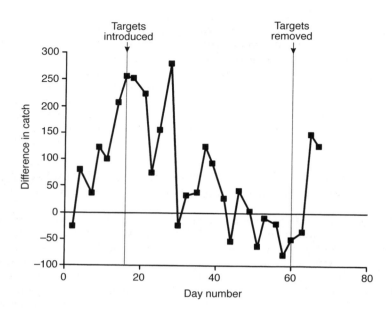

Fig. 7.8. Difference in *Musca domestica* catch at each sample date, between treatment house one and the control house. In treatment house one, 50 triflumuron impregnated, sugar baited targets were suspended between days 14 and 59, where day 1 = 15 March 1995. In the control house, 50 targets, baited with sugar only, were suspended over the same period. The deviation from zero (solid line) shows the change in the *Musca domestica* population in the treatment house relative to that in the control (after Howard and Wall, 1996a).

The number of *M. domestica* caught in all three houses showed significant changes over time (F = 15.89, df = 9.81, $P < 0.001$) and was significantly different between the control house and the treatment houses (F = 1.85, df = 18.81, $P < 0.03$). To illustrate this, the difference between the catch in the control and each of the treatment houses has been plotted in Fig. 7.8. Targets were introduced on day 14, following which the population in the treatment house dropped steadily relative to the control, until it was lower than that in the control house. Subsequently, the targets were removed on day 59, following which the housefly population increased and exceeded that in the control house.

7.9 Discussion

Cultural control encompasses techniques such as crop rotations and intercropping that have a long history of successful use. These are techniques that have been tested and refined over generations by trial and error and passed on to each new generation as the conventional accepted practice. However, the trend towards large scale monoculture and high input farming has led to a widespread abandoning of cultural control as the techniques appeared unnecessary when compared with the high level of control that could frequently be obtained by insecticides alone (Coaker, 1987). For instance crop rotation is a traditional technique used to both maintain soil fertility and reduce the incidence of pests, but the availability and use of chemical fertilizers and insecticides reduced the need for rotations and hence they have gradually become a less important component of intensive farming systems. However, cultural control is not just about the use and reintroduction of 'old' traditional techniques, although this may be relevant in many situations (especially where there is concern about excessive use of fertilizers and pesticides). There are more modern cultural/agronomic practices that can enhance productivity and have an important role in IPM, the use of reduced tillage and direct drilling techniques, for instance, as alternatives to traditional land cultivation. These can be used to reduce the time and effort required for land preparation, provide greater control over sowing time and also enable a larger area to be cropped (Webster and Wilson, 1980). Hence, there is a need to look more widely than just traditional methods for possible techniques that can be utilized in cultural control while evaluating new ways of incorporating the essential elements of traditional approaches into more modern systems. The subjects in cultural control that have been considered in this chapter (condition of the host, modification of the physical environment, crop rotation, tillage and planting date, intercropping and sowing/planting density) do not cover all types of cultural control. The subjects covered here were chosen because there are some well documented examples of each, but more importantly because they illustrate the point that links exist between different areas of cultural control. For instance, host condition cannot be separated from manipulation of the physical environment or intercropping effects from plant density, despite the fact that they are too often considered in isolation.

In order to get at the underlying principles, there is a need for a greater understanding of the ecology of pest/host interactions. This necessitates more detailed studies of the processes contributing to the infestation and exploitation of the host by the pest insect, for instance the role of host odour and other stimuli, and the influence of these on rates of immigration and subsequent dispersal. Such studies take time and at present there are too few researchers considering the problems. It is this long-term approach rather than short-term comparisons of various cultural treatments that will yield the general principles that will unify the concepts of cultural control and make the approach more widely acceptable. Whether the technique orientated approach of cultural control will gain sufficient credence without the driving force of 'product development' still remains to be seen.

It is interesting to note, when considering these broader issues of ecological principles in IPM, that the concepts considered important in cultural control, i.e. the processes of host location, colonization, reproduction, survival and dispersion, are the same ones that are important in host plant resistance, and in a slightly different way in the location and predation/parasitism of pests by natural enemies. In host plant resistance, the problem is one of genetic differences between individual plants of the same crop species and the effects this may or may not have on pest infestation, etc. The situation in cultural control is on a different scale, genetic heterogeneity over space and time between different crop species, e.g. crop rotation, intercropping, mixed cropping, and yet the location, colonization, reproduction, survival and dispersal of hosts by the pest remain the essential components of the interaction. These components are also evident in natural enemy/pest interactions especially in pest/parasitoid interactions. It is upon these three types of control option (host plant resistance, natural enemies and cultural control) that traditional farming systems have been based and it is perhaps within these areas that we should search for the unifying concepts in IPM, within which more modern control techniques such as chemical insecticides, behaviour modifying chemicals and biotechnology can be placed in context. However, before this can really happen our understanding of the processes involved in cultural control techniques must be brought up to the level of that of host plant resistance and natural enemy studies, without which cultural control will continue to provide the weak link in the chain.

8

Legislation, Codes of Conduct and Conventions

8.1 Introduction

Legislation, regulations and voluntary codes are used at national and international levels to ensure good practice in pest management, to govern the safe use and availability of pest control products, the quality of food and other products that require treatment against pests and the introduction of exotic organisms across national boundaries. Such legislation, etc. takes many forms and has a major impact on farming practices, the agribusiness and the consumer.

The legislation passed by governments can be one of two types, it is either passed as direct acts, orders or regulations or it is enabling legislation that authorizes other bodies (usually Ministries of Health and Agriculture) to issue rules, orders or directives. In most cases the legislation produced by governments will not intrude on the day-to-day practice of IPM but in specific cases it can have a significant and major impact on the general approach taken. One example of the role of legislation, that of defining quarantine regulations, is discussed below. This is one form of legislation that is generally applied throughout the world but other laws may be more specific to particular situations and circumstances. For instance, in East Africa there is legislation to ensure that there is a closed season for cotton growing in order to prevent a population build-up

of the pink bollworm (*Pectinophora gossypiella*), a monophagous pest of cotton (Hill and Waller, 1982). Legislation can also be used to ban the application of insecticides during specific periods, such as the ban imposed in Colombia on the use of synthetic pyrethroids to control the tobacco budworm *Heliothis virescens* on cotton during periods of maximum flowering (Sawaki and Denholm, 1989). This was carried out as part of a strategy to manage insecticide resistance in this cropping system. Such control by government through legislation will generally only occur where the imposition of certain changes is sure to guarantee major national economic or environmental benefits. The circumstances under which such legislation occurs will be dependent on the situation, and no general point can be made other than the obvious need to ensure that the legislation imposed by government can actually be put into practice.

Farmers who do not implement weeding of *Echinocloa* spp. (alternate hosts of stemborers *Chilo* spp.) in Burkina Faso are fined (Zethner, 1995) whereas in the case of the legislation in East Africa for the control of the pink bollworm, the law stipulates that all cotton plants should be uprooted and destroyed by a certain date in order to kill the diapausing larvae. However, many peasant farmers do not bother to destroy old plants by the appointed date and so in some areas there is considerable survival of diapausing lar-vae (Hill and Waller, 1982). In this situa-tion, the legislation reflects a genuine need but the lack of a means to implement and police it effectively reduces its chances of success.

The legislation, codes of conduct and conventions that deal with issues relevant to IPM are mainly concerned with quaran-tine regulations, registration of hazardous materials (pesticides and genetically modi-fied organisms), food quality standards and the exchange and exploitation of biological diversity.

8.2 Quarantine Regulations

The increasing levels of trade and transport links between countries and continents has meant an increased risk of accidental intro-duction and spread of pest insects (Dowell and Gill, 1989; Singh, 1988; Zimmerman, 1990; Section 6.7). Species that were previ-ously capable of dispersal only over short distances by natural processes can now be transferred inadvertently from country to country or from one geographical region to another by means of fast international transport links. The threat of the accidental introduction of a major pest species which could potentially cause massive problems for agriculture, forestry or public health has encouraged governments to take steps to reduce the likelihood of such introduc-tions occurring. Generally this is achieved through the enforcement of what are gener-ally referred to as quarantine regulations.

Quarantine regulations are produced by governments to reduce the chances of pests being introduced on articles imported from foreign countries. The legal basis of such quarantine is either legislation passed as acts, orders and regulations or enabling legislation that authorizes a body, usually the Ministry of Agriculture, to issue rules, orders or directives specifying which arti-cles are prohibited for importation, the exceptions to this ruling, and the condi-tions under which such imports can be made (including the need for permits,

inspections, treatments and quarantines (Kahn, 1982)). The articles in question may be plants, plant parts or materials, agricul-tural cargoes, soil, containers, packing material, plant growing media, baggage or mail; any article that could potentially har-bour a novel pest may need to be prohib-ited from entry. Pests such as viruses and bacteria that are transmitted in plant cut-tings and soil are particularly difficult to find and exclude. Insect pests, although a problem, probably represent less of a con-cern for the regulatory authorities.

In a survey of pests named in quaran-tine regulations in 125 countries, 614 were species of insects and mites (Kahn, 1982). Of these, the ten most frequently cited pests are listed in Table 8.1. Although major pests, these ten insects are not likely to be moved by natural means between regions, except perhaps for species such as *Leptinotarsa decemlineata* and fruit flies which can be dispersed on wind currents. Generally there is little point in regulating against pests if their dispersive ability will inevitably lead to their introduction, although quarantine measures may slow down their rate of spread.

The importation of living plants or plant parts represents the greatest hazard for introduction of exotic insects. For instance, imported germplasm, used to improve national crop breeding programmes, has a very great potential for facilitating the transfer of exotic pests (Kahn, 1983; Verma *et al.*, 1988). It is all a question of balanc-ing the risk associated with an importation of the germplasm with the potential bene-fits from the new plant material. The bene-fits from the imported material must exceed the economic or environmental costs should an exotic pest be transported and become established. Risk is the princi-pal element considered in the development of plant quarantine regulations (Kahn, 1977). The pest risk is the actual or per-ceived threat of moving pests of quarantine significance along man-made pathways, and risk is defined by an entry status which reflects the regulations, policies, procedures or decision made by quarantine

Table 8.1. The ten most frequently cited insects in the quarantine regulations of 125 countries (after Kahn, 1982).

Insects		Number of countries citing organisms
Scientific name	Common name	
Quadraspidiotus perniciosus	San José scale	44
Leptinotarsa decemlineata	Colorado potato beetle	40
Ceratitis capitata	Mediterranean fruit fly	39
Rhagoletis pomonella	Apple maggot	33
Bactrocera dorsalis	Oriental fruit fly	28
Popillia japonica	Japanese beetle	28
Anthonomus grandis grandis	Boll weevil	21
Anastrepha ludens	Mexican fruit fly	20
Rhagoletis cerasi	Cherry maggot	20
Phthorimaea operculella	Potato tuberworm	19

officials (Kahn, 1983). The entry status of a pest is represented by a continuum with conservative policies (and hence restrictive) at one end of the scale and liberal and unrestrictive policies at the other. A biologically sound position is reached when the pest risk is matched with the attitude towards entry status. Thus, a sound position would occur when a high pest risk is matched by a conservative entry status, or a low pest risk with liberal entry status (Kahn, 1983). However, positions other than those based on a match between risk and entry status may be taken. An economically based position would be influenced by the relative costs of effective quarantine compared with control if a pest entered the country. A conservative entry status might be adopted even for pests of low risk because the benefits from safe importation exceed the potential control costs of an inadvertent introduction. Alternatively, a high pest risk and a liberal entry status might be employed by governments importing grain during periods of famine or food shortage (Kahn, 1983). In general, the optimal strategy to be taken by the regulatory authorities would be to adopt a biologically sound position where the

resources allocated for the entry status of a pest are matched by the pest risk.

Internationally, regulations usually specify risk categories and safeguards with imported plant germplasm. If the pest is categorized as prohibited, the risks of its introduction are great, and the existing safeguards to prevent entry are probably considered inadequate. Under such prohibitions importation will probably also not be permitted even to government services. Post-entry quarantine procedures will be adopted for pests in the high risk categories for which adequate safeguards to prevent entry can be provided by transit through a government quarantine station. A restricted entry category for a pest would mean that a permit is required for the importation of material likely to harbour the pest. Plants will be subject to inspection and treatment on arrival. The no permit restricted category is similar to the restricted category but without the requirement for a permit. Some plants may fall within the unrestricted category which would allow unrestricted entry without the need for permits or inspection, etc.

Plants are grown in quarantine at special quarantine stations when it is not possible to certify the imported plant material as healthy and free from all pests and diseases. The quarantine stations have facilities where plants can be grown in biological isolation under high phytosanitary conditions until freedom from pests and diseases is assured. This is usually for a period long enough to break the life cycle of harmful pests or to extend beyond the duration of the normal resting stage of the insect. Third country quarantine is also sometimes used. Temperate countries are often used for the quarantine of tropical clonal or perennial crops because it is unlikely that the exotic tropical pest could survive if it escaped. If pests are found on imported material they may then be sent to a third country for the purposes of pest control (Hill and Waller, 1982).

Plant material under the restricted category will be inspected on arrival and treated if there is any suspicion of pest contamination. With insect pests insecticides often as fumigants are the common form of treatment (e.g. Yokoyama *et al.*, 1992, 1993a) but excision of plant parts involving removal of lateral branches and leaves (e.g. for leaf miners) high or low temperature treatments, irradiation (Benschoter, 1984; Burditt, 1986; Burditt and Balock, 1985; Couey *et al.*, 1985; Leibee, 1985) or even compression may be considered appropriate (e.g. Yokoyama *et al.*, 1993b). Heat treatment may be administered by hot water, air or vapour heat. The plant material is raised to a temperature that will kill the pest but not the plant. It has been used to control insects and mites in narcissus bulbs and strawberry runners (Kahn, 1977), the Caribbean fruit fly (*Anastrepha suspensa*) in mangoes (Sharp, 1986; Sharp *et al.*, 1989) and *Dacus* spp. in papaya (Couey *et al.*, 1985). Such measures should, however, be unnecessary when the exporting country has checked the material to certify that it meets the health standards for the importing country, although countries which lack well developed quarantine capability are more inclined to accept material certified elsewhere than countries having extensive services.

The ability to evaluate the hazard posed by the accidental introduction of exotic organisms also applies to those agents which are introduced for the purposes of biological control. Although biological control is generally thought to offer an environmentally benign alternative to pesticides, this does not mean they will have no environmental impact. Once an alien biological control agent has been introduced and established, it is often very difficult or impossible to eradicate. The growing interest in developing and utilizing biological control agents, particularly for augmentative releases (mainly macrobiologicals; Section 6.9) and biopesticides (pathogens; Section 6.8) has meant a larger number of organizations are now involved in the introduction of potentially alien species. Such trends have identified the need for generally agreed standards for procedures for the introduction of non-indigenous natural enemies and a promotional effect to ensure both practitioners and regulatory authorities are well informed of the risks as well as the opportunities (Waage, 1996). The FAO Council ratified a Code of Conduct for the Import and Release of Exotic Biological Control Agents in 1995 which recommends that governments establish authorities responsible for regulating introductions and ensuring that risks are properly evaluated. The Code specifies that organizations intending to introduce an alien biocontrol agent should prepare a dossier that provides information on both the pest and the natural enemy and indicates the risk involved to non-target species. The value of this Code is not so much in its precise wording but the opportunity it offers to harmonize safety procedures for biological control introductions across regulatory authorities worldwide (Waage, 1996).

8.3 Regulation of Pesticides

The regulation of pesticides, which in the context of pest management often refers to both chemical and biological pesticides, is usually carried out by a government agency. In the USA, these responsibilities are held by the Environmental Protection Agency (EPA) and the Food and Drug Administration (FDA), while in the UK, the role of providing regulatory oversight is held by the Pesticides Safety Directorate (PSD). At an international level, the Organisation of Economic Co-operation and Development (OECD; an inter-governmental organization with 29 member countries), United Nations Food and Agriculture Organisation (FAO) and the World Health Organisation (WHO) play important roles in the harmonization of pesticide procedures and regulations and to reduce the negative impact of pesticide use.

The FAO have produced an International Code of Conduct on the distribution and use of pesticides (FAO, 1986). Article 6.11 states: 'governments should take action to introduce the necessary legislation for regulation, including registration of pesticides and make provisions for its effective enforcement, including the establishment of appropriate educational, advisory, extension and health care services'. Registration should take full account of local needs, social and economic conditions, level of literacy, climatic conditions and availability of pesticide application equipment. The registration of pesticides by government authorities according to stipulated regulations provides the main method by which the distribution and use of pesticides can be controlled by legislation. Each new pesticide should be approved before it can be imported and sold within a country as this is the only real means by which governments can ensure that only those products deemed suitable to their particular needs and circumstances are made available. Two methods of restricting availability can be exercised: (i) not registering the product; or (ii) imposing conditions on registration and restricting the availability to certain groups or users in accordance with national assessments of the hazards involved in the use of the products (Article 7.3; FAO, 1986). Thus, the registration procedure and the regulations

governing the amount and type of information that needs to be produced by the industry in support of their application for product registration must be appropriate to ensure that only those products that are suitable are accepted. Hence governments provide legislation to ensure that the registration procedures for insecticides are stringent enough to reduce the hazard to the users and general public (thereby allaying public fears), objective enough to satisfy scientific opinion, while not placing such a heavy burden on insecticide development that they are uneconomical to produce. Whether the legislation produced to date has achieved a balance between these opposing influences is a matter for debate.

There are three terms that are commonly used in reference to pesticide safety: toxicity, hazard and risk. Toxicity refers to the capacity of an insecticide to cause harm, while hazard is the state of affairs that can lead to harm and risk is the probability of that particular event occurring (Barnes, 1976; Copplestone, 1985; Berry, 1994).

The toxicity of a substance is dependent on the dose administered or degree of exposure and each insecticide will produce a dose-response relationship with the organisms against which it is tested. The degree of toxicity is usually measured as the proportion of test organisms that die in a given sample. The dose at which 50% of a large population of test organisms are killed is called the LD_{50} (LD = lethal dose), expressed in milligrams of substance per kilogram of body weight. An insecticide having a low LD_{50} is more toxic than one having a high LD_{50}. Hazards, by contrast are evaluated on the reverse scale. A high hazard represents a greater problem than a low hazard. Hazard is very rarely assessed numerically, instead it is considered in relative terms, according to the effects on a particular organism under specified circumstances. For instance, a hazard would be given a low rating when organisms are exposed infrequently to very low doses of a microbial insecticide, having a low toxicity, via a route of entry that would occur in only unusual circumstances.

The route by which an insecticide enters the body is particularly important in assessing the level of hazard because some routes are more likely to cause poisonings than others. There are four recognized routes: oral, dermal absorption, inhalation and inoculation. Insecticides that are taken in orally and through dermal absorption represent the greatest hazard. Ingestion of an insecticide may occur if a spray operator drinks, eats or smokes during application, or, more rarely, directly swallows the insecticide. Dermal absorption usually occurs when areas of skin are exposed to spray or dust particles during preparation for application or during application itself. Dermal toxicity is lower than oral toxicity but it is enhanced when skin is warm and sweating, conditions that may occur in a temperate summer or in tropical countries (Copplestone, 1985). The hazard posed by the inhalation of insecticides is usually quite low because most droplets or dust particles used in insecticides are too large to be readily inhaled. The entry of insecticides through inoculation is the situation in which they are taken up through broken skin, cuts, scrapes and rashes; provided wounds are adequately covered inoculation does not present a problem. However, in developing countries there may be a greater risk because suitable dressings may not be available and farmers often walk barefoot with cracked and sore feet. In these circumstances insecticide poisoning through inoculation could be considered a hazard.

The different elements of toxicity, hazard and routes of entry have provided the basis for safety testing insecticides. The types of study undertaken as part of the hazard identification process in pesticide regulation is given in Table 8.2. From these data, an acceptable daily intake (ADI) for use in consumer risk assessment is derived (Harris, 1998). The ADI is defined as the amount of a chemical which can be consumed every day of an individual's entire lifetime in the practical certainty, on the basis of all known facts, that no harm will result. The ADI is based on data of no-observed adverse effect levels in a representative sensitive animal species or humans.

Another measure that is used by regulatory authorities to advance acceptable standards of safety for pesticides is the maximum residue level (MRL). The US EPA sets food tolerance levels for residues for each chemical insecticide to ensure that any residues remaining on produce will be below what is considered to be safe levels

Table 8.2. Toxicology studies assessed as part of the hazard identification process in pesticide registration.

Type of study	Used to assess
Pharmacokinetics and metabolism	Absorption, distribution, metabolism and excretion of pesticides
Acute single dose	Effects of high dose exposure after gastrointestinal, skin or eye exposure
Repeat dose (short-term to life-time)	Effects ranging from behavioural changes to alterations at the biochemical level (e.g. enzyme and hormone changes) and macroscopic/microscopic level changes (e.g. cancers)
Mutagenicity	Effects on chromosomes and DNA
Reproductive (teratogenicity/multi-generation)	Effects on reproduction and offspring neurotoxicity effects on the nervous system
Dermal absorption	The rate and amount of absorption of pesticides through skin
Human surveillance data and biomonitoring	Effects on operators or less frequently volunteers

(Zalom and Jones, 1994). MRLs are primarily used as a trading standard for bulked or composite samples rather than individual crop items (Harris, 1998). The regulatory authority is responsible for ensuring that any variation of residues within a bulked sample which conforms to the appropriate MRL does not give rise to an unacceptable risk to the consumers when individual items are consumed. A refinement of the MRL has been the Supervised Trial Median Residue (STMR) level. The STMR takes into account that 70% of samples analysed do not contain residues and consumer eating patterns to provide what is thought to be a more realistic measure to estimate theoretical maximum daily intake (TMDI) or more recently the national estimated daily intake (NEDI). However, even these more refined measures do not take into account the fact that children receive greater exposures (on a mg kg^{-1} of body weight basis), consume widely varying levels of food, multiple residue impacts and non-dietary exposure (Luijk *et al.*, 1998). Thus, there are still opportunities for refining the regulatory standards to ensure that pesticides are safely used. Some would argue that the costs of maintaining such regulatory systems and of managing pesticide risk is too high (estimated at $17 billion (1971–1995)), about 7.4% of gross pesticide sales in the USA (Benbrook *et al.*, 1996), and that a shift is needed away from reducing pesticide problems to finding safer effective pest control methods. Even though biopesticides may provide a solution to part of this problem, it is unrealistic to believe that pest management will at some future time not require chemical insecticides. It is just a matter of whether the chemicals can be made more selective and less damaging to the environment. There is every indication that this is the direction that the pesticide industry is moving.

8.4 Regulation of GMOs

A large number of organizations and individuals have expressed concerns about the potential environmental impacts of genetically modified organisms (GMOs). In terms of regulatory issues, concerns relate to the need to consider indirect and cumulative effects of GM crops (in particular their use in conjunction with chemicals), the possibility of gene transfer from GM crops to wild relatives and 'contamination' through cross pollination of non-GM crops by GM crops (Hill, 1998).

The basis for environmental safety of genetically modified organisms in the USA and Europe is one of 'substantial equivalence', where the novel product is compared to a closely related product that has an accepted standard of safety (WHO, 1991, 1995; OECD, 1993; FAO/WHO, 1996). The assessment occurs in two steps, the first being agronomic characteristics and the composition of the novel product and of the traditional counterpart are compared to assess equivalence within the limits of natural variation. Secondly, the introduced change is characterized and assessed to ensure there is no harm to human health and the environment (König, 1998). This process involves a molecular characterization and safety assessment of the introduced proteins.

The regulatory system for GMOs in the USA is under the overall guidance of the EPA through the Federal Insecticide, Fungicide and Rodenticide Act (FIFRA) and the USDA regulate import, movement, field trials and commercial release of GMOs under the Federal Plant Pest Act and the Plant Quarantine Act, administered by the Animal and Plant Health Inspection Service (APHIS). In Europe, the European Union implemented Directive 90/220/EEC in 1990 deals with the release and commercialization of GMOs. The Directive outlines the procedures for environmental safety evaluations for Europe, but approvals for field trials are issued by national authorities. However, approval for commercialization of a GMO product can only be gained by a complex authorization procedure in which all EU member states participate (Landsmann, 1998). This process is inevitably time consuming so that in Europe it can take between 17 months and two years to gain product approval compared to the USA where the whole process takes on average 5 months (König, 1998). The biotechnology industry consider that GMOs represent new hope for feeding the world's growing population (Stübler and Kern, 1998). The extent to which the industry is allowed to meet that goal will depend on convincing the regulatory authorities that their products indeed offer a safe, low risk opportunity for invigorating agricultural production and crop protection.

8.5 The Convention on Biological Diversity

The most important international event of recent years has been the United Nations Conference on Environment and Development (UNCED) held in Rio de Janeiro in 1992 which produced a document, the Convention on Biological Diversity, that was later ratified in 1995 by 142 countries around the world. The Rio-Conference, commonly known as 'Agenda 21', includes a chapter on 'Promoting Sustainable Agriculture and Rural Development' (Agenda 21: Chapter 14) that deals exclusively with the problems of pesticide overuse and the need to promote effective alternatives under the general banner of Integrated Pest Management. Agenda 21 states:

> Chemical control of agricultural pests has dominated the scene but its overuse has adverse effects on farm budgets, human health and the environment, as well as international trade. Integrated Pest Management, which combines biological control, host plant resistance and appropriate farming practices and minimises the use of pesticides, is the best option for the future, as it guarantees yields, reduces costs, is environmentally friendly and contributes to the sustainability of agriculture.

While some see the Convention as no more than an expression of goodwill (Pincus *et al.*, 1999), others have recognized its relevance to the future of pest management, particularly biocontrol (Waage, 1996), the structure and focus of international research organizations and the regulation of pesticide use (van Emden and Peakall, 1996). However, as van Emden and Peakall (1996) point out the target deadlines that were set in 1992 now seem hopelessly unrealistic. For instance, the Convention recommendation 'to establish operational and interactive networks among farmers, researchers and extension services to promote and develop integrated pest management' not later than 1998, has already passed with little progress made towards achieving this objective. In other areas, though, the Convention has had a far reaching impact – particularly relating to the exchange and exploitation of genetic resources. The signatories to the Convention may have access to genetic resources (e.g. pathogen isolates for developing biopesticides) in other countries party to the agreement but only with prior informed consent of the country concerned, and on the condition that any studies carried out involve the scientists from that country and that benefits derived from exploitation of the genetic resources will be shared in a fair and equitable manner. Such global commitments inevitably change the way in which issues such as biological diversity are viewed, and as much as national regulatory legislation for pesticide registration procedures, has a place in influencing the way pest management develops and the options that are ultimately available.

8.6 Discussion

Legislation and regulations impact on many areas of pest management, especially where products, quality standards and safety are concerned. Environmental concerns are now considered on a global scale which is having a similarly worldwide influence on how pest management measures are perceived. A number of controversies exist which need to be dealt with at a regulatory level. For instance, the legislation and regulations needed for registration of a pesticide require tests to be carried out on animals to determine how safe they are. There are many who object to such use of animals, and most people would agree that more appropriate means of assessing microbial insecticides, for instance, the use of an oral LD_{50} test with a *B. thuringiensis* active ingredient are not applicable since *B. thuringiensis* will not kill the animals; even an acute single dose toxicity test does not cause the animals harm, yet animals still have to be sacrificed and autopsied to satisfy registration procedures. Clear, pertinent protocols need to be established to cope with these new situations. As has already been discussed, the big challenge for the government agencies in the next few years is to ensure that the protocols for the development and release of genetically engineered organisms are stringent enough to alleviate concern and the possibility of major environmental disasters.

The controversy that surrounds the subject of quarantine is concerned with whether or not quarantine procedures are effective in keeping out exotic pests. The argument comes from the fact that it is not possible to measure how many pests would have entered a new region had quarantine procedures not been in operation (Kahn, 1982). Certainly pests do enter new regions in spite of quarantines but there is likely to be no answer to the controversy unless a comparison can be made between the number of man-made introductions prior to quarantine procedures and the number introduced since their establishment. It is unlikely that any nation would be willing to withdraw quarantine procedures just to test the hypothesis, so we will have to be satisfied with the fact that the perceived need for quarantine measures is high.

9

Programme Design, Management and Implementation

9.1 Introduction

Agricultural systems rarely remain static, rather they undergo a continuous series of changes which are influenced by government policy, economic forces and technological advances (Norton, 1993). Within this framework new pest problems arise which may be the result of a change in agricultural practice or introduction of a technology, e.g. secondary pest outbreaks due to overuse of chemical pesticides. Any change introduced into a crop system that is widely adopted by farmers may have 'knock-on' effects for another component of that system (Dent, 1995). If this effect is a negative one then a new problem will have been created. Over time any number of changes may combine to create a particular suite of pest or agronomic problems. In these circumstances it is self evident that to clearly define the extent, nature and cause of a new pest problem it is necessary to take an holistic view taking into account historical influences, the current control measures as well as the ecology of the pest and the cropping system. Time spent on the systematic, directed and focused definition of a pest problem, identifying the key factors influencing pest status, damage and control (Norton and Mumford, 1993) is a vital step in the process of devising appropriate solutions to pest problems (Dent, 1995). Solutions may be dependent on a programme of R&D, advice and education

expressed through an appropriate means of dissemination (e.g. extension service) or through farmer experimentation and adoption of their own or introduced practices.

9.2 Defining the Problem

A number of techniques have been developed that have proved useful for identifying key components and relationships associated with pest problems, including historical profiles, seasonal or damage profiles, interaction matrices and decision trees (Norton and Mumford, 1993).

9.2.1 Historical profile

An historical profile is most usefully constructed in a workshop situation involving representatives of all relevant stakeholders (e.g. scientists, farmers, extension personnel, input retailers/wholesalers/manufacturers) who can provide a range of different viewpoints and perspectives (Norton 1987, 1990, 1993; Norton and Mumford, 1993). An historical profile is constructed by consideration of the major factors which may influence the pest problem. The changes in these factors over time are graphed over a relevant period to the present (Norton, 1990; Fig. 9.1) and the contribution of the different factors assessed. In Fig. 9.1 the major developments in the Sudan Gezira (1925–1998) are mapped out. From this profile, it was possible to deduce that the

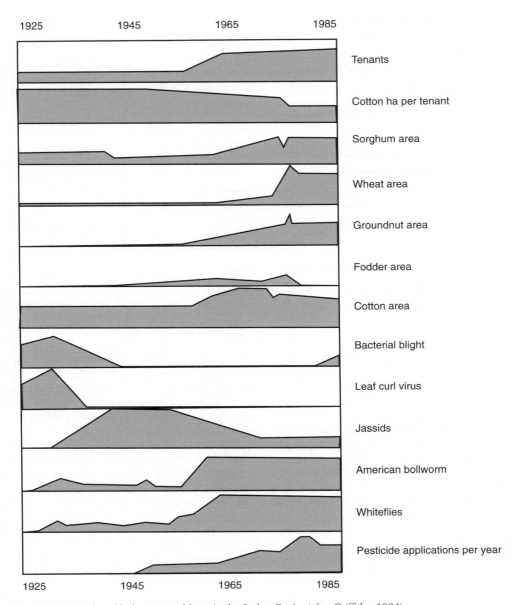

Fig 9.1. Historical profile for pest problems in the Sudan Gezira (after Griffiths, 1984).

changes that occurred in the area and rotation of crops in the 1950s probably contributed to the increase in American bollworm which led to a subsequent increase in insecticide use. The consequence of this was an increase in whitefly populations (Griffiths, 1984). In addition, cotton production and protection has been influenced by institutional and financial arrangements which affected farmers' incentives to grow good cotton. Through highlighting these changes and associations over time the historical profile provides a structured means of collating

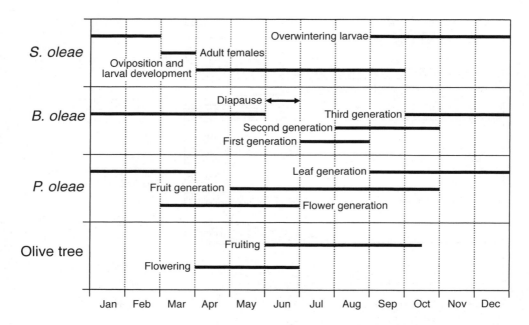

Fig. 9.2. A seasonal profile for olive tree pests in Spain (Claridge and Walton, 1992).

diverse information, brings a range of expertise to bear on a problem, and raises questions to hypotheses relevant to developing solutions (Norton, 1987).

9.2.2 Seasonal profiles and damage matrices
Seasonal profiles and damage matrices provide a simple but effective means of describing the structure of the pest–crop system and identifying where information is lacking or unavailable. A seasonal profile (Fig. 9.2) maps out the presence of different

life stages of pests, or location of major pests on the host throughout the cropping season (Norton, 1979, 1982b). They can also be used to describe events such as the timing of agronomic practices or application of control measures.

A damage matrix (Table 9.1) is used to identify which plant parts are damaged by a complex of pests. It effectively classifies the pest problems according to the type of damage they cause (Norton, 1982a).

Table 9.1. A damage matrix for pests of apples (after Norton, 1982a).

Tree components	Codling/ tortix	Winter moth	Sawfly	Apple aphid	Woolly aphid	Mites
Vegetative buds				*		
Leaves				*	*	*
Wood						
Fruit buds		*		*		
Blossom trusses	*	*		*	*	*
Fruitlet	*		*			
Fruit	*					

9.2.3 Interaction matrices

Interaction matrices identify how the pest interacts with the physical environment (climate, weather, soil type), the biotic environment (the host, natural enemies, competing species) as well as all the relevant information about the pest itself (life cycle, host preferences, mobility, fecundity etc.). A matrix is constructed to allow the primary effect of any column component on any row component to be shown (Norton, 1990; Fig. 9.3). A relationship between components is identified with a dot and where the cell is left blank then no direct relationship is thought to exist. The dots in a column indicate the primary effects that component has on a system while the dots in the rows indicate the various factors that can have an effect on that component. Interaction matrices have been used as a preliminary step in the development of simulation models (Holt *et al.*, 1987; Norton, 1990; Day and Collins, 1992) where they are used to identify the possible relationships and interactions between major components of a system. They also have an important role to play as a means of summarizing available information on the biology and ecology of a pest.

9.2.4 Decision trees

Decision trees provide a means of summarizing the range of options that can be adopted in managing pest problems. When used as descriptive tools, decision trees can identify constraints that prevent certain options being adopted, determine the time at which pest management decisions need to be made and ensure all possible combinations of strategies are considered (Norton and Mumford, 1993).

Each branch of a decision tree (Fig. 9.4) represents a particular pest management strategy, with the nodes of each branch indicating decision points. As the season progresses the number of options diminish while at the same time information on the level of pest attack potentially increases. Thus farmers are in the invidious position of having increased information as time progresses but a reduced number of

options with which to implement any form of control. In this way decision trees can play a dual role, of problem specification and as a means of studying problems of implementation, particularly those associated with the design of control recommendations.

9.2.5 Understanding the farmer

Insect pest damage is just one production constraint among many that a farmer has to consider. There are numerous factors within a farmer's family and the farming system that can affect the farmer's perception of a pest problem, the pest control decision-making process and the likelihood of adoption of new technologies and techniques. It is important that the problem definition stage of developing an IPM system includes and involves farmers; those individuals who may be the beneficiaries of 'technology' or participants in the process of development.

Farmers may be categorized according to characteristics, knowledge, beliefs, attitudes, behaviour goals (Beal and Sibley, 1967). Farmers' characteristics are the known facts about sex, age, education, literacy, and ethnic background, while knowledge includes what farmers know about alternative farming practices, cropping systems, inputs and markets. Farmers' attitudes can be described in terms of feelings, emotions and sentiments that may influence, for instance, decisions on whether or not to accept new ideas and adopt new technologies. The behaviour of a farmer is an important variable because knowledge of the way farmers have behaved in the past may well help predict how they will behave in the future. This is particularly useful in prediction of the utility of credit facilities, adoption of new technology, use of inputs and pest control decision making. Beliefs may influence both the farmers' attitudes and behaviour, and to some extent their goals. Beliefs are what the farmer thinks is true, whether correct or not, and they are based on the farmer's own experience and knowledge. What a farmer believes is attainable will

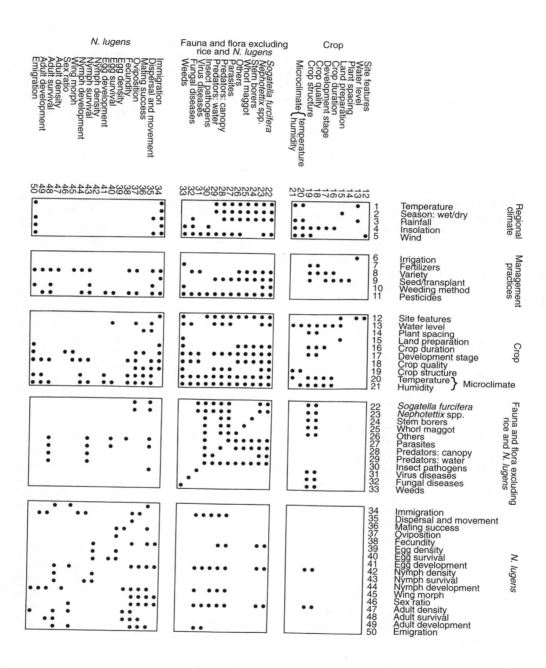

Fig. 9.3. An interaction matrix for *Nilaparvata lugens* (after Holt *et al.*, 1987).

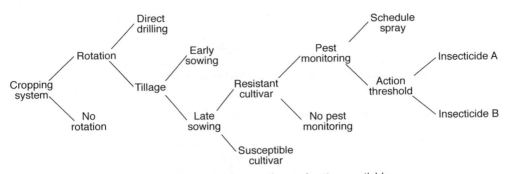

Fig. 9.4. A decision tree showing a sequence and range of control options available.

influence the farmers' and their families' goals and desires, whether they be wealth, property or status.

A number of techniques have been used to gain information about farmers, including participant observation (Shaner *et al.*, 1982), surveys (Byerlee *et al.*, 1980) (Table 9.2) and interviews (Reichelderfer, 1989). Discussions with the farmers will also allow the researcher to become acquainted with the farmer's terms, concepts and ideals, which can only lead to a much deeper understanding of a farmer's circumstances, farming system, reasoning and decision-making process (Shaner *et al.*, 1982). These informal, largely unstructured interviews and observations made by the researcher (condition of crops, animals, farm buildings, tools, pest damage) should provide an overview of the farmer's circumstances. This information can then be used to provide direction for research, highlight potential areas for change and identify gaps in the researcher's knowledge. It will also enable the social scientist to place a farmer into a 'recommendation domain', a group of farmers whose climate, soils, pest problems and socio-economic circumstances are sufficiently similar that a single recommendation would be applicable to that group (Reichelderfer, 1989).

9.3 Programme Design

The problem specification, definition phase of a pest management programme, if carried out well, is a rather involved process requiring a large number of inputs utilizing multidisciplinary expertise. However, to a large extent IPM programmes are devised without this comprehensive approach to problem framing. There are two fundamental reasons for this: the first, is that there is no theoretical framework to aid in the design process of a pest management programme; and the second is that there are no organizational structures available in which the necessary definition phase of development can be considered. The lack of a theoretical framework is evident when questions are asked along the lines of 'on what basis should I select a technique for the control of this pest?', 'is this control strategy sustainable?', or 'which combinations of control techniques are compatible for this pest?' At present there exists only a rudimentary theoretical basis on which to answer such questions (e.g. Southwood, 1977a). This need for a theoretical framework in IPM and the form it should take is considered further throughout this chapter. The second reason why an appropriate definition phase of a project rarely takes place is concerned with the types of funding provided for research and the use of inappropriate organizational structures for IPM research.

The emphasis on specialization in research has meant that research projects are considered for funding on a piecemeal basis and on individual merit. There is little or no coordination between projects and in some cases different institutes working on similar projects do not know of the

Table 9.2. The general type of information that should be collected from an exploratory survey (Shaner *et al.*, 1982).

A. Dates of data collection .... **Collector** ....

B. Farm access
Distance to nearest road usable by: e.g. motorcycle, 4-wheel drive vehicle, truck
....
Distance to nearest all-weather road ....
Distance to other transportation facilities: e.g. river, canal, airfield, railroad
....

C. Farm land status (hectares)		**D. Farm land use** (hectares)	
Privately owned	....	Crops	....
Rented	....	Other tillable purposes	....
Other: e.g. village	....	Pasture	....
or tribal land	....	Forest	....
Total	....	Other: e.g. fish ponds	....
		Wasteland	....
Note: Totals for C and D should be equal		Total	....

E. Other information about land
Number and sizes of land parcels in farm ....
Distances from farmhouse to field ....
Percentages of total farmland suitable for:
    Motorized equipment ....
    Irrigation ....
Access to other land and resources: e.g. community pasture and forest, roadside pasture ....
....

F. Farm enterprises:	Major enterprise	Minor enterprise
Crops: e.g. species, varieties, principal uses.		
Cereals	....	....
Root crops	....	....
Vegetables	....	....
Fruit	....	....
Other tree crops	....	....
Other: e.g. sugar cane, sisal	....	....
Animals: e.g. breeds, sex,		
number, principal uses		
Cattle	....	....
Buffalo	....	....
Sheep	....	....
Goats	....	....
Swine	....	....
Poultry	....	....
Non-agricultural enterprises:		
e.g. spinning, weaving, pottery making.	....	....

G. Farm labour ....

H. Power, equipment and tools
Power   ....
Equipment   ....
Tools   ....

I. Farm buildings and facilities
Storage facilities   ....
Processing facilities   ....
Livestock housing and yards ....
Irrigation facilities   ....

Table 9.2. Continued

Other ....

J. Marketing of output ....

K. Acquisition of inputs ....

L. Estimates of income, expenditure and savings
Farm income ....
Farm expenditure ....
Off-farm income ....
Savings ....

M. Type of household: e.g. nuclear or extended
Ethnic background ....
Number in household ....
Rights and obligations of members by age and sex ....
Characteristics of members:
Literacy and education ....
Health ....
Knowledge: e.g. farming and off-farm experiences ....
Beliefs: e.g. what the person thinks is true ....
Attitudes: e.g. feelings, emotions, sentiments ....
Behaviour: e.g. past actions ....
Goals ....
Other ....

N. Miscellaneous: e.g. help from others, obligations ....

other's work. While this approach to funding research may be highly appropriate for promoting innovative, specialist projects within a single discipline it is much less suitable for promoting the objectives of a coordinated multidisciplinary research effort such as that required for a research programme in IPM. Even if the separate components needed for an integrated approach were funded, there would still be little hope for integration because no organizational structure exists to coordinate the efforts and direct them towards a common goal. This lack of an appropriate organizational structure for management of research projects has important implications during both the definition and the research phases of a programme. During the definition phase it is important that a common paradigm of the problem be developed to integrate the individual goals of each research group and to provide a common objective (Dent, 1992, 1995). In the absence of this, each group will carry out their own research in isolation from every other group. This will result in a number of separate solutions to a number of unrelated problems, because each group will perceive their objectives differently according to their own perceptions and expertise. Thus, organizational structures have an important role to play in pest management research to provide an appropriate institutional framework within which integration can take place.

9.3.1 Research status analysis

Part of the process of programme design requires an analysis of the current status of research, in order that a programme of R&D can be properly defined and built on existing know-how. A good understanding of the research process enables R&D to be directed more effectively towards the successful development of IPM systems (Dent, 1997b). Research on especially well established forms of control, e.g. pesticide development and host plant resistance, tend to follow a roughly similar sequence or pathway. Such pathways arise through accumulated experiences of the work of many scientists and often provide an optimum pathway that can

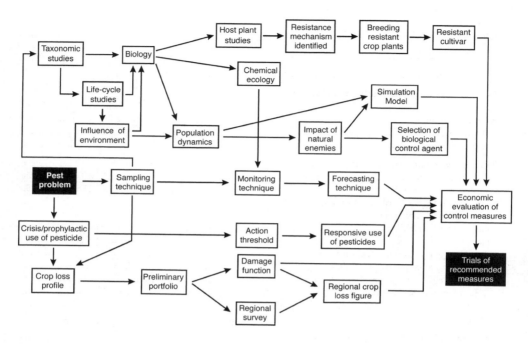

Fig. 9.5. A hypothetical example of a research pathway for development of an IPM programme.

be devised only with hindsight (Dent, 1995). Research pathways are available for pheromone monitoring systems (see Fig. 2.11), classical biological control (Pschorn-Walcher, 1977), host plant resistance (Gallun and Khush, 1980) and biopesticide development (Baldwin, 1986). They are also implicit in most of the R&D processes in pest management. Diagrams of research pathways provide a simple, visual means of describing the research process and in addition allow a systematic appraisal of progress (Fig. 9.5). The research pathway can be taken one stage further and a formal Planning Evaluation and Review Technique (PERT) which utilizes a critical path analysis in order to identify critical tasks and the amount of slack available if a programme of research is to be completed on time (Dent, 1997b). Such approaches provide an overview of the status of research and the likelihood of the availability of a 'new tech-

nology'.

9.3.2 Selection of control measures – pest types

The manager now has information about the pest and the control options available, but the question then arises, how to select appropriate control options for the pest under consideration? Knowledge of the techniques that have previously been used successfully will certainly be useful in this context, but what if the techniques were not successful or have been successful in slightly different circumstances? On what basis should a manager consider the different options for control? An outbreak classification system such as that proposed by Berryman (1987) (Table 9.3) can be of some value. For example, suppression with insecticides may be considered a useful tactic against pests with sustained or pulse eruptions, while cultural practices aimed at improving host resistance are useful strategies for lowering the probability of all types of outbreaks (Berryman, 1987). However, what is really required is a theoretical framework that will specifically combine a particular pest type with appropriate forms of control option. Few attempts have been made at this and the work that has been done (Conway, 1976; Southwood, 1975, 1977a; Stenseth, 1981, 1987) provide only conflicting conclusions. This work does, however, provide the basis for further studies and forms part of the much wider need for a more rigorous formal theoretical framework within insect pest management.

The work of Conway (1976), Southwood (1977a) and Stenseth (1981, 1987) has revolved around the concept of r- and K-selection and, particularly, the extremes of this r–K continuum. r-selected insects have evolved under highly uncertain and variable environments; they are characterized by a high rate of increase, strong dispersal and host-finding ability and a small size relative to other members of Insecta. These r-selected insects are particularly good at exploiting temporary habitats which means they are ideally suited for effective colonization of short-lived annual crops. The damage they cause is often a function of the actual number of infesting insects and the ability to overwhelm the host. By contrast the K pests acquire pest status largely because of the character or quality of the damage they cause (Conway, 1984a). The K pest insect tends to have greater size than an r pest, a lower potential rate of increase, a greater competitive ability and a tendency for more specialized food preferences. Conway (1976) cites sheep ked, codling moth, rhinoceros beetle and tsetse fly as examples of K pests and the desert locust, black bean aphid, housefly and black cutworm as examples of r pests.

K- and r-selection is a generalized concept that does not apply to every situation; r and K strategies are only relative and not absolute phenomena, i.e. a species is an r strategist or a K strategist only relative to

Table 9.3. A general scheme for classifying insect pest outbreaks (from Berryman, 1987).

Pest outbreak	Class
1. Outbreaks do not spread from local epicentres to cover large areas	**Gradient**
(a) Short delays in the response of density-dependent regulating factors	
(i) Outbreaks restricted to particular kinds of environments (site or space dependent)	Sustained gradient
(ii) Outbreaks follow changes in environmental conditions (time dependent)	Pulse gradient
(b) Long delays in the response of density-dependent regulating factors	
(i) Outbreaks restricted to particular environments (site or space dependent)	Cyclical gradient
(ii) Outbreaks follow changes in environmental conditions (time dependent)	Pulse gradient
2. Outbreaks spread out from local epicentres to cover large areas	**Eruptive**
(a) Short delay in the density-dependent response at the low-density equilibrium	
(i) Short delay in the density-dependent response at the high-density equilibrium	Sustained eruption
(ii) Long delay in the density-dependent response at the high-density equilibrium	Pulse eruption
(b) Long delay in the density-dependent response at the low-density equilibrium	
(i) Short delay in the density-dependent response at the high-density equilibrium	Permanent eruption
(ii) Long delay in the density-dependent response at the high-density equilibrium	Cyclical eruption

some other species. To complicate this further, the strategy of individuals within a population is population specific and not species specific (Stenseth, 1987). This is because a species may behave as an r or a K strategist depending on the environment under which a particular population has evolved and is living. Obviously, on the basis of these qualifications it is important that something is known about the habitat of the species as well as the characteristics of the population itself.

Southwood (1977b) described the habitat in terms of durational stability and favourableness. The K strategist will be selected for in habitats having a high abiotic predictability, low biotic predictability and a high habitat favourability (Stenseth, 1987). The biotic environment (competitors, host defences, natural enemies) plays the dominating role in the evolution of the K strategists. Hence, you would expect to

find K strategists in more diverse environments than r strategists which are favoured by higher biotic predictability not low abiotic predictability. In an attempt to account for differences in population regulation between r and K strategists, Southwood (1975, 1977a) and Southwood and Comins (1976) developed a synoptic population model based on habitat stability, population density and population growth (Fig. 9.6). The model is derived from variations of the natality curve, natural enemy action and intraspecific competition as expressed by nine parameters. The response surface of this model produces some interesting features that shed some light on the relative value of different control options. The major features of the response surface are the endemic ridge, the epidemic ridge and the natural enemy ravine (Fig. 9.6).

The entire population of an extreme r strategist occurs on the epidemic ridge,

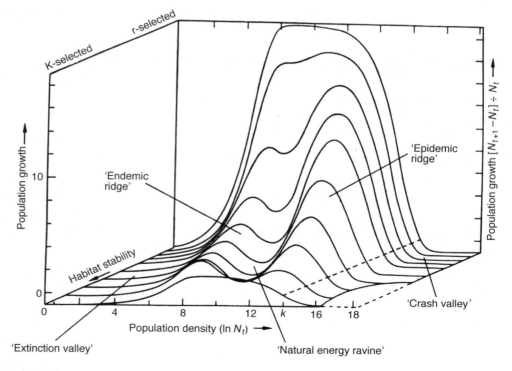

Fig. 9.6. The response surface of a synoptic model of population growth (Southwood and Comins, 1976).

with the marked fluctuations of the boom and bust dynamics between generations a consequence of the high r values. Examples of species suffering such marked changes may be some aphid species and phytophagous mites (Horn, 1988). The extreme K strategist lies exclusively within the endemic ridge where only small amplitude population fluctuations occur and the population is characterized by the high survival rate of offspring but a low recruitment rate. The population is often in a region of equilibrium determined by competition (Southwood and Comins, 1976).

The natural enemy ravine is the area in the model where natural enemies are important in the population dynamics of a pest population. The synoptic model has shown that where the ravine drops below the zero growth contour there is a lower equilibrium point maintained by predation and a release point 'R' beyond which the population becomes epidemic. The model also shows that at both extremes of the r–K continuum natural enemies will have very little influence on population regulation. At the r end of the ravine the numerical response of the natural enemies is limited; they are often polyphagous predators, large relative to their prey, and have equilibrium points at relatively low densities (Southwood and Comins, 1976). The short-

age of time in the temporary habitat limits the numerical response and the r strategists simply outbreed predation (Horn, 1988). At the K end of the ravine the saturation point of the numerical response and the handling time are higher and the attack rate may be limited by prey size and other defensive features of K strategists.

From these differences between r and K strategists it was possible to draw up a framework for potential control options to be used against particular types of pest (Table 9.4; Southwood, 1977a). Natural enemies will obviously be of most use in the intermediate position on the r–K continuum and should be the preferred strategy against such pests, e.g. many forest Lepidoptera (Horn, 1988). Natural enemy regulation of insect pests of this type may be enhanced through the use of conservation and augmentation techniques. Classical introductions of natural enemies and use of selective insecticides could also prove effective against the r–K intermediate pests.

The rapid development of outbreaks of r pests can make them very difficult to control. Natural enemies can rarely respond quickly enough to provide a suitable level of control, but because r pests can rebound after even very heavy mortalities, insecticides have to be used more or less continuously (Conway, 1984a). Polygenic resistant

Table 9.4. The principal control measures appropriate for different pest strategies (Conway, 1984a).

Control measure	r pests	Intermediate pests	K pests
Insecticides	Early wide-scale applications based on forecasting	Selective insecticides	Precisely targeted applications based on monitoring
Biological control		Introduction of enhancement of natural enemies	
Cultural control	Timing, cultivation, sanitation and rotation		Changes in agronomic practice, destruction of alternative hosts
Resistance	General, polygenic resistance		Specific, monogenic resistance
Genetic control			Sterile mating technique

crops (utilizing horizontal resistance) and animals can be used to reduce overall rates of increase and cultural control may be of help if it serves to reduce the size of immigration, e.g. early or late planting.

K strategists are theoretically suitable targets for genetic male-sterile control because they can be forced to low population levels where such control techniques will have a major impact. Insecticides may also be appropriate where small populations cause high damage such as fruit blemishes, but timing of application will be critical, e.g. codling moth control. Monogenic host resistances can be used effectively against K strategists because a small change in their habitat or niche can have a major effect on population dynamics. For the same reason cultural control may be effective, since it reduces the effective size of the pest niche (Conway, 1981).

The recommendations for changes in habitat structure as an effective preventative control measure is one area in which the analysis of Stenseth (1981) coincides with that of Southwood (1977b) and Conway (1976, 1981). In other ways the analyses produce conflicting scenarios and hence different types of recommendations for control. The approach used by Stenseth (1981) was based on the equilibrium theory of island biogeography (MacArthur and Wilson, 1967). The basis of the approach was to aim to have the potential pest species present for as short a time span as possible. The environment was assumed to remain constant, which excluded predators and parasitoids from the analysis. Stenseth (1981) came to the following conclusions about control strategies based on r and K strategies:

1. If there are methods for reducing immigration into empty patches by almost 100% then, regardless of the demography of the pest, most of the resources should be used on this. The techniques used must be efficient and should be applied with great spatial variability between patches.

2. If no particularly effective methods for reducing immigration exist then separate predictions for r and K strategists in the same environment and habitat complex should be as follows:

(a) For r strategists resources should be devoted to influencing extinction rates through mortality and birth. A greater emphasis should be placed on reducing the birth rate rather than increasing mortality. The techniques should be applied with as little variation between patches as possible, but as variable as possible over time.

(b) For K strategists, resources should in most cases be devoted to reducing dispersal, and techniques should be applied in space and time as in (1). Remaining resources should then, in most cases, be devoted to increasing mortality.

(c) These recommendations provided by Stenseth's (1981) analysis differ from those advocated on the basis of the Southwood and Comins (1976) synoptic model. The latter's approach advocates application of insecticides (affecting mortality) for r-selected species as well as for some K-selected species. The analysis of Stenseth (1981), however, suggests that insecticides may only be appropriate in some K-selected species. Stenseth (1981) argues that since most pest species are r-selected this apparent difference in strategy for insecticide use may explain why insecticides have proven so inefficient in protecting the world's food supply. Thus these differences represent quite a contentious issue.

The difficulty at present is how to resolve the differences between these two analyses to the benefit of pest management. Stenseth (1981, 1987) suggests that the differences are primarily due to different assumptions made about the patch dynamics and the need in Stenseth's model for effective instantaneous measures for control. Stenseth (1981) also assumes that the r and K pests exist in the same environment and habitat complex while Southwood (1977a) stressed the different dynamics of the habitats of r and K species. Any preference for either model and analysis will then be dependent on how the habitat template (Southwood, 1977b) is viewed. It is obvious that there is a need for greater understanding of the dynamics of the habitat template if this dichotomy of opinion is to be resolved (Stenseth, 1987). Further work is needed on this subject and the more general relation between successful control options and different pest types. The theoretical studies carried out to elucidate principles will ultimately also require supportive field data if the subject is really to make headway. This surely must represent one of the major study areas in IPM in future years.

9.3.3 Selection of control measures – types of control

Control measures may be selected because they either reduce the initial pest numbers infesting the crop (N_t) or they are used to reduce the rate of population increase (r) after colonization. An IPM programme involving a number of control measures would aim to reduce both N_t and r. Measures that reduce N_t include quarantine procedures, eliminating alternative hosts, crop rotations and vertical resistance – anything that impacts on host finding and establishment of colonies. Control measures that reduce r can affect mortality, reproductive rate, development or generation times will have an effect on the intrinsic rate of natural increase. Antibiotic resistance can influence all of these factors whereas chemical pesticides are normally used simply to increase mortality of a particular development stage. Natural enemies will also affect mortality but chronic infection by pathogens can slow development rates or affect reproduction. Insect growth regulators will act in a similar way while the sterile insect technique will reduce r through decreasing mating success and hence, reproduction. The choice of control measures should depend on the relative impact on N_t and r that is required. However, more often decisions are made on a more pragmatic basis depending upon what measures are available, practical, fea-

sible, economically desirable, environmentally and socially acceptable and politically advantageous (Norton, 1987; Bomford 1988; Norton and Mumford 1993). Control measure can be assessed for each of these characteristics in a feasibility table (Bonford, 1988; Prinsley, 1987). A control attribute table can also assist in the appraisal of control measures to ensure the most appropriate measures are introduced (Table 9.5) (Dent, 1995).

9.3.4 Selection of control measures – operational factors

A large number of operational factors will influence how a control measure is deployed, used and its subsequent efficacy. Control measures may be used in either a prophylactic or a responsive way, although most are limited by their nature and attributes to one or other of these. A control measure that is used prophylactically is deployed without any assessment of need or economic benefits. This usually means it is deployed before a farmer has knowledge of whether pest populators are present or likely to reach damaging levels. Examples include use of resistant cultivar, cultural control techniques, and calendar spraying of insecticides. A responsive deployment of a control measure occurs on the basis of knowledge or need and likely economic gain by the farmer (Vandermeer and Andow, 1986). In practice, deployment of a control measure in a responsive way depends upon availability of reliable information from a pest monitoring and forecasting system. Ideally use of a responsive measure would be tied to an economic threshold model, although this is too rarely the case. Mumford and Norton (1987) reviewed the conditions that favour use of prophylactic and responsive control measures (Table 9.6).

9.3.5 Objectives and strategies

The setting of goals/objectives is an essential component in the design of an IPM programme. Goals must define what it is that the programme is attempting to achieve. If the goal(s) are inappropriate

Table 9.5. Control attributes table – a hypothetical example.

Control measure	Chemical insecticide	Microbial insecticide
Cost: H, high; M, medium; L, low		
Product purchase	M	M
Application	L	L
Equipment/machinery	H	H
Operating/maintenance	M	M
Time	L	L
Labour	L	L
Control characteristics: 1, good; 5, poor		
Reliability	2	4
Specificity	3	2
Ease of use	2	2
Efficacy		
Speed of action	2	4
Toxicity to target organ	1	2
Compatibility (natural mortality factors)	5	2
Duration of control	3	4
Hazard: H, high; M, medium; L, low		
Operator	H	L
Family	H	L
Consumer	L	L
Environment	H	L
Infrastructure support requirements		
Extension services	H	M
Diagnostic laboratories	M	M
Credit facilities	M	M

Table 9.6. Relative conditions which favour use of prophylactic and responsive control measures (from Mumford and Norton, 1987).

Prophylactic	Responsive
Regular pest attack (in space and time)	Irregular pest attack
Frequent pest attack	Infrequent pest attack
Endogenous pests	Exogenous pests
Multiple pest complexes	Single major pest
High rate pest increase	Slow rate of increase
High damage cost	Low damage cost
Less effective control	More effective control
Long lasting control	Short duration control
Few bad secondary control effects (pollution, resistance, etc.)	Serious bad secondary effects
Low control cost	High control cost
Limited choice of controls	Wide choice of controls
Poor natural control	Good natural control
Monitoring technique difficult	Monitoring technique easy
High managerial cost	Low managerial cost
Poor access to advice	Good access to advice
More risk aversion	Less risk aversion

then *de facto* the programme will fail. A goal is the benchmark against which the success of the programme will be judged. A goal is a generic term combining the aspirations of the programme in terms of its vision and levels of ambition (Karlöf, 1987). 'Vision' refers to the image of the future that encompasses current knowledge, hopes, aspirations and opportunities The level of ambition places this in the context of the ability or will to perform, to achieve a specified standard. Thus, these two aspects of 'vision and level of ambition' combine to provide a goal. Each goal is linked to the activities necessary to achieve them. A strategy identifies the means (the activities) by which the goal will be achieved. Karlöf (1987) considered goals and strategies to be linked in a hierarchical way such that the goal of one level becomes the strategy of the level above it. Conversely, the strategy at one level becomes the goal of the level below it. Dent

(1995) used this to provide a descriptive summary of the goal/strategy relationship, as a strategic tree (Fig. 9.7). In designing an IPM programme the role of different organizations, which may include governments, research institutes, extension services, farmers, agrochemical companies, can be clearly defined using a strategic tree.

9.4 Programme Management

Programme or project management refers to the techniques used by those individuals leading R&D teams involved in the development of pest control products, techniques or IPM programmes. It is through the process of management that the efforts of individuals within and between research teams are coordinated, directed and motivated towards the achievement of particular goals. Ineffective management will reduce the likelihood of a research project

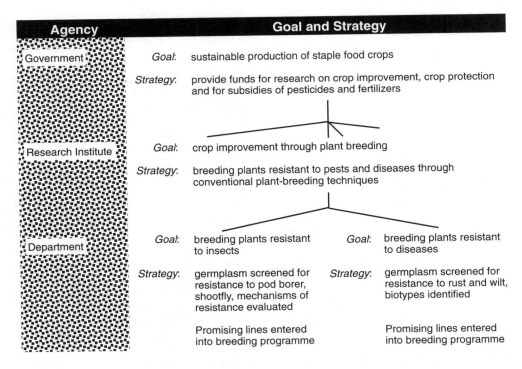

Fig. 9.7. A strategic tree for sustainable production of staple food crops.

obtaining its objectives. A neglect of research management has to some extent been responsible for the slow progress made in the development of interdisciplinary IPM systems. This is disappointing, as much because the techniques of management are well established and are readily applicable to research and IPM, as for the lost opportunities for developing successful IPM programmes. There are many books on management that are relevant and should be read by anyone interested in becoming a research manager (e.g. Gilman, 1992). Dent (1995) specifically addresses issues relating to IPM programme planning and management. The subject is only given cursory coverage here dealing with organizational structures and management.

9.4.1 Organizational structures

Structure within an organization is created by management for the purpose of division of labour and coordination of activities towards common objectives. The structure will be used to define research tasks and assign responsibilities as well as to develop clear channels for the flow of information. The structure may be formal or informal. Formal structures tend to be prevalent in government institutes and commercial companies – the hierarchy is clearly delineated and formally adhered to – whereas in universities less formal structures exist and the division between levels is less rigidly enforced. Both have their advantages and disadvantages as far as coordinating research is concerned. With more formal structures clear lines of authority exist and managers can if they wish actively direct and control the research that is carried out. By contrast in universities few researchers have direct authority over peers and coordination of multidisciplinary teams is dependent on cooperation and mutual agreement. Hence, in universities there may be situations in which it is less easy to coordinate and direct the research activities towards specific goals.

The emphasis on research for the last 25 years has been for increasing specialization and this had seemed particularly necessary with pest management where detailed specialized information has been required. This specialization, combined with a coincidental decline in resources, has caused research organizations to concentrate on what they are recognized as doing well, reducing any opportunity for maintaining a broad based multidisciplinary approach to research. Universities with traditional areas of expertise have increasingly narrowed their research interests so that few have the necessary range of expertise to cater for an integrated multidisciplinary approach to pest management. The result has been that few organizations now have the required breadth of expertise to independently carry out research of this kind. Where the expertise has existed the pervading doctrine of specialization has meant projects have been taken on piecemeal, and little attempt has been made to develop programmes of interdisciplinary research. Pest management problems have for far too long been considered on this very narrow basis and if the process is to be reversed attention will need to be paid to structural organization. The problem is one of creating suitable organizational structures that promote integration of pest management research. One solution has been the increased emphasis on collaboration between organizations that have complementary expertise (Fig. 9.8).

The hierarchical organization depicted in Fig. 9.8 could also be used in an organization where all the necessary expertise existed at the one location. An alternative, however, would be a matrix organization which is a combination of functional research groups that integrate various activities of different functional groups on a project team. The functional research groups provide a stable base for any specialized activities and a permanent location for individual members of the research group. A project team will be set up, seconding individuals from appropriate functional groups, as a separate unit on a temporary basis for the purpose of solving a specific research problem. The matrix organization forms a two-way flow of

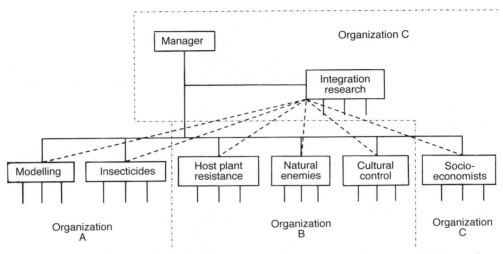

Fig. 9.8. An organizational scheme for the type of hierarchical relationships that may exist for a research programme carried out between three collaborating organizations, A, B and C; (———) formal relationships, (-------) functional relationships.

authority and responsibility (Fig. 9.9) and can be very effective if the project team has a very clear objective and the composition of the team is carefully chosen (Mullins, 1989). It also provides a flexible approach to problem solving and integration of expertise, as well as good opportunities for individual development and initiative. However, management skills or matrix structures are highly advanced and demanding. For this reason alone, they are probably not appropriate for organizations such as universities where managers traditionally have no management training. In other organizations, use of a matrix structure should be engaged only with caution ensuring the good managers hold the key positions as opposed to leading scientists. It is too often assumed that a good scientist makes a good manager, while in practice this is rarely true (Dent, 1996).

Appropriate organizational structures are fundamental to good management. Without them, integration of research will be less likely to occur and at a different level it may be found to affect motivation, innovation, morale and decision making, and exacerbate conflict and poor coordination. Much more care needs to be given to

devising appropriate organizational structures for research programmes in pest management.

9.4.2 Management as an integrating activity

Management is not a separate discrete function, it combines a variety of activities that include planning, organizing, directing, coordinating and motivating.

Planning is an activity that requires the manager to think ahead, to decide what needs to be achieved and the order of priorities for action. In applied research it is not enough to trust to good fortune to bring you to a point that will provide a satisfactory conclusion to the work. The end objective needs to be carefully selected and defined and the means of achieving it fully outlined.

Of course the plan for the research work may have to be changed, according to information gained during the course of the research programme, but if this is the case then the plan would have to be revised and the work developed along these new lines. Without a definite plan it is very easy for research to lose direction and while this is perfectly admissible in pure research it is less easy to justify in an applied subject

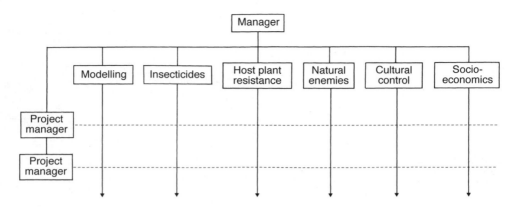

Fig. 9.9. An example of a matrix organization with a two-way flow of authority and responsibility via the project manager and the manager of each functional research group.

such as pest management research. There will always be exceptions to this rule, the uncovering of a promising new area of research being one of them, but generally, unless a manager can plan an effective research programme, it is unlikely that suitable goals will be attained.

A number of techniques exist to assist in this process. The most commonly used techniques are activity or Gantt charts (Dent, 1992), logistic frameworks but in addition implementation plans that define the operating procedures, performance standards, and decision rules for a research programme have proven of value in multi-institute multidisciplinary R&D programmes (Dent, in press). Even with such aids, providing direction and focus of teams of scientists will not be an easy task for a manager. The natural inclination and training of scientists creates independence of thought and action and hence differences in rates of progress and the uncovering of novel ideas, techniques, etc. will all confound efforts to keep the research task goal orientated. This task of keeping the programme on course will be easier in some circumstances than others. Where direct lines of control and authority are permissible the manager has some recourse to these for use in directing work. In universities, however, where lecturers working in a collaborative programme have no formal authority or influence over one another, other than that of a title such as manager, the task of directing the research will be dependent on the goodwill of individuals and the leadership skills of the manager. These skills will need to be greater with programmes involving a number of different disciplines where research from the different sources will need to be effectively coordinated.

The coordination of research, especially interdisciplinary research, is made difficult by differences in perception of problems, the approach that should be used and the interpretation of experimental results. The need for a manager to understand how different people perceive different issues, etc. has been mentioned above, but the role of this both for managers and individual members working in interdisciplinary research programmes has to be stressed.

Everyone has their own perceptions of events and will interpret what they experience in the light of accumulated knowledge and wisdom. Thus, scientists trained in one discipline may perceive things differently from a scientist trained in another, or an economist or sociologist. In the realms of interdisciplinary research these differences in perception can have tragic consequences for a research programme. Consider the picture in Fig. 9.10: is it a rabbit or a duck, or both? What would happen if your discipline allowed you to see a duck while another's discipline allowed

Fig. 9.10. A question of perception: is this a rabbit or a duck, or both?

him to see a rabbit and you did not realize the difference (Dent, 1992)? The chances are that the other's view may be considered foolish, or, further, misunderstandings could accrue when each blames the other for lack of intelligence, deceptiveness or insincerity (Swanson, 1979). There are a number of ways of avoiding these problems. The first makes it necessary for individuals to understand and acknowledge that different disciplines have different cognitive maps (Petrie, 1976). A cognitive map is taken to include basic concepts, modes of enquiry, what counts as a problem, observational categories, representation techniques, standards of proof, and types of explanation, and generate ideals of what constitutes a discipline (Petrie, 1976). Once this fact has been acknowledged it is then necessary for each to learn at least part of other disciplinary maps. This can be done by learning about the observational categories of the other discipline and by learning the meanings of key terms in the other's discipline (Dent, 1992, 1994). The aim being to understand enough to allow one to interpret the problem in the other's terms. It is this aspect of interdisciplinary research that has been overlooked in the past but is absolutely necessary if the work of the group is to be successful (Petrie, 1976).

The coordination of research activities also requires that the form of integration of research that is to be achieved is defined.

Rossini *et al.* (1978) considered there to be four types of integration and four socio-cognitive frameworks for integration. The four types of integration were termed editorial, conceptual/terminological, systemic, and theoretical. The editorial form of integration involved the organization and ordering of written material obtained from different disciplinary scientists. Devices to integrate the work might include use of introductions and conclusions. The conceptual/terminological form of integration took a further step and ensured that a consistent vocabulary and terminology was used throughout the written material. These two forms of integration may often merge in practice, with systemic integration a common view or representation permeating the whole study. This may be formalized by use of a conceptual or mathematical model to help achieve integration. These are the four types of integration that could potentially be used in pest management research. In practice an emphasis occurs on the lower order types of integration although the increased interest in modelling has focused attention on systemic integration. However, it is usually considered as an isolated technique and is not often used as a mechanism for integrating research programmes.

The four processes that can be used to achieve integration, the socio-cognitive frameworks of Rossini *et al.* (1978), can be referred to as common group learning, negotiation among experts, modelling, and integration by leader. These represent idealized mutually exclusive frameworks which in practice rarely exist, but they do provide a structure for comparison. Components of each will inevitably be found in existing research programmes.

A common group learning approach requires the members of a research team to decide on a definition and boundaries of the problem as a group exercise. A division of effort is determined and each team member carries out their allotted tasks. The analyses of each member's data are commented on by every other member and after discussion, reports are written by members who are not specialists in the area. The final report is then the common intellectual property of the group (Swanson, 1979).

The negotiation-among-experts approach is similar to that of group learning except that there is a greater emphasis on the role of individual experts within their given discipline. The assignments are allocated to each expert who then brings the full power of their appropriate discipline to bear on the problem. The group then discusses the result of the work focusing on the overlaps and links between the different components. After these negotiations the individual team experts rewrite the report of their work bearing in mind the results of these negotiations. Non-experts do not write a common report.

Both the group learning and the negotiation among experts approaches can suffer from an initial failure to identify the problem and its boundaries. If this is combined with poor definition of individual tasks and assignments then the approach will be prone to disintegration of effort because each member will tend to do what they think is appropriate irrespective of the partner's goals and requirements. This problem arises because few groups can generate the conciliatory atmosphere and the necessary self-discipline to allow individuals to discuss their research open and frankly. Negotiation among experts can even be said to run contrary to standard research training since it involves the internal tampering with discipline analysis to include other intellectual perspectives (Rossini et al., 1978); not a practice scientists will readily accept unless among close colleagues with whom they share absolute confidence and respect. Such situations do occur, but it can take several years before a team can fully interact and cooperate in this way (Barfield et al., 1987; Dent, 1992). The common group learning approach has the advantage that if any confrontation situation does develop over a particular issue then the approach places the burden of confronting the expert upon the whole.

Modelling, as a socio-cognitive framework, has the advantage in that the use of a model tends to depersonalize any confrontation in favour of forcing individuals to meet the information demands of the model. The modelling approach provides a definite focus for participation, especially in the early stages of problem definition and framing. Not all the members of the team need participate in the model construction but ideally all should agree on its form and should contribute data. The use of a model in this way will have the added advantage of also providing the team with a shared paradigm. It has the disadvantage of sometimes providing too narrow a focus and excluding aspects of the problem that do not really fit within the model framework. For this reason mathematical models may be of most use for problems that are well defined and lie within a fairly narrow range of closely related disciplines. Conceptual models on the other hand can have wide ranging application.

The integration by leader may appear an attractive proposition, but in the situation defined by Rossini et al. (1978) (also see Swanson, 1979) the leader functions as the sole integrator and interacts with each member and the members do not interact between themselves. In this situation the success or failure of the approach is totally dependent on the skills of the leader. A poor leader will not be able to integrate the work of the individual members and even a good leader would have problems with understanding, assimilating and making effective use of the individual's contribution from all but a relatively small research team. Hence, the excessive demands placed on the leader for larger groups will make this approach impractical.

Each of the four frameworks discussed above has its strong and weak points, with approaches being more relevant to some situations than to others and rarely with one approach being used exclusively throughout. Rossini et al. (1978) recommended that strategies for integration might best include a combination of approaches. In the past, a research team has only rarely deliberately selected in advance the intellectual and social components that determine the particular socio-cognitive framework for their collaborative programme; more often the organization evolves into a stable pattern by trial and error (Swanson, 1979). However, a more positive and decisive approach to selecting a suitable combination of frameworks may speed up the coordination of the research, especially during the important definition phase.

A scheme such as that depicted in Fig. 9.11 may be considered generally appropriate for combining different frameworks within an R&D programme. Firstly a conceptual model could be produced by a few members of the programme, e.g. the team leader, a modeller and a socio-economist. This should aim to incorporate the main components of the system. This model should then be discussed by the group to

eliminate basic differences and misunderstandings and then each should take the general model away and consider it carefully in the light of their own knowledge and experience. They should make any alterations they consider necessary, but must be able to discuss and justify their additions and/or omissions to the model to the other team members. The model is now openly discussed and everyone attempts to explain their own components as well as to understand those of the others in the team. Thus the model is eventually accepted as the common paradigm of the group. It will have been used to both define the problem and its boundaries and to promote individual and group participation in its formulation, aiding the process of integration. During the research phase the model defines the research and data requirements and hence can be used to direct the research effort. In this task,

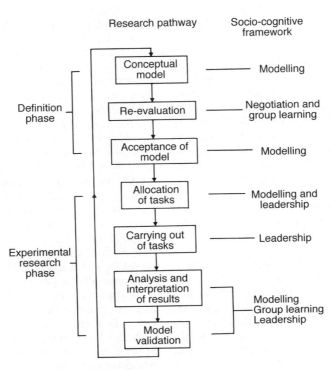

Fig 9.11. A scheme for the type of socio-cognitive framework appropriate for each stage of a research programme. See text for further explanation.

and also to promote continued communication and integration of approach the team leader will encourage and maintain links between the different research groups. The leader's role will here differ from that of the leader integrator of Rossini *et al.* (1978) where all information passed only through the leader. Here, the leader will be responsible for maintaining interactions between all team members, as well as him/herself.

Periodically it will be necessary and important that groups meet to discuss progress towards their mutual and individual goals. The frequency of such meetings will be dependent on many things, such as the location of each of the research groups, the completion of work according to seasons, etc., but the more often the groups can meet and the more informally discussions can take place, the greater will be the integration achieved (Barker, 1994; Clarke, 1994). During these meetings each of the group experts will make a report of their work which will be discussed by the group. The leader will write a report of the programme's progress, integrating all the different components. The future direction of the research will be considered and discussed in the presence of all the members, with the leader taking an active part to encourage and direct the work towards the common goal. The process will effectively start again when the model reaches the validation stage, or the research moves to the field integration stage. The results of the validation and the experiments will be used to reassess the whole programme and to define new targets, changes to the model, etc.

9.4.3 Management training for scientists

There are two basic concerns about our present ability to produce scientific leaders capable of managing pest management research groups. The first is that the training received in science does not equip young people for management and the education/qualification system promotes and rewards only specialist research work (Dent, 1996).

Scientists need to be trained first through a Bachelor of Science Degree and then either a Masters' or a Doctor of Philosophy or both, after which the post-doctoral scientist will work on a specific research problem for a number of years. At no point during this training period does a scientist receive any training in management, but from the post-doctoral level onwards, throughout the scientist's career, management will become an increasingly greater component of the scientist's work. The need for management skills will start modestly enough, perhaps with the need to supervise only a technician or post-graduate student, but as the scientist progresses to section leader, lecturer, head of department, director of the research institute, etc., the need for management skills will become overwhelmingly important. It might be argued that during the progress up the management ladder the scientist acquires the necessary skills, or only those who obtain or have the skills ascend to higher management positions. That this were only true, but in reality it is not the situation. Scientists are largely promoted on scientific grounds rather than on management ability (Dent, 1996). This situation may be acceptable in much of the scientific community but within the context of pest management research it takes on a further incongruity. Pest management research involves the integration of multidisciplinary teams that can only be achieved with careful leadership and management; management for which few in science have been trained. Now more than ever, there is a need for research managers in pest management research, but the only real way that such people can gain the necessary scientific and management training will be with specially taught courses that combine these two components. Courses that are run at present do not fulfil this role, although most of them provide scientists with the broad based knowledge required by such managers. This need has long been recognized (Smith,1978) and even though a number of courses do assist in developing the skills of generalist scientist there is still a tendency to regard specialist work as somehow better than the more generalized (Thorne, 1986). Such attitudes do little for the progress of integrating pest management research and will hopefully in the course of time be replaced by a respect and understanding for the specialists' skills, and the abilities of the generalist insect pest manager.

9.5 Delivery of Research Results

The outputs of research in pest management may be classified as information, techniques or products (Dent, 1997b). Products are characterized by having commercial value and that they are usually purchased as off-farm inputs (Dent, 1993, 1995). They may be available as devices, materials or substances (which may be living organisms, chemicals or plant material) which is usually manufactured, produced, formulated or packaged for sale. Products include a diverse range of control agents such as biopesticides, chemical insecticides, resistant crop cultivars, macrobiological control agents and monitoring and trapping devices. In contrast to products a technique is a form of procedure, skill or method that may be utilized by farmers from available on-farm resources (Dent, 1993). Techniques are often based on traditional agronomic or husbandry practices such as crop rotation, tillage or planting dates that are specific to at least the level of the cropping system. Both techniques and products have information components in that they require both conceptual and technical know-how to ensure they are used properly (Dent, 1997b). Information can be 'formatted' as decision support systems, e.g. decision rules based on monitoring and forecasting and expert systems (Norton and Mumford, 1993). These decision tools are used for the analysis and delivery of research outputs, being concerned with the design of control recommendations, providing pest forecasts or delivery specific information and advice (Norton, 1993) and include database systems, computer models particularly simulation models, expert systems and goal, linear and dynamic programming.

The provision of information and advice, the transfer of technology, education and training is traditionally the remit of the extension service. Extension methods may target individuals, groups or mass audiences according to the type of dialogue required and the number of farmers involved (Adams, 1982). In addition the use of farmer field schools have provided alternative means of empowering farmers, incorporating external and indigenous know-how with a strong dissemination element to its message.

9.5.1 Simulation models

In the USA the evolution of computer simulation models occurred concurrently with the evolving concept of IPM (Logan, 1989). This coincidence provided the IPM researcher with a tool that could help deal with the complexity of IPM. Simulation methodology also provides an unequalled ability for producing realistic models of systems behaviour for a reasonable investment of time and money (McKinion, 1989). These models can also be readily used by non-mathematicians because they do not necessarily require a high degree of mathematical sophistication to implement (Getz and Gutierrez, 1982). The real advantage of a simulation model is, however, that it permits study of the real system without actual modification of that system. This means that a system simulated on the computer can be subjected to a series of alternative modifications in a controlled and systematic manner and the consequences of these modifications studied (McKinion, 1989). These attributes and the common claim that simulation models are a preeminently useful research tool because they help identify gaps in our knowledge (Conway, 1984a) has made the technique highly acceptable as an approach consistent with IPM and IPM needs and objectives.

The technique does, however, have a number of recognized pitfalls (Getz and

Gutierrez, 1982). A lack of awareness of certain important processes may lead to their omission from the model. By their very nature it may not be readily apparent what impact such omissions will have on the model, but eventually such models will fail. There is also the problem that sometimes the construction of the complete model becomes an end in itself.

It is not possible to relate here the techniques used in simulation models (see Rabbinge *et al.*, 1989; Holt and Norton, 1993; Holt and Cheke, 1997) but a brief description of the principles involved may provide some understanding of the type of relations and processes that drive such models. Simulation models should all start as a relational diagram (Fig. 9.12). Relational diagrams are qualitative models that contain the most important elements and relationships of the system. They make use of a special form of notation and convention, developed to depict industrial systems (Forrester, 1961). An understanding of this notation (Table 9.7) means that the relational diagrams and the general functioning of the model can be understood. The modelling approach using this notation has gained wide acceptance and is known as the state variable approach. State variables are quantities such as number of species, or number of individuals or biomass and each state variable is associated with a rate variable. Rate variables characterize the rate of change of a state variable over a given length of time and they include such variables as development rate, mortality rate and reproductive rate. The value of a state variable will change over time according to the rate variable which in turn can influence the value of the rate variable. In this way positive or negative feedback loops are developed. The form the loop takes will determine whether a state variable continues to increase, decrease or reaches a steady state. A positive feedback loop is characterized by a positive relation between the rate and state variable which will lead to the rate enhancing the state and vice versa so that both continue to increase, e.g. exponential

growth. In a negative feedback loop the rate may be either positive or negative but will decrease as a function of the state variable (Leffelaar and Ferrari, 1989). Unlike the positive feedback loop, negative feedback causes the system to approach some goal. Such an equilibrium state is stable since any departure will cause a return to the equilibrium position. The state variables and the rate variables can be influenced by driving variables which tend to characterize the effect of the environment on the system, e.g. temperature, rainfall, but they can also be immigration and emigration or sex ratio. A simulation model is constructed using state variables, rate variables and driving variables (also parameters, see Table 9.7, Fig. 9.12). These variables are all related in a model with the use of difference equations.

Although the use of simulation modelling is most often associated with research, they provide an excellent means of testing and developing management strategies and for providing a basis for pest management advice (Holt and Norton, 1993). The ability of simulation models to predict means they can be used to either derive general rules or in real time operation, and given appropriate field inputs, they can provide short term tactical advice.

Simulation models have now been produced for a range of pest species and systems, e.g. cereal aphids (Carter and Rabbinge, 1980; Rabbinge *et al.*, 1979), larch moths (van den Bos and Rabbinge, 1976), planthoppers in rice (Cheng and Holt, 1990), spider mites in fruit trees (Rabbinge, 1989), potato leafhopper in alfalfa (Onstad *et al.*, 1984) and *Helicoverpa amigera* in pigeonpea (Holt *et al.*, 1990). They are used in tactical models to predict pest outbreaks in tandem with management models (Cheng and Holt, 1990) or to determine optimum insecticide spray timing (Baumgaertner and Zahner, 1984; Cheng *et al.*, 1990; Holt *et al.*, 1990) while in strategic models they have been used to simulate plant growth and pest damage (Boote *et al.*, 1983; Gutierrez *et al.*, 1988a,b; see also Rabbinge *et al.*, 1989) and strategies for

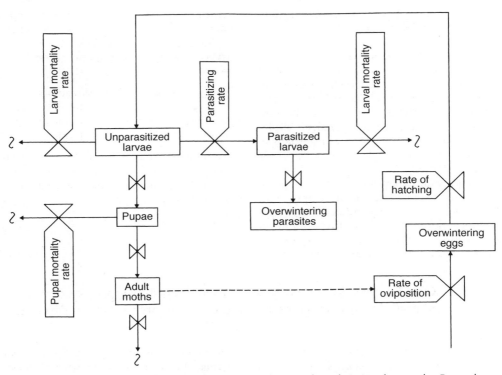

Fig. 9.12. A relational diagram for the grey larch bud moth (*Zieraphera diniaria*) (after van den Bos and Rabbinge, 1976).

Table 9.7. The basic elements of relational diagrams, the symbols of Forrester (1961) notation (Leffelaar and Ferrari, 1989).

Symbol	Meaning
▭	A state variable, or integral of the flow; the final result of what has happened
⟶	Flow direction of an action by which an amount, or state variable is changed. These flows always begin or end at a state variable, and may connect two state variables
∘⟶	Flow and direction of information derived from the state of the system. Dotted arrows always point to rate variables, never to state variables. The use of information does not affect the information source itself. Information may be delayed and as such be a part of the process itself
▭◁	Valve in a flow, indicating calculation of a rate variable takes place; the lines of incoming information indicate the factors upon which the rate depends
⌇	Source and sink of quantities in whose content one is not interested. This symbol is often omitted
⊶∘⊸	A constant or parameter
◯	Auxiliary or intermediate variable in the flow of material or of information
+, −	Sometimes placed next to a flow of information, to indicate whether a loop involves a positive or negative feedback

insecticide use (Holt *et al.*, 1992) and biological control (Hearn *et al.*, 1994).

9.5.2 Expert systems

Expert systems are processes that mimic
the way in which humans diagnose problems and dispense advice (Mumford and
Norton, 1989, 1993). They are usually computer based although for practical implementation it is often more appropriate to
transfer the expert system to a paper or
manual format. The steps involved in
developing an expert system include problem structuring, knowledge acquisition,
knowledge engineering and encoding, veri-
fication, validation and testing (Mumford
and Norton, 1993). The basis of expert systems is a set of linked logical relationships
of rules with the structure '*If* some condi-
tion(s) *then* some conclusion(s) *else* other
conclusion(s)' (Holt and Cheke, 1997).

A very simple programme could have
one rule:

If pest population is above threshold
                                    (quality value)
then spray   (choice)
else do not spray   (choice)

A number of expert systems have been
developed for pest management use (Jones
et al., 1990; Compton *et al.*, 1992; Knight *et
al.*, 1992; Tang *et al.*, 1994).

Case Study: GABY: a computer-based decision support system for integrated pest management in Dutch apple orchards (van den Ende *et al.*, 1996)

Historically Dutch fruit growers have controlled pests of apple by broad spectrum insecticides applied at regular intervals. However, the occurrence of insecticide resistant insects and problems with the control of apple rust mite, which
led to the introduction of predatory mites as a control option, forced a reconsid-
eration of insecticide use strategies. A computerized advisory system (GABY)
was developed to support decision making of individual fruit growers (de Visser,
1991; Helsen and Blommers, 1992; Mols *et al.*, 1992; van den Ende, 1994a,b)

Table 9.8. Insects considered in GABY.

Apple leaf gall midge (*D. mali*)
Apple blossom weevil (*A. pomorum*)
Apple sawfly (*H. testudinea*)
Mussel scale (*I. ulmi*)
Woolly apple aphid (*E. lanigerum*)
Rosy apple aphid (*D. plantaginea*)
Rosy leaf curling aphid (*D. devecta*)
Green apple aphid (*A. pomi*)
Apple grass aphid (*R. insertum*)
Summer fruit tortrix (*A. orana*)
Leaf-rollers (*Tortricidae* spp.)
Codling moth (*C. pomonella*)
Winter moth (*O. brumata*)
Clouded drab moth (*Orthosia* spp.)
Green apple capsid (*L. pabulinus*)
Rose tortrix moth (*A. rosana*)
Fruit tree red spider mite (*P. ulmi*)
Apple rust mite (*A. schlechtendali*)
Phytoseiid mite (*T. pyri*)

which would provide a holistic integration of control tactics for almost a complete set of apple pests in a region (Table 9.8). Separate decision models for each pest or pest group form the core of GABY. The system requires that the farmers carry out monitoring and record treatments in order to regularly update GABY. In this way with a series of inputs (orchard characteristics, temperature and pest numbers) GABY generates recommendations appropriate for a particular pest. Recommendations to spray insecticides are in accordance with Dutch IFP guidelines and are only generated if control thresholds are exceeded. The type and dosage of the most appropriate insecticide is recommended, and in turn the type, dosage and date of treatment is entered into GABY since this will affect future recommendations.

GABY was developed as a stand-alone system. Its value is dependent on the input of properly collected data which requires initial on-farm support (van den Ende, 1994a). In addition GABY needs to be regularly updated if it is to continue to be relevant to farmers' needs, for instance any disappearance of insecticides will change the decision schemes. Hence, although of great value to farmers, the maintenance requirements by scientists and advisory experts are high.

9.5.3 Linear programming

Linear programming is a technique that has long been used by economists as a general purpose technique for determining the best allocation of scarce resources. It can be readily applied to any pest management problem that needs to optimize the use of two or more resources provided that certain conditions are met. The most important of these is the need for the objective function and the constraints to be linear in the decision variables. The objective function is the method used for measuring how good an allocation is, while the decision variables are the ways in which scarce resources are allocated. The resource constraints are the limitations placed on the decision variables to reflect the resource

scarcity (Rossing, 1989). The requirement for linearity means that a change of one unit in the decision variables results in a constant change of the objective function and the resource constraints.

Linear programming (LP) problems are solved using an algorithm. A graphical analysis of an LP problem is used here to illustrate the technique but such graphical analyses are not really practical because they can only be used where there are only two decision variables. It does, however, provide an introduction to the structure of the algebraic methods used to solve and analyse a general LP problem (Markland and Sweigart, 1987). Further details on linear programming can be found in Gass (1975) or Walsh (1985).

Case Study: A hypothetical linear programming problem

Suppose that a farmer normally has to apply seven applications of insecticide to a particular crop each season to ensure control of an insect pest. The farmer has to purchase all of the insecticide that will be needed at one time as part of a credit deal. There is the choice of two insecticides for the job, one of these is 0.6 of the price of the other but is unpleasant to apply and is more harmful to beneficial insects. Hence, the farmer prefers to keep the use of this insecticide as low as economically possible, refusing to apply it more than three times during a season. If the two insecticides have equal efficacy and the farmer wants to optimize purchase, then the question of how many applications should be made with each insecticide (and hence how much of each insecticide should be purchased) needs to be answered.

The constraints are:

1. The farmer does not want to apply the insecticides more than seven times.
2. The farmer does not want to apply the cheaper insecticide (X_1).
3. The farmer cannot afford more than five applications with the expensive insecticide (X_2).

Hence:

1. $X_1 + X_2 < 7$
2. $X_1 < 3$
3. $X_2 < 5$
4. $X_1 > 0$
5. $X_2 > 2$

Let the objective function be:

$$Z = 2X_1 + 3X_2 \qquad (9.1)$$

The graphical method for analysing this problem is shown in Fig. 9.13. The first step is to draw in the constraint lines equations 1, 2, and 3 which then demarcate the area known as the 'feasible region'. This is the area in which the optimal solution to the problem can be found. There are an infinite number of feasible solutions and so a systematic procedure for determining the best feasible solution is required.

The first step is to select an arbitrary profit level and to determine all feasible solutions that yield the selected profit or objective value, for instance a value of $Z = 6$.

So X_1 and X_2 must satisfy the linear relationship:

$$2X_1 + 3X_2 = 6 \qquad (9.2)$$

The question 'are there any objective values greater than 6?' then needs to be answered. It is instructive to rewrite the objective function:

$$Z = 2X_1 + 3X_2 \qquad (9.1)$$
$$\text{as: } 3X_2 = 2X_1 + Z \qquad (9.3)$$
$$\text{then: } X_2 = \tfrac{2}{3}X_1 + \tfrac{1}{3}Z \qquad (9.4)$$

The coefficient of X is the slope of the line and $-\tfrac{2}{3}$ and $(\tfrac{1}{3})$ Z is the intercept (i.e. 2). The slope is independent of the value given for Z, thus for any specified value of Z the profit line has a slope of $-\tfrac{2}{3}$. For example, for $Z = 6$:

$$X_2 = -\tfrac{2}{3}X_1 + 2 \qquad (9.5)$$

where $-\tfrac{2}{3}X$ is the slope and 2 the intercept.

As the value of Z is increased the profit line moves further from the origin, although remaining parallel with each other. If the profit line is moved out from the origin it will eventually leave the feasible region. The point in the feasible region that lies on the profit line, with the largest value of Z, is the optimal solution for the linear programme (Fig. 9.13). Note that the optimal solution lies on the constraint lines. Determining the exact values from the graph is not advisable (Markland and Sweigart, 1987). To find these coordinates it is best to solve the constraint equations simultaneously:

6. $X_1 + X_2 = 7$
7. $\qquad X_2 = 5$

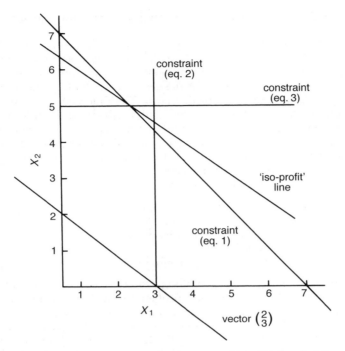

Fig. 9.13. A graphical illustration of a linear programming problem. The two decision variables are the amounts of an inexpensive insecticide (X_1) and of an expensive insecticide (X_2). Further explanation in the text.

8. $X_1 + 5 = 7$
9. $X_1 = 2$

Hence, the farmer would optimize his purchase of insecticides by buying five applications of X_2 and two applications worth of insecticide X_1.

The graphical method can only be used where there are only two decision variables then the simplex algorithm is used. Despite its name the method is too complex to be considered here, Gass (1975) or Walsh (1985) provide useful reviews.

9.5.4 Goal programming

Goal programming is similar to linear programming except that it permits the inclusion of a number of conflicting objectives and provides a solution to the problem on the basis of specified goal priorities. Where the linear programming method aims to provide an optimal solution the goal programming technique attempts to reach a satisfactory level of achievement given several different objectives (Markland and Sweigart, 1987). In this way goal programming perhaps better reflects the way actual decisions may be made, since few decisions are made with only a single objective in mind. In goal programming the decision maker can priority rank the goals, which are incorporated into the model as constraints. The technique is applicable to pest management problems and particularly to research management where choices between a number of conflicting options may need to be made, e.g. selection of an area for field evaluation trials on the basis of such variables as farm size, area grown to specific crops, farmer income, contribution of livestock etc.

Case Study: A hypothetical goal programming problem

Suppose a farmer has two fodder crops (X_1 and X_2) and a total of 10 ha of land available on which to grow them. Crop X_1 can be used as a cash crop whereas X_2 is needed to maintain his own cattle stock; a minimum of 4 ha is required for this. The labour requirement for the cash crop is high so the farmer cannot afford to grow more than 5 ha of this crop. The financial return from the two crops is different, 1 ha of X_1 gives two and a half times the return of crop X_2 and the farmer wishes to return a profit at least equal to 30. There is one further constraint: very occasionally there is a pest that attacks crop X_1 and can cause complete devastation. Since pest control options are too expensive, it only pays to grow about 2 ha of this crop.

The priority order of goals of the farmer are:

1. Maximal use of 10 ha of land.
2. Must grow a minimum of 4 ha of crop X_2.
3. Cannot afford to grow more than 5 ha of X_1.
4. Return must at least equal 30.
5. During outbreak years it is only economical to grow 2 ha of X_1.

This two-dimensional goal programming problem can be solved graphically by repeatedly solving the optimization problem, each time adding an extra goal (Fig. 9.14). The solution space remains unchanged, or if new goals turn out to be constraints, is decreased. Eventually a goal may be added that cannot be met by any of the solutions satisfying the higher priority goals (Fig. 9.14). The solution feasible with respect to the higher priority objectives which deviates least from the unsatisfied goal is designated the optimum solution to the problem.

The fifth goal cannot be achieved at any of the points in the solution space, hence a solution has to be found that minimizes the deviation from this fifth constraint and is still feasible. The dotted line moved towards the solution space first comes into contact with point P and this is the first solution encountered which is feasible with respect to the first four goals. Thus, the optimum solution point P is to grow 3.5 ha of crop X_1 and 6.5 ha of crop X_2. In outbreak years the farmer will have minimized the losses due to pest outbreak with respect to optimizing in non-outbreak years.

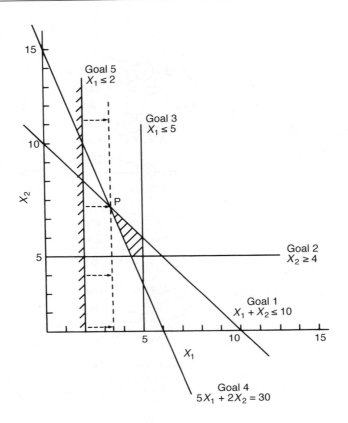

Fig. 9.14. A hypothetical example of a simple goal programming problem. The hatched area is the solution space. See text for explanation.

9.5.5 Decision trees

Where a number of sequential decisions need to be made in pest management then decision trees can be useful: for instance, where a number of options for control exist that can be applied at different times of the year (see Fig. 9.4). The decision tree presents the decisions and their outcomes in sequence that is followed by the decision-making process. If a probability and a value on an outcome can be determined then a decision tree can be used to study the effects of different strategies. The sensitivity of choices to changes in probabilities or assumptions concerning control effectiveness can be assessed so that the most robust strategies can be selected (Mumford and Norton, 1984).

Decision trees should be constructed using a set of decision nodes (□; at which one of several alternatives may be chosen) and state of nature nodes (○; at which a chance event occurs) (Fig. 9.15). The probabilities assigned to the state of nature nodes are based on historical records and reflect the frequency of occurrence of states of nature (Mumford and Norton, 1984).

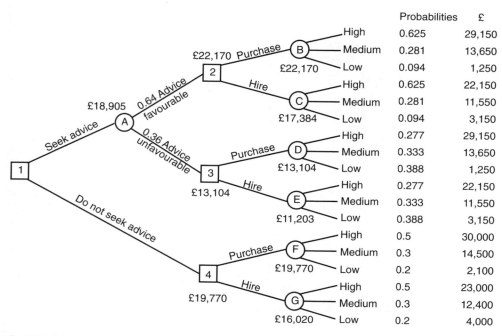

Fig. 9.15. A hypothetical decision tree analysis where a farmer has to decide whether or not to: (i) purchase or hire a piece of equipment and (ii) seek advice on this from a consultant, given a high, medium or low probability of insect pest attack. □ = decision nodes, ○ = state of nature nodes. See text for further explanation.

Case Study: A hypothetical decision tree analysis

The market for a particular crop is increasing and hence a number of farmers who have not normally produced the crop are now turning to it as an alternative source of income. Farms that grow the crop have suffered losses due to a high incidence of an insect pest which can be controlled by insecticides. One of the farmers adopting the new crop needs to obtain use of spraying equipment suitable for applying insecticides against the insect pest. The farmer believes that the equipment will be used often, but since there is some doubt as to the likely incidence of an outbreak on his particular farm he does not know whether to hire or purchase the equipment. The farmer can choose to consult an independent advisor but this will cost £850. What decision should the farmer take?

To carry out a decision tree analysis a great deal of information is required. Firstly, a set of conditional probabilities must be determined based on the demand for the spraying equipment by other farmers (who had or had not taken the advice of the independent consultant) (Table 9.9). Secondly, a set of prior probabilities is required that identifies the actual demand for the sprayer by farmers who did not pay for independent advice (Fig. 9.15). These prior and conditional probabilities are then used to determine a set of posterior probabilities which are used in the sequential decision tree analysis.

At the chance node *A* the overall probabilities associated with the advice

Table 9.9. The conditional probabilities of sprayer demand when the advice from the consultant was either favourable or unfavourable.

Advice	Actual situation					
	High demand (H)		Moderate demand (M)		Low demand (L)	
Favourable (F)	$P(F\|H)$	0.8	$P(F\|M)$	0.6	$P(F\|L)$	0.3
Unfavourable (U)	$P(U\|H)$	0.2	$P(U\|M)$	0.4	$P(U\|L)$	0.7

being either favourable or unfavourable are computed. These probabilities are computed using the prior probabilities (0.5, 0.3, 0.2) and the conditional probabilities (Table 9.9), i.e. where the advice was favourable:

$$P_F = (0.8)(0.5) + (0.6)(0.3) + (0.3)(0.2) \tag{9.6}$$
$$P_F = P(F\mid H).P(H) + P(F\mid M).P(M) + P(F\mid L).P(L)$$
$$P_F = 0.4 + 0.18 + 0.06$$
$$P_F = 0.64$$

The probability of 0.36 (0.1 + 0.12 + 0.14) was similarly obtained for the situation in which the advise was unfavourable. The next step in the analysis is to consider the probabilities associated with the chance nodes – the conditional probabilities. These are computed using Baye's theorem, the general form being:

$$P(Ai)B = \frac{P(Ai) \cdot P(B)Ai}{\sum\limits_{i=1}^{n} P(Ai) \cdot P(B \mid Ai)} \tag{9.7}$$

where:
Ai = set of n mutually exclusive and exhaustive events
B = a known end effect, or outcome of an experiment
$P(Ai)$ = the prior probability for event i
$P(B \mid Ai)$ = the conditional probability of end effect B given the occurrence of Ai.

The posterior probability is thus a revision of the prior probability using new or additional information. At each chance node the posterior probability $P(Ai \mid B)$ is the probability of the event 'high, medium or low demand' given the outcome of either favourable or unfavourable advice (new information). At each chance node (B–E) the probability of high, moderate or low demand given that the advice was favourable would be obtained by:

$$P(F \mid H) = \frac{P(F \mid H) \cdot P(H)}{P_F} \tag{9.8}$$

$$P(F \mid H) = \frac{0.4}{0.64} = 0.625$$

The probabilities for all other conditions (F|M), (F|U), (U|H), (U|M) and (U|L) were calculated similarly replacing P_F with P_U as appropriate. Using these conditional probabilities the expected monetary value (MEV) at each chance node can be determined by:

$$\text{EMV}_B = (0.625) \ (29,150) + (0.281) \ (13,650) + (0.094) \ (1250)$$
$$= 13,843 + 3245 + 296$$
$$= 22,170$$
$$\text{EMV}_C = (0.625) \ (22,150) + (0.281) \ (11,550) + (0.094)(3150)$$
$$= 13,843 + 3245 + 296$$
$$= 17,384$$

etc. for each chance node.

From the expected monetary values shown at chance nodes B and C the largest EMV £22,170 is selected for the decision node 2 and £13,104 for decision node 3. Using the EMVs at decision nodes 2 and 3 the EMV at chance node A can be computed:

$$\text{EMV}_A = (0.64) \ (22,170) + (0.36) \ (13,104)$$
$$= 14,188 + 4717$$
$$= 18,905$$

From the results of this decision tree analysis it can be concluded that the best strategy for the farmer would be to not seek any advice from the independent consultant but to purchase the spray equipment.

9.5.6 Dynamic programming

Dynamic programming (Bellman, 1957) developed from studying a sequence of decision problems that arose in inventory control theory (Markland and Sweigart, 1987). It is a technique similar to decision tree analysis in that it involves breaking problems down into stages and the decision at one stage can affect the decision and outcome at the next stage, but it is a complex technique that requires quite extensive modelling expertise. There are no standard dynamic programming procedures and the major challenge with the technique in relation to pest management is to develop a mathematical model of low dimension which can adequately describe the impact of weather and management decisions on the dynamics of an agricultural system (Shoemaker, 1982). The models may be dynamic, non-linear and stochastic but essentially the computation begins at the final stage of the decision process and then works backwards along an optimal pathway to the beginning (Conway, 1984b). This means that all combinations of route do not have to be checked which makes the technique quite efficient. Another advantage of the technique is that if for any reason the optimal path is not followed between two decision periods, a new optimal path is pro-

vided. The dynamic programming technique does have a major disadvantage in that it is unable to deal with more than four state variables. This tends to limit its usefulness and partly accounts for the difficulties involved at the model formulation stage.

The dynamic programming solution approach involves four major steps (Markland and Sweigart, 1987):

1. The problem is divided into subproblems called stages.
2. The final stage of the problem is analysed and solved for all possible conditions or states.
3. Each preceding stage is solved working backwards from the final stage of the problem. This involves making an optimal policy decision for the intermediate stages of the problem with each intermediate stage being linked to its preceding stage by a recursion relationship. It should be stressed that the exact form of the recursion relationship will vary according to the dynamic programming problem but it will always take the general form:

$$\text{fn}^*(Sn) = \text{max/min}\{\text{fn}(Sn, Xn)\} \qquad (9.9)$$

where the function $\text{fn}(Sn, Xn)$ is the value associated with the best overall policy for the remaining stages of the problem, given that the system is in a state (Sn) with n

stages to go and the decision variable (Xn) is selected. This recursion relationship involves a return at each stage of the problem. The return at a particular stage is due to the policy decision selected and its interaction with the state of the system.

4. The initial stage of the problem is solved, and when it has been achieved the optimal solution is obtained (Markland and Sweigart, 1987).

Dynamic programming has largely been used in pest management as a means of optimizing pesticide use (Valentine *et al.*, 1976; Shoemaker, 1979, 1982, 1984; Dudley *et al.*, 1989).

Case Study: An illustration of a dynamic programming approach using a hypothetical shortest route problem

Figure 9.16 is a diagrammatic representation of the shortest route problem. The solution to the problem will be the shortest route between 1 and 7. The problem is broken down into three stages (Tables 9.10, 9.11 and 9.12). The stage 3 table

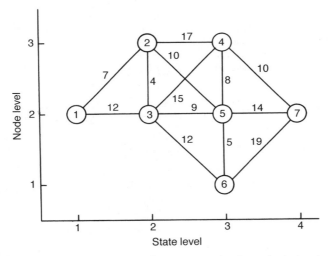

Fig. 9.16. An example of the dynamic programming approach using a shortest route problem. The node number is the number in the circle and the distances between nodes are the numbers adjacent to each line. See text and Tables 9.10, 9.11 and 9.12 for explanation.

Table 9.10. The stage 1 table for a shortest route, dynamic programming problem. See text and Fig. 9.16 for explanation.

Node number	Route	Total distance	Minimum distance (route)
4	4→7	10	10 (4→7)
	4→5→7	22	
	4→5→6→7	32	
5	5→7	14	14 (5→7)
	5→4→7	18	
	5→6→7	24	
6	6→7	19	19 (6→7)
	6→5→7	19	19 (6→5→7)
	6→5→4→7	23	

(Table 9.12) indicates that when starting at node 1 the minimum distance is traversed by going to node 2. In the stage 2 table (Table 9.11), looking at node 2 the shorter route will be via node 5 and in the stage 1 table (Table 9.10) at node 5 the minimum distance to node 7 is 14. Hence, the shortest route from node 1 to node 7 is achieved by passing from node 1 to 2 to 5 to 7 (7 + 10 + 14 = 31).

Table 9.11. The stage 2 table for a shortest route, dynamic programming problem. See text and Fig. 9.16 for explanation.

Node number	Route	Total distance	Minimum distance (route)
2	2→4	17	17 (2→4)
	2→5→4	18	
	2→3→4	19	
2	2→5	10	10 (2→5)
	2→3→5	13	
	2→4→5	25	
2	2→3→6	16	
	2→5→6	15	15 (2→5→6)
	2→3→5→6	18	
	2→4→5→6	30	
3	3→4	15	15 (3→4)
	3→2→4	21	
	3→5→4	17	
3	3→5	9	9 (3→5)
	3→4→5	23	
	3→6→5	17	
	3→2→4→5	29	
	3→6	12	12 (3→6)
	3→5→6	14	
	3→4→5→6	18	

Table 9.12. The stage 3 table for a shortest route, dynamic programming problem. See text and Fig. 9.16 for explanation.

Node number	Route	Total distance	Minimum distance (route)
1	1→2	7	7 (1→2)
	1→3→2	16	
1	1→3	12	
	1→2→3	11	11 (1→2→3)

9.6 Implementation and Adoption

The development of an implementation strategy should not be left to the final phases of a research programme. It should initially be considered during problem formulation and then continually readdressed throughout the research phase of the programme. Since it is often at the point of implementation that many pest management programmes fail (Reichelderfer *et al.*, 1984; Heong, 1985) it is important that the processes involved in implementation and adoption be considered in further detail.

9.6.1 Conditions for change

A farmer who is satisfied with a control measure or a pest management strategy is unlikely to be receptive to change (Norton, 1982a) unless there is some alteration in circumstances that makes a current practice less attractive. Conditions that are necessary to make a farmer reconsider the value on a current pest management strategy are depicted in Fig. 9.17 (Norton, 1982b). These trigger events refer to two basic categories of change: big-bang implementation, and incremental implementation. There is a third form of implementation, off-farm implementation, where a change of strategy occurs because outside agencies undertake decision making and implementation on behalf of the farmer. This sort of off-farm implementation occurs in plantation crops or when field crops covering large areas can be treated economically as a homogeneous unit by a parastatal body, e.g. cotton in Sudan Gezira (Eveleens, 1983).

Big-bang implementation occurs when a trigger event produces an unsatisfactory outcome as a result of an increase in potential yield loss, increased pest attack, the availability of a revolutionary new technology, or the failure of existing control measures (particularly as a result of insecticide resistance and the development of secondary pests) (Norton, 1982a). It has been argued that the adoption of pest management strategies appears only to be associated with the failure of insecticide use (May, 1978; Brader, 1979; Gutierrez, 1995; van Lenteren, 1995). In contrast, incremental implementation occurs as a gradual shift from one control strategy to another, usually through the intense activity of extension workers to promote the adoption of the new technology. The flow of information is of crucial importance in this type of implementation.

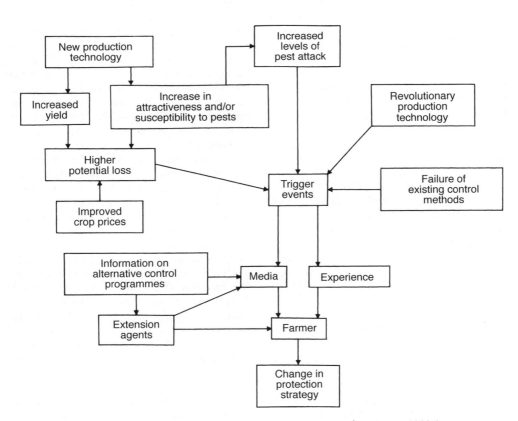

Fig 9.17. Major factors that influence changes in pest control strategies (after Norton, 1982a).

9.6.2 Reasons for failure to adopt new technologies

Even when a new technology is both appropriate and feasible there will still be occasions when a particular pest management strategy will fail to be adopted by farmers. The reasons for this may be caused by a communication gap, social or cultural constraints, or it could be that some form of incentive is required.

A communication gap may occur when information about the new technology is not being relayed to the farmer because of lack of resources, extension workers or use of inappropriate communication channels. Most small farmers in developing countries get their information from neighbours, friends and relatives and often these sources are more important than the extension service (Kenmore *et al.*, 1985). Hence, emphasis solely on extension through these agencies may not produce a desired result. A range of information delivery systems may have to be considered, from pamphlets and media articles on the radio and television, to the use of computer based systems. Improvements in understanding the actual communication networks used by farmers may alleviate problems caused by communication gaps.

Of course if the farmer is receiving the information but is not convinced by the content, its meaning or relevance, then a reception gap has developed between the researchers and their target farmers (Norton, 1990). This may occur as a result of too little farmer participation during the research phase of the programme. Such participation is important and can have a marked effect on the uptake of pest management strategies (Saud *et al.*, 1985). Farmers' understanding of what is involved and required to encourage adoption and their contributions towards coordination of effort among themselves should never be underestimated. Successful implementation and adoption of a new technology will both be assured but will certainly be improved given greater levels of farmer participation in the implementation process. Such involvement may also prove useful where there are difficulties with adoption due to lack of education among farmers. If farmers lack sufficient education concerning the effects of pests on crop production then they are unlikely to adopt an approach they do not understand (Reichelderfer *et al.*, 1984). If there are also too few extension workers to aid farmers in using a new or complex practice then widespread adoption will not take place unless farmers are intimately involved with the process and can pass on relevant information.

9.6.3 Extension services

In general, extension includes: (i) the transfer of technology; (ii) provision of information and advice; (iii) problem solving; (iv) education and training; (v) strengthening the organization base of farmers; and (vi) supplying inputs, credit and technical services (Garforth, 1993). The establishment and maintenance of an extension service that provides all of these services is a costly exercise, prohibitively so in developing countries especially because little foreign aid is allocated for this purpose (Goodell, 1984). Resource availability for implementing IPM systems has also been cited as an obstacle to IPM adoption (Wearing, 1988; Dent, 1995) and will certainly have an impact on the extension methods that can be used. Methods may be classified on the basis of scale, from individual to group to mass methods.

Individual, one-to-one farmer-extension discussions, are without doubt highly valued by farmers (Wardlow, 1992); however, such methods are on the decline mainly due to the need to improve levels of cost effectiveness (Wearing, 1988; Allen and Rajotte, 1990; Dent, 1995). Individual one-to-one extension may be feasible where there are a few large farms in an area having similar requirements (e.g. Percy-Smith and Philpsen, 1992) but most services concentrate on providing extension at group levels or to mass 'audiences'.

Group methods involve extensive workers in addressing a number of farmers at one time at specified dates, study tours and method demonstration or informal discussions with no fixed agenda (Garforth, 1993). Group methods rarely allow in-depth analysis of specific problems but demonstrations and study tours allow farmers to see new technologies with their own eyes or to carry out new techniques for themselves, often under supervision. Mass methods of dissemination seek to reach very large numbers of farmers raising awareness of general issues and to shape attitudes and influence approaches used. Mass methods include radio and television, film, video, audio cassettes, drama, newspapers and other print material (Stone, 1992; Huus-Brown, 1992). More recently use of the internet and computer software have become available (Teng, 1990). Allen and Rajotte (1990) predict that traditional methods of extension are likely to decline in importance in relation to computers and mass electronic media.

9.6.4 Farmers' Field Schools

Education and technical complexity of IPM are thought to have contributed to the slow rates of adoption of IPM (Wearing, 1988; Dent, 1995). The complex nature of IPM necessitates a certain degree of knowledge and some understanding of the basic concepts of biology and ecology. A farmer may recognize a spider or a ladybird but may not have the accompanying concept of 'predator' or 'beneficial insect' to enable him/her to make sense of any advice on how not to kill them (Garforth, 1993). The Farmer Field School approach has been based on the premise that by assisting farmers to know and understand their crop ecosystem they are better placed to make rational decisions about pest management. The approach places farmers at the centre of the IPM process empowering them as the key pest management decision maker, independent of constant extension advice (Chambers *et al.*, 1989; Conway and Barbier, 1990; Matteson *et al.*, 1994).

The Field Schools work by bringing together farmers who utilize their existing knowledge complemented by a 'trainer' in an agroecosystem analysis which involves the trainees working in small groups and observing the state of a crop. The plant is the central focus and drawings are used to analyse plant change, pests, natural enemies and environmental changes (e.g. water levels) (Matteson *et al.*, 1994). Group members then go through a set of prepared questions which cover agronomic and pest problems, on the basis of which management decisions are made – thus integrating their skills and knowledge. In addition, in the agroecosystem analysis there are activities that address plant physiology, insects, rats, population dynamics, economics and issues concerning pesticides. Insect life cycles are studied by rearing insects in what are referred to as 'Insect Zoos'.

The IPM Farmers' Field Schools approach has the ability to empower farmers – understanding the ecology of their own fields means they are less susceptible to exploitation by representatives of agrochemical companies. As one graduate of an Indonesian IPM farmer's field school put it, 'After following the field school I have peace of mind. Because I know how to investigate, I am not panicked anymore into using pesticides so soon as I discover some (pest damage) symptoms' (van de Fliert, 1992; Matteson *et al.*, 1994). The philosophy guiding the successfully held schools in Indonesia is that farmers are very capable of training other farmers in a season-long process (Escalada and Heong, 1993). Farmers graduating from the Field Schools receive certificates which allow them to train other farmers; thus the process is disseminated.

9.7 Discussion

There is no detailed formula for the development of an IPM programme that can be applied to every situation. This is as much because of the variety of possible starting points for a programme as for the diverse range of pests/cropping systems that exist. All that can really be considered here is an idealized scheme identifying procedures that should be generally applicable. It will always be the responsibility of the pro-gramme manager and the research team to identify the best and most appropriate pro-cedures to suit their particular circum-stances.

The problem definition stage of a pest management programme involves bringing together all available information relevant to the problem. This will involve literature searches and synthesis of information from their sources, and it will also involve the use of workshops, bringing together the rel-evant extension workers and researchers to discuss the problem (Norton, 1987) as well as carrying out a socio-economic evalua-tion. This combination of studies and dis-cussions should provide all the necessary background information, provided suffi-cient is known about pest ecology and yield losses. If insufficient is known about these things then there will need to be a prelimi-nary phase of the work dedicated to obtain-ing the necessary information. In some well studied systems such as cotton or maize a great deal of information may already be available and this will shorten the time needed for the problem definition phase of a programme.

The problem definition stage is vitally important to the likely success of the whole programme. The information obtained from the different sources needs to be carefully analysed so that it forms the basis of selec-tion between control options and provides a framework for research and incorporation of new information (Heong, 1985). The major components of the system need to be identified and key processes described. From this it should be possible to formulate key questions that relate to the ecological, socio-economic and technical information required for the management of pests. At this point it should be possible to produce a conceptual model of the system and the problem to be dealt with (Fig. 9.18).

Decisions will need to be made as to whether management will be best achieved by modification of existing options or whether development and testing of innova-tive technologies is required. The time and research commitments needed will be greater for the latter. Close attention will also need to be paid to proposals for how the management programme is to be imple-mented. The role of farmers in the pro-gramme and the importance of identifying appropriate channels of communication need to be addressed. Since the implementa-tion procedures provide the ultimate goal of the programme it is essential that these are clearly defined and act as a focus and com-mon objective for the research.

The needs of the research/experimental phase of the programme will be determined by the key questions that have been identi-fied and the type of control options selected. Obviously the choice of control options is going to affect the type of specialist input required for the research. Ideally a number of key specialists would contribute to almost every pest management programme; these would include a socio-economist, a mod-eller, an agronomist and specialists in cul-tural control, natural enemies, host resistance and insecticides. Additional expertise may be required in the form of an ecologist and statistician and specialists in semiochemicals, genetic manipulation and quarantine. All of the above may form a per-manent team for the duration of the pro-gramme or provide advice at appropriate times as work progresses. For instance, the input of the socio-economist may be needed most at the definition and implementation stages of a programme whereas a statistician will largely only be needed during the research/experimental phase. Each specialist will have their own projects which should be directed to achieve goals required for the success of the programme. The research will undoubtedly develop at different rates but as techniques and products become available their field compatibility will need to be assessed. Both quantitative models and experimental procedures can be used for these evaluations. Where integrative experi-mental trials are required then a specialist group of researchers may need to be brought together. A statistician and an economist will play important roles here. When on-farm trials are carried out an agronomist may need to be added to the group.

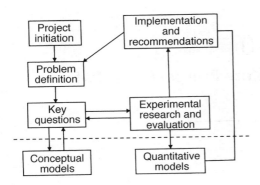

Fig 9.18. A procedure for the organization of an insect pest management research programme (after Heong, 1985).

If the formulation phase of an insect pest management programme has been properly conducted then the implementation phase should be a well coordinated and organized formality. The necessary techniques, materials and communication channels to be used will all have been identified in advance, the involvement of the farmers throughout the programme ensuring that the management programme is appropriate. Sadly this is rarely the case but, even if it were, it is unlikely that there will be no problems with the recommendations and implementation phases of a programme. Where difficulties arise then time should be spent to formulate the problem properly and identify the key questions and then the cycle should continue as before.

10

Driving Forces and Future Prospects for IPM

10.1 Introduction

The direction taken by pest management over the last 50 years has been influenced by a variety of factors, including the availability of pest control technologies, priorities of governments to provide adequate food supplies, the growth of agribusiness and the value placed on scientific endeavour. These factors and their impact have been tempered by increasing public concern for food safety, 'fads' championed by various interest groups (scientists, politicians, funding agencies) and the global realization of the need to use resources more sustainably. This combination of driving forces has shaped IPM, the type of strategies that are currently used and they also form the foundation for pest management in the future. This final chapter considers the current status of pest management from a number of different perspectives, firstly by describing three crop/pest management systems and a few of the lessons that can be learned from these. This is followed by an analysis of the influence of the 'panacea mentality' on the progress of pest management, its influence in the future and of the impact of 'paradigms'. The future of pest control technologies, systems approaches, the IT revolution and the role of the general public are then addressed in an attempt to place IPM in the context of modern developments in agriculture and the wider world.

10.2 Working IPM Systems

Three crop/pest management systems are described: soybean, cotton and greenhouse crops. Each system addresses a complex of key insect pests, and problems associated with chemical insecticide use. The range of other control measures available and how they are integrated to provide an IPM system are considered in each case.

10.2.1 Pest management in soybean

The soybean *Glycine max* is grown in at least 45 countries with a production of 113,069,000 tonnes. The major producers are the USA (44%), Brazil (21%), China (12%) and Argentina (11%) (Soyatech Inc, 1995). To produce a soybean crop that yields economically anywhere in the world farmers must control pathogens, insect pests and weeds (Sinclair *et al.*, 1997). For many years from the 1940s onwards, pest control in soybean relied on chemical insecticides, but for the last two decades concerns about farm worker safety, environmental pollution, pest resistance to agrichemicals and secondary pest outbreaks have contributed to the acceptance of IPM as the approach of choice among soybean growers. The IPM approach advocated by Sinclair *et al.* (1997) involves the components outlined in Fig. 10.1 which is based on the active participation of the growers. Although these guidelines for IPM in soybean are set out for pathogens and

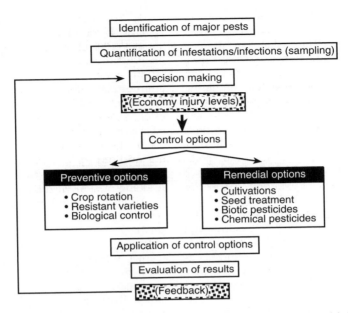

Fig. 10.1. Components of a basic integrated pest management programme for an annual field crop (from Sinclair *et al.*, 1997).

weeds as well as insect pests, only those relating to insects are considered here.

Soybean crops worldwide attract a large and diverse insect fauna (Kogan, 1988) but no more than 15 insect species normally account for up to 99% of the pest induced damage to soybean in most regions (Kogan, 1980). Insect induced injury can occur at any time during the crop development cycle: stand establishment, vegetative growth and flower and pod growth. Stand establishment may be affected by cutworms that cut seedlings at ground level and open gaps in the stand or bean flies that bore into young stems and cause seedling death. Soybean can, however, compensate for minor stand reductions but rows with linear gaps of 25–100 cm may suffer yield reductions of between 5 and 10%. Lepidopterous caterpillars (e.g. *Anticarsia gemmatalis*), coleopteran adults and larvae (e.g. *Epilactina varivestis*) and grasshoppers destroy leaf tissue while aphids (*Aphis glycine*), whiteflies, planthoppers and stinkbugs (e.g. *Nezara viridula*, *Acrosternum hilare*) interfere with photo-synthate production and translocation and in addition many can act as vectors of plant pathogens. Root pests such as white grubs can cause extensive root pruning leading to plant wilting. Injury to the reproductive plant parts results in loss or shedding of blossom and young pods, abortion of seeds, seed malformation and changes in the quality of seeds. Pests such as *Helicoverpa* spp., *Heliothis* spp., *Maruca testualis* and *Matsumuraeses phaseoli* larvae feed and cause damage by scraping or boring through pod walls into developing seeds. However, stinkbugs are the most serious pests of soybean because extensive pod feeding can result in pod abortion. In addition stinkbugs can transmit *Nematospora coryli*, the causal agent of the yeast spot disease (Sinclair *et al.*, 1997).

The management tactics employed in the implementation of IPM in soybean are based on preventive measures including pest exclusion, host plant resistance, biological and cultural control, combined with remedial use of augmentative biocontrol, biopesticides and chemical insecticides.

The preventive tactics are those associated with legislative embargoes, inspections and quarantines that are used to prevent the introduction of soybean pests into new areas. Genotypes of resistant soybean have been identified for defoliating, pod sucking and some stem boring genera. Most commercial cultivars in North and South America express resistance to deter colonization by leaf hoppers in the form of trichomes on leaves, stems and pods. Genes for resistance to leaf chewing insects (*Helicoverpa zea*, *Epilachna varivestis*, *Cerotoma trifurcata*, *Epicauta* spp.) have been found in the genotypes Kasamame (PI 17451), Sodendaizu (PI 229358) and Niyaka White (PI 227687) (Clark *et al.*, 1972; Van Duyn *et al.*, 1972). Cultural control practices are used to reduce all pest categories (insects, disease and weeds) and are listed in Table 10.1.

The damage caused by insect pests can be affected by planting dates, e.g. the bean leaf beetle *Cerotoma trifurcata* (Pedigo and Zeiss, 1996) and seed corn maggots, *Delia platura* (Hammond, 1995). The late planting of soybean in Brazil has been shown to reduce population size of thrips which normally peak early in the season (Almeida *et al.*, 1988). Trap crops in soybean have been used successfully for stinkbugs, the Mexican bean beetle and the bean leaf beetle (Hokkanen, 1991). Early planted, early maturing soybean varieties near the main soybean plantings have been successfully used in the USA (McPherson and Newsom,

1984; Todd and Schumann, 1988) and Nigeria (Jackai, 1984). A trap crop area of about 1–10% of the main crop proves sufficient to attract up to 85% of the pest population which can then be treated with appropriately timed insecticide applications (Kobayashi and Cosenza, 1987; Todd *et al.*, 1994). Use of trap crops in this way conserves natural enemy populations that occur in soybean at levels capable of holding most pest species in check. Under some circumstances more than 90% of the stinkbug *Nezara viridula* eggs and nymphs are killed by natural enemies and weather factors before reaching adult stage (Pitre, 1983).

Remedial control measures used against insect pests in soybean include augmentative releases of *Trichogramma* spp. against lepidopteran pests, NPV for the velvetbean caterpillar *Anticarsia gemmatalis* (Funderburk *et al.*, 1992; Moscardi and Sosa-Gomez, 1992) and use of chemical insecticides (Table 10.2). In 1992 chemical insecticide use in soybean was predicted to increase at 5% per annum, particularly in South America (Rosier, 1992). All over South America soybean receives the largest amount of insecticide (Campanhola *et al.*, 1995). However, the introduction of various IPM programmes has seen a reduction in some areas, for instance in Argentina an IPM programme focusing on control of defoliating caterpillars and green stinkbugs (*N. viridula*) has reduced the number of insec-

Table 10.1. Cultural practices for control of insect, disease and weed pests in soybean (Sinclair and Backman, 1989).

Avoidance of water deficit or excess
Minimization of plant stress by providing optimum nutrients and soil pH
Proper harvest date
Proper harvest to avoid seed injury
Proper storage conditions
Use of non-host buffer crop(s)
Use of trap crops
Crop rotation, optimum nutrients and soil pH
Soil tillage prior to planting
Planting of high quality seeds
Proper planting date
Proper row spacing

Table 10.2. Insecticide use in soybeans (from Rosier, 1992).

Country	Market (US$ millions)
Argentina	18
Brazil	57
USA	132
Others	13
Total	220

ticide treatments from 2–3 per season to an average 0.3 treatments, corresponding to savings of US$1.2 million per year in pesticides and application costs. Similarly in Brazil an IPM programme which has been adopted by about 40% of soybean farmers has achieved savings of over US$200 million annually due to reduced use of insecticides, labour, machinery and fuel (Iles and Sweetmore, 1991).

In Brazil and Paraguay the reduction of chemical use has been achieved through the availability and use of the NPV biopesticide *Baculovirus anticarsia* for control of the velvetbean caterpillar. The production of this virus in Brazil was initially done by EMBRAPA/CNSPO but has since been transferred to four private sector companies. In Paraguay, growers' cooperatives are presently producing large amounts of *B. anticarsia* for use (Campanhola *et al.*, 1995). In the 1988/89 season 19,000 ha were applied with the biopesticide in Paraguay, while in Brazil it is estimated that 1 million hectares are sprayed annually (Moscardi and Sosa-Gomez, 1992).

Parasitoids, particularly *Trichogramma* spp., are also widely used for control of lepidopteran pests. Over 20 private organizations produce parasitoids in Columbia and 12 in Venezuela. *A. gemmatalis*, *Omiodes indicata* and *Semiothisa abydata* are significantly attacked by naturally occurring *Trichogramma* species and timely release of mass reared individuals can maintain these pests below the economic damage level (Campanhola *et al.*, 1995; Table 3.4; Section 3.9.3).

Economic damage levels have been determined for all major insect pests of soybean

and a variety of grower sampling and monitoring techniques have been devised (Sinclair *et al.*, 1997). Numerous models, developed to describe the population dynamics of insect pests and their natural enemies, coupled with soybean growth models, have been developed to assist in decision making for soybean pest management, e.g. SOYGROW (Mishoe *et al.*, 1984) and AUSIMM (Herbert *et al.*, 1992). However, AUSIMM, which was developed as a decision aid for growers in Alabama, USA was not taken up by growers. The reasons given for this were: lack of funds for implementation, technical complexity of the product, inadequate collaboration with extension specialists, institutional inertia towards research/extension cooperation, an inability to demonstrate a great increase in net profits and the requirement of a computer for growers to use the software (Mack, 1992).

10.2.2 Pest management in cotton

Cotton (primarily *Gossypium hirsutum* but also *G. barbadense*) is grown in over 70 countries and is the most important fibre crop grown worldwide (Gutierrez, 1995; Matthews, 1997b). Annual global production is approximately 19×10^6 t lint (Gillham *et al.*, 1995) with China leading production followed by the USA, Central Asia, India, Pakistan, Brazil and Egypt. The need for irrigation and the dominant influence of pest problems in most systems greatly affects potential profits by increasing production costs and reducing yield (Luttrell *et al.*, 1994). Insects are a major constraint and cotton was one of the first crops on which insecticide use reached unacceptable levels. The problems that subsequently arose had a dramatic impact on the cotton production industry and led to a concerted effort to find alternative systems of pest management.

In the USA misuse of chemical pesticides to control *Lygus hesperus* in the 1960s in the San Joaquin Valley induced other more serious pests, such as the cotton bollworm (*Heliothis zea*) and a number of defoliating noctuids (*Spodoptera exigua*, *Trichoplusia ni* and *Estigmene acraea*;

Gutierrez, 1995). In the Imperial Valley of California the pesticide induced outbreaks of whiteflies, mites and insecticide resistance to *Heliothis virescens* caused economic ruin in the 1980s and the industry effectively collapsed, falling from *c.* 44,000 ha of production to *c.* 6000 ha. In the Sudan, the use of aerial applications of chemicals so reduced the impact of natural enemies that the cotton suffered major outbreaks of whiteflies, *Bemisia tabaci*, the honeydew from which encouraged sooty moulds to the extent that the lint was downgraded (Eveleens, 1983). During the 1950s in Peru extensive aerial application including over 30 sprays per season of DDT, parathion and toxaphene caused the death of and poisoned thousands of people (Matthews, 1997b). Overall, the overuse and misuse of chemicals in cotton pest management represents an abject lesson in what not to do with a toxicant in an agricultural system, creating more problems than it solved.

Wherever cotton is grown in the world there exists an impressive array of pest insects that are associated with reductions in yield. There are pests from five different orders, Lepidoptera, Hemiptera, Coleoptera, Thysanoptera and Acarina, although a complex of Lepidoptera including the cotton bollworms *Heliothis* and *Helicoverpa*, spiny bollworms *Earias* spp. and red bollworm *Diparopsis* spp. are most well known. The cotton boll weevil *Anthomonas grandis*, the tobacco budworm *Heliothis virescens* and plant bugs such as jassids, whiteflies and aphids (particularly *Bemisia tabacci* and *Aphis gossypii*) can cause excessive fruit loss. The *Lygus* bugs attack very small fruits hence the plant has time to compensate for this damage unless the rate of loss is high (Gutierrez, 1995). However, boll weevils and bollworms attack more mature fruits making it more unlikely that the plants can compensate, hence damage tends to be greater from these pests. The pink bollworm *Pectinophora gossypiella* similarly causes most damage when it attacks mature bolls and yield losses accrue through

destruction of seed and reduced lint quality (Gutierrez *et al.*, 1977; Stone and Gutierrez, 1986a,b).

Mites, particularly the red spider mite *Tetranychus urticae*, kill cells in photosynthetically active leaves and hence have an impact on yield. Insect defoliators such as *Spodoptera exigua* and *Trichoplusia ni* also reduce availability of photosynthetically active area but only have a significant effect on yield when they attack young leaves (Gutierrez, 1995). Yield may also be reduced by stem borers such as *Eutinobrothrus brasiliensis* that kill whole plants and stunt the growth of others; this can be prevented by application of seed treatments (Dos Santos *et al.*, 1989).

Chemcial insecticides remain the major tactic for insect pest control in cotton. Most cotton is routinely examined for insect pest damage with private consultants providing scouting services in the USA and Australia (Fitt, 1994; Luttrell *et al.*, 1994). In Australia, scouts examine up to 60 plants per 100 ha, 2–3 times per week whereas in the USA most fields are scouted 1–2 times per week and samples are generally based on examination of 100 terminal buds and fruiting forms per 20–40 ha. The use of scouting in Andhra Pradesh cotton by both state and federal agencies has reduced the number of pesticide applications from as many as 20 to 3–6 (Raheja, 1995). Cotton consumes 50% of the insecticides used annually in India even though it occupies only 5% of the cultivated area; 80% of synthetic pyrethroid consumption is confined to cotton alone. Hillocks (1995) lists the insecticides used in cotton in Africa including: Endosulphan for *Lygus*, American bollworm, spiny bollworm and cotton leafworm control, Pirimicarb for *Aphis gossypii* control, Amitrax for red spider mite and whitefly control and Dimethoate for cotton stainers.

Although economic threshold levels for pesticide application against key pests have been known since the 1950s in countries such as Sudan they were not modified over decades which resulted in the application of insecticides at lower densities than was actually necessary (Zethner, 1995). A wide variation in threshold values exists across and within different cotton systems for specific pests. Thresholds for *Heliothis* species in Brazil range from 10 to 40% of the squares infested by larvae relative to crop phenological development (Table 10.3). Assuming a density of 10 plants m^{-2} and 5 fruits/plant, thresholds are 4–20 larvae per 100 plants, 5–25 larvae per 100 plants, 10–20 larvae per 100 plants and 50–200 larvae per 100 plants in the USA, the Commonwealth of Independent States (former USSR), Australia and Brazil respectively.

The use of thresholds has not prevented the need for insecticide resistance management strategies in some countries to extend the useful life of some chemical insecticides. The most successful has been the pyrethroid resistance management strategy adopted in Australia (see Case Study in Chapter 4, p. 110) which confined the use of pyrethroids to a defined period each year on all crops (Forrester *et al.*, 1993) in order to reduce the selection pressure on the major pest *Helicoverpa armigera*. While this strategy enabled farmers to continue growing cotton, an alternative IPM programme is urgently needed and research on transgenic cotton and other tactics is currently being developed (Matthews, 1997b).

The transgenic cotton Bollgard® containing a *Bt* gene was first commercialized in the USA in 1996 on 729,000 ha which then increased to just over one million hectares a year later, followed in 1998 by planting of around 2 million hectares (Merritt, 1998). During 1998 China planted 53,00 ha, Mexico 40,500 ha, Australia 81,000 ha and South Africa 12,000 ha. In a three year study in Mississippi, USA where *Bt* cotton was compared with conventional cotton, *Bt* cotton received on average 6.7 sprays compared with 11.7 sprays on conventional fields (Stewart *et al.*, 1998). The *Bt* cotton costs, at $61.48 per acre (which includes a $32 technology), were lower than the insecticide costs of $68.15 per acre in conventional cotton. In addition, in some *Bt* cotton crops higher levels of beneficial insects were found than in conventional crops (Stewart *et al.*, 1998) while in others no differences have been demonstrated (Van Tol and Lentz, 1998).

Natural enemy populations have also been shown to be enhanced when mating disruption is used for the control of *P. gossypiella* (Campion *et al.*, 1989). Today, almost all pink bollworm management is achieved by using the synthetic pheromone Gossyplure (Howse *et al.*, 1998). With the development of Gossyplure combined with appropriate slow release formulations (Campion *et al.*, 1989; Critchley *et al.*, 1989) cotton growing countries affected by pink bollworm seized eagerly on the new mating disruption strategy. The Egyptian Ministry of Agriculture, more than any other, has made a significant contribution to the use of pheromones for pink

Table 10.3. Economic thresholds for cotton pests in Brazil (from Ramalho, 1994).

Species	Critical period (days)	Economic threshold	Sampling site
Aphis gossypii	10–60	71% plants infested	All of plant
Frankliniella sp.	10–25	6 nymphs and/or adults per plant	All of plant
Anthomonas grandis	40–120	10% infested squares	Medium squares – upper $\frac{1}{2}$ of plant
Heliothis virescens	50–60	40% infested squares by larvae	1/3 grown squares from upper $\frac{1}{2}$ of plant
Heliothis virescens	80–120	10% infested squares by larvae	1/3 grown squares from upper $\frac{1}{2}$ of plant
Pectinophora gossypiella	75–120	11% hard bolls damaged	First hard bolls from top to plant bottom

bollworm control with 300,000 acres of cotton (36% of the country's cotton) treated in 1994.

The use of mating disruption as a control technique tended also to reduce the severity of outbreaks of secondary pests, probably due to the conservation of natural enemy populations. However, it is unlikely that natural enemy populations can control all secondary pests; this is certainly thought to be the case of *Helicoverpa* spp. (Titmarsh, 1992). Augmentative control of *Helicoverpa* using the egg parasitoid *Trichogramma* spp. suggests that the approach is not economically viable in US cotton systems (King and Coleman, 1989). However, in Usbekistan where there are now over 700 factories producing *Trichogramma pinoi* and to a lesser extent *Bracon herbetos*, the approach is considered a highly effective means of controlling *H. armigera* and *Agrotis* moths (Matthews 1997b). Other natural enemies such as the entomopathogens, *Bt* and NPVs have been used against *H. virescens* and *H. zea* and although they are not widely applied, interest is increasing (Luttrell, 1994).

A variety of other control techniques including lure and kill for the boll weevil *A. grandis* (McKibben *et al*, 1990; Smith *et al.*, 1994), host plant resistance obtained through various morphological and biochemical traits (Fisbie *et al.*, 1989; King and Phillips, 1983) and cultural controls such as destruction of stubble to reduce over-wintering pupal populations of *H. armigera* (Fitt and Daly, 1990) are used in different parts of the world. The impacts of a variety of control measures have been incorporated into complex management systems utilizing simulation models and expert systems (Luttrell, 1994) but in most situations simple decision rules have been devised to integrate measures (Fig. 10.2). In Asia, a farmer participatory approach to cotton IPM has been tested (Anon., 1999). An IPM approach evaluated at Wanshong, China included a range of control measures (Box 10.1) that provide cotton yields comparable with conventional farmer practice and reduced the number of insecticide applications necessary by 29%. This reduced total input costs leading to profits 7.5% higher per hectare than was achieved

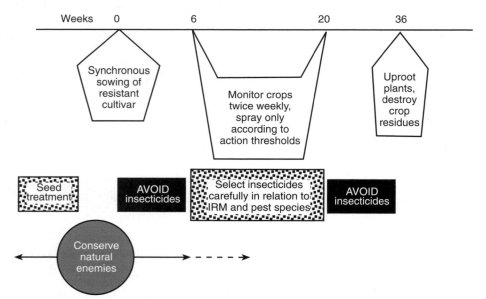

Fig. 10.2. Integration of some of the tactics that can be used in an IPM programme. In some countries it is possible to deploy pheromones for confusion or 'lure and kill' as well as monitoring pest populations (after Matthews, 1997b).

Box 10.1. Methods used in the IPM cotton plots at Wenshang in China (Anon., 1999).

Ploughing and irrigation of the fields after harvest for control of overwintering *Helicoverpa* pupae

Minimal use of organophosphate insecticides on wheat to encourage build up of natural enemies which later migrate to cotton

Irrigation during March–May to increase humidity to help control red spider mites

Dipping growing tips of seedlings in insecticides for aphid control in place of sprays

Use of *Bacillus thuringiensis* against bollworms

Topping and bud thinning to increase boll shedding rate and number of fruiting points per plant

Minimal insecticide use

in the standard practice plots. Although such approaches are yet to be tested on a large scale, the involvement of farmers in such studies increases the likelihood of widespread adoption of similar approaches. New initiatives to implement IPM are now underway in cotton growing areas most affected by problems of insecticide resistance in China, India and Pakistan (Matthews, 1997b).

10.2.3 Pest management in greenhouse crops
The global greenhouse (plastic plus glasshouses) industry covers an area of 279,000 ha (Wittwer and Castilla, 1995) but the monetary value of the industry well exceeds what might be expected from such

Box 10.2. Pest insects of glasshouse vegetable and ornamental crops.

Trialeurodes vaporariorum
Tetranychus urticae
Bemisia tabaci
Liriomyza spp.
Thrips tabaci
Frankliniella occidentalis
Otiorrhynchus sulcatus
Myzus persicae
Aphis gossypii
Macrosiphum euphorbiae

a relatively small hectarage. In the Netherlands only 0.5% of the area in use for agriculture is covered with glasshouses (9300 ha) but on this small area, 17% of the total value of agricultural production is realized (van Lenteren, 1995). In 1992, the value of Dutch greenhouse vegetables was US$31.7 million (Snyder, 1993). Crops grown under glass/plastic are thus high value crops such as tomato, cucumbers, sweet peppers, aubergines and ornamentals (including flowers). Growing vegetables in a protected environment under glass is rather expensive and pest damage cannot be tolerated. For ornamentals, the situation is even more serious because the presence of even extremely low numbers of pests may prevent export and therefore a zero-tolerance may be enforced (van Lenteren, 1992b). Given this situation, chemical insecticides with quick knock-down would seem the obvious first choice as a control measure, especially as chemicals are easy and relatively inexpensive to apply. However, the development of resistance to chemical insecticides in several key pests in greenhouses has led to an increasing interest in biological control.

Small scale application of biological control in glasshouses started in 1968 with the use of the predatory mite *Phytoseilius persimilis* and the whitefly parasitoid *Encarsia formosa* has been used commercially since 1970 (van Lenteren, 1992b; 1993). Currently, biological control of the two key pests in greenhouses; whitefly (*Trialeurodes vaporarium*) and spider mite (*Tetranychus urticae*), is now applied in more than 20 out of a total of 35 countries with a greenhouse industry (van Lenteren, 1995). The other major pests of greenhouse crops are listed in Box 10.2; in addition *Sciaridae* and some Lepidoptera are also a problem.

A range of pest control measures are used in protected crops including cultural methods, chemical insecticides, biopesticides, host plant resistance and a limited use of pheromones. Pesticides are more widely used for pest control in ornamentals than in vegetables largely because there are more available, the whole plant is marketed so no leaf damage is allowed and there is a zero-tolerance to damage for export material (van Lenteren, 1995). Non-pesticide control measures for ornamentals tend to be only used for non-export crops.

In vegetable crops several of the aphid pest species can be controlled with the selective insecticide pirimicarb but *Aphis gossypii* which is resistant to pirimicarb occurs frequently on cucumber. Organophosphates may be used to kill thrips or alternatively application of polybutenes mixed with deltamethrin can be applied topically on plastic sheeting that covers the soil and is picked up and kills thrips when they leave the plant (van Lenteren, 1995). Thrips (*Frankliniella occidentalis* and *Thrips tabaci*) can also be controlled by releases of *Amblyseius* species leading to a rapid increase in their use in IPM programmes.

Soil sterilization is used in crops such as tomato largely to control diseases but it also has an impact on *Liriomyza* spp. (leafminers). Unfortunately red spider mite (*T. urticae*) and the tomato looper (*Chrysadeixis chalcites*) tend to survive soil sterilization. The looper can be controlled using *Bt* while *T. urticae* requires a combination of chemical control and the use of *Phytoseiulus persimilis* (Table 10.4). Tomato cultivars that are resistant to whiteflies are available (de Ponti *et al.*, 1990). Host plant resistance is a relatively new approach to insect pest control in glasshouses. Levels of resistance in cucum-

ber have been identified for control of spider mites and thrips (*F. occidentalis*; de Ponti, 1982; Mollema *et al.*, 1993).

IPM has been promoted and supported in the glasshouse crops industry by a combination of grower participation, IPM labels and the availability of suitable decision support systems. Several expert systems have been developed for greenhouse applications, providing advice on a range of factors such as plant nutrition and temperature control (Gohler, 1989) as well as pest control (Clarke *et al.*, 1994). In general, however, IPM has gained favour with growers working in greenhouses because:

• Of chemical resistance in key pests.
• Biological control is cheaper than chemical control.
• The release of biocontrol agents is more pleasant and safer than chemicals.
• Biocontrol agents do not require a safety period prior to harvesting.
• The release of biocontrol agents usually occurs after the planting period when the grower has plenty of time to check for successful development of natural enemies, thereafter it is reliable for many months, whereas chemical control requires continuous attention (van Lenteren, 1995).

10.2.4 Lessons learnt

While it is acknowledged that the types of lessons that can be learnt from case studies of pest management are to a large extent dependent on the systems selected, the case studies used here do highlight a number of factors of general importance.

Table 10.4. Commercially applied IPM programme for tomato crops (after van Lenteren, 1995).

Pests	IPM programme
Trialeurodes vaporariorum	*Encarsia formosa*
Tetranychus urticae	Chemical control; *Phytoseuilus persimilis*
Liriomyza bryoniae; *L. trifolii; L. huidobrensis*	*Dacnusa sibirica; Diglyphus isaea*
Noctuid spp.	*Bacillus thuringiensis*
Aphid spp.	*Aphidoletes aphidimyza; Aphidius matricariae; Aphidius colemani*

The first of these is that the force for change in each of these three crop/pest systems has been the development of pest resistance to chemical insecticides. In each case this has led to the failure of insecticide to control the key pest, decimation of natural enemy populations and secondary pest outbreaks. The provision of information and training has proved important in convincing growers and farmers of the value of alternatives to chemicals, whether provided in terms of sophisticated expert systems or through farmer participatory approaches. The alternative control measures adopted vary widely and the relative mix used is dependent on the cropping systems, particularly with regard to cultural control methods. A number of technologies may not be as effective as the chemicals they replace but all those adopted have proved economic. Where chemicals constitute part of an IPM system then selective use of insecticides combined with thresholds for application have been used. An understanding of the crop yield and physiology particularly with regard to yield compensation for damage, combined with a good knowledge of insect pest ecology have also proved essential in optimizing interventions.

Each of the above systems – soybean, cotton and greenhouse crops – are well studied. A great deal of research effort has been dedicated to solving the problems posed by insect pests to these crops. Effort has also been applied to redressing problems created by the use of inappropriate solutions to their control that have been tried over the years. The process by which we have tackled pest management problems and the means by which we have derived 'solutions' can perhaps tell us as much about how to develop more successful approaches in the future as can the detail of what has been achieved in terms of control. A better understanding of the process by which we can develop successful IPM systems may ensure we make fewer mistakes in the future.

10.3 Panaceas, Paradigms and Pragmatism

The history of pest management (Chapter 1) is a history of panaceas and paradigms; some of the greatest failings of pest control have come about quite simply because of the limitations imposed by both. A panacea may be described as a 'universal medicine' a solution to all ills, whereas a paradigm is a common way of thinking, approach, goal or established scientific opinion that is shared by particular scientific communities (Dent, 1995). Examples of paradigms within pest management include IPM (Perkins, 1982; Morse and Buhler, 1997), deterministic models of natural enemy population dynamics, and the pursuance of vertical resistance for insect resistance breeding (Dent, 1995). The use of analytical models to study natural enemy dynamics during the 1970s and early 80s was carried out under the guise of providing a greater understanding of factors influencing the success of biological control but there is little evidence that such studies contributed much to the actual practice of biological control. In fact, the results of theoretical studies were used as a basis to narrow down the attributes of natural enemies considered appropriate for biological control – characteristics which mitigated against the potential benefits of predators in favour of parasitoids. The basis for this selection has been the theoretical ability of host specific parasitoids to depress pest population levels to new, low and stable equilibrium levels (Beddington et al., 1978; Hassell 1978, 1980, 1982; Waage, 1983). However, the depression of pest populations to a new but lower stable equilibrium level is actually an unnecessary pre-requisite for achieving effective control. Provided a pest population can be maintained below its economic threshold it is irrelevant whether or not the pest population is stable or not. Low stable equilibria have always been sought because it had been assumed that if the pest population became extinct then so would the natural enemy and that this would be undesirable. If you remove the need to obtain low stable equilibria then it opens up the possibility of predators being suitable biocontrol agents, largely because when pest populations become extinct, predators can switch to alternative prey and hence, be preserved in the absence of the pest. Thus, the use of the theoretical models limited what was actually attempted in practice.

A similar situation is evident in the host plant resistance paradigm where the predominance of the vertical resistance, Mendelian school of genetics in plant breeding has essentially precluded development of cultivars utilizing horizontal resistance (Chapter 5; Robinson, 1991). Given the only limited success of breeding resistance to insect pests it is quite surprising that the conventional approach was not more rigorously questioned. This may have been because the debate was deflected by entomologists becoming preoccupied with the mechanisms rather than the genetics of resistance.

Paradigms that influence pest management are not just those associated with research and development. The subjects of anthropology and socio-economics have had a major impact on the approaches taken in pest management particularly farmer participatory approaches to IPM. For many years the main emphasis in both the research and practice of insect pest management has been the development and use of various insect control techniques. Such a focus of attention implies a single-interest strategy that excludes the pest from its total natural and social environment (Gabriel, 1989). This separation of insect pest management from the social and human aspects of the pest problem inevitably leads to poor targeting of research, the development of inappropriate technology and, subsequently, to poor rates of adoption. The work and reappraisal of the role of farmers/communities in development undertaken by Chambers (1983) created a paradigm shift, through the realization that the failure to properly involve farmers in development issues contributed to failure in development programmes.

The situation with regard to pest management can be best described using the terminology of Andrews et al. (1992) who categorized pest management systems by scale and the degree to which farmers participated in implementation. Essentially, there are two models, the 'Farmer by-pass model' and the 'Farmer as protagonist model'. The former applies to a regional pest management system where there is a minimal farmer and extension involvement (e.g. SIT, quarantine and classical biological control). Such approaches are biased towards straightforward research programmes uncomplicated by outreach and extension. The 'Farmer as protagonist model' takes into account the ecological and social heterogeneity of farming systems and the need for site specific decision making, selection and use of control measures. Further subdivision of these models is possible in terms of whether research and implementation is participatory or non-participatory.

The approach prior to the paradigm shift (Chambers, 1983) was very much non-participatory research, non-participatory implementation where research was undertaken in isolation of farming systems and implementation was also carried out totally independent of farmers who were seen simply as the beneficiaries of the outcome. A common model that still largely persists is the non-participatory research, participatory implementation model, i.e. the top down approach where scientists involved in fundamental research develop new concepts, devices etc. that are adapted by applied scientists and tested by extension scientists in collaboration with influential farmers. These models have traditionally been used for the development and adoption of pest control products such as chemical insecticides that can be packaged in simple ways to promote dissemination and adoption (Dent, 1995). However, the approach is less effective when information rather than inputs are to be disseminated and has proven useful for techniques such as conservation of natural enemies and habitat modification (Andrews et al., 1992).

The paradigm shift changed the emphasis from these non-participatory research approaches to one of participatory research, participatory implementation which takes as its central premise that engaging the farmer in the research process will ensure that the research is pertinent to farmers' needs and hence will mean that there is more likelihood that the products of such research will be adopted by farmers. *Participatory research – participatory implementation* more readily suits the complex needs of IPM systems where on-farm research is used to identify the constraints to performance of 'baskets' of new technologies (Sumberg and Okali, 1989). The farmer field school approach is the epitome of the participatory research and implementation model where farmers carry out their own ecosystem analysis, determine which control measures to assess and then carry out their own evaluations.

The value of paradigms to science is that they focus effort and create a framework for scientists who are practitioners of a scientific speciality to make up a discrete community in the knowledge that each shares a common approach, goals and opinions. The paradigm has a key role in the furtherance of scientific endeavour because research progresses within a framework of paradigms until their limitations and constraints become overwhelmingly apparent. Then through the efforts of extraordinary scientists (here I refer also to social scientists) or scientific phenomena, the prevailing paradigm is supplanted by another (Musgrave, 1980; Shapere, 1980; Medawar, 1981, 1986). Herein lies the problem with the paradigms that have shaped pest management (including IPM) to this day. Paradigms can limit scientific progress and they do this by: (i) creating problems that do not exist by adherence to a particular division, polarization and conceptualizations; (ii) acting as conceptual traps or prisons which prevent a more useful arrangement of information; and (iii) through blocking by adequacy (de Bono, 1970). Thus, paradigms can actually limit the scope for solving pest management

problems, as is apparent from the examples above, for analytical models and Mendelian approaches to host plant resistance breeding. Paradigms, however, have an impact beyond the scientific community: paradigms tend to be readily adopted by politicians and funding agencies as a means of prioritizing support for research – hence, the emphasis given to the funding of IPM socio-economic approaches, gender issues and more recently farmer participatory and farmer field schools in pest management. This is not to say that these are not valid or valuable areas for funding but the problem comes when equally valid or valuable activities are ignored just because they do not fit within the current orthodoxy or paradigm. The impact of this is that you do not necessarily obtain the funding support for the most appropriate solution to a problem but just the one that is currently in vogue. Sadly, the history of pest management is littered with such examples.

The situation is no different for 'panaceas' in pest control – the 'new' pest management technologies that have been developed over the last 50 years that have been heralded as the 'new solution' to pest control are numerous, e.g. SIT, pheromones, high yielding dwarf cultivars – different classes of insecticides, organochlorines, organophosphates, carbamates, pyrethroids, *Bt*, NPVs etc. Each of these product based control measures, at their time of introduction, received funding support, and were greeted with great enthusiasm until the limitations in their development and more often their use, became apparent. Biotechnology represents a current paradigm and transgenic crops represent the current predominant panacea receiving disproportionate amounts of research funding and support compared to many other equally valuable approaches and products. Paradigms and panaceas limit the ability of stakeholders to fund the most appropriate solutions to pest management problems. It is important for decision makers to be aware of this fact and to understand that the only logical and sensible solution is to adopt a more pragmatic

approach to problem solving that avoids the error of 'locking in' to a particular development path (Norton, 1993). This can be achieved through a decision tool approach which provides techniques to analyse the factors leading to a particular problem and considers all of the available options before selecting the most appropriate solution for solving the problem (Norton and Mumford, 1993). Thus, a more objective approach to the problem of improving pest management is provided and avoids preconceived notions, based on predominant paradigms of how improvements can be best achieved (Norton, 1993).

10.4 Models, Information Technology and Communication Technology in IPM

Throughout this book reference has been made to and examples given of many different types of model. There have been examples of models used to forecast pest abundance (Chapter 2), models to simulate plant growth and pest damage (Chapter 3), models to explain the interactions between pests and their natural enemies (Chapter 6). These models have been of different types and forms but the fact that they are now used in most areas of insect pest management says something for their utility.

Models are representations of a system. They attempt to mimic the essential features of a particular system where a system is taken as a limited part of reality. Rabbinge and de Wit (1989) distinguished four main levels of system in agriculture, pathosystems, cropping systems, farming systems and agroecosystems. As discussed in Chapter 5, a pathosystem includes a single host and parasite population and their interactions. These may be described by population genetics, dynamics and crop physiology (Rabbinge and de Wit, 1989). A cropping system may have a number of pathosystem components but will only deal with a single crop or intercrop. Agronomic and pest management practices

and their economics are also encompassed within the cropping system.

A farming system consists of a number of different crops, their interactions, economics and management, and is also a subsystem of an agroecosystem. Agroecosystems are ecosystems adapted by mankind to serve their own needs and they are characterized by their component recognizable structure. This structure will consist of functional relations among the different components (Rabbinge and de Wit, 1989). Most models that have been formulated in insect pest management deal with the pathosystem and cropping system levels mainly because at these levels the amount of available information makes models less speculative.

Models then, seek to mimic an existing system. If the literature were studied it might be thought that the only way to represent such systems would be through a series of mathematical formulae, and while it has to be accepted that most models require a certain level of mathematical dexterity, they all start at a conceptual level that requires no mathematical knowledge or ability. It is perhaps at this conceptual level that modelling can make one of its greatest contributions to insect pest management. This is because, if generally adopted as a method, scientists would have to define more closely the system with which they were working and, by doing so, focus their attention on the problem, in a way that can only improve scientific method.

Insect pest management is an apparently complex subject which is often too difficult to comprehend fully. Even separate components of the subject present diverse and complicated interactions, for instance, the relations between weather, insect population dynamics, damage, yield loss and economic application of an insecticide. If the interactions of natural enemies and the effects of agronomic practices or use of resistant plant cultivars are also included then it is clearly a complex situation in which it is difficult to envisage the role and effect of each interaction. When dealing with such complexity, unless the system is formalized in some way, it is very difficult to maintain a balanced and holistic perspective. This can lead to biases in emphasis, a narrowing of approach and adopted options, poor decision making and communication. If a system is formalized as a conceptual model the complexity is placed in a more objective context. Just the exercise of assembling and formulating all known information about a system forces clear and concise thinking. It also helps to identify key processes or decisions and gaps in our knowledge, as well as highlighting untested assumptions and emphasizing the potential importance of elements of the system and their interconnection which, without the modelling process, may have seemed of little importance (McKinion, 1989). Such conceptual models are qualitative and are usually produced as a diagram where each of the components are identified and their inter-relations depicted as connecting lines or arrows (Fig. 10.3). Southwood (1978) referred to these as dioristic models, where dioristic means serving

to define. Although only an initial step in the development of a model it is an important process since it forces the modeller to consider and define the objectives of the model, the system boundaries and inter-relations of the different components of the system. It is this aspect of modelling that clearly relates to the basic tenets of good scientific method and which, if adopted, whether at the level of strategic planning in a research institute or the design and execution of a simple laboratory experiment, will greatly improve research in insect pest management. The complexity and diversity of IPM and insect pest management have, especially at some higher levels, encouraged some woolly thinking which is something a modelling approach would help to reduce.

The defining of objectives makes the modeller consider the present status of a situation 'where are we now?' and the position that is needed at the end of the exercise 'where do we want to get to?' (Tait, 1987). These are important questions! The

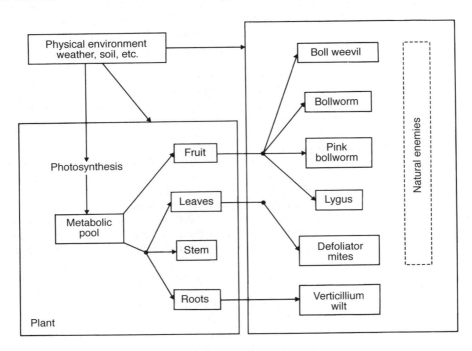

Fig. 10.3. A simple conceptual model of the California cotton product system (after Gutierrez, 1987).

first is usually considered by all entomologists since it has an immediate influence on the work they currently propose to carry out and any good scientist should be well informed on the current status of their subject. The second question is, however, one that is only rarely considered and, even less often, clearly and unambiguously defined. The end objective and the final form of the programme are rarely considered at the problem formulation stage. If insect pest management is to pass successfully from research and on to implementation then greater emphasis needs to be placed, at the outset, on how this is to be achieved. The objectives considered must extend beyond initial questions to the implications of their solutions. Conceptual models can help to clarify objectives because in formulating a model an end point has to be considered and defined. It may become apparent during the process that the model reflects only a small part of a larger system, providing the opportunity to define the role of this sub-model and place it in the context of the objectives of a larger systems model. This is, however, the nub of the problem. There are major drawbacks associated with any form of problem framing because the objectives can only be based on a person's initial perception of the problem. If the person involved does not perceive the relations and interconnections that exist with other much larger systems then this could place the analysis of the problem in jeopardy simply because the wider implications of the larger system are not taken into account (Tait, 1987). This is where the involvement of scientists from a range of disciplines in insect pest management and IPM can have a definite benefit. The perspective provided by a multidisciplinary team can only serve to broaden the group's perception of a problem with the result that the larger systems of interconnecting problems should be more readily identified. Once identified their importance and influence can be evaluated.

The decision as to what to include in the model is part of the process of defining the system's boundaries. On the one hand there is the problem of the need to include all relevant aspects of the system, on the other is the need to keep the scope of the model within manageable proportions. At the start, the number of variables will be increased and as a result the model will begin to reflect reality, but a point will be reached where the addition of new variables and an increase in the number of interactions will divert attention away from more important variables already present (Rabbinge and de Wit, 1989). There is some argument as to whether models should be complex and a top down approach be adopted, i.e. a complex model is reduced in size and complexity until it can be reasonably validated, or the contrasting bottom up approach used, i.e. where the simplest explanation is initially considered and complexity is increased if the model is unsatisfactory (Vansteenkiste, 1984). There is, of course, no single correct solution to this argument. Each model and system will be different and the boundaries of the system and the complexity of the model will invariably be determined by the objectives of the study. The aim during any modelling exercise must be to produce sufficient checks in the procedure to ensure that the model becomes neither too simplistic nor too complex, but that it contains descriptions of the key processes and provides an adequate reflection of the system.

The conceptual elements of model formulation, of problem framing, system description and identification of objectives and constraints represent only the initial stages in the systems analysis (Fig. 10.4). Next, it is important to identify the route(s) by which the objective may be obtained and then to formulate measures of performance by which the model can be assessed. The measures of performance are the criteria by which the effectiveness of the model is judged. These may be based on the degree to which the model agreed with accepted theory, actual events, or economic criteria, but in each case work has to be carried out to test the model, to validate it relative to these performance criteria. In

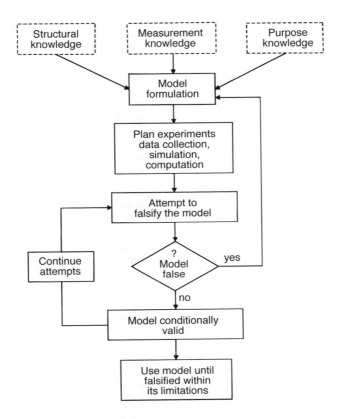

Fig. 10.4. The methodological steps used in model building (after Vansteenkiste, 1984).

this sense a model is only a complex hypothesis that, like any other hypothesis, needs to be falsified by experiment. A great deal of modelling methodology should involve attempts to falsify the model and any model that is validated is only useful until conditions, situations or information arise that falsify it. Once a model has been conditionally validated it can then be implemented, in that its results/output can be put to use. These steps in model building are depicted in Fig. 10.4. They are generally applicable for most model types, whether they are used to plan research, experiments or management.

Models and their use in ecological and agricultural entomology have recently been reviewed by Holt and Cheke (1997). Models can be taken forward to develop computer decision support systems

(Walton, 1998; Section 9.5). Uptake among farmers has, however, tended to be poor even though in some cases there have been clear financial benefits to using the systems (e.g. Lucey *et al.*, 1997). One of the disadvantages of using computer decision support systems has been the cost and inconvenience of revising the system and copying changes onto suitable media and distributing them to users. The problem can now be easily overcome through use of the internet. The advantage of internet based decision support systems is that only a single version of the software is implemented on a computer server at any one time and users have to access the programme remotely from their own computers. Any changes to the software are made to a single programme and can be implemented quickly and cheaply (Morgan *et al.*,

1998). Such use of information and communication technology and the world wide web will have a major impact on how information is accessed, synthesized and used by all stakeholders in pest management in the near future.

Farmers and the general public already have access via the internet to a wide range of informed and uninformed opinion, information and data, relating to pest management issues. Scientists and vested interest groups will be able to receive a personalized update of information from a range of pre-selected sources dealing with subject matter relevant to their specific interests. Access to information on matters relating to food and environmental safety (e.g. on pesticide residues, GMOs) could potentially lead to a better informed and aware general public. In an ideal world this would contribute to a general public empowered to make more rational assessments of personal risk. This in turn could reduce the need for over-protective regulation by government authorities. Governments and associated agencies may benefit from an ability to canvas public opinion either through the internet or interactive television. The strength of public opinion will be gauged by politicians from results of on-line opinion polls or television debates, providing a mechanism for feedback to changes in legislation or policies. Hence, public opinion will have an increasingly direct impact on government policies and decisions, changing forever the way government works in developed countries.

10.5 A Question of Scale?

Scale adds an important dimension to pest management, which can be attempted at the level of individual farms, across a number of farms, regions, countries or continents. It can involve the use of a control measure used on a small scale by a large number of farmers or the control measure can be area-wide in its application or influence, e.g. classical biological control or

SIT. The scale at which activities are attempted has important implications for pest management.

10.5.1 Sustainable use of control measures
The widespread and frequent use of chemical insecticides of organochlorines, organophosphates and pyrethroids has consistently led to the development of insecticide resistance. The problem is particularly acute when the product persists in the environment and hence maintains a high selection pressure. Resistance has been slower to develop to *Bt*, probably due to its less widespread use and short persistence but resistance has developed (Harris, 1997b). However, increasing sales have led to resistance in the Diamondback moth *Plutella xylostella* on crucifer crops in South-East Asia (Harris, 1995). The widespread use of most biological or chemical based products for insect control are potentially at risk if used on a sufficiently widespread basis that they cause sufficient selection pressure. The breakdown of host plant resistance is a common enough phenomenon where resistant cultivars are widely planted (Russell, 1978; Johnson and Gilmore, 1980; Robinson, 1987) and now with genetically modified crops there are real concerns that their widespread adoption in cotton, soybean and potato will for instance quickly lead to the development of resistance. In the case of transgenic *Bt* crops this could also mean the loss of *Bt* as a useful control agent as a foliar spray. If other biopesticides based on fungi, viruses or nematodes are also used sufficiently, resistance can potentially develop. Recently, the first suspected case of selection of an insect strain capable of 'overcoming' a crop rotation was reported (Gray *et al.*, 1996). The insect corn rootworm (*Diabrotica virgifera virgifera*) has been controlled for decades in some parts of the USA through use of an annual crop rotation of maize with soybean. Severe damage in maize caused by *D. virgifera* in 1995 was investigated and it was discovered that corn rootworm adults were laying eggs in soybean fields. The work of Gray *et al.*

(1996) suggests that growers may have selected inadvertently for a new strain of western corn rootworm because of routine practice of rotating maize with soybean. Such are the salutary lessons of pest management.

10.5.2 Levels of integration

In Section 10.4 the four main levels of system in agriculture – pathosystems, cropping systems, farming systems and agroecosystems (Rabbinge and de Wit, 1989) – were described. The main emphasis in insect pest management is on the pathosystem and to a more limited extent, the cropping system; this book considers insect pest management within the context of IPM. However, insect pest management is just one component of IPM which may be conceived of as an interactive system at three possible levels of interaction (Kogan, 1988):

1. Level I: The integration of control methods for single species or species complexes (species/population level integration).
2. Level II: The integration of the impact of multiple pest categories (insects, pathogens, weeds) on the crop methods and their control (community level integration).
3. Level III: The integration of multiple pest impacts and the methods for their control within the context of the entire cropping system (ecosystem level of integration).

A more current philosophy of agricultural production systems classifies the different levels of integration as integrated pest management, integrated crop management (ICM) and integrated farming systems (IFS). These different levels of integration are of great importance for IPM which is considered the lowest level of integration. The significance of this is that although stakeholders involved in IPM consider pest management to be of primary concern and the need to understand the detail of its dynamics and economics highly important, when you start to evaluate problems from a larger scale at the levels of ICM or IFS, then pests just become one of a number of production constraints that have to be taken

into account, and their management one other factor that has to be balanced against a whole range of others competing for resources and attention by the farmer. Although a sobering thought for those of us involved in pest management, the reality of the situation is that there is a tendency to address problems in agriculture and development on a wider and wider scale, moving from the pathosystem to the ecosystem. The implication of such approaches is that the level of detail and understanding that is required in order to make pragmatic decisions (by a farmer) is not that supposed by researchers attempting to manipulate a single pest in a single crop. Integrated crop management is moving towards lower inputs and a return to use of cultural control methods that remove the major constraints to production. With such approaches the significance of individual pest problems is reduced. If the farmer deals macro level problems by manipulating the crop/farm environment it is arguable that detailed understanding of each pest and its control is unnecessary. This in turn calls into question the value of and need for detailed research at the pest/crop level. With governments keen to cut back on public spending such justification could be used to reduce research on IPM related topics.

10.6 Technological Advances and Commerce

In Section 10.3 the role of paradigms both as a process by which science advances and at the same time is constrained, was discussed. Overlaying this process are the constraints to progress presented by the limitations of particular methodologies (Dent, 1997b; Walton and Dent, 1997). Areas of potential advancement open up when new methodologies are developed which make possible previously impossible or difficult experiments. Walton and Dent (1997) described this as a 'vertical leap' in capability and contrasted it with a 'horizontal expansion' in capability, the process by which the 'vertical leap' in methodology is more widely exploited and used across a wide range of conditions and circumstances. The 'vertical leap' in capability is a relatively rare event in science, but its impact can be substantial, creating whole new possibilities in technology or knowhow. Take for instance, the discovery of Polymerase Chain Reaction (PCR). The process of PCR and its associated enzyme DNA polymerase were named by *Science* as the 1989 'Molecule of the Year' due to its likely effect on future science. The Nobel Prize for Chemistry was also awarded to the inventor of the PCR process in 1993 (Symondson and Hemmingway, 1997). The technique has made possible DNA diagnostics for insecticide resistance (Hemingway *et al.*, 1995), the study of insect taxonomy and insect population genetics (Symondson and Hemmingway, 1997). Instrumental to this is rapid amplification of polymorphic DNA (RAPD) which was first reported in 1990 (Welsh and McClelland, 1990) – a simple technique for amplifying non-specific fragments of genomic DNA for which no sequence is available. RAPDs have been used to investigate taxonomic differences between aphid species (Cenis *et al.*, 1993; Robinson *et al.*, 1993) as well as clonal variation within a single species (Black *et al.*, 1992; Puterka *et al.*, 1993). Such DNA and biochemical advances will eventually result in 'litmus paper', 'pregnancy testing' type kits that can be used in the field to identify insect pest species, and insecticide resistant insect populations, including those resistant to *Bt* transgenic crops. In addition they will be used as indicators to test for the presence of insecticide residues or the presence of proteins produced in transgenic crop plants. Such technologies will continue to revolutionize pest management and the ability of consumers and retailers to make decisions about food and its origins prior to purchase. Traceability of food products through the availability of such technologies is likely to have a major impact on the future use of and markets for chemical insecticides (Dent and Waage, 1999).

The advances that have been made in developing transgenic crops with resistance to insect pests expressing toxins from *Bt* (first successfully reported in 1987; Vaeck *et al.*, 1987), inhibitors of proteases and amylases, lectins and enzymes such as chitinases and lipoxygenases (Gatehouse *et al.*, 1998) are increasingly either available for farmers to plant or are likely to be so within the next few years. Inevitably the development of transgenic biocontrol agents such as viruses, bacteria and fungi are also likely in the future (Winstanley *et al.*, 1998) – having enhanced host range and speed of kill, storage and persistence.

Each of these technological advances provide potential new products for agribusiness to commercialize, market and sell. The pest management agribusiness has been built on pest control agents/technologies particularly chemical pesticides and today these and transgenic crops represent the largest share of the business. However, the structure of the industry has changed since the late 1960s when, driven by IPM, new and different pest control technologies were sought, e.g. semiochemicals and biopesticides, which over the years have spawned a multitude of smaller companies specializing in different aspects of pest management. Today there are companies that specialize in pheromone monitoring and mating disruption, target systems, barriers and mulches, decision support software, as well as production and sale of predators, parasitoids and biopesticides based on fungi, entomopathogenic nematodes, bacteria and viruses. In addition to these are the support industries providing packaging, adjuvants, sprayers, inert carriers, diagnostic systems, tracers and meteorological equipment.

Pest management has also produced a wide range of service industries including consultants dispensing advice, independent providers of toxicological, ecotoxicological and registration support, as well as contract research and product development. The pest management business is a multinational, multi-billion dollar industry that provides employment, wealth, taxes and by most accounts contributes to agricultural production. However, the signs are that the pest management aspect of agribusiness could be in decline. A number of multinational agrochemical companies are retrenching, reducing their chemical pesticide product portfolios, diversifying into seed companies and transgenic crops and in addition there is growing evidence in some cropping systems, e.g. rice and cotton, that pest control inputs (particularly chemical pesticides) are not necessary for sustainable crop production. It is unlikely that the pest management agribusiness is in terminal decline, rather that the major players (the multinational companies) foresee the prospect of lower returns and smaller markets. This is a process driven by farmers – responding to the needs of consumers who are increasingly discriminating in choice of food products that are considered 'safe'. The result will be an increasingly diverse industry with fewer but safer chemicals available, complemented by a larger number of pest management products manufactured by small, flexible and specialist pest management companies meeting more local pest management product needs. Given this, however, does not imply that the industry as a whole will be any less important in wealth and employment generation, just that there will be a shift in where and how the wealth and employment is generated.

10.7 Politics, the Public and the Environment

The policies of governments can have a significant effect on insect pest management in a number of different ways. They can affect the availability of funds for research and extension, or the pricing of products and the availability of subsidies, its implementation and adoption. The overall philosophy of IPM will also only continue to gain political credence if the climate of public opinion is appropriate to allow this to occur.

There are a number of factors that make IPM more politically attractive now than at any previous time. These factors are mainly relevant to the economically developed nations of Europe and USA although, if anything, these countries have until relatively recently been the bastions of pest control dominated by the use of chemical pesticides. The reduced pressure on arable land (due to surpluses) and greater public concern for the environment have been the main driving forces for this change. In Europe enormous stockpiles of grain, meat and dairy products have prompted changes to reduce over-production. The growing public concern about the misuse of pesticides and the general interest in a cleaner, safer environment have pushed the politicians towards an acceptance that environmental issues will influence the way people vote. The factors that influence votes in a democratically run government will subsequently influence policy and thus, the evolution of 'green politics', a political force for environmental issues. IPM, a philosophy of pest management that seeks to reduce the use of harmful chemical pesticides and to promote the integration of natural or environmentally friendly control methods, is uniquely suited to the new green era. IPM and green politics are well matched.

Public opinion is now a powerful force for change in agriculture and is hence being taken seriously by governments. However, because public opinion is never homogeneous, it is not always easy to be sure exactly who is represented when 'public opinion' is quoted. Pressure groups often appear to represent public opinion but it may be dangerous to assume they do so, especially if they hold extreme views (Spedding, 1998), or have vested political interests or social engineering at heart (Hillman, 1998). The media plays a major part in the public's perception of scientific issues, for instance the use of transgenic crops. Since the media is arts dominated, and all too often artistic licence embellishes scientific observation with a view to making the news rather than just reporting it, the presentation of the issues is rarely properly dealt with. Given this, however, public perceptions are a reality even when they are ill-founded and erroneous, and can have sufficient influence to affect government policy as well as demand for agricultural products (Spedding, 1998).

The public perceive risks differently from experts. Scientists and regulators focus on measurable, quantifiable attributes of risk whereas the public as consumers emphasize more qualitative, value laden attributes of risk such as fairness and controllability which experts tend to ignore (Groth, 1991). Risk perception by the public strongly depends on whether the risk is involuntary; out of an individual's control, whether it is inequitably distributed and whether or not it has the potential for catastrophic consequences (Slovic, 1987). The more that risk falls into these categories the higher the perceived risk, hence the pressure for the management and reduction of this perhaps through government regulations (Luijk *et al.*, 1998). In the UK, the BSE situation has damaged public confidence in government's ability to understand and manage risk – to the extent that new approaches to policy making are urgently needed (König, 1998).

Part of the problem faced by governments is that there is generally a poor knowledge, standard of education and understanding of the science behind the issues that are perceived as problems by the public. Often the public have a rather 'vague notion' of what is at issue and such notions may be driven by generally held beliefs and 'truisms' such as a need for the 'balance of nature between species' or 'humans need to live in harmony with nature'. Even perceptions of what is meant by 'the environment' can be very vague. All of which may be used to justify arguments or support whatever action a particular group advocates. Sweeping generalizations about systems and methods may be invalid and do not help debate at all (Spedding, 1996, 1998).

Whereas in the past the concerns of the public in relation to pest management have been associated with chemical pesticide use, the present major subject of debate is the use of transgenic crops and the potential negative impact they could have on 'the environment' and our food. In contrast to healthcare the application of biotechnology to food and the environment has tended to be greatly influenced by the level of education, perceived social and ethical issues as well as reaction and frequent irrational responses towards non-medical sciences (Hillman, 1998). None the less a balance point has to be reached, taking into account illogical fears, genuine damage to 'the environment', healthcare and the need for science. To this end governments will need to ensure greater public disclosure (product labelling) and tighter regulation (encompassing more systematic research and evaluation of risk and hazard before and after release). In fact such requirements are more or less becoming politically inescapable, at least in Europe. Thus, governments and their policies will continue to have a major impact on IPM. In developing countries governments are the single most important influence, because of their policies which can affect availability and regulation of biocontrol agents, subsidies to pesticides and other control measures, but also because governments are usually the single most important investor, employer and controller of resources.

Governments can control the availability of insecticides through the provision of tax exemptions and reductions, low cost credit or subsidized prices for the purchase of both sprayers and chemicals. Hence, farmers are encouraged to purchase and use insecticides because they are inexpensive. By reducing the cost of insecticides governments hope to increase crop production either for home use or to generate revenue from exporting the products from cash crops. The level of subsidy as a percentage of the retail price of the pesticide is usually high, for instance 89% in Senegal, 83% in Egypt, 87% in Ghana, 44% in Colombia (Bifani, 1986) and between 50 and 75% in Pakistan (Ahmad, 1987). These subsidies cost the governments a substantial amount, especially as most of the pesticides consumed in developing countries are imported. Such subsidies are bound to encourage pesticide use (van Emden and Peakall, 1996), in Indonesia the consumption of subsidized pesticides grew from 5100 tonnes in 1978 to 15,100 tonnes in 1982 (Bifani, 1986).

Credit facilities may also be extended to help farmers purchase spray equipment or insecticides. The credit may be obtained from banks, merchants or government supported agricultural co-operatives and the security for the loan will often have to be the productive capacity of the crop, if the farmers have few other assets. Hence, the lender receives both the return on the capital and interest from the proceeds of the harvested crop. There are disadvantages to some credit schemes, other than the obvious problems with defaulting. Some agricultural credit programmes involve packages of inputs, so that a prescribed cocktail of seeds and pesticides must be purchased at the time the load is drawn (Goodell, 1984). In doing this farmers may gain by taking advantage of special low interest rates, but they may also lose out by purchasing pesticides they may not need at higher than open market prices. Such packages preclude farmer experimentation and discourage use of other forms of control, a situation contrary to that required for the adoption of IPM.

The control of resources and the policies for their allocation provides governments with the capability to promote IPM within a country. Governments create and control the institutions important to IPM such as national agricultural research stations, extension services, regulatory agencies and have the ability to establish mechanisms for coordinating the activities of the complex of government, non-governments (NGOs, private sector) and international organizations that are required to provide IPM in a relevant form for the farmers. Governments also need to establish appropriate mechanisms to engage a broader range of people than those in the usual circles of scientific and official advisors in risk management. Only in this way will they regain public confidence in the

process of decision making regarding use of novel technologies for pest management (König, 1998). The public and their concerns about food and the environment will continue to be a major driving force in agriculture. It will be through developing better partnerships between all relevant parties, coordinated and led by government that IPM can remain at the forefront of this process thereby alleviating unwarranted concerns and reducing harm to the environment.

10.8 Conclusion

The driving forces in pest management are clearly many and varied. The future prospects of IPM will be influenced by each to differing degrees and it is difficult to predict the extent to which any one will predominate. However, in bringing together the elements of pest management, its history, the control options, the principles and information that relate so much to what is researched and implemented, it is important that the future prospects of IPM are properly considered. At risk, of course, is the fact that in time everything written in this section may prove totally incorrect. In making this attempt the context in which pest management is given real meaning needs to be outlined – the problem of pest management in meeting the needs of the world's population for a sustainable food supply. It is within this context that the important future trends in pest management can then be addressed.

10.8.1 The problem

The problem starts with population – life span has surged this century from an average 39 years to 65 years and between 1950 and 1990 2.8 billion people were added to the world population (Nelson-Smith, 1998). Modern medicine and increased food availability has lowered the world's death rates which has produced a one-off population growth surge (Avery, 1998). In addition, fertility rates tend to reduce with increased food security. Births per woman in developing countries have already fallen from 6.1 births in 1965 to 3.1 at this time. The long term population equilibrium for affluent, urban societies seems to be 1.7 births per woman. As a result, demographic trends now indicate a peak world human population of approximately 8.5 billion people that will be reached around 2035 (Seckler and Cox, 1994).

Obtaining world food security is a complex and difficult task depending on many interrelated factors encompassing political, social and technical agendas (Ives *et al.*, 1998). Agricultural research and development has over the last 30 years met the needs for improving agricultural production sufficient to barely meet requirements for feeding the world's growing population. This has largely been achieved through use of fertilizers, chemical pesticides and plant breeding. It is anticipated that further agricultural research and development will be able to meet the challenge to provide the required levels of food to meet continued population growth – at least up to a level where the world population stabilizes around 2035.

Worldwide agricultural production faces many constraints, one of which are the losses to yield caused by pest damage. This is commonly quoted to be around 20–40% yield loss (e.g. Walgate, 1990). Improved levels of pest management obviously provides one of a number of options for addressing the problem of world food shortages. It is in this context that the future of insect pest management ultimately needs to be viewed.

10.8.2 A framework for the future

The simplest and most concise means of presenting a framework for the future of pest management is to make a number of unsubstantiated statements. The arguments for these are largely to be found or can be inferred from the main body of the book.

1. There will be a number of influential paradigms that will both guide and constrain the development of pest management. These will be: (i) the need to obtain sustainability in agriculture and by implication sustainable pest management; (ii) biotechnology; and (iii) pressure for scientists to commercially exploit the intellectual property generated by their research.

2. The panacea for agriculture will be transgenic crops even though ultimately such approaches will be acknowledged as ecologically unsustainable for pest management. Despite concerns among pressure groups and the general public transgenic crop products will become more widely available but diagnostic techniques will enable consumers to differentiate and make informed choices about GM food products.

3. There will be a proliferation of smaller agribusiness companies as the multinationals reduce product ranges of chemical pesticides to meet only major markets and concentrate on biotechnology. The smaller niche markets will be exploited by more specialized companies meeting local and regionally based markets. There will be a continued diversification of IPM products.

4. There will be increasing financial pressure to manage public sector research and development more cost effectively, leading to greater collaboration and accountability. Also decision-makers will be pressured to understand and account for the scientific process as well as its outputs. More emphasis will thus be placed on generating income from research outputs (materials, devices and know how) which will inevitably mean greater effort directed towards pest management products rather than techniques. Reinvestment of income derived from sale of pest management products into the institutes and universities that generated them will allow public finances to fund 'public good' research such as pest management techniques.

5. The environment will increasingly be seen as a social issue. Information sources will expand and diversify, improving understanding and education of the general public which could balance the likelihood of the need for a more stringent regulatory environment. The ability for mass audiences to vote on major issues through television or the internet will significantly

affect the way in which governments develop policy and evaluate perception of risk associated with new technologies.

6. Frameworks to promote international public/private sector collaboration will be established to better coordinate and identify the most effective means of promoting sustainable agriculture for major crop commodities on a global scale.

10.8.3 And finally

IPM has had a significant impact on the way pest control is viewed, studied and practised. From being just a concept in the early 1960s it has become a political and social force for the beginning of the new millennium. The philosophy and practice of IPM will continue to change and develop. It will remain dynamic simply because of its diverse nature and the range of individuals, disciplines and organizations involved in its study, development and implementation. Whether IPM can meet the real challenges that lie ahead with practical and sustainable solutions remains to be seen, but it is likely that, because it always achieves or partly achieves its aims, IPM will continue as a valuable means of addressing pest management problems for many years hence.

References

A'Brook, J. (1964) The effect of planting date and spacing on the incidence of groundnut rosette disease and of the vector, *Aphis craccivora* Koch, at Mokwa, Northern Nigeria. *Annals of Applied Biology* 54, 199–208.

A'Brook, J. (1968) The effect of plant spacing on the numbers of aphids trapped over the groundnut crop. *Annals of Applied Biology* 61, 298–294.

Acock, B., Reddy, V.R., Whisler, F.D., McKinion, J.M., Hodges, H.F. and Boote, K.J. (1983) The soybean crop simulator, GLYCIM: model documentation 1982. 002 in the series "Response of vegetation to carbon dioxide". US Department of Energy and US Department of Agriculture, pp. 18–27.

Acreman, T.M. and Dixon, A.F.G. (1986) The role of awns in the resistance of cereals to the grain aphid, *Sitobion avenae*. *Annals of Applied Biology* 109, 375–381.

Adams, A.J. and Hall, F.R. (1989) Influence of bifenthrin spray deposit quality on the mortality of *Trichoplasis ni* (Lepidoptera: Noctuidae) on cabbage. *Crop Protection* 8, 206–211.

Adams, A.J. and Hall, F.R. (1990) Initial behavioural responses of *Aphis gossypii* to defined deposits of bifenthrin on chrysanthemum. *Crop Protection* 9, 39–43.

Adams, J.M. and Schulten, G.G.M. (1976) Losses caused by insects, mites and microorganisms. In: Harris, K.L. and Lublad, C.J. (eds) *Postharvest Grain Loss Assessment Methods*, Meeting of the American Association of Cereal Chemists 1976, 8–10 September 1976, Slough. The League for International Food Education, The Tropical Products Institute (England), Food and Agriculture Organisation, UN, pp. 83–93.

Adams, M.E. (1982) *Agricultural Extension in Developing Countries*. Longman Scientific & Technical, London.

ADAS (1976) The utilisation and performance of field crop sprayers. *Farm Mechanisation Studies* No. 29. Ministry of Agriculture, Fisheries and Food, pp. 32.

Agarwal, R.A. (1969) Morphological characteristics of sugarcane and insect resistance. *Entomologia Experimentalis et Applicata* 12, 767–776.

Ahmad, M. (1987) Marketing pesticides in Pakistan in relation to legal and other controls. In: Tait, J. and Napompeth, B. (eds) *Management of Pests and Pesticides, Farmers' Perceptions and Practices*. Westview Press, Boulder and London, pp. 79–85.

Aldridge, C.A. and Hart, A.D.M. (1993) Validation of the EPPO/CoE risk assessment scheme for honeybees. In: *Proceedings of the Fifth International Symposium on the Hazards of Pesticides to Bees*, 26–28 October 1993. Plant Protection Service, Wageningen, pp. 37–41.

Alford, D.V., Carden, P.W., Dennis, E.B., Gould, H.J. and Vernon, J.D.R. (1979) Monitoring codling and tortix moths in United Kingdom apple orchards using pheromone traps. *Annals of Applied Biology* 91, 165–178.

Allard, R.W. (1960) *Principles of Plant Breeding*. John Wiley & Sons, New York and London.

Allen, J.C. (1976) A modified sine wave method for calculating degree days. *Environmental Entomology* 5, 388–396.

Allen, J.C. (1981) The citrus rust mite game: a simulation model of pest losses. *Environmental Entomology* 10, 171–176.

Allen, W.A. and Rajotte, E.G. (1990) The changing role of extension entomology in the IPM era. *Annual Review of Entomology* 35, 379–397.

Allsopp, P.G. (1981) Development, longevity and fecundity of the false wireworms *Pterohelaeus darlingensis* and *P. alternatus* I. Effects of constant temperatures. *Australian Journal of Zoology* 29, 605–619.

Allsopp, P.G. (1994) An artificial diet suitable for testing antimetabolic products against sugarcane whitegrubs (Coleoptera: Scarabaeidae). *Journal of the Australian Entomological Society* 34, 135–137.

Allsopp, P.G. and McGhie, T.K. (1996) Snowdrop and wheatgerm lectins and avidin as antimetabolites for the control of sugarcane whitegrubs. *Entomologia Experimentalis et Applicata* 80, 409–414.

Allsopp, P.G., Daglish, G.J., Taylor, M.F.J. and Gregg, P.C. (1991) Measuring development of *Heliothis* species. In: Zalucki, M.P. (ed.) *Heliothis: Research Methods and Prospects*. Springer-Verlag, Berlin, pp. 90–108.

Almeida, A.M.R., Corso, I.C. and Machado, N.F. (1988) Estudos epidemiologicos com o virus da queima-do-broto da soja. *Resultados de Pesquisa de Soja, 1987–88*. EMBRAPA/CNP Soja, Londrina, Brazil.

Amatobi, C.I., Apeji, S.A. and Oyidi, O. (1988) Effects of farming practices on populations of two grasshopper pests (*Kraussaria angulifra* Krauss and *Oedaleus senegalensis* Krauss (Orthoptera: Acrididae)) in Northern Nigeria. *Tropical Pest Management* 34, 173–179.

Anderson, R.G. (1961) The inheritance of leaf rust resistance in seven varieties of common wheat. *Canadian Journal of Plant Science* 41, 342.

Andow, D.A. (1991) Vegetational diversity and arthropod population response. *Annual Review of Entomology* 36, 561–586.

Andow, D.A. (1992) Fate of eggs of first generation *Ostrinia nubilalis* (Lepidoptera: Pyralidae) in three conservation tillage systems. *Environmental Entomology* 21, 388–393.

Andrewartha, H.G. and Birch, L.C. (1954) *The Distribution and Abundance of Animals*. University of Chicago Press, Chicago.

Andrews, D.J. and Kassam, A.H. (1976) Importance of multiple cropping in increasing world food supplies. In: Papendick, R.I., Sanchez, P.A. and Triplet, G.B. (eds) *Multiple Cropping*. American Society of Agronomy Special Publication No. 27, pp. 1–10.

Andrews, K.L., Bentley, J.W. and Cave, R.D. (1992) Enhancing biological control's contributions to integrated pest management through appropriate levels of farmer participation. *Florida Entomologist* 75, 429–439.

Annan, I.B., Schaefers, G.A. and Tingey, W.M. (1996) Impact of density of *Aphis craccivora* (Aphididae) on growth and yield of susceptible and resistant cowpea cultivars. *Annals of Applied Biology* 128, 185–193.

Anon (1976) Control of insects affecting livestock. *USDA Agricultural Research Service Natural Research Program*. No. 2048.

Anon (1993) *Reference Volume of the Agrochemical Service*. Wood Mackenzie, May 1993.

Anon (1997) Bt com: seed supplies in the USA. *Agricultural Chemical News 15 Nov. 1997*, 218, 5.

Anon (1999) *Farmer Participatory Cotton IPM Source Book*. CABI Bioscience Technical Support Group to the Global IPM Facility, Swiss Agency for Development and Co-operation.

APO (Asian Productivity Organisation) (1993) *Pest Control in Asia and the Pacific*. APO, Tokyo.

Appleyard, W.T., Williams, J.T. and Davie, R. (1984) Evaluation of three synthetic pyrethroids in the control of sheep headfly disease. *Veterinary Record* 114, 214–215.

Arbogast, R.T. (1979) The biology and impact of the predatory bug *Xylocoris flavipes*. *Proceedings of the 2nd International Working Conference on Stored Product Entomology, Ibadan, Nigeria, 10–16 September 1978*. Savannah, Georgia, pp. 91–105.

Artigues, M., Avilla, J., Sarasua, M.J. and Albajes, R. (1992) Egg laying and host stage preference at constant temperatures in *Encarsia tricolor*. *Entomphaga* 37, 45–53.

Ashby, J.W. (1974) A study of arthropod predation of *Pieris rapae* L. using serological and exclusion techniques. *Journal of Applied Ecology* 11, 419–425.

Asher, M.J.C. and Dewar, A.M. (1994) Control of pests and diseases in sugar beet by seed treatments. *British Crop Protection Council Monograph No. 57. Seed Treatment: Progress and Prospects*. British Crop Protection Council, Farnham, UK, pp. 151–158.

Askew, R.R. (1971) *Parasitic Insects*. Heinemann Educational Books Limited, London.

Atlegrim, O. (1989) Exclusion of birds from bilberry stands: impact on insect larval density and damage to the bilberry. *Oecologia* 79, 136–139.

Audemard, H. (1971) Crop loss assessment in *Malus* and *Pyrus* spp. of damage caused by *Zeuzera pyrina*. No. 38 Crop Loss Assessment Methods. *FAO Manual on the Evaluation and Prevention of Losses by Pests, Disease and Weeds.*

Aveling, C. (1977) The biology of Anthocorids (Heterophera: Anthocoridae) and their role in the integrated control of the damson-hop aphid (*Phorodon humuli* Schrank). PhD Thesis, University of London.

Aveling, C. (1981) Action of mephosfolan on anthocorid predators of *Phorodon humuli*. *Annals of Applied Biology* 97, 155–164.

Avery, D. (1998) 1997 Saving the planet with pesticides, biotechnology and European farm reform. In: Lewis, T. (ed.) *The Bawden Memorial Lectures.* British Crop Protection Council, Farnham, UK, pp. 309–323.

Axtell, R.C. and Arends, J.J. (1990) Ecology and management of arthropod pests of poultry. *Are* 35, 101–126.

Bach, C.E. (1980) Effects of plant diversity and time of colonisation on an herbivore-plant interaction. *Oecologia* 44, 319–326.

Bailey, J.C. (1986) Infesting cotton with tarnished plant bug (Heteroptera: Miridae) nymphs reared by improved laboratory rearing methods. *Journal of Economic Entomology* 79, 1410–1412.

Bailey, N.T.J. (1981) *Statistical Methods in Biology*, 2nd edn. Hodder and Stoughton, London, UK.

Baker, D.N., Lambert, J.R. and McKinion, J.M. (1983) GOSSYM: a simulator of cotton growth and yield. *South Carolina Agricultural Experiment Station Technical Bulletin* No. 1089.

Baker, P.B., Shelton, A.M. and Andaloro, J.T. (1982) Monitoring of diamondback moth (lepidoptera: yponomeutidae) in cabbage with pheromones. *Journal of Economic Entomology* 75, 1025–1028.

Baker, R.H.A. (1994) The potential for geographical information systems in analysing the risk posed by exotic pests. *Brighton Crop Protection Conference: Pests and Diseases* Volume 1. British Crop Protection Council, Farnham, UK, pp. 159–165.

Bakke, A., Sether, T. and Kvamme, T. (1983) Mass-trapping of the spruce bark beetle *Ips typographus*: Pheromone and trap technology. *Meddelelser fra Norsk Institut for Skogforskning* 38, 1–35.

Baldwin, B. (1986) Commercialisation of microbially produced pesticides. *The World Biotech Report 1986: Proceedings of Biotech' 86 Volume 1, Applied Biotechnology.* Online Publications, London, New York, pp. 39–49.

Baliddawa, C.W. (1985) Plant species diversity and crop pest control: an analytical review. *Insect Science Applications* 6, 479–487.

Barcelo, P. and Lazzeri, P.A. (1998) Direct gene transfer: chemical, electrical and physical methods. In: Lindsey, K. (ed.) *Transgenic Plant Research.* Harwood Academic, Amsterdam, pp. 35–56.

Bardner, R. and Fletcher, K.E. (1974) Insect infestations and their effects on the growth and yield of field crops: a review. *Bulletin of Entomological Research* 64, 141–160.

Bardner, R., Fletcher, K.E. and Stevenson, J.H. (1978) Pre-flowering and post-flowering insecticide applications to control *Aphis fabae* on field beans: their biological and economic effectiveness. *Annals of Applied Biology* 88, 265–271.

Barfield, C.S., Cardelli, D.J. and Boggess, W.G. (1987) Major problems with evaluating multiple stress factors in agriculture. *Tropical Pest Management* 33, 109–118.

Barker, K. (1994) Management of collaboration in ECR & D projects. In: *Management of Collaborative European Programmes and Projects in Research, Education and Training.* University of Oxford, Oxford, UK.

Barlow, C.A. (1961) On the biology and reproductive capacity of *Syrphus corollae* in the laboratory. *Entomologia Experimentalis et Applicata* 4, 91–100.

Barlow, F. (1985) Chemistry and formulation. In: Haskell, P.T. (ed.) *Pesticide Application: Principles and Practice.* Clarendon Press, Oxford, UK, pp. 1–34.

Barnard, D.R. (1985) Injury thresholds and production loss functions for the lone star tick, *Amblyomma americanum* (Acari: Ixodidae) on pastured, preweaner beef cattle, *Bos taurus*. *Journal of Economic Entomology* 78, 852–855.

Barnard, D.R., Ervin, R.T. and Epplin, F.M. (1986) Production system-based model for defining economic thresholds in pre-weaner beef cattle, *Bos taurus*, infested with the lone star tick, *Amblyomma americanum* (Acari: Ixodidae). *Journal of Economic Entomology* 79, 141–143.

Barnes, D.K., Ratcliffe, R.H. and Hanson, C.H. (1969) Interrelationship of three laboratory screening procedures for breeding alfalfa resistant to the alfalfa weevil. *Crop Science* 9, 77–79.

Barnes, J.M. (1976) Hazards to people. In: Gunn, D.L. and Stevens, J.G.R. (eds) *Pesticides and Human Welfare*. Oxford University Press, pp. 181–192.

Barnes, M.M., Millar, J.G., Kirsch, P.A. and Hawks, D.C. (1992) Codling moth (Lepidoptera: Tortricidae) control by dissemination of synthetic female sex pheromone. *Journal of Economic Entomology* 85, 1274–1277.

Barnett, E.A. and Fletcher, M.R. (1998) The poisoning of animals from the negligent use of pesticides. *The 1998 Brighton Conference: Pests and Diseases* Volume 1. British Crop Protection Council, Farnham, UK, pp. 279–284.

Barrett, J.A. (1983) Exploitation and conservation of germplasm for crop protection. In: *Plant Protection for Human Welfare, Proceedings of the 10th International Congress of Plant Protection 1983, Brighton* Volume 1, pp. 820–827.

Bartlett, P.W. and Murray, A.W.A. (1986) Modelling adult survival in the laboratory of diapause and non-diapause Colorado beetle *Leptinotarsa decemlineata* (Coleoptera: Chrysomelidae) from Normandy, France. *Annals of Applied Biology* 108, 487–501.

Bateman, R. (1999) Delivery systems and protocols for biopesticides. In: Hall, F.R. and Menn, J.J. (eds) *Methods in Biotechnology, Volume 5: Biopesticides: Use and Delivery*. Humana Press, Totowa, New Jersey, USA, pp. 509–528.

Bathan, H. and Glas, M. (1982) The distribution of the cereal leaf roller, *Cnephasia pumicana* Zell. (Lepidoptera: Tortricidae) in the German Federal Republic. First results of a pheromone trap survey in 1982. *Nachrichtenblatt des Deutschen Pflanzenschutzdienstes* 35, 81–86.

Baumgaertner, J. and Zahner, P. (1984) Simulation experiments with stochastic population models as a tool to explore pest management strategies in an apple-tree mite system (*Panonychus ulmi* Koch, *Tetranychus urticae* Koch). In: Cavalloro, R. (ed.) *Statistical and Mathematical Methods in Population Dynamics*. A.A. Balkema, Rotterdam and Boston, pp. 166–178.

Beal, G.M. and Sibley, D.N. (1967) Adoption of agricultural technology by the Indians of Guatemala. *Rural Society Report No. 62*. Department of Sociology and Anthropology, Iowa State University, Ames, Iowa, USA.

Beck, S.D. (1965) Resistance of plants to insects. *Annual Review of Entomology* 10, 207–232.

Beckman, J.S. and Soller, J. (1983) Restriction fragment length polymorphisms in genetic improvement: methodologies, mapping and costs. *Theoretical and Applied Genetics* 67, 35–43.

Beddington, J.R., Free, C.A. and Lawton, J.H. (1975) Dynamic complexity in predator-prey models framed in difference equations. *Nature* 225, 58–60.

Beddington, J.R., Free, C.A. and Lawton, J.H. (1978) Characteristics of successful natural enemies in models of biological control of insect pests. *Nature* 273, 513–519.

Beeden, P. (1972) The pegboard – an aid to cotton pest scouting. *Pans* 18, 43–45.

Begon, M. and Mortimer, M. (1981) *Population Ecology*. Blackwell Scientific Publications, Oxford, UK.

Beirne, B.P. (1975) Biological control attempts by introductions against pest insects in the field in Canada. *Canadian Entomologist* 107, 225–236.

Bell, M.R. and Hayes, J.L. (1994) Areawide management of cotton bollworm and tobacco budworm (Lepidoptera: Noctuidae) through application of a nuclear polyhedrosis virus on early-season alternate hosts. *Journal of Economic Entomology* 87, 53–57.

Bellman, R.E. (1957) *Dynamic Programming*. Princeton University Press, Princeton, New Jersey, USA.

Bellotti, A. and Kawano, K. (1980) Breeding approaches in cassava. In: Maxwell, F.G. and Jennings, P.R. (eds) *Breeding Plants Resistant to Insects*. John Wiley & Sons, Chichester, Brisbane and Toronto, pp. 313–316.

Bellows, T.S. Jr., van Driesche, R.G. and Elkinton, J.S. (1989) Extensions to Southwood and Jepson's graphical method of estimating numbers entering a stage for calculating mortality due to parasitism. *Researches on Population Ecology* 31, 169–184.

Ben Saad, A.A. and Bishop, G.W. (1976) Effect of artificial honeydews on insect communities in potato fields. *Environmental Entomology* 5, 453–457.

Benbrook, c. M., Groth, E., Halloran, J.M. and Hansen, M.K. (1996) *Pest Management at the Crossroads*. Consumer Union, Yonkers, New York, USA.

Bence, J.R. (1988) Indirect effects and biological control of mosquitoes by mosquito fish. *Journal of Applied Ecology* 25, 505–521.

Benedict, M.R. (1953) *Farm Policies of the United States 1790–1950*. Twentieth Century Fund, New York.

Bennet, S.R., McClelland, G.A.H. and Smilanick, J.M. (1981) A versatile system of fluorescent marks for studies of large populations of mosquitoes (Diptera: Culicidae). *Journal of Medical Entomology* 18, 173–174.

Bennett, F.D. and Yaseen, M. (1980) Investigations on the natural enemies of cassava mealybug (*Phenacoccus* spp.) in the Neotropics. *Unpublished Report*. International Institute for Biological Control, Ascot, UK.

Bennett, J., Cohen, M.B., Katiyar, S.K., Ghareyazie, B. and Khush, G.S. (1997) In: Carozzi, N. and Koziel, M. (eds) *Advances in Insect Control: the Role of Transgenic Plants*. Taylor and Francis, London, pp. 75–93.

Benschoter, C.A. (1984) Low-temperature storage as a quarantine treatment for the Caribbean fruit fly (Diptera: Tephritidae) in Florida citrus. *Journal of Economic Entomology* 77, 1233–1235.

Benson, J.F. (1973) Population dynamics of cabbage root fly in Canada and England. *Journal of Applied Ecology* 10, 437–446.

Bentley, B.R. and Clements, R.O. (1989) Impact of time of sowing on pest damage to direct-drilled grass and the mode of attack by dipterous stem borers. *Crop Protection* 8, 55–62.

Benz, G. (1987) Integrated pest management in material protection, storage, and food industry. In: Delucchi, V. (ed.) *Integrated Pest Management, Protection Integrée: Quo Vadis? An International Perspective*. Parasitis 86, Geneva, pp. 31–69.

Bergman, M.K. and Turpin, F.T. (1984) Impact of corn planting date on the population dynamics of corn rootworms (Coleoptera: Chrysomelidae). *Environmental Entomology* 13, 898–901.

Bernays, E.A., Chapman, R.F. and Woodhead, S. (1983) Behaviour of newly hatched larvae of *Chilo partellus* (Swinhoe) (Lepidoptera: Pyralidae) associated with their establishment in host-plant, sorghum. *Bulletin of Entomological Research* 73, 75–83.

Berry, C. (1994) Conversion of hazard to risk: a high-cost mistake. *Eighth International Congress of Pesticide Chemistry Proceedings, USA*.

Berry, N.A., Wratten, S.D. and Frampton, C. (1995) The role of predation in carrot rust fly (*Psila rosae*) egg loss. *Proceedings of the 48th New Zealand Plant Protection Conference*, 297–301.

Berry, N.A., Wratten, S.D. and Frampton, C. (1997) Effects of sowing and harvest dates on carrot rust fly (*Psila rosae*) damage to carrots in Canterbury, New Zealand. *New Zealand Journal of Crop and Horticultural Science* 25, 109–115.

Berryman, A.A. (1987) The theory and classification of outbreaks. In: Barbosa, P. and Schultz, J.C. (eds) *Insect Outbreaks*. Academic Press, San Diego and London, pp. 3–30.

Bifani, P. (1986) Socioeconomic aspects of technological innovation in food production systems. In: Marini Bettolo, G.B. (ed.) *Towards a Second Green Revolution. From Chemicals to New Biological Technologies in Agriculture in the Tropics. Developments in Agriculture and Managed Forest Ecology, 19*. Elsevier, Oxford, UK, pp. 177–221.

Bintcliffe, E.J.B. and Wratten, S.D. (1982) Antibiotic resistance in potato cultivars to the aphid *Myzus persicae*. *Annals of Applied Biology* 100, 383–391.

Birch, N. and Wratten, S.D. (1984) Patterns of aphid resistance in the genus *Vicia*. *Annals of Applied Biology* 104, 327–338.

Black, W.C. (IV), DuTeau, N.M., Puterka, G.J., Nechols, J.R. and Pettorini, J.M. (1992) Use of the random amplified polymorphic DNA polymerase chain reaction (RAPD-PCR) to detect DNA polymorphisms in aphids (Homoptera: Aphididae). *Bulletin of Entomological Research* 82, 151–159.

Blackshaw, R.P. (1994) Sampling for leatherjackets in grassland. *Aspects of Applied Biology* 37, 95–102.

Blackshaw, R.P. (1995) Effect of tank-mixing on economic thresholds for pest control. *Annals of Applied Biology* 41, 41–49.

Blackshaw, R.P. and Thompson, D. (1993) Comparative effects of bark and peat based composts on the occurrence of vine weevil larvae and the growth of containerised polyanthus. *Journal of Horticultural Science* 68, 725–729.

Blackshaw, R.P., Stewart, R.M., Humphreys, I.C. and Coll, C. (1994) Preventing leatherjacket damage to cereals. *Aspects of Applied Biology* 37, 189–196.

Blanford, S., Thomas, M.B. and Langewald, J. (1998) Behavioural fever in a population of the

Senegalese grasshopper, *Oedaleus senegalensis*, and its implications for biological control using pathogens. *Ecological Entomology* 23, 9–14.

Blanford, S., Thomas, M.B. and Langewald, J. (in press) Thermal ecology of *Zonocerus variegatus* and its effect on biocontrol using pathogens. *Agricultural and Forest Entomology*.

Blight, M.M., Hick, A.J., Pickett, J.A., Smart, L.E., Wadhams, L.J. and Woodcock, C.M. (1995) Oilseed rape volatiles cueing host-plant recognition by the seed weevil, *Ceutorhynchus assimilis*: chemical, electrophysiological and behavioural studies. *Proceedings of the Ninth International Rapeseed Congress: Rapeseed Today and Tomorrow* 3, 1031–1033.

Boller, E.F. (1972) Behavioural aspects of mass-rearing of insects. *Entomophaga* 17, 9–25.

Boller, E.F. (1987) Genetic control. In: Burn, A.J., Coaker, T.H. and Jepson, P.C. (eds) *Integrated Pest Management*. Academic Press, London, pp. 161–187.

Bomford, M. (1988) Effect of wild ducks on rice production. In: Norton, G.A. and Pech, R.P. (eds) *Vertebrate Pest Management in Australia*. CSIRO, Division of Wildlife and Ecology, Canberra, Australia, pp. 53–57.

Bonnemaison, L. (1951) Contributio a l'edtude des facteurs provoquant l'apparition des formes ailees et sexuees chez les Aphidinae. *Annals Epiphyties* 2, 1–380.

Boote, K.J., Jones, J.W., Mishoe, J.W. and Berger, R.D. (1983) Coupling pests to crop growth simulators to predict yield reductions. *Phytopathology* 73, 1581–1587.

Borden, J.H. (1990) Use of semiochemicals to manage coniferous tree pests in western Canada. In: Ridgway, R.L., Silverstein, R.M. and Inscoe, M.N. (eds) *Behaviour-modifying Chemicals for Insect Management. Applications of Pheromones and other Attractants*. Marcel Dekker Inc., New York and Basel, pp. 281–315.

Boreham, P.F.L. and Ohiagu C.E. (1978) The use of serology in evaluating invertebrate prey-predator relationships: a review. *Bulletin of Entomological Research* 68, 171–194.

Bottrell, D.G., Aguda, R.M., Gould, F.L., Theunis, W., Demayo, C.G. and Magalit, V.F. (1992) Potential strategies for prolonging the usefulness of *Bacillus thuringiensis* in engineered rice. *Korean Journal of Applied Entomology* 31, 247–255.

Boulter, D., Edwards, G.A., Gatehouse, A.M.R., Gatehouse, J.A. and Hilder, V.A. (1990) Additive protective effects of incorporating two different higher plant derived insect resistance genes in transgenic tobacco plants. *Crop Protection* 9, 351–354.

Bowden, J. (1973) The influence of moonlight on catches of insects in light-traps in Africa. Part I. The moon and moonlight. *Bulletin of Entomological Research* 63, 113–128.

Bowden, J. (1982) An analysis of factors affecting catches of insects in light-traps. *Bulletin of Entomological Research* 72, 535–556.

Bowden, J. and Church, B.M. (1973) The influence of moonlight on catches of insects in light-traps in Africa. Part II. The effect of moon phase on light-trap catches. *Bulletin of Entomological Research* 63, 129–142.

Bowden, J. and Morris, M.G. (1975) The influence of moonlight on catches of insects in light-traps in Africa. Part III. The effective radius of a mercury-vapour light-trap and the analysis of catches using effective radius. *Bulletin of Entomological Research* 65, 303–348.

Bowers, R.C. (1982) Commercialisation of microbial biological control agents. In: Charudattan, R. and Walker, H.L. (eds) *Biological Control of Weed with Plant Pathogens*. John Wiley, New York, USA, pp. 157–173.

Bowers, W.S. (1971) Juvenile hormones. In: Jacobson, M. and Crosby, D.G. (eds) *Naturally Occurring Insecticides*. Marcel Dekker Inc., New York.

Boyetchko, S.M., Pederson, E., Punja Z.K. and Reddy, M.S. (1999) Formulations of biopesticides. In: Hall, F.R. and Menn, J.J. (eds) *Methods in Biotechnology, Volume 5: Biopesticides: Use and Delivery*. Humana Press, Totowa, New Jersey, USA, pp. 487–508.

Brader, L. (1979) Integrated pest control in the developing world. *Annual Review of Entomology* 24, 225–254.

Bradshaw, N.J., Parham, C.J. and Croxford, A.C. (1996) The awareness, use and promotion of integrated control techniques of pests, diseases and weeds in British agriculture and horticulture. In: *Proceedings 1996 British Crop Protection Conference – Pests and Diseases*. British Crop Protection Council, Farnham, UK, pp. 591–596.

Breese, M.H. (1960) The infestibility of stored paddy by *Sitophilus sasakii* (Tak.) and *Rhyzopertha dominica* (F.). *Bulletin of Entomological Research* 51, 599–630.

Briggs, K.G. and Shebeski, L.H. (1968) Implications concerning the frequency of control plots in wheat breeding nurseries. *Canadian Journal of Plant Science* 48, 149–153.

Briggs, K.G. and Shebeski, L.H. (1970) Visual selection for yielding ability of F$_3$ lines in a hard red spring wheat breeding programme. *Crop Science* 10, 400–402.

Bringhurst, R.S. and Galleta, G.J. (1990) Strawberry management. In: Galleta, G.J. and Himmilrick, D.G. (eds) *Small Fruit Crop Management*. Prentice-Hall, Englewood Cliffs, New York, USA.

Britton, R.J. (1988) Physiological effects of natural and artificial defoliation on the growth of young crops of lodgepole pine. *Forestry* 61, 165–175.

Brodbeck, B. and Strong, D. (1987) Amino acid nutrition of herbivorous insects and stress to host plants. In: Barbosa, P. and Schultz, J.C. (eds) *Insect Outbreaks*. Academic Press, San Diego and London, pp. 347–363.

Brodsgaard, H.F. (1994) Effect of photoperiod on the bionomics of *Frankliniella occidentalis*. *Zeitschrift fur Angewandte Entomologie* 117, 498–507.

Brodsgaard, H.F. and Enkegaard, A. (1997) Interactions among polyphagous anthocorid bugs used for thrips control and other beneficials in multi-species biological pest management systems. *Recent Research Developments in Entomology* 1, 153–160.

Broughton, S., Gleeson, P., Hancock, D. and Osborne, R. (1998) Eradicating *Bactrocera papayae* Drew & Hancock from North Queensland. *Proceedings of the 5th International Symposium – Fruitflies of Economic Importance*. Penang, Malaysia.

Browde, J.A., Pedigo, L.P., Owen, M.D.K. and Tylka, G.L. (1994a) Soybean yield and pest management as influenced by nematodes, herbicides and defoliating insects. *Agronomy Journal* 86, 601–608.

Browde, J.A., Pedigo, L.P., Owen, M.D.K., Tylka, G.L. and Levene, B.C. (1994b) Growth of soybean stressed by nematodes, herbicides and simulated insect defoliation. *Agronomy Journal* 86, 968–974

Browde, J.A., Tylka, G.L., Pedigo, L.P. and Owen, M.D.K. (1994c) A method for infesting small field plots with soybean cyst nematode. *Agronomy Journal* 86, 585–587.

Brown, K.C. (1998) The value of field studies with pesticides and non-target arthropods. *The 1998 Brighton Conference: Pests and Diseases* Volume 2. British Crop Protection Council, Farnham, UK, pp. 575–582.

Brown, T.M. and Brogdon, W.G. (1987) Improved detection of insecticide resistance through conventional and molecular techniques. *Annual Review of Entomology* 32, 145–162.

Bryant, E.H., Combs, L.M. and McCommas, S.A. (1986a) Morphometric differentiation among experimental lines of the housefly in relation to a bottleneck. *Genetics* 114, 1213–1223.

Bryant, E.H., McCommas, S.A. and Combs, L.M. (1986b) The effect of an experimental bottleneck upon quantitative genetic variation in the housefly. *Genetics* 114, 1191–1214.

Buckner, C.H. (1966) The role of vertebrate predators in the biological control of forest insects. *Annual Review of Entomology* 11, 449–470.

Buckner, C.H. (1967) Avian and mammalian predators of forest insects. *Entomophaga* 12, 491–501.

Buddenhagen, I.W. (1983) Plant breeding or pesticides to narrow the Yield Gap? *10th Annual Congress of Plant Protection 1983, Plant Protection for Human Welfare*. Volume 2. British Crop Protection Council, Farnham, UK, pp. 803–809

Budenberg, W.J. (1990) Honeydew as a contact kairomone for aphid parasitoids. *Entomologia Experimentalis et Applicata* 55, 139–148.

Burditt, A.K. Jr. (1986) γ irradiation as a quarantine treatment for walnuts infested with codling moths (Lepidoptera: Tortricidae). *Journal of Economic Entomology* 79, 1577–1579.

Burditt, A.K. Jr. and Balock, J.W. (1985) Refrigeration as a quarantine treatment for fruits and vegetables infested with eggs and larvae of *Dacus dorsalis* and *Dacus cucurbitae* (Diptera: Tephritidae). *Journal of Economic Entomology* 78, 885–887.

Burgess, H.D. (ed.) (1981) *Microbial Control of Pests and Plant Diseases, 1970–1980*. Academic Press, London.

Burgess, H.D. (1982) Control of insects by bacteria. *Parasitology* 84, 79–117.

Burgess, H.D. (1998) *Formulation of Microbial Biopesticides*. Kluwer Academic Publishers, Dordrecht.

Burn, A.J. (1984) Life-tables for the carrot fly *Psila rosae*. *Journal of Applied Ecology* 21, 891–902.

Busvine, J.R. (1971) *A Critical Review of the Techniques for Testing Insecticides*. Commonwealth Institute of Entomology, London.

Busvine, J.R. (1976) Pest resistance to pesticides. In: Gunn, D.L. and Stevens, J.G.R. (eds) *Pesticides and Human Welfare*. Oxford University Press, pp. 193–205.

Byerlee, D., Collinson, M.P., Perrin, R.K., Winkelmann, D.L., Biggs, S., Moscardi, E.R., Martinez, J.C., Harrington, L. and Benjamin, A. (1980) *Planning Technologies Appropriate to Farmers: Concepts and Procedures*. CIMMYT, El Batan, Mexico.

Cadogan, B.L., Retnakaran, A. and Meating, J.H. (1997) Efficacy of RH5992, a new insect growth regulator against spruce budworm (Lepidoptera: Tortricidae) in a boreal forest. *Journal of Economic Entomology* 90, 551–559.

Caltagirone, L.E. (1985) Identifying and discriminating among biotypes of parasites and predators. In: Hoy, M.A. and Herzog, D.C. (eds) *Biological Control in Agricultural IPM Systems*. Academic Press, Orlando and London, pp. 189–200.

Campanhola, C., José de Moraes, G. and de Sá, L.A.N. (1995) Review of IPM in South Africa. In: Mengech, A.N., Saxena, K.N. and Gopalan, H.N.B. (eds) *Integrated Pest Management in the Tropics*. John Wiley & Sons, Chichester, UK, pp. 121–152.

Campbell, A., Frazer, B.D., Gilbert, N., Gutierrez, A.P. and Mackauer, M. (1974) Temperature requirements of some aphids and their parasites. *Journal of Applied Ecology* 11, 431–438.

Campion, D.G. (1972) Insect chemosterilants: a review. *Bulletin of Entomological Research* 61, 577–635.

Campion, D.G. (1989) Semiochemicals for the control of insect pests. *British Crop Protection Council Monograph No. 43. Progress and Prospects in Insect Control*. British Crop Protection Council, Farnham, UK, pp. 119–127.

Campion, D.G., Bettany, B.W. and Steedman, R.A. (1974) The arrival of male moths of the cotton leafworm *Spodoptera littoralis* (Boisd.) (Lepidoptera, Noctuidae) at a new continuously recording pheromone trap. *Bulletin of Entomological Research* 64, 379–386.

Campion, D.G., Bettany, B.W., McGinnigle, J.B. and Taylor, L.R. (1977) The distribution and migration of *Spodoptera littoralis* (Boisduval) (Lepidoptera, Noctuidae), in relation to meteorology on Cyprus, interpreted from maps of pheromone trap samples. *Bulletin of Entomological Research* 67, 501–522.

Campion, D.G., Critchley, B.R. and McVeigh, L.J. (1989) Mating disruption. In: Jutsum, A.R. and Gordon, R.F.S. (eds) *Insect Pheromones in Plant Protection*. John Wiley & Sons, Chichester, UK, pp. 89–119.

Canning, E.U. (1982) An evaluation of protozoal characteristics in relation to biological control of pests. *Parasitology* 84, 119–149.

Cannon, R.J.C. (1986) Summer populations of the cereal aphid *Metopolophium dirhodum* (Walker) on winter wheat: three contrasting years. *Journal of Applied Ecology* 23, 101–114.

Cantelo, W.W. and Sanford, L.L. (1984) Insect population response to mixed and uniform plantings of resistant and susceptible plant material. *Environmental Entomology* 13, 1443–1445.

Capinera, J.L., Epsky, N.D. and Thompson, D.C. (1986) Effects of adult western corn rootworm (Coleoptera: Chrysomelidae) ear feeding on irrigated field corn in Colorado. *Journal of Economic Entomology* 79, 1609–1612.

Cardé, R.T. (1990) Principles of mating disruption. In: Ridgway, R.L., Silverstein, R.M. and Inscoe, M.N. (eds) *Behaviour-modifying Chemicals for Insect Management. Applications of Pheromones and other Attractants*. Marcel Dekker Inc., New York and Basel, pp. 47–71.

Carson, R. (1962) *Silent Spring*. Riverside Press, Boston.

Carter, A.D. and Heather, A.I.J. (1995) Pesticides in Groundwater. In: Best, G.A. and Ruthven, A.D. (eds) *Pesticides – Developments, Impacts, and Controls*. The Royal Society of Chemistry, London, UK, pp. 113–123.

Carter, N. and Rabbinge, R. (1980) Simulation models of the population development of *Sitobion avenae*. *IOBC/WPRS Bulletin 1980*. 111, 93–98.

Caudwell, R.W. and Gatehouse, A.G. (1994) Extruded starch contact baits for the formulation of grasshopper and locust entomopathogens. *British Crop Protection Council Conference: Pests and Diseases 1994, Volume 1*. British Crop Protection Council, pp. 67–74.

Cenis, J.L., Perez, P. and Fereres, A. (1993) Identification of aphid (Homoptera: Aphididae) species and clones by random amplified polymorphic DNA. *Annals of the Entomological Society of America* 86, 545–550.

Chambers, R. (1983) *Rural Development: Putting the Last First*. Longman Scientific & Technical, Harlow, UK.

Chambers, R.J., Sunderland, K.D., Wyatt, I.J. and Vickerman, G.P. (1983) The effects of predator exclusion and caging on cereal aphids on winter wheat. *Journal of Applied Ecology* 20, 209–224.

Chambers, R., Pacey, A. and Thrupp, L.A. (1989) *Farmer First. Farmer Innovation and Agricultural Research*. Intermediate Technology Publications, London.

Chandler, L.D., Gilstrap, F.E. and Browning, H.W. (1988) Evaluation of the within-field mortality of *Liriomyza trifolii* (Diptera: Agromyzidae) on bell pepper. *Journal of Economic Entomology* 81, 1089–1096.

Chandler, L.D., Pair, S.D. and Harrison, W.E. (1992) RH5992 a new insect growth regulator active against corn earworm and fall armyworm (Lepidoptera: Noctuidae). *Journal of Economic Entomology* 85, 1099–1103.

Chang, K.-S. and Morimoto, N. (1988) Life table studies of the walnut leaf beetle, *Gastrolina depressa* (Coleoptera: Chrysomelidae), with special attention to aggregation. *Research in Population Ecology* 30, 297–313.

Chapman, C.G.D. (1996) Using molecular markers to breed for pest and disease resistance. *Brighton Crop Protection Conference: Pests and Diseases* Volume 3. British Crop Protection Council, Farnham, UK, pp. 783–788.

Chapple, A.C. and Bateman, R.P. (1997) Application systems for microbial pesticides: necessity not novelty. In: Evans, H.F. (ed.) *British Crop Protection Council Symposium Proceedings No. 68. Microbial Insecticides: Novelty or Necessity?* British Crop Protection Council, Farnham, UK, pp. 243–252.

Charmillot, P.J. (1990) Mating disruption technique to control codling moth in Western Switzerland. In: Ridgway, R.L., Silverstein, R.M. and Inscoe, M.N. (eds) *Behaviour-modifying Chemicals for Insect Management. Applications of Pheromones and other Attractants*. Marcel Dekker, New York and Basel, pp. 65–182.

Chaudhary, R.C. and Khush, G.S. (1990) Breeding rice varieties for resistance against *Chilo* spp. of stem borers in Asia and Africa. *Insect Science Applications* 11, 659–669.

Cheng, J.A. and Holt, J. (1990) A systems analysis approach to brown planthopper control on rice in Zhejiang Province, China. I. Simulation of outbreaks. *Journal of Applied Ecology* 27, 85–99.

Cheng, J.A., Norton, G.A. and Holt, J. (1990) A systems analysis approach to brown planthopper control on rice in Zhejiang Province, China. II. Investigations of control strategies. *Journal of Applied Ecology* 27, 100–112.

Chiang, H.C. and Wallen, V.R. (1971) Detection and assessment of crop diseases and insect infestations by aerial photography. In: Chiarappa, L. (ed.) *Crop Loss Assessment Methods*. FAO Manual on the Evaluation and Prevention of Losses by Pests, Disease and Weeds. Commonwealth Agricultural Bureau, Slough, UK.

Church, B.M. (1971) The place of sample survey in crop-loss estimation. In: Chiarappa, L. (ed.) *Crop Loss Assessment Methods, FAO Manual on the Evaluation and Prevention of Losses by Pest, Disease and Weeds*. Commonwealth Agricultural Bureau, Slough, UK.

Cilgi, T., Jepson, P.C. and Unal, G. (1988) The short term exposure of non-target invertebrates to pesticides in the cereal canopy. *British Crop Protection Council Conference Pests and Diseases, Volume 2*. British Crop Protection Council, pp. 759–764.

Claridge, M.F. and Den Hollander, J. (1980) The 'biotypes' of the rice brown planthopper, *Nilaparvata lugens*. *Entomologia Experimentalis et Applicata* 27, 23–30.

Claridge, M.F. and Den Hollander, J. (1983) The biotype concept and its application to insect pests of agriculture. *Crop Protection* 2, 85–95.

Claridge, M.F. and Walton, M.P. (1992) The European Olive and its pests – management strategies. In: Haskell, P.T. (ed.) *Research Collaboration in European IPM Systems*. British Crop Protection Council Monograph No. 52, British Crop Protection Council, Farnham, UK, pp. 3–12.

Clark, W.J., Harris, F.A., Maxwell, F. G and Hartwig, E.E. (1972) Resistance of certain soybean cultivars to beanleaf beetle, blister beetle and bollworm. *Journal of Economic Entomology* 65, 1669–1672.

Clarke, A.M. (1994) The human factors of project management. In: *Management of Collaborative European Programmes and Projects in Research, Education and Training*. University of Oxford, Oxford, UK.

Clarke, N.D., Shipp, J.L., Jarvis, W.R., Papadopoulos, A.P. and Jewett, T.J. (1994) Integrated management of greenhouse crops – a conceptual and potentially practical model. *HortScience* 29, 846–849.

Clausen, C.P. (ed.) (1978) *Introduced Parasites and Predators of Arthropod Pests and Weeds: A World Review*. USDA Handbook No. 480.

Clements, A.N. and Paterson, G.D. (1981) The analysis of mortality and survival rates in wild populations of mosquitoes. *Journal of Applied Ecology* 18, 373–399.

Clunies-Ross, T. and Hildyard, N. (1992) *The Politics of Industrial Agriculture: A Report by the Ecologist*. Earthscan, London.

Coaker, T.H. (1965) Further experiments on the effect of beetle predators on the numbers of the cabbage root fly, *Erioishia brassicae* (Bouche), attacking brassica crops. *Annals of Applied Biology* 56, 7–20.

Coaker, T.H. (1987) Cultural methods: the crop. In: Burn, A.J., Coaker, T.H. and Jepson, P.C. (eds) *Integrated Pest Management*. Academic Press, London, pp. 69–88.

Coaker, T.H. (1990) Intercropping for pest control. *British Crop Protection Council Monograph No. 45. Crop Protection in Organic and Low Input Agriculture. Options for Reducing Agrochemical Usage*. British Crop Protection Council, Farnham, UK, pp. 71–76.

Cobley, L.S. and Steele, W.M. (1976) *An Introduction to the Botany of Tropical Crops*, 2nd edn. Longman, London and New York.

Cochran, W.G. (1977) *Sampling Techniques*. John Wiley & Sons, New York.

Cochran, W.G. and Cox, G.M. (1957) *Experimental Designs*, 2nd edn. John Wiley & Sons, New York.

Cogburn, R.R. and Bollich, C.N. (1980) Breeding for host plant resistance to stored rice insects. In: Harris, M.K. (ed.) *Biology and Breeding for Resistance to Arthropods and Pathogens in Agricultural Plants*. Texas A & M University, pp. 355–358.

Cohen, M.B., Alam, S.N., Medina, E.G. and Bernal, C.C. (1997) Brown planthopper, *Nilaparvata lugens*, resistance in rice cultivar IR64: mechanism and role in successful *N. lugens* management in Central Luzon, Philippines. *Entomologia Experimentalis et Applicata* 85, 221–229.

Cole, N.A., Guillot, F.S. and Purdy, C.W. (1984) Influence of *Psoroptes ovis* (Heing) (Acari: Psoroptidae) on the performance of beef steers. *Journal of Economic Entomology* 77, 390–393.

Collier, R.H. and Finch, S. (1985) Accumulated temperatures for predicting the time of emergence in the spring of the cabbage root fly, *Delia radicum* (L.) (Diptera: Anthomyiidae). *Bulletin of Entomological Research* 75, 395–404.

Collier, R.H., Davies, J., Roberts, M., Leatherland, M., Runham, S. and Blood Smyth, J. (1994) Supervised control of foliar pests in Brussels sprouts and calabrese crops. In: Finch, S. and Brunel, E. (eds) *International Control in Field Vegetable Crops*. IOBC wprs Bulletin Vol. 17, pp. 31–40.

Collins, S.A. (1994) Strategic management in the development of integrated pest and disease management programmes. *Proceedings 1994 British Crop Protection Conference – Pests and Diseases*. British Crop Protection Council, Farnham, UK, pp. 115–124.

Comins, H.N. and Wellings, P.W. (1985) Density-related parasitoid sex-ratio: influence on host-parasitoid dynamics. *Journal of Animal Ecology* 54, 583–594.

Compton, J.A.F., Tyler, P.S., Mumford, J.D., Norton, G.A., Jones, T.H. and Hindmarsh, P.S. (1992) Potential for an expert system for pest control in tropical grain stores. *Tropical Science* 32, 295–303.

Conway, G.R. (1976) Man versus pests. In: May R.M. (ed.) *Theoretical Ecology*. Blackwell Scientific Publications, Oxford, UK, pp. 356–386.

Conway, G.R. (1981) Man versus pests. In: May R.M. (ed.) *Theoretical Ecology*, 2nd edn. Blackwell Scientific Publications, Oxford, UK, pp. 356–386.

Conway, G.R. (1984a) Strategic models. In: Conway, G.R. (ed.) *Pest and Pathogen Control: Strategic, Tactical and Policy Models*. John Wiley and Sons, Chichester, pp. 15–28.

Conway, G.R. (1984b) Tactical models. In: Conway, G.R. (ed.) *Pest and Pathogen Control: Strategic, Tactical and Policy Models*. John Wiley and Sons, Chichester, pp. 209–220.

Conway, G.R. and Barbier, E.B. (1990) *After the Green Revolution: Sustainable Agriculture for Development*. Earthscan Publications Ltd, London, UK.

Cook, S.P., Hain, F.P. and Smith, H.R. (1994) Oviposition and pupal survival of gypsy moth (Lepidoptera: Lymantriidae) in Virginia and North Carolina pine-hardwood forests. *Environmental Entomology* 23, 360–366.

Cook, S.P., Smith, H.R., Hain, F.P. and Hastings, F.L. (1995) Predation by gypsy moth (Lepidoptera: Lymantriidae) pupae by invertebrates at low small mammal population densities. *Environmental Entomology* 24, 1234–1238.

Copping, L. (1999) *The Biopesticides Manual*. British Crop Protection Council, Farnham, UK.

Copplestone, J.F. (1985) Safe use of pesticides. In: Haskell, P.T. (ed.) *Pesticide Application: Principles and Practice*. Clarendon Press, Oxford, pp. 202–216.

Couey, H.M., Armstrong, J.W., Hylin, J.W., Thornburg, W., Nakamura, A.N., Linse, E.S., Ogata, J. and Vetro, R. (1985) Quarantine procedure for Hawaii papaya, using a hot-water treatment and high-temperature, low-dose ethylene dibromide fumigation. *Journal of Economic Entomology* 78, 879–884.

Coulson, J.R., Klaasen, W., Cook, R.J., King, E.G., Chiang, H.C., Hagen, K.S. and Yendol, W.G. (1982) Notes on biological control of pests in China, 1979. In: *Biological Control of Pests in China*. United States Department of Agriculture, Washington, D.C., USA.

Cowgill, S.E., Wratten, S.D. and Sotherton, N.W. (1993) The effect of weeds on the number of hover-fly (Diptera: Syrphidae) adults and the distribution and composition of their eggs in winter wheat. *Annals of Applied Biology* 123, 499–515.

Cox, J. and Williams, D.J. (1981) An account of cassava mealybug with a description of a new species. *Bulletin of Entomological Research* 71, 247–258.

Craig, J.B. (1955) The natural history of sheep blowflies in Britain. *Annals of Applied Biology* 42, 197–207.

Cremlyn, R. (1978) *Pesticides: Preparation and Mode of Action*. John Wiley & Sons, Chichester.

Critchley, B.R., Campion, D.G., McVeigh, L.J., Hunter-Jones, P., Hall, D.R., Cork, A., Nesbitt, B.F., Marks, G.J., Justun, A.R., Hosny, M.M. and Nasr, E-S.A. (1983) Control of pink bollworm *Pectinophora gossypiella* (Saunders) (Lepidoptera: Gelechiidae), in Egypt by mating disruption using an aerially applied microencapsulated pheromone formulation. *Bulletin of Entomological Research* 73, 289–299.

Critchley, B.R., Campion, D.G. and McVeigh, L.J. (1989) Pheromone control in the integrated pest management of cotton. In: Green, M.B. and de B. Lyon, D.J. (eds) *Pest Management in Cotton*. Ellis Horwood, Chichester, UK, pp. 83–92.

Croll, B.T. (1995) The removal of pesticides during drinking water treatment. In: Best, G.A. and Ruthven, A.D. (eds) *Pesticides – Developments, Impacts, and Controls*. The Royal Society of Chemistry, London, UK, pp. 125–134.

Cromatie, W.J. (1991) The environmental control of insects using crop diversity. In: Pimentel, D. (ed.) *Handbook of Pest Management in Agriculture*, Volume 1, 2nd edn. CRC, Boca Raton, Florida, USA, pp. 182–216.

Crook, N.E. and Sunderland, K.D. (1984) Detection of aphid remains in predatory insects and spiders by ELISA. *Annals of Applied Biology* 105, 413–422.

CTIC (1995) Conservation Impact. *Conservation Technology Information Center* 12, 1–6.

Cunnington, A.M. (1976) The effect of physical conditions on the development and increase of some important storage mites. *Annals of Applied Biology* 82, 175–201.

Daly, J.C. (1993) Ecology and genetics of insecticide resistance in *Helicoverpa armigera*: interactions between selection and gene flow. *Genetica* 90, 217–226.

Daly, J.C. and Fisk, J.H. (1992) Inheritance of metabolic resistance to the synthetic pyrethroids in Australian *Helicoverpa armigera* (Lepidoptera: Noctuidae). *Bulletin of Entomological Research* 82, 5–12.

Daly, J.C. and Fisk, J.H. (1995) Decrease in tolerance to fenvalerate, in resistant *Helicoverpa armigera* after pupal diapause. *Entomologia Experimentalis et Applicata* 77, 217–222.

Daly, J.C. and McKenzie, J.A. (1986) Resistance management strategies in Australia: the *Heliothis* 'Wormkill' programmes. *Proceedings 1986 British Crop Protection Conference – Pests and Diseases*. British Crop Protection Council, Farnham, UK, pp. 951–959.

Daly, J.C. and Trowell, S. (1996) Biochemical approaches to the study of ecological genetics: the role of selection and gene flow in the evolution of insecticide resistance. In: Symondson, W.O.C. and Liddell, J.E. (eds) *The Ecology of Agricultural Pests*. Chapman & Hall, London, pp. 73–91.

Davidson, G. (1974) *Genetic Control in Insect Pests*. Academic Press, London, UK.

Davies, E.T. and Dunford, W.J. (1962) *Some Physical and Economic Considerations of Field Enlargement*. University of Exeter; Department of Agricultural Economics Publication No. 133.

Dawson, G.W., Doughty, K.J., Hick, A.J., Pickett, J.A., Pye, B.J., Smart, L.E. and Wadhams, L.J. (1993) Chemical precursors for studying the effects of glucosinolate catabolites on diseases and pests of oilseed rape (*Brassica napus*) or related plants. *Pesticide Science* 39, 271–278.

Dawson, G.W., Griffiths, D.C., Hassanali, A., Pickett, J.A., Plumb, R. T, Pye, B.J., Smart, L.E. and Woodcock, C.M. (1986) Antifeedants: a new concept for control of barley yellow dwarf virus in winter cereals. *Proceedings 1986 Crop Protection Conference – Pests and Diseases*. British Crop Protection Council, Farnham, UK, pp. 1001–1008.

Day, K.R. and Leather, S.R. (1997) Threats to forestry by insect pests in Europe. In: Watt, A.D., Stork, N.E. and Hunter, M.D. (eds) *Forests and Insects*. Chapman & Hall, London, pp.177–206.

Day, R.K. and Collins, M.D. (1992) Simulation modelling to assess the potential value of formulation development of lambda-cyhalothrin. *Pesticide Science* 15, 45–61.

de Bono, E. (1970) *Lateral Thinking: A Textbook of Creativity*. Penguin Books, Aylesbury, UK.

de Lima, C.P.F. (1979) Ecology and the integrated control approach under tropical conditions. *Proceedings of the 2nd International Working Conference on Stored Product Entomology, Ibadan, Nigeria*. Savannah, Georgia, USA, pp. 44–48.

de Ponti, O.M.B. (1982) Resistance to insects: a challenge to plant breeders and entomologists. In: Visser, J.H. and Minks, A.K. (eds) *Proceedings of the 5th International Symposium on Insect Plant Relationships*. Pudoc, Wageningen, The Netherlands, pp. 337–348.

de Ponti, O.M.B., Romanov, L.F. and Berlinger, M.J. (1990) Whitefly-plant relationships: plant resistance. In: Gerling, D. (ed.) *Whiteflies: their Bionomics, Pest Status and Management*. Intercept, Andover, pp. 91–106.

de Visser, P. (1991) Kennissysteem helpt fruiteler bij gewasbescherming. *Kennissystemen* 6, 3–5.

Deacon, J.W. (1983) *Microbial Control of Plant Pests and Diseases*. Van Nostrand Reinhold (UK) Company Limited, Wokingham, UK.

DeBach, P. (1974) *Biological Control by Natural Enemies*. Cambridge University Press, Cambridge, UK.

DeBach, P. and Hagen, K.S. (1964) Manipulation of entomophagous species. In: DeBach, P. (ed.) *Biological Control of Insect Pests and Weeds*. Chapman and Hall, London, pp. 283–304.

DeBach, P. and Huffaker, C.B. (1971) Experimental techniques for evaluation of the effectiveness of natural enemies. In: Huffaker, C.G. (ed.) *Biological Control*. Plenum Press, New York and London, pp. 113–140.

DeGrooyer, T.A., Pedigo, L.P. and Giles, K.L. (1995) Population dynamics of the alfalfa weevil (Coleoptera: Curculionidae) in Central and Southern Iowa. *Journal of the Kansas Entomological Society* 68, 268–278.

Delanney, X., Lavallee, B.J., Proksch, R.K., Fuchs, R.L., Sims, S.R., Greenplate, J.T., Marrone, P.G., Dodson, R.B., Augustine, J.J., Layton, J.G. and Fischhoff, D.A. (1989) Field performance of transgenic tomato plants expressing the *Bacillus thuringiensis* var. *kurstaki* insect control protein. *Bio/Technology* 7, 1265–1269.

Delobel, A.G.L. (1981) Effects of sorghum density on oviposition and survival of the sorghum shoot fly, *Atherigona soccata*. *Entomologia Experimentalis et Applicata* 31, 170–174.

Delrio, G. (1992) Integrated control in olive groves. In: van Lenteren, J.C., Minks, A.K. and de Ponti, O.M.B. (eds) *Biological Control and Integrated Crop Protection: Towards Environmentally Safer Agriculture*. Pudoc Scientific Publishers, Wageningen, The Netherlands, pp. 67–76.

Dempster, J.P. (1960) A quantitative study of the predators on the eggs and larvae of the broom beetle, *Phytodecta olivacea* Forster, using the precipitin test. *Journal of Animal Ecology* 29, 149–167.

Dempster, J.P. (1967) The control of *Pieris rapae* with DDT. I. The natural mortality of the young stages of *Pieris*. *Journal of Applied Ecology* 4, 485–500.

Den Hollander, J. and Pathak, P.K. (1981) The genetics of the 'biotypes' of the rice brown planthopper, *Nilaparvata lugens*. *Entomologia Experimentalis et Applicata* 29, 76–86.

Deng, G.R., Yang, H.H. and Jin, M.X. (1987) Augmentation of coccinellid beetles for controlling sugarcane woolly aphid. *Chinese Journal of Biological Control* 3, 166–168.

Denholm, I., Birnie, L.C., Kennedy, P.J., Shaw, K.E., Perry, J.N. and Powell, W. (1998) The complementary roles of laboratory and field testing in ecotoxicological risk assessment. *The 1998 Brighton Conference: Pests and Diseases* Volume 2. British Crop Protection Council, Farnham, UK, pp. 583–590.

Dennis, P. and Wratten, S.D. (1991) Field manipulation of populations of individual staphylinid species in cereals and their impact on aphid populations. *Ecological Entomology* 16, 17–24.

Dennis, P., Thomas, M.B. and Sotherton, N.W. (1994) Structural features of field boundaries which influence the overwintering densities of beneficial arthropod predators. *Journal of Applied Ecology* 31, 361–370.

Dent, D.R. (1986) Resistance to the aphid *Metopolophium festucae cerealium*: effects of the host plant on flight and reproduction. *Annals of Applied Biology* 108, 577–583.

Dent, D.R. (1990) *Bacillus thuringiensis* for the control of *Heliothis armigera*: bridging the gap between the laboratory and the field. In: *Aspects of Applied Biology, 24. The Exploitation of Micro-organisms in Applied Biology*. Association of Applied Biologists, Wellesbourne, UK, pp. 179–185.

Dent, D.R. (1992) Scientific programme management in collaborative research. In: Haskell, P.T. (ed.) *Research Collaboration in European IPM Systems*. British Crop Protection Council Monograph No. 52, British Crop Protection Council, Farnham, UK, pp. 69–76.

Dent, D.R. (1993) Product versus techniques in community-based pest management. In: Saini, R.K. and Haskell, P.T. (eds) *Community-based and Environmentally Safe Pest Management*. ICIPE Science Press, Nairobi, Kenya, pp. 169–180.

Dent, D.R. (1994) Strategic Management of an Olive IPM Programme. *International Symposium on Crop Protection, Gent University*.

Dent, D.R. (1995) *Integrated Pest Management*. Chapman & Hall, London.

Dent, D.R. (1996) Research specialisation: a constraint to integration. In: Waibel, H. and Zadoks, J.C. (eds) *Institutional Constraints to IPM*. The Pesticide Policy Project, Hannover, Germany, pp. 21–26.

Dent, D.R. (1997a) Quantifying insect populations: estimates and parameters. In: Dent, D.R. and Walton, M.P. (eds) *Methods in Ecological and Agricultural Entomology*. CAB International, Wallingford, UK, pp. 57–110.

Dent, D.R. (1997b) The research dynamic in the development of IPM systems. In: Allsopp, P.G., Rogers, D.J. and Robertson, L.N. (eds) *Soil Invertebrates in 1997*. Bureau of Sugar Experiment Stations, Brisbane, Australia, pp. 86–91.

Dent, D.R. (1997c) Integrated pest management and microbial insecticides. In: Evans, H.F. (ed.) *British Crop Protection Council Symposium Proceedings No. 68. Microbial Insecticides: Novelty or Necessity?*. British Crop Protection Council, Farnham, UK, pp. 127–139.

Dent, D.R. and Pawar, C.S. (1988) The influence of moonlight and weather on catches of *Helicoverpa armigera* Hübner (Lepidoptera: Noctuidae) in light and pheromone traps. *Bulletin of Entomological Research* 78, 365–377.

Dent, D.R. and Spencer, J. (1993) A simulation study of variables involved in spray application procedures using a hydraulic knapsack sprayer. *International Journal of Pest Management* 39, 139–143.

Dent, D.R. and Waage, J. (1999) Wanted: investors in biological control. *Pesticide News* 45, 10–11.

Dent, D.R. and Wratten, S.D. (1986) The host-plant relationships of apterous virginoparae of the grass aphid *Metopolophium festucae cerealium*. *Annals of Applied Biology* 108, 567–576.

Dewar, A.M. (1992a) The effect of pellet type and insecticides applied to pellets on plant establishment and pest incidence in sugar beet in Europe. *Proceedings of the 54th Winter Congress of the Institut International de Recherches Betteravieres*, pp. 89–112.

Dewar, A.M. (1992b) The effects of imidacloprid on aphids and virus yellows in sugar beet. *Pflanzenschutz-Nachrichten Bayer* 45, 423–442.

Dewar, A.M. (1996) The effect of soil pests on seedling root growth and yield of sugar beet. *Proceedings of the 59th IIRB Congress, February 1996*, pp. 229–237.

Dewar, A.M. and Asher, M.J.C. (1994) A European perspective on pesticide seed treatments in sugar beet. *Pesticide Outlook* June 1994, 11–17.

Dewar, A.M., Dean, G.J. and Cannon, R. (1982) Assessment of methods for estimating the numbers of aphids (Hemiptera: Aphididae) in cereals. *Bulletin of Entomological Research* 72, 675–685.

Dewar, A.M., Read, L., Prince, J. and Ecclestone, P. (1993) Profile on imidacloprid – another new seed treatment for sugar beet pest control. *Beet Review* 61, 5–8.

Dickerson, W.A. (1986) Grandlure: use in boll weevil control and eradication programmes in the United States. *Florida Entomologist* 69, 147–153.

Diehl, S.R. and Bush, G.L. (1984) An evolutionary and applied perspective of insect biotypes. *Annual Review of Entomology* 29, 471–504.

Dina, S.O. (1976) Effect of insecticidal application at different growth phases on insect damage and yield of cowpea. *Journal of Economic Entomology* 69, 186–188.

Dingle, H. (1972) Migration strategies of insects. *Science* 175, 1327–1335.

Dobrin, G.C. and Hammond, R.B. (1983) Residual effect of selected insecticides against the adult Mexican bean beetle (Coleoptera: Coccinellidae) on soybeans. *Journal of Economic Entomology* 76, 1456–1459.

Dobrin, G.C. and Hammond, R.B. (1985) The antifeeding activity of selected pyrethroids towards the Mexican bean beetle (Coleoptera: Coccinellidae). *Journal of the Kansas Entomological Society* 58, 422–427.

Dombrowski, N. and Lloyd, T.L. (1974) Atomization of liquids by spinning cups. *Chemical Engineering Journal* 8, 63–81.

Dos Santos, W.J., Gutierrez, A.P. and Pizzamiglio, M.A. (1989) Evaluating the economic damage caused by the cotton stem borer *Eutinobothrus brasiliensis* (Hambleton 1937) in cotton in southern Brazil. *Pesquisa Agropecuaria Brasileira* 24, 337–345.

Dowell, R.V. and Gill, R. (1989) Exotic invertebrates and their effects on California. *Pan-Pacific Entomologist* 65, 132–145.

Down, R.E., Gatehouse, A.M.R., Hamilton, W.D.O. and Gatehouse, J.A. (1996) Snowdrop lectin inhibits development and decreases fecundity of the glasshouse potato aphid (*Aulacorthum solani*) when administered *in vitro* and via transgenic plants both in laboratory and glasshouse trials. *Journal of Insect Physiology* 42, 1035–1045.

Drake, V.A. (1990) Methods for studying adult movement in *Heliothis*. In: Zalucki, M.P. (ed.) *Heliothis: Research Methods and Prospects*. Springer-Verlag, Berlin, pp. 109–121.

Drummond, R.O., George, J. E and Kunz, S.E. (1988) *Control of Arthropod Pests of Livestock: A Review of Technology*. CRC Press Inc., Boca Raton, Florida.

Dubus, I.G., Hollis, J.M., Brown, C.D., Lythgo, C. and Jarvis, J. (1998) Implications of a first-step environmental exposure assessment for the atmospheric deposition of pesticides in the UK. *The 1998 Brighton Conference: Pests and Diseases* Volume 1. British Crop Protection Council, Farnham, UK, pp. 273–278.

Dudley, N.J., Mueller, R.A.E. and Wightman, J.A. (1989) Application of dynamic programming for guiding IPM on groundnut leafminer in India. *Crop Protection* 8, 349–357.

Duncan, D.P. and Hodson, A.C. (1958) Influence of the forest tent caterpillar upon the aspen forests of Minnesota. *Forest Science* 4, 71–93.

Durno, J. (1989) Recipes for fish with rice. *New Scientist* 124, 47–49.

Easwaramoorthy, S. and Nandagopal, V. (1986) Life tables of internode borer, *Chilo sacchariphagus indicus* (k.), on resistant and susceptible varieties of sugarcane. *Tropical Pest Management* 32(3), 221–228.

Edgar, W.D. (1970) Prey and feeding behaviour of adult females of the wolf spider *Pardosa amentata* (Clerk.). *Netherlands Journal of Zoology* 20, 487–491.

Edwards, A.J. (1970) Field size and machinery efficiency. In: Hooper, M.D. and Holdgate, M.W. (eds) *Hedges and Hedgerow Trees*. Monks Wood Symposium No. 4. The Nature Conservancy, Peterborough, UK.

Edwards, C.A. (1973) *Persistent Pesticides in the Environment*, 2nd edn. CRC Press, Cleveland, Ohio.

Edwards, C.A. (1987) The environmental impact of pesticides. In: Delucchi, V. (ed.) *Integrated Pest Management, Protection Integrée Quo vadis? An International Perspective*. Parasitis 86, Geneva, Switzerland, pp. 309–329.

Edwards, C.A., Huelsman, M.F., Yardim, E.N. and Shuster, W.D. (1992) Cultural inputs into integrated crop management and minimising losses of processing tomatoes to pests in the US. *Brighton Crop Protection Conference: Pests and Diseases* Volume 2. British Crop Protection Council, Farnham, UK, pp. 597–602.

Edwards, C.A., Huelsman, M.F., Yardim, E.N. and Shuster, W.D. (1996) Cultural inputs into integrated management and minimising losses of processing tomatoes to pests in the US. *Brighton Crop Protection Conference – Pests and Diseases 1996*, 18–21 November, Vol. 2, pp. 597–602.

Edwards, C.A., Knacker, T. and Pokarzhevskii, A. (1998) The prediction of the fate and effects of pesticides in the environment using tiered laboratory soil microcosms. *The 1998 Brighton Conference: Pests and Diseases* Volume 1. British Crop Protection Council, Farnham, UK, pp. 267–272.

Edwards, P.J. and Wratten, S.D. (1980) *Ecology of Insect-Plant Interactions*. Edward Arnold, London.

Egwuatu, R.I. and Ita, C.B. (1982) Some effects of single and split applications of carbofuran on the incidence of and damage by *Locris maculata, Busseola fusca* and *Sesamia calamistis* on maize. *Tropical Pest Management* 34, 210–214.

Eigenbrode, S.D. and Bernays, E.A. (1997) Evaluation of factors affecting host plant selection, with an emphasis on studying behaviour. In: Dent, D.R. and Walton, M.P. (eds) *Methods in Ecological and Agricultural Entomology*. CAB International, Wallingford, UK, pp. 147–170.

El-Adl, M.A., Hosny, M.M. and Campion, D.G. (1988) Mating disruption for the control of pink bollworm *Pectinophora gossypiella* (Saunders) in the delta cotton growing area of Egypt. *Tropical Pest Management* 34, 210–214.

Elkinton, J.S. and Wood, D.L. (1980) Feeding and boring behaviour of *Ips paraconfusus* on the bark of a host and a non-host tree species. *Canadian Entomologist* 112, 797–809.

Elliott, M., James, N.F. and Potter, C. (1978) The future of pyrethroids in insect control. *Annual Review of Entomology* 23, 443–469.

Elliott, N.C., Kieckhefer, R.W. and Walgenbach, D.D. (1990) Binomial sequential sampling methods for cereal aphids in small grains. *Journal of Economic Entomology* 83, 1381–1387.

Ellis, B.R. (1993) Advances in pest control. *Alpine Garden Society Quarterly Bulletin* 61, 98–102.

Ellis, P.R., Saw, P.L. and Crowther, T.C. (1991) Development of carrot inbreds with resistance to carrot fly using a single seed descent programme. *Annals of Applied Biology* 119, 349–357.

Ellis, P.R., Singh, R., Pink, D.A.C., Lynn, J.R. and Saw, P.L. (1996) Resistance to *Brevicoryne brassicae* in horticultural brassicas. *Euphytica* 88, 85–96.

Emehute, J.K.U. and Egwuatu, R.I. (1990) Effects of field populations of cassava mealybug, *Phenacoccus manjhoti* on cassava yield and *Epidinocarsis lopezi* at different planting dates in Nigeria. *Tropical Pest Management* 36, 279–281.

Entwistle, J.C. and Dixon, A.F.G. (1986) Short-term forecasting of peak population density of the grain aphid (*Sitobion avenae*) on wheat. *Annals of Applied Biology* 109, 215–222.

Ericsson, A., Hellqvist, C., Långström, B., Larsson, S. and Tenow, O. (1985) Effects on growth of simulated and induced shoot pruning by *Tomicus piniperta* as related to carbohydrate and nitrogen dynamics in Scots pine. *Journal of Applied Ecology* 22, 105–124.

Escalada, M.M. and Heong, K.L. (1993) Communication and implementation of change in crop protection. In: *Crop Protection and Sustainable Agriculture*. John Wiley & Sons, Chichester, UK, pp. 191–207.

Ester, A., Embrechts, A., Vlaswinkel, M.E.R. and de Moel, C.P. (1994) Protection of field vegetables against insect attacks by covering the plot with polyethylene nets. *International Symposium on Crop Protection*. University of Gent, p.64.

Evans, A.V. (1936) The physiology of the sheep blow-fly *Lucilia sericata* Meig. (Diptera). *Transactions Royal Entomological Society* 85, 363–378.

Evans, D.E. (1987) Stored products. In: Burn, A.J., Coaker, T.H. and Jepson, P.C. (eds) *Integrated Pest Management*. Academic Press, London, pp. 425–461.

Evans, G.C. (1972) *The Quantitative Analysis of Plant Growth, Studies in Ecology Vol. 1*. Blackwell Scientific Publications, Oxford.

Eveleens, K.G. (1983) Cotton-insect control in the Sudan Gezira: analysis of a crisis. *Crop Protection* 2, 273–287.

FAO (1986) *International Code of Conduct on the Distribution and Use of Pesticides*. Food and Agriculture Organisation of the United Nations, Rome, Italy.

FAO/WHO (1996) Biotechnology and Food Safety. *Report of a Joint FAO/WHO Consultation. FAO Food and Nutrition Paper 61*. Food and Agriculture Organisation of the United Nations, Rome, Italy.

Farrell, G., Hodges, R.J. and Golob, P. (1996) Integration of control methods for stored product pests in East Africa. In: Farrell, G., Greathead, D.J., Hill, M.G. and Kibata, G.N. (eds) *Management of Farm Storage Pests in East and Central Africa*. Proceedings of the East and Central African Storage Pest Management Workshop, Naivasha, Kenya.

Farrell, J.A.K. (1976) Effects of groundnut crop density on the population dynamics of *Aphis craccivora* Koch (Hempitera, Aphididae) in Malawi. *Bulletin of Entomological Research* 66, 317–329.

Ferro, D.N. (1987) Insect pest outbreaks in agroecosystems. In: Barbosa, P. and Schultz, J.C. (eds) *Insect Outbreaks*. Academic Press, San Diego and London, pp. 195–215.

Finch, S. and Skinner, G. (1974) Some factors affecting the efficiency of water-traps for capturing cabbage root flies. *Annals of Applied Biology* 77, 213–226.

Finch, S., Skinner, G. and Freeman, G.H. (1976) The effect of plant density on populations of the cabbage root fly on four cruciferous crops. *Annals of Applied Biology* 83, 191–197.

Finney, D.J. (1971) *Profit Analysis*. Cambridge University Press, Cambridge, UK.

Fisbie, R.F., El-Zik, K.M. and Wilson, L.T. (1989) *Integrated Pest Management Systems and Cotton Production*. John Wiley and Sons, Chichester, UK.

Fisher, K.J., Cromartie, T.H., Kanne, D.B., Leadbetter, M.R., Haag, W.G. and Broadhurst, M D. (1996) Imidate insecticides: a new class of broad spectrum insecticides. *Brighton Crop Protection Conference: Pests and Diseases* Volume 2. British Crop Protection Council, Farnham, UK, pp. 467–472.

Fisher, M.H. (1990) Novel avermectin insecticides and miticides. In: Crombie, L. (ed.) *Recent Advances in Chemistry of Insect Control II*. The Royal Society of Chemistry, London, UK, pp. 52–68.

Fisher, R.A. and Ford, E.B. (1947) The spread of a gene in natural conditions in a colony of the moth *Panaxia dominula* L. *Heredity* 1(II), 143–174.

Fisher, W.F. and Wright, F.C. (1981) Effects of the sheep scab mite on cumulative weight gains in cattle. *Journal of Economic Entomology* 74, 234–237.

Fitt, G.P. (1994) Cotton pest management, Part 3. An Australian perspective. *Annual Review of Entomology* 39, 543–562.

Fitt, G.P. and Daly, J.C. (1990) Abundance of overwintering pupae and the spring generation of *Helicoverpa* spp. (Lepidoptera: Noctuidae) in northern New South Wales, Australia: consequences for pest management. *Journal of Economic Entomology* 83, 1827–1836.

Fitzgerald, J.D., Solomon, M.G. and Murray, R.A. (1986) The quantitative assessment of arthropod predation rates by electrophoresis. *Annals of Applied Biology* 109, 491–498.

Fletcher, M.R. and Grave, R.C. (1992) Post registration surveillance to detect wildlife problems arising from approved pesticides. *The 1998 Brighton Conference: Pests and Diseases* Volume 2. British Crop Protection Council, Farnham, UK, pp. 793–798.

Flinn, P.W., Hagstrum, D.W. and McGaughey, W.H. (1996) Suppression of beetles in stored wheat by augmentative releases of parasitic wasps. *Environmental Entomology* 25, 505–511.

Flinn, P.W., Hagstrum, D.W. and Muir, W.E. (1997) Effects of time of aeration, bin size and latitude on insect populations in stored wheat: a simulation study. *Journal of Economic Entomology* 90, 646–651.

Flor, H.H. (1942) Inheritance of pathogenicity in *Melampsora lini*. *Phytopathology* 32, 653–669.

Fluckiger, C.R. (1989) Cotton simulation models as research tools for insect control. In: McFarlane, N.R. (ed.) *British Crop Protection Council Monograph No. 43. Progress and Prospects in Insect Control*. British Crop Protection Council, Farnham, UK, pp. 217–231.

Foley, D.H. (1981) Pupal development rate of *Heliothis armigera* under constant and fluctuating temperatures. *Journal of Australian Entomological Society* 20, 13–20.

Follett, P.A., Cantelo, W.W. and Roderick, G.K. (1996) Local dispersal of overwintered Colorado potato beetle (Chrysomelidae: Coleoptera) determined by mark and recapture. *Environmental Entomology* 25, 1304–1311.

Forbes, V.E. and Forbes, T.L. (1994) *Ecotoxicology in Theory and Practice*. Chapman & Hall, London.

Ford, M.G. and Salt, D.W. (1987) Behaviour of insecticide deposits and their transfer from plant to insect surfaces. In: Cottrell, H.J. (ed.) *Pesticides on Plant Surfaces*, Critical Reports on Applied Chemistry Volume 18. John Wiley & Sons, Chichester, pp. 26–81.

Forrester, J.W.O. (1961) *Industrial Dynamics*. M.I.T. Press, Cambridge, Massachusetts, USA.

Forrester, N.W., Cahill, M., Bird, L.J. and Layland, J.K. (1993) Management of pyrethroid and endosulfan resistance in *Helicoverpa armigerai* (Lepidoptera: Noctuidae) in Australia. *Bulletin of Entomological Research* Supplement 1.

Frampton, G.K., Cilgi, T., Fry, G.L.A. and Wratten, S.D. (1995) Effects of grassy banks on the dispersal of some carabid beetles (Coleoptera: Carabidae) on farmland. *Biological Conservation* 71, 347–355.

France, J. and Thornley, J.H.M. (1984) *Mathematical Models in Agriculture*. Butterworth and Co., London, pp. 161–174.

Frazer, B.D. and Gilbert, N. (1976) Coccinellids and aphids: a quantitative study of the impact of adult ladybirds (Coleoptera: Coccinellidae) preying on field populations of pea aphids (Homoptera: Aphididae). *Journal of the Entomological Society of British Columbia* 73, 33–56.

Frazer, B.D., Gilbert, N., Nealis, V. and Raworth, D.A. (1981) Control of aphid density by a complex of predators. *Canadian Entomologist* 113, 1035–1041.

Free, J.B., Pickett, J.A., Feguson, A.W., Simpkins, J.R. and Smith, M.C. (1985) Repelling foraging honey bees with alarm pheromones. *Journal of Agricultural Sciences, Cambridge* 105, 255–260.

French, N.M., Follett, P., Nault, B.A. and Kennedy, G.G. (1993) Colonization of potato fields in eastern North Carolina by Colorado potato beetle. *Entomologia Experimentalis et Applicata* 68, 247–256.

French, N.P., Wall, R., Cripps, P.J. and Morgan, K.L. (1992) Prevalence, regional distribution and control of blowfly strike in England and Wales. *Veterinary Record* 131, 337–342.

French, N.P., Wall, R. and Morgan, K.L. (1995) The seasonal pattern of sheep blowfly strike in England and Wales. *Medical and Veterinary Entomology* 9, 1–8.

Funderburk, J., Maruniak, J., Boucias, D. and Garcia-Canedo, A. (1992) Efficacy of baculoviruses and their impact on pest management programs. In: Copping, L.G., Green, M.B. and Rees, R.T. (eds) *Pest Management in Soybean*. Elsevier Applied Science, London and New York, pp. 88–97.

Gabriel, T. (1989) Pest control, pest management and the 'human factor'. *Tropical Pest Management* 35, 254–256.

Gallun, R.L. (1980) Breeding for resistance in wheat. In: Harris, M.K. (ed.) *Biology and Breeding for Resistance to Arthropods and Pathogens in Agricultural Plants*. Texas A & M University, pp. 245–262.

Gallun, R.L. and Khush, G.S. (1980) Genetic factors affecting the expression and stability of resistance. In: Maxwell, F.G. and Jennings, P.R. (eds) *Breeding Plants Resistant to Insects*. John Wiley and Sons, Chichester, pp. 63–86.

Gardner, C.O. (1961) An evaluation of effects of mass selection and seed irradiation with thermal neutrons on yield of corn. *Crop Science* 1, 241–245.

Gardner, S.M. and Dixon, A.F.G. (1985) Plant structure and the foraging success of *Aphidius rhopalosiphi* (Hymenoptera: Aphididae). *Ecological Entomology* 10, 171–179.

Garforth, C. (1993) Extension techniques for pest management. In: Norton, G.A. and Mumford, J.D. (eds) *Decision Tools for Pest Management*. CAB International, Wallingford, UK, pp. 247–264.

Garguillo, P.M., Berisford, C.W., Canalos, C.G., Richmond, J.A. and Cade, S.C. (1984) Mathematical descriptions of *Rhyacionia frustrana* (Lepidoptera: Tortricidae) cumulative catches in pheromone traps, cumulative eggs hatching, and their use in timing of chemical control. *Environmental Entomology* 13, 1681–1685.

Gash, A.F., Carter, N. and Bale, J.S. (1996) The influence of nitrogen fertiliser applications on the cereal aphids *Metopolophium dirhodum* and *Sitobion avenae*. *Brighton Crop Protection Conference: Pests and Diseases* Volume 1. British Crop Protection Council, Farnham, UK, pp. 209–214.

Gass, S.I. (1975) *Linear Programming*. McGraw-Hill Book Company, New York.

Gatehouse, A.M.R., Hilder, V.A. and Boulter, D. (1992) *Plant Genetic Manipulation for Crop Protection*. CAB International, Wallingford, UK.

Gatehouse, A.M.R., Davison, G.M., Newell, C.A., Merryweather, A., Hamilton, W.D.O., Burgess, E.P.J., Gilbert, R.J.C. and Gatehouse, J.A. (1997) Transgenic potato plants with enhanced resistance to the tomato moth, *Lacanobia oleracea*: growth room trials. *Molecular Breeding* 3, 49–63.

Gatehouse, A.M.R., Brown, D.P., Wilkinson, H.S., Down, R.E., Ford, L., Gatehouse, J.A., Bell, H.A. and Edwards, J.P. (1998) The use of transgenic plants for the control of insect pests. In: Kerry, B.R. (ed.) *British Crop Protection Council Symposium Proceedings No. 71. Biotechnology in Crop Protection: Facts and Fallacies*. British Crop Protection Council, Farnham, UK, pp. 25–33.

Gebre-Amlak, A., Sigvald, R. and Pettersson, J. (1989) The relationship between sowing date, infestation and damage by the maize stalk borer, *Busseola fusca* (Noctuidae), on maize in Awassa, Ethiopia. *Tropical Pest Management* 36, 279–281.

Gednalske, J.V. and Walgenbach, D.D. (1984) Effect of tillage practices on the emergence of *Smicronyx fulvus* (Coleoptera: Curculionidae). *Journal of Economic Entomology* 77, 522–524.

Geissbuhler, H. (1981) The agrochemical industry's approach to integrated pest control. *Philosophical Transactions of the Royal Society London B* 295, 111–123.

Georghiou, G.P. and Mellon, R.B. (1983) Pesticide resistance in time and space. In: Georghiou, G.P. and Saito, T. (eds) *Pest Resistance to Pesticides*. Plenum Press, New York, pp. 1–46.

Georghiou, G.P. and Saito, T. (eds) (1983) *Pest Resistance to Pesticides*. Plenum Press, New York.

Georgis, R. (1997) Commercial prospects of microbial insecticides in agriculture. In: Evans, H.F. (ed.) *British Crop Protection Council Symposium Proceedings No. 68. Microbial Insecticides: Novelty or Necessity?* British Crop Protection Council, Farnham, UK, pp. 241–252.

Gerson, U. and Smiley, R.L. (1990) *Acarine Biocontrol Agents, an Illustrated Key and Manual.* Chapman and Hall, New York, USA.

Getz, W.M. and Gutierrez, A.P. (1982) A perspective on systems analysis in crop production and insect pest management. *Annual Review of Entomology* 27, 447–466.

Gibson, R.W. (1971) Glandular hairs providing resistance to aphids in certain wild potato species. *Annals of Applied Biology* 68, 113–119.

Gilkeson, L.A. (1990) Cold storage of the predatory midge *Aphidoletes aphidimyza* (Diptera: Cecidomyiidae). *Journal of Economic Entomology* 83, 965–970.

Gilkeson, L.A. (1992) Mass rearing of phytoseiid mites for testing and commercial application. In: Anderson, T.E. and Leppla, N.C. (eds) *Advances in Insect Rearing for Research and Pest Management.* Westview Press, Boulder, Colorado, USA, pp. 489–506.

Gillham, F., Bell, T., Arin, T., Matthews, G., le Rumeur, C. and Hearn, B. (1995) *Cotton Production Prospects for the Next Decade.* World Bank, Washington DC.

Gilman, J.J. (1992) *Inventivity The Art and Science of Research Management.* Van Nostrand Reinhold, New York, USA.

Givovich, A., Weibull, J. and Pettersson, J. (1988) Cowpea aphid performance and behaviour on two resistant cowpea lines. *Entomologia Experimentalis et Applicata* 49, 259–264.

Givovich, A., Morse, S., Cerda, H., Niemeyer, H.M., Wratten, S.D. and Edwards, P.J. (1992) Hydroxamic acid glucosides in honeydew of aphids feeding on wheat. *Journal of Chemical Ecology* 18, 841–846.

Glen, D.M. and Brain, P. (1982) Pheromone-trap catch in relation to the phenology of codling moth (*Cydia pomonella*). *Annals of Applied Biology* 101, 429–440.

Goddard, R.E., Hollis, C.A., Kok, H.R., Rockwood, D.L. and Strickland, R.K. (1973) Co-operative forest genetics research programme. *Research Report No. 21. 15th Annual Report. University of Florida, School of Forest Resources and Conservation,* pp. 1–19.

Godfray, H.C.J. and Waage, J.K. (1991) Predictive modelling in biological control: the mango mealy bug (*Tastrococcus invadens*) and its parasitoids. *Journal of Applied Ecology* 28, 434–453.

Gohler, A. (1989) Expert system cucumber. *Acta Horticulture* 248, 453–457.

Goodell, G. (1984) Challenges to international pest management research and extension in the third world: do we really want IPM to work? *Bulletin of the Entomological Society of America* 30, 18–26.

Gouck, H.K., Meifert, D.W. and Gahan, J.B. (1963) A field experiment with apholate as a chemosterilant for the control of house flies. *Journal of Economic Entomology* 56, 445–446.

Gouge, D.H. and Hague, N.G.M. (1993) Effects of *Steinernema feltiae* against sciarids infesting conifers in a propagation house. *Annals of Applied Biology* 112(Supplement), 184–185.

Gouge, D.H., Lee, L.L., van Berkum, J.R., Henneberry, T.J., Smith, K.A., Payne, C. and Ortega, D. (1997) Control of pink bollworm, *Pectinophora gossypiella* (Saunders) (Lepidoptera: Gelechiidae), with biocontrol and biorational agents. *1997 Proceedings: Beltwide Cotton Conference.* National Cotton Council of America, Memphis, USA, pp. 1066–1072.

Gould, H.J., Parr, W.J., Woodville, H.C. and Simmonds, S.P. (1975) Biological control of glasshouse whitefly (*Trialeurodes vaporariorum*) on cucumbers. *Entomophaga* 20, 285–292.

Grace, J.K., Yates, J.R., Tome, C.H.M. and Oshiro, R.J. (1996) Termite-resistant construction: use of a stainless steel mesh to exclude *Coptotermes formosanus* (Isoptera: Rhinotermitidae). *Sociobiology* 28, 365–372.

Graf, B., Baumgärtner, J. and Gutierrez, A.P. (1990) Modelling agroecosystem dynamics with the metabolic pool approach. *Itteilungen der Schweizerischen Entomologischen Gesellschaft* 63, 465–476.

Graf, B., Lamb, R., Heong, K.L. and Fabellar, L. (1992) A simulation model for the population dynamics of rice leaf-folders (Lepidoptera: Pyralidae) and their interactions with rice. *Journal of Applied Ecology* 29, 558–570.

Graham, J., Gordon, S.C. and Williamson, B. (1996) Progress towards the use of transgenic plants as an aid to control soft fruit pests and diseases. *Brighton Crop Protection Conference 1996: Pests and Diseases,* Volume 3. British Crop Protection Council, Farnham, UK, pp. 777–782.

Graham-Bryce, J.J. (1977) Crop protection: a consideration of the effectiveness and disadvantages of current methods and of the scope for improvement. *Philosophical Transactions of the Royal Society London B* 281, 163–179.

Gray, M.E., Levine, E. and Steffey, K.L. (1996) Western corn rootworms and crop rotation: have we selected a new strain? *Brighton Crop Protection Conference: Pests and Diseases,* Volume 2. British Crop Protection Council, Farnham, UK, pp. 653–660.

Greathead, D.J. (1984) Biological control constraints to agricultural production. In: Hawksworth, D.L. (ed.) *Advancing Agricultural Production in Africa. Proceedings of Commonwealth Agricultural Bureaux's 1st Scientific Conference.* CAB International, Wallingford, UK, pp. 200–206.

Greathead, D.J. (1986) Parasitoids in classical biological control. In: Waage, J. and Greathead, D. (eds) *Insect Parasitoids.* Academic Press, London, pp. 289–318.

Greathead, D.J. and Waage, J.K. (1983) Opportunities for biological control of agricultural pests in developing countries. *World Bank Technical Paper No. 11.*

Greaves, M.P. and Marshall, E.J.P. (1987) Field margins: definitions and statistics. In: Way, J.M. and Greig-Smith, P.W. (eds) *British Crop Protection Council Monograph No. 35: Field Margins.* British Crop Protection Council, Farnham, UK, pp. 3–11.

Greenstone, M.H. (1996) Serological analysis of arthropod predation: past, present and future. In: Symondson, W.O.C. and Liddell, J.E. (eds) *The Ecology of Agricultural Pests.* Chapman & Hall, London, pp. 265–300.

Greenstone, M.H. and Morgan, C.E. (1989) Predation on *Heliothis zea*: an instar-specific ELISA for stomach analysis. *Annals of the Entomological Society of America* 84, 457–464.

Grégoire, J.-C., Baisier, M., Merlin, J. and Naccache, Y. (1990) Interactions between *Rhizophagus grandis* (Coleoptera: Rhizophagidae) and *Dendroctonus micans* (Coleoptera: Scolytidae) in the field and the laboratory. Their application for the biological control of *D. micans* in France. In: Kulhavy, D. and Miller, M.C. (eds) *The Potential of Biological Control of* Dendroctonus *and* Ips *Bark Beetles.* The Stephen Austin University Press, Nagocdoches, pp. 95–108.

Grégoire, J.-C., Raty, L., Drumont, A. and de Windt, N. (1997) Pheromone mass trapping: does it protect windfalls from attack by *Ips typographus* L. (Coleoptera: Scolytidae)? In: Grégoire, J.C., Liebhold, A.M., Stephen, F.M., Day, K.R. and Salom, S.M. (eds) *Integrating Cultural Tactics into the Management of Bark Beetle and Reforestation Pests.* USDA, Forest Service General Technical Report NE-236, pp. 1–8.

Griffiths, E. (1982) The carabid *Agonum dorsale* as a predator in cereals. *Annals of Applied Biology* 101, 143–203.

Griffiths, E., Wratten, S.D. and Vickerman, G.P. (1985) Foraging by the carabid *Agonum dorsale* in the field. *Ecological Entomology* 10, 181–189.

Griffiths, W.T. (1984) A review of the development of cotton pest problems in the Sudan Gezira. PhD Dissertation, University of London.

Groth, E. (1991) Communicating with consumers about food safety and risk issues. *Food Technology* 45, 245–253.

Guillon, M. (1997) Production of biopesticides: scale up and quality assurance. In: Evans, H.F. (ed.) *British Crop Protection Council Symposium Proceedings No. 68. Microbial Insecticides: Novelty or Necessity?.* British Crop Protection Council, Farnham, UK, pp. 151–162.

Gunning, R.V. (1991) Measuring insecticide resistance. In: Zalucki, M.P. (ed.) *Heliothis: Research Methods and Prospects.* Springer-Verlag, Berlin, pp. 151–156.

Gunning, R.V., Easton, C.S., Balfe, M.E. and Ferris, I.G. (1991) Pyrethroid resistance mechanisms in Australian *Helicoverpa armigera*. *Pesticide Science* 33, 473–490.

Gurr, G.M., Wratten, S.D. and van Emden, H.F. (1998) Habitat manipulation and natural enemy efficiency: implications for the control of pests. In: Barbosa, P. (ed.) *Conservation Biocontrol.* Academic Press, New York.

Guthrie, W.D. (1980) Breeding for resistance to insects in corn. In: Harris, M.K. (ed.) *Biology and Breeding for Resistance to Arthropods and Pathogens in Agricultural Plants.* Texas A & M University, pp. 290–302.

Gutierrez, A.P. (1987) Systems analysis in integrated pest management. In: Delucchi, V. (ed.) *Integrated Pest Management, Protection Integrée: Quo Vadis? An International Perspective.* Parasitis 86, Geneva, pp. 71–84.

Gutierrez, A.P. (1995) Integrated pest management in cotton. In: Dent, D.R. (ed.) *Integrated Pest Management.* Chapman & Hall, London.

Gutierrez, A.P., Butler, G.D. Jr., Wang, Y.H. and Westphal, D.F. (1977) The interaction of pink boll-worm (Lepidoptera: Gelichiidae) cotton and weather: a detailed model. *Canadian Entomologist* 4, 125–136.

Gutierrez, A.P., Pizzamiglio, M.A., Dos Santos, W.J., Tennyson, R. and Villacorta, A.M. (1984) A general distributed delay time varying life table plant population model, cotton (*Gossypium hirsutum* L.) growth and development as an example. *Ecological Modelling* 26, 131–149.

Gutierrez, A.P., Schulthess, F., Wilson, L.T., Villacorta, A.M., Ellis, C.K. and Baumgärtner, J.U. (1987) Energy acquisition and allocation in plants and insects, a hypothesis for the possible role of hormones in insect feeding patterns. *Canadian Entomologist* 119, 109–129.

Gutierrez, A.P., Wermelinger, B., Schulthess, F., Baumgärtner, J.U., Herren, H.R., Ellis, C.K. and Yaninek, J.S. (1988a) Analysis of biological control of cassava pests in Africa. I. Simulation of carbon, nitrogen and water dynamics in cassava. *Journal of Applied Ecology* 25, 901–920.

Gutierrez, A.P., Neuenschwander, P., Schulthess, F., Herren, H.R., Baumgärtner, J.U., Wermelinger, B., Löhr, B. and Ellis, C.K. (1988b) Analysis of biological control of cassava pests in Africa. II. Cassava mealybug *Phenococcus manihoti*. *Journal of Applied Ecology* 25, 921–940.

Gutierrez, A.P., Yaninek, J.S., Wermelinger, B., Herren, H.R. and Ellis, C.K. (1988c) Analysis of biological control of cassava pests in Africa. III. Cassava green mite *Mononychellus tanajoa*. *Journal of Applied Ecology* 25, 941–950.

Gutierrez, A.P., Neuenschwander, P. and van Alphen, J.J.M. (1994) Factors affecting biological control of cassava mealybug by exotic parasitoids: a ratio-dependent supply-demand driven model. *Journal of Applied Ecology* 30, 706–721.

Gutridge, C.G. (1958) The effects of winter chilling on the subsequent growth and development of the cultivated strawberry. *HortScience* 33, 119–127.

Hadaway, A.B. and Barlow, F. (1950) Studies of aqueous suspensions of insecticides. *Bulletin of Entomological Research* 41, 603–622.

Hagen, A.F. (1982) *Labops hesperius* (Hemiptera: Miridae) management in crested wheatgrass by haying: an eight-year study. *Journal of Economic Entomology* 75, 706–707.

Hagen, J.S., Sawall, E.F. Jr. and Tassan, R.L. (1971) Use of food sprays to increase effectiveness of entomophagous insects. *Proceedings of the Tall Timbers Conference on Ecological Animal Control by Habitat Management*. Tallahassee, Florida, USA, pp. 59–81.

Hagler, J.R., Brower, A.G., Zhijian Tu, Byrne, D.N., Bradley-Dunlop, D. and Enriques, F.J. (1993) Development of a monoclonal antibody to detect predation of the sweetpotato whitefly, *Bemisia tabaci* (Gennadius). *Entomologia Experimentalis et Applicata* 68, 321–326.

Hagley, E.A.C. and Simpson, C.M. (1981) Effect of food sprays on numbers of predators in an apple orchard. *Canadian Entomologist* 113, 75–77.

Hagstrum, D.W. and Flinn, P.W. (1990) Simulations comparing insect species differences in response to wheat storage conditions and management practices. *Journal of Economic Entomology* 2469–2475.

Hågvar, E.B. and Höfsvang, T. (1991) Aphid parasitoids (Hymenoptera: Aphididae): Biology, host selection and use in biological control. *Biocontrol News and Information* 12, 13–41.

Hall, R.A. and Papierok, B. (1982) Fungi as biological control agents of arthropods of agricultural and medical importance. *Parasitology* 84, 205–240.

Hall, R.W. and Ehler, L.E. (1979) Rate of establishment of natural enemies in classical biological control. *Bulletin of the Entomological Society of America* 25, 280–282.

Hallahan, D.L., Pickett, J.A., Wadham, L.J., Wallsgrove, R.M. and Woodcock, C.M. (1992) Potential of secondary metabolites in genetic engineering of crops for resistance. In: Gatehouse, A.M.R., Hilder, V.A. and Boulter, D. (eds) *Plant Genetic Manipulation for Crop Protection*. CAB International, Wallingford, UK, pp. 215–247.

Hammond, R.B. (1995) Timing of plowing and planting: effects on seedcorn maggot populations in soybean. *Crop Protection* 14, 471–477.

Hammond, R.B. (1996) Residual activity of Lambda-cyhalothrin against bean leaf beetle (Coleoptera: Chrysomelidae) in soybeans. *Journal of Agricultural Entomology* 13, 365–373.

Hammond, R.B. (1997) Long-term conservation tillage studies: impact of no-till on seed corn maggot (Diptera: Anthomyiidae). *Crop Protection* 16, 221–225.

Hammond, R.B., Pedigo, L.P. and Poston, F.L. (1979) Green clover-worm leaf consumption on greenhouse and field soybean leaves and development of a leaf-consumption model. *Journal of Economic Entomology* 72, 102–107.

Hammond, R.B., Bledsoe, L.W. and Anwar, M.N. (1995) Maturity and environmental effects on soybeans resistant to Mexican bean beetle (Coleoptera: Coccinellidae). *Journal of Economic Entomology* 88, 175–181.

Haniotakis, G., Kozyrakis, M. and Fitsakis, T. (1991) An effective mass trapping method for the control of *Dacus oleae* (Diptera: Tephritidae). *Journal of Economic Entomology* 84, 564–569.

Hanover, J.W. (1980) Breeding forest trees resistant to insects. In: Maxwell, F.G. and Jennings, P.R. (eds) *Breeding Plants Resistant to Insects.* John Wiley and Sons, Chichester, pp. 487–512.

Hanson, C.H., Busbice, T.H., Hill, R.R., Hunt, O.J. and Oaks, A.J. (1972) Directed mass selection for developing pest resistance and conserving germplasm in alfalfa. *Journal of Environmental Quality* 1, 106–111.

Harder, H.H., Riley, S.L., McCann, S.F. and Irving, S.N. (1996) DPX-MP062: a novel broad-spectrum, environmentally soft, insect control compound. *Brighton Crop Protection Conference: Pests and Diseases,* Volume 2. British Crop Protection Council, Farnham, UK, pp. 449–454.

Hardie, J., Isaacs, R., Pickett, J.A., Wadhams, L.J. and Woodcock, C. M (1994) Methyl salicylate and (-)-(1R,5S)-myrtenal are plant-derived repellents for black bean aphid, *Aphis fabae* Scop. (Homoptera: aphididae). *Journal of Chemical Ecology* 20, 2847–2855.

Hardin, G. (1962) The tragedy of the commons. *Science* 162, 1243–1248.

Harlan, J.R. and Starks, K.J. (1980) Germplasm resources and needs. In: Maxwell, F.G. and Jennings, P.R. (eds) *Breeding Plants Resistant to Insects.* John Wiley and Sons, Chichester, pp. 253–275.

Harris, C.A. (1998) Food safety and pesticides residues: is there a problem? A regulator's perspective. *The 1998 Brighton Conference: Pests and Diseases* Volume 1. British Crop Protection Council, Farnham, UK, pp. 465–470.

Harris, F.A. (1997a) Transgenic *B.t.* cotton in the Mississippi Delta. *IBC's 2nd Annual Conference on Biopesticides and Transgenic Plants.* International Business Communications, Southborough, MA, USA.

Harris, J. and Dent, D.R. (1999) *Priorities in Biopesticide R&D in Developing Countries.* CAB International, Wallingford, UK.

Harris, J.G. (1997b) Microbial insecticides – an industry perspective. In: Evans, H.F. (ed.) *British Crop Protection Council Symposium Proceedings No. 68. Microbial Insecticides: Novelty or Necessity?* British Crop Protection Council, Farnham, UK, pp. 41–52.

Harris, J.G. (1995) The efficacy of different strains of *Bacillus thuringiensis* for Diamondback Moth control in south east Asia, and their strategic usage to combat resistance to chemical insecticides, especially acyl urea compounds. In: Feng, T.-Y. (ed.) Bacillus thuringiensis *Biotechnology and Environmental Benefits*, Volume 1. Hua Shinag Yuan Publishing Company, Taipei, Taiwan, pp. 259–268.

Harris, M.O., Rose, S. and Malsch, P. (1993) The role of vision in the host plant-finding behaviour of the Hessian fly. *Physiological Entomology* 18, 31–42.

Harris, P. (1990) Environmental impact of introduced biological control agents. In: Mackauer, M., Ehler, L.E. and Roland, J. (eds) *Critical Issues in Biological Control.* Intercept, Andover, pp. 289–300.

Haskell, P.T. (1987) Natural selection in crop protection. In: Brent, K.J. and Atkin, R.K. (eds) *Rational Pesticide Use.* Cambridge University Press, pp. 1–13.

Hassan, S.A. (1986) Side effects of pesticides to entomophagous arthropods. In: Franz, J.M. (ed.) *Biological Plant and Health Protection. Biological Control of Plant Pests and of Vectors of Human and Animal Diseases.* Gustav Fischer Verlag, Stuttgart and New York, pp. 89–94.

Hassan, S.A. (1989) Testing methodology and the concept of the AIOBC/WPRS Working Group. In: Jepson, P.C. (ed.) *Pesticides and Non-target Invertebrates.* Intercept, Wimborne, Dorset, pp. 1–18.

Hassell, M.P. (1978) *The Dynamics of Arthropod Predator-Prey Systems.* Princeton University Press, Princeton, New Jersey.

Hassell, M.P. (1980) Foraging strategies, population models and biological control: a case study. *Journal of Animal Ecology* 49, 603–628.

Hassell, M.P. (1982) Patterns of parasitism by insect parasitoids in patchy environments. *Ecological Entomology* 7, 365–377.

Hassell, M.P. (1984) Host-parasitoid models and biological control. In: Conway, G.R. (ed.) *Pest and Pathogen Control: Strategic, Tactical and Policy Models.* John Wiley & Sons, Chichester, pp. 73–92.

Hassell, M.P. and Anderson, R.M. (1984) Host susceptibility as a component in host-parasitoid systems. *Journal of Animal Ecology* 53, 611–621.

Hassell, M.P. and May, R.M. (1973) Stability in insect host-parasite models. *Journal of Animal Ecology* 42, 693–726.

Hassell, M.P. and Rogers, D.J. (1972) Insect parasite responses in the development of population models. *Journal of Animal Ecology* 41, 661–676.

Hassell, M.P. and Varley, G.C. (1969) A new inductive population model for insect parasites and its bearing on biological control. *Nature* 223, 1133–1160.

Hassell, M.P. and Waage, J.K. (1984) Host-parasitoid population interactions. *Annual Review of Entomology* 29, 89–114.

Hassell, M.P., Waage, J.K. and May, R.M. (1983) Variable parasitoid sex ratios and their effect on host-parasitoid dynamics. *Journal of Animal Ecology* 52, 889–904.

Hayes, J.L. and Bell, M.R. (1994) Evaluation of early-season baculovirus treatment for suppression of *Heliothis virescens* and *Helicoverpa zea* (Lepidoptera: Noctuidae) over a wide area. *Journal of Economic Entomology* 87, 58–66.

Hayes, J.L. and Strom, B.L. (1994) 4-allylanisole as an inhibitor of bark beetle (Coleoptera: Scolytidae) aggregation. *Journal of Economic Entomology* 87, 1586–1594.

Hayes, J.L., Strom, B.L., Roton, L.M. and Ingram, L.L. (1994) Repellent properties of the host compound 4-allylanisole to the southern pine beetle. *Journal of Chemical Ecology* 20, 1595–1615.

Hayes, J.L., Meeker, J.R., Foltz, J.L. and Strom, B.L. (1996) Suppression of bark beetles and protection of pines in the urban environment: a case study. *Journal of Arboriculture* 22, 67–74.

Head, G., Hoy, C.W. and Hall, F.R. (1995) Influence of permethrin droplets on movement of larval *Plutella xylostella* (Lepidoptera: Plutellidae). *Pesticide Science* 45, 271–278.

Head, J., Baker, R.H.A. and Jarvis, C.H. (1998) Utilising computer models to determine the risk of outbreaks of gypsy moth, *Lymantria dispar*, to the UK amenity tree industry. *Proceedings 1998 British Crop Protection Conference – Pests and Diseases*. British Crop Protection Council, Farnham, UK, pp. 823–828.

Hearn, J.W., van Coller, L.M. and Conlong, D.E. (1994) Determining strategies for the biological control of a sugarcane stalk borer. *Ecological Modelling* 73, 117–133.

Heidari, M. and Copland, M.J.W. (1993) Honeydew: a food resource or arrestant for the mealybug predator *Cryptolaemus montrouzieri*? *Entomophaga* 38, 63–68.

Heinrichs, E.A. (1986) Perspectives and directions for the continued development of insect-resistant rice varieties. *Agricultural Ecosystems and Environments* 18, 9–36.

Heinz, K.M. and Parrella, M.P. (1994) Biological control of *Bemisia argentifolii* (Homoptera: Aleyrodidae) infesting *Euphorbia pulcherrima*: evaluations of releases of *Encarsia luteola* (Hymenoptera: Aphelinidae) and *Delphastus pusillus* (Coleoptera: Coccinellidae). *Environmental Entomology* 23, 1346–1353.

Heinz, K.M. and Zalom, F.G. (1995) Variation in trichome-based resistance to *Bemisia argentifolii* (Homoptera: Aleyrodidae) oviposition on tomato. *Journal of Economic Entomology* 88, 1495–1502.

Helm, C.G., Kogan, M., Onstad, D.W., Wax, L.M. and Jeffords, M.R. (1992) Effects of velvetleaf competition and defoliation by soybean looper (Lepidoptera: Noctuidae) on yield of indeterminate soybean. *Journal of Economic Entomology* 85, 2433–2439.

Helsen, H. and Blommers, L.H.M. (1992) Integrated decision models in apple IPM: the codling moth as an example. *Acta Phytopathological Entomology* 27, 271–276.

Hemingway, J., Lindsay, S.W., Small, G.J., Jawara, M. and Collins, F.H. (1995) Insecticide susceptibility status in individual species of the *Anopheles gambiae* complex where pyrethroid impregnated bednets are used extensively for malaria control. *Bulletin of Entomological Research* 85, 229–234.

Henshaw, M.D., Brown, J.E. and Griffey, W.A. (1991) Use of reflective mulches in control of mosaic viruses in summer squash. *Proceedings of the 23rd National Agricultural Plastics Congress*, pp. 78–87.

Heong, K.L. (1985) Systems analysis in solving pest management problems. In: Lee, B.S., Loke, W.H. and Heong, K.L. (eds) *Integrated Pest Management in Malaysia*. The Malaysian Plant Protection Society, Kuala Lumpur, pp. 133–50.

Heong, K.L., Aquino, G.B. and Barrion, A.T. (1991) Arthropod community structures of rice ecosystems in the Philippines. *Bulletin of Entomological Research* 81, 407–416.

Herbert, D.A., Mack, T.P., Backman, P.A. and Rodriguez-Kabana, R. (1992) Validation of a model for estimating leaf feeding by insects in soybean. *Crop Protection* 11, 27–33.

Herren, H.R. and Neuenschwander, P. (1991) Biological control of cassava pests in Africa. *Annual Review of Entomology* 36, 257–284.

Hessayon, D.G. (1983) Risk – do the public, press and politicians really care? *10th International Congress of Plant Protection 1983, Plant Protection for Human Welfare*, Brighton, Volume 2, pp. 663–669.

Herzog, D.C. and Funderburk, J.E. (1986) Ecological bases for habitat manipulation. In: Kogan, M. (ed.) *Ecological Theory and Integrated Pest Management Practice*. John Wiley, New York, USA, pp. 217–250.

Hickman, J.M., Lövei, G.L. and Wratten, S.D. (1995) Pollen feeding by adults of the hoverfly *Melanostoma fasciatum* (Diptera: Syrphidae). *New Zealand Journal of Zoology* 22, 387–392.

Higley, L.G. and Wintersteen, W.K. (1992) A new approach to environmental risk assessment of pesticides as a basis for incorporating environmental costs into economic injury levels. *American Entomologist* 38, 34–39.

Hilder, V.A., Gatehouse, A.M.R., Sheerman, S.E., Barker, R.F. and Boutler, D. (1987) A novel mechanism of insect resistance engineered in tobacco. *Nature* 300, 160–163.

Hill, D.S. (1983) *Agricultural Insect Pests of the Tropics and their Control*, 2nd edn. Cambridge University Press, Cambridge.

Hill, D.S. and Waller, J.M. (1982) *Pests and Diseases of Tropical Crops. Volume 1: Principles and Methods of Control*. Longman, London and New York.

Hill, J.E. (1998) Public concerns over the use of transgenic plants in the protection of crops from pests and diseases and government responses. In: Kerry, B.R. (ed.) *British Crop Protection Council Symposium Proceedings No. 71. Biotechnology in Crop Protection: Facts and Fallacies*. British Crop Protection Council, Farnham, UK, pp. 57–65.

Hill, L. (1993) Colour in adult *Helicoverpa punctigera* (Wallengren) (Lepidoptera: Noctuidae) as an indicator of migratory origin. *Journal of the Australian Entomological Society* 32, 145–151.

Hillman, J.R. (1998) 1993 Bio-engineering – intellect, enterprise and opportunity. In: Lewis, T. (ed.) *The Bawden Memorial Lectures*. British Crop Protection Council, Farnham, UK, pp. 249–261.

Hillocks, R.J. (1995) Integrated management of insect pests, diseases and weeds of cotton in Africa. *Integrated Pest Management Reviews* 1, 31–47.

Hodges, R.J. and Surendro (1996) Detection of controlled atmosphere changes in CO_2-flushed sealed enclosures for pest and quality management of bagged milled rice. *Journal of Stored Product Research* 32, 97–104.

Hokkanen, H.M.T. (1991) Trap cropping in pest management. *Annual Review of Entomology* 36, 119–138.

Hokkanen, H.M.T. and Pimentel, D. (1984) New approach for selecting biological control agents. *Canadian Entomologist* 116, 1109–1121.

Holt, J. and Cheke, R.A. (1997) Modelling. In: Dent, D.R. and Walton, M.P. (eds) *Methods in Ecological and Agricultural Entomology*. CAB International, Wallingford, UK, pp. 351–378.

Holt, J. and Norton, G.A. (1993) Simulation models. In: Norton, G.A. and Mumford, J.D. (eds) *Decision Tools for Pest Management*. CAB International, Wallingford, UK, pp. 119–146.

Holt, J. and Wratten, S.D. (1986) Components of resistance to *Aphis fabae* in faba bean cultivars. *Entomologia Experimentalis et Applicata* 40, 35–40.

Holt, J., Cook, A.G., Perfect, T.J. and Norton, G.A. (1987) Simulation analysis of brown planthopper (*Nilaparvata lugens*) population dynamics on rice in the Philippines. *Journal of Applied Ecology* 24, 87–102.

Holt, J., King, A.B.S. and Armes, N.J. (1990) Use of simulation analysis to assess *Helicoverpa armigera* control on pigeonpea in southern India. *Crop Protection* 9, 197–206.

Holt, J., Wareing, D.R. and Norton, G.A. (1992) Strategies of insecticide use to avoid resurgence of *Nilaparvata lugens* (Homoptera: Delphacidae) in tropical rice: A simulation analysis. *Journal of Economic Entomology* 85, 1979–1989.

Hommes, M., Hurni, B., Van de Steen, F. and Vanparys, L. (1994) Action thresholds for pests of leek – results from the co-operative experiment. In: Finch, S. and Brunel, E. (eds) *International Control in Field Vegetable Crops*. IOBC wprs Bulletin Vol. 17, pp. 67–74.

Hong, T.D., Jenkins, N.E., Ellis, R.H. and Moore, D. (1998) Limits to the negative logarithmic relationship between mositure content and longevity in conidia of *Metarhizium flavoviride*. *Annals of Botany* 81, 625–630.

Hopper, K.R., Roush, R.T. and Powell, W. (1993) Management of genetics of biological control introductions. *Annual Review of Entomology* 38, 27–51.

Horn, D.J. (1988) *Ecological Approach to Pest Management*. Elsevier Applied Science Publishers, London.

Horsch, R.B., Fry, J.E., Hoffman, N.L., Eichholtz, D., Rogers, S.G. and Fraley, R.T. (1985) A simple and general method for transferring genes into plants. *Science* 227, 1229–1231.

Hörstadius, S. (1974) Lennaeus, animals and man. *Biological Journal of the Linnaean Society* 6, 269–275.

Howard, J. and Wall, R. (1996a) Autosterilization of the house fly, *Musca domestica* (Diptera: Muscidae) in poultry houses in north-east India. *Bulletin of Entomological Research* 86, 363–367.

Howard, J. and Wall, R. (1996b) Control of the house fly, *Musca domestica*, in poultry units: current techniques and future prospects. *Agricultural Zoology Reviews* 7, 247–265.

Howard, L.O. (1930) *A History of Applied Entomology*. Smithsonian Institute, Washington, D.C., USA.

Howell, J.F., Knight, A.L., Unruh, T.R., Brown, D.F., Krysan, J.L., Sell, C.R. and Kirsch, P.A. (1992) Control of codling moth in apple and pear with sex pheromone-mediated mating disruption. *Journal of Economic Entomology* 85, 918–925.

Howse, P.E., Stevens, I.D.R. and Jones, O.T. (eds) (1998) *Insect Pheromones and their Use in Pest Management*. Chapman and Hall, London, UK.

Hoy, J.B., Kauffman, E.E. and O'Berg, A.G. (1972) A large-scale field test of *Gambusia affinis* and chlorpyrifos for mosquito control. *Mosquito News* 32, 161–171.

Hoy, M.A. (1985) Improving establishment of arthropod natural enemies. In: Hoy, M.A. and Herzog, D.C. (eds) *Biological Control in Agricultural IPM Systems*. Academic Press, Orlando and London, pp. 151–166.

Hsiao, T.H. (1969) Chemical basis of host selection and plant resistance in oligophagous insects. *Entomologia Experimentalis et Applicata* 12, 777–788.

Huffaker, C.B. (1958) Experimental studies on predation: dispersion factors and predator-prey oscillations. *Hilgardia* 27, 343–383.

Huffaker, C.B. (1985) Biological control in integrated pest management: an entomological perspective. In: Hoy, M.A. and Herzog, D.C. (eds) *Biological Control in Agricultural IPM Systems*. Academic Press, Orlando and London, pp. 13–24.

Huffaker, C.B. and Kennett, C.E. (1969) Some aspects of assessing efficiency of natural enemies. *Canadian Entomologist* 101, 425–447.

Huffaker, C.B., Messenger, P.S. and DeBach, P. (1971) The natural enemy component in natural control and the theory of biological control. In: Huffaker, C.G. (ed.) *Biological Control*. Plenum Press, New York and London, pp. 16–67.

Hughes, G. (1988) Models of crop growth. *Nature* 332, 16.

Hull, L.A. and Beers, E.H. (1985) Ecological selectivity: modifying chemical control practices to preserve natural enemies. In: Hoy, M.A. and Herzog, D.C. (eds) *Biological Control in Agricultural IPM Systems*. Academic Press, Orlando and London, pp. 103–122.

Hull, L.A. and Starner, R. van (1983) Effectiveness of insecticide applications timed to correspond with the development of rosy apple aphid (Homoptera: Aphididae) on apple. *Journal of Economic Entomology* 76, 594–598.

Hulspas-Jordan, P.M. and van Lenteren, J.C. (1989) The parasite-host relationship between *Encarsia formosa* (Hymenoptera: Aphelinidae) and *Trialeurodes vaporariorum* (Homoptera Aleyrodidae). XXX. Modelling population growth of greenhouse whitefly on tomato. *Agricultural University of Wageningen Paper* 89, 1–54.

Hunt, R. (1978) *Plant Growth Analysis*, Studies in Biology No. 96. Edward Arnold, London.

Hunt, R. (1982) *Plant Growth Curves, the Functional Approach to Plant Growth Analysis*. Edward Arnold, London.

Hunter, K. (1995) The poisoning of non-target animals. In: Best, G.A. and Ruthven, A.D. (eds) *Pesticides – Developments, Impacts, and Controls*. The Royal Society of Chemistry, London, UK, pp. 74–86.

Husain, M.A. and Lal, K.B. (1940) The bionomics of *Empoasca devastans* Distant on some varieties of cotton in the Punjab. *Indian Journal of Entomology* 2, 123–136.

Hussey, N.W. and Bravenboaer, L. (1971) Control of pests in glasshouse culture by the introduction of natural enemies. In: Huffaker, C.G. (ed.) *Biological Control*. Plenum Press, New York and London, pp. 195–216.

Hutchins, S.H. and Pedigo, L.P. (1989) Potato leafhopper-induced injury on growth and development of alfalfa. *Crop Science* 29, 1005–1011.

Huus-Brown, T. (1992) Field vegetables in Denmark: an example of the role of the Danish Advisory Service in reducing insecticide use. In: van Lenteren, J.C., Minks, A.K. and de Ponti, O.M.B. (eds) *Biological Control and Integrated Crop Protection: Towards Environmentally Safer Agriculture*. Pudoc Scientific Publishers, Wageningen, The Netherlands, pp. 201–207.

ICRISAT (International Crops Research Institute for the Semi-Arid Tropics) (1989) *Annual Report 1988*. ICRISAT, Patancheru, India.

Ignoffo, C.M. and Couch, T.L. (1981) The nucleopolyhedrosis virus of *Heliothis* species as a microbial insecticide. In: Burgess, H.E. (ed.) *Microbial Control of Pests and Plant Diseases 1970–1980*. Academic Press, London, pp. 330–362.

Iles, M.J. and Sweetmore, A. (eds) (1991) *Constraints on the Adoption of IPM in Developing Countries – A Survey*. Natural Resources Institute, Kent, UK.

Ilse, D. (1937) New observations on responses to colours in egg laying butterflies. *Nature* 40, 544–545.

Ingram, W.R. (1980) Studies of the pink bollworm, *Pectinophora gossypiella* on sea island cotton in Barbados. *Tropical Pest Management* 26, 118–137.

Ishaaya, I., Yablonski, S., Mendelson, Z., Mansour, Y. and Horowitz, A.R. (1996) Novaluron (MCW-275), a novel benzoylphenyl urea, suppressing developing stages of lepidopteran, whitefly and agromyzid leafminer pests. *The 1996 Brighton Conference: Pests and Diseases* Volume 3. British Crop Protection Council, Farnham, UK, pp. 1013–1020.

Ishaaya, I., Damme, N. and Tirry, L. (1998) Novaluron, optimisation and use for the control of the beet armyworm and the greenhouse whitefly. *The 1998 Brighton Conference: Pests and Diseases* Volume 1. British Crop Protection Council, Farnham, UK, pp. 49–56.

Ishimoto, M., Sato, T., Chrispeels, M.J. and Kitamura, K. (1996) Bruchid resistance of transgenic azuki bean expressing seed a-amylase inhibitor of common bean. *Entomologia Experimentalis et Applicata* 79, 309–315.

Ismail, A.B. and Valentine, J. (1983) The efficiency of visual assessment of grain yield and its components in spring barley rows. *Annals of Applied Biology* 102, 539–549.

Ives, C.L., Bedford, B.M. and Maredia, K.M. (1998) The agricultural biotechnology for sustainable productivity project: a new model in collaborative development. In: Ives, C.L. and Bedford, B.M. (eds) *Agricultural Biotechnology in International Development*. CAB International, Wallingford, UK, pp. 1–14.

Ives, P.M. (1981) Estimation of coccinellid numbers and movement in the field. *The Canadian Entomologist* 113, 981–997.

Jackai, L.E.N. (1982) A field screening technique for resistance of cowpea (*Vigna unguiculata*) to the pod-borer *Maruca testulalis* (Geyer) (Lepidoptera: Pyralidae). *Bulletin of Entomological Research* 72, 145–156.

Jackai, L.E.N. (1984) Using trap plants in the control of insect pests of tropical legumes. *Proceedings of the International Workshop on Integrated Pest Control of Grain Legumes, Goiania, Goias, Brasil*. Dep. Difus. Technol. EMBRAPA, Brasilia, Brasil, pp. 101–112.

Jackai, L.E.N. (1991) Laboratory and screenhouse assays for evaluating cowpea resistance to the legume pod borer. *Crop Protection* 10, 48–52.

Jackson, R.D. and Lewis, W.J. (1981) Summary of significance and employment strategies for semiochemicals. In: Nordlund, D.A., Jones, R.L. and Lewsi, W.J. (eds) *Semiochemicals, their Role in Pest Control*. John Wiley & Sons, Chichester, pp. 283–296.

Jager, C.M. de, Butôt, R.P.T., Klinkhamer, P.G.L., Jong, T.J. de, Wolff, K. and Meijden, E. van der (1995) Genetic variation in chrysanthemum for resistance to *Frankliniella occidentalis*. *Entomologia Experimentalis et Applicata* 77, 277–287.

Jager, C.M. de, Butôt, R.P.T., Meijden, E. van der and Verpoorte, R. (1996) The role of primary and secondary metabolites in chrysanthemum resistance to *Frankliniella occidentalis*. *Journal of Chemical Ecology* 22, 1987–1999.

James, H.G. (1966) Insect predators of univoltine mosquitoes in woodland pools of the Recambrian shield in Ontario. *Canadian Entomologist* 98, 550–555.

Jay, C.N. and Cross, J.V. (1998) Monitoring and predicting the development of summer fruit tortrix moth, *Adoxophyes orana*, larvae in spring in the UK as an aid to the timing of fenoxycarb applications. *Proceedings 1998 British Crop Protection Conference – Pests and Diseases*. British Crop Protection Council, Farnham, UK, pp. 833–836.

Jenkins, N.E., Heviefo, G., Langewald, J., Cherry, A.J. and Lomer, C.J. (1998) Development of mass production technology for aerial conidia of mitosporic fungi for use as mycopesticides. *Biocontrol News and Information* 19, 21–31.

Jepson, L.R., McMurty, J.A., Mead, D.W., Jesser, M.J. and Johnson, H.G. (1975) Toxicity of citrus pesticides to some predaceous phytoseiid mites. *Journal of Economic Entomology* 68, 707–710.

Jepson, P.C. (1987) An experimental rationale for the quantitative evaluation of pesticide side effects on beneficial insects in cereal crops. *IOBC/WPRS Bulletin* 10, 206–215.

Jervis, M.A. and Copeland, M.J.W. (1996) The life cycle. In: Jervis, M.A. and Kidd, N.A.C. (eds) *Insect Natural Enemies*. Chapman & Hall, London, pp. 63–161.

Jervis, M.A. and Kidd, N.A.C. (1986) Host feeding strategies in hymenopteran parasitoids. *Biological Reviews* 61, 395–434.

Jin, W-G., Sun, G-H. and Xu, Z-M. (1996) A new broad-spectrum and highly active pyrethroid ZXi 8901. *Brighton Crop Protection Conference: Pests and Diseases*, Volume 2. British Crop Protection Council, Farnham, UK, pp. 455–460.

John, J.A. and Quenouilille, M.H. (1977) *Experiments: Design and Analysis*. Charles Griffin and Company Limited, London.

John, P.W.M. (1971) *Statistical Design and Analysis of Experiments*. Macmillan Company, New York.

Johnson, C.G., (1950) The comparison of suction trap and tow-net for the quantitative sampling of small airborne insects. *Annals of Applied Biology* 37, 268–285.

Johnson, D.A. and Gilmore, E.C. (1980) Breeding for resistance to pathogens in wheat. In: Harris, M.K. (ed.) *Biology and Breeding for Resistance to Arthropods and Pathogens in Agricultural Plants*. Texas A & M University, pp. 263–275.

Johnson, D.L. and Worsbec, A. (1988) Spatial and temporal computer analysis of insects and weather: grasshoppers and rainfall in Alberta. *Memorandum of the Entomological Society of Canada* 146, 33–48.

Johnson, J.E. and Blair, E.C. (1972) Cost, time and pesticide safety. *Chemical Technology* (Nov), 666–669.

Johnson, J.W. and Teetes, G.L. (1980) Breeding for arthropod resistance in sorghum. In: Harris, M.K. (ed.) *Biology and Breeding for Resistance to Arthropods and Pathogens in Agricultural Plants*. Texas A & M University, pp. 168–180.

Johnson, R., Narvaez, J., An, G. and Ryan, C. (1989) Expression of proteinase inhibitors I and II in transgenic tobacco plants: Effects on natural defence against *Manduca sexta* larvae. *Proceedings National Academy Sciences, USA* 86, 9871–9875.

Johnson, T.B., Turpin, F.T., Schreiber, M.M. and Griffith, D.R. (1984) Effects of crop rotation, tillage, and weed management systems on black cutworm (Lepidoptera: Noctuidae) infestations in corn. *Journal of Economic Entomology* 77, 919–921.

Johnstone, D.R. (1973) Physics and meteorology. In: Haskell, P.T. (ed.) *Pesticide Application: Principles and Practice*. Clarendon Press, Oxford, UK, pp. 35–67.

Jokela, J.J. (1966) Incidence and heritability of *Melampsora* rust in *Populus deltoides*. In: Gerhold, H.D. (ed.) *Breeding Pest Resistant Trees*. Pergamon, Oxford, pp. 111–117.

Jones, H.G. (1983) *Plants and Microclimate. A Quantitative Approach to Environmental Physiology*. Cambridge University Press, Cambridge.

Jones, K.A. and Burgess, H.D. (1998) Principles of formulation. In: Burges, H.D. (ed.) *Formulation of Microbial Biopesticides, Beneficial Micro-organisms and Nematodes*. Chapman & Hall, London, UK.

Jones, O. and Langley, P. (1998) Target technology – bring the insect to the insecticide not the insecticide to the insect. *The 1998 Brighton Conference: Pests and Diseases* Volume 1. British Crop Protection Council, Farnham, UK, pp. 433–440.

Jones, T.H., Young, J.E.B., Norton, G.A. and Mumford, J.D. (1990) An expert system for the management of wheat bulb fly *Delia coaractata* (Diptera: Anthomyiidae) in the United Kingdom. *Journal of Economic Entomology* 83, 2063–2072.

Josephson, L.M., Bennett, S.E. and Burgess, E.E. (1966) Methods of artificially infesting corn with the corn earworm and factors influencing resistance. *Journal of Economic Entomology* 59, 1322–1324.

Joyce, R.J.V. and Roberts, P. (1959) The determination of the size of plot suitable for cotton spraying experiments in the Sudan Gezira. *Annals of Applied Biology* 47, 287–305.

Judenko, E. (1972) The assessment of economic losses in yield of annual crops caused by pests, and the problem of the economic threshold. *PANS* 18, 186–191.

Kahn, R.P. (1977) Plant quarantine: principles, methodology and suggested approaches. In: Hewitt, W.B. and Chiarappa, L. (eds) *Plant Health and Quarantine in International Transfer of Plant Genetic Resources.* CRC Press, Boca Raton, USA, pp. 289–307.

Kahn, R.P. (1982) The host as a vector, exclusion as a control. In: Harris, K. and Maramorsch, K. (eds) *Pathogens, Vectors, and Plant Diseases.* Academic Press, New York, USA, pp. 123–149.

Kahn, R.P. (1983) Safeguarding the international exchange of plant germplasm. *10th International Congress of Plant Protection, Plant Protection for Human Welfare, Brighton*, Volume 2. British Crop Protection Council, Farnham, UK, pp. 517.

Kakehashi, N., Suzuki, Y. and Iwasa, Y. (1984) Niche overlap of parasitoids in host-parasitoid systems: its consequence to single versus multiple introduction controversy in biological control. *Journal of Applied Ecology* 21, 115–131.

Karlöf, B. (1987) *Business Strategy in Practice.* John Wiley & Sons, New York, USA.

Kay, C.A.R., Veaszey, J.N. and Whitcomb, W.H. (1977) Effects of date of soil disturbance on numbers of adult field crickets (Orthoptera: Gryllidae) in Florida. *Canadian Entomologist* 109, 721–726.

Kaya, H.K. (1985) Entomogenous nematodes for insect control in IPM systems. In: Hoy, M.A. and Herzog, D.C. (eds) *Biological Control in Agricultural IPM Systems.* Academic Press, Orlando and London, pp. 283–302.

Kaya, H.K. (1987) Diseases caused by nematodes. In: Fuxa, J.R. and Tanada, Y. (eds) *Epizootiology of Insect Diseases.* John Wiley & Sons, Chichester, pp. 453–470.

Keerthisinghe, C.I. (1982) Economic thresholds for cotton pest management in Sri Lanka. *Bulletin of Entomological Research* 72, 239–246.

Kenmore, P.E., Heong, K.L. and Putter, C.A.J. (1985) Political, social and perceptual factors in integrated pest management programmes. In: Lee, B.S., Loke, W.H. and Heong, K.L. (eds) *Integrated Pest Management in Malaysia.* The Malaysian Plant Protection Society, Kuala Lumpur, pp. 47–66.

Kennedy, G.G., Gould, F. de Ponti, O.M.B. and Stinner, R.E. (1987) Ecological, agricultural, genetic, and commercial considerations in the deployment of insect resistant germplasm. *Environmental Entomology* 16, 327–338.

Kennedy, J.S. (1985) Migration, behavioural and ecological. In: Rankin, M.A. (ed.) *Migration: Mechanisms and Adaptive Significance.* Contributions in Marine Science, 27 (Supplement), 5–26.

Kennedy, J.S., Booth, C.O. and Kershaw, W.J.S. (1961) Host finding by aphids in the field (III) Visual attraction. *Annals of Applied Biology* 49, 1–21.

Kfir, R. (1981) Fertility of the polyembryonic parasite *Copidosoma koehleri*, effect of humidities on life length and relative abundance as compared with that of *Apanteles subandinus* in potato tuber moth. *Annals of Applied Biology* 99, 225–230.

Khan, Z.R., Ampong-Nyarko, K., Chiliswa, P., Hassanali, A., Kimani, S., Lwander, W., Overholt, W.A., Pickett, J.A., Smart, L.E., Wadhams, L.J. and Woodcock, C.M. (1997) Intercropping increases parasitism of pests. *Nature* 38, 631–632.

Khush, G.S. (1980) Breeding for multiple disease and insect resistance in rice. In: Harris, M.K. (ed.) *Biology and Breeding for Resistance to Arthropods and Pathogens in Agricultural Plants.* Texas A & M University, pp. 341–354.

Khush, G.S. (1992) Selecting rice for simply inherited resistance. In: Stalker, H.T. and Murphy, J.P. (eds) *Plant Breeding in the 1990s.* CAB International, Wallingford, UK, pp. 303–346.

Kidd, N.A.C. and Jervis, M.A. (1989) The effects of host-feeding behaviour on the dynamics of parasitoid-host interactions, and the implications for biological control. *Research in Population Ecology* 31, 211–250.

Kidd, N.A.C. and Jervis, M.A. (1996) Population dynamics. In: Jervis, M.A. and N.A.C. Kidd (eds) *Insect Natural Enemies*. Chapman & Hall, London, pp. 293–374.

Kimmins, F.M. (1989) Electrical penetration graphs from *Nilaparvata lugens* on resistant and susceptible rice varieties. *Entomologia Experimentalis et Applicata* 50, 69–79.

King, A.B.S. and Saunders, J.L. (1984) *The Invertebrate Pests of Annual Food Crops in Central America. A Guide to their Recognition and Control*. Overseas Development Administration, London.

King, E.G. and Coleman, R.J. (1989) Potential for biological control of *Heliothis* spp. *Annual Review of Entomology* 34, 53–75.

King, E.G. and Phillips, J.R. (eds) (1983) *Cotton Insects and Mites: Characterisation and Management*. Cotton Foundation Reference Book Series No. 3. National Cotton Council, Memphis, USA.

King, E.G., Hopper, K.R. and Powell, J.E. (1985) Analysis of systems for biological control of crop arthropod pests in the U.S. by augmentation of predators and parasites. In: Hoy, M.A. and Herzog, D.C. (eds) *Biological Control in Agricultural IPM Systems*. Academic Press, Orlando and London, pp. 201–228.

Kips, R.H. (1985) Environmental aspects. In: Haskell, P.T. (ed.) *Pesticide Application: Principles and Practice*. Clarendon Press, Oxford, pp. 190–201.

Kirby, R.D. and Slosser, J.E. (1984) Composite economic threshold for three lepidopterous pests of cabbage. *Journal of Economic Entomology* 77, 725–733.

Kiritani, K. and Dempster, J.P. (1973) Different approaches to the quantitative evaluation of natural enemies. *Journal of Applied Ecology* 10, 323–330.

Kishaba, A.N., Toba, H.H., Wolf, W.W. and Vail, P. (1970) Response of laboratory-reared male cabbage looper to synthetic sex pheromone in the field. *Journal of Economic Entomology* 63, 178–181.

Knight, J.D., Tatchell, G.M., Norton, G.A. and Harrington, R. (1992) FLYPAST. An information management system for the Rothamsted aphid database to aid pest control research and advice. *Crop Protection* 11, 419–426.

Kobayashi, T. and Cosenza, G.W. (1987) Integrated control of soybean stink bugs in the Cerrados. *Japanese Agricultural Research Q.* 20, 229–236.

Kogan, M. (1980) Insect problems of soybean in the United States. In: Corbin, F.T. (ed.) *World Soybean Research*. Westview Press, Boulder, CO, USA, pp. 303–325.

Kogan, M. (1988) Integrated pest management theory and practice. *Entomologia Experimentalis et Applicata* 49, 59–70.

Kogan, M. (1994) Plant resistance in pest management. In: Metcalf, R.L. and Luckmann, W.H. (eds) *Introduction to Insect Pest Management*, 3rd edn. John Wiley & Sons, pp. 73–128.

Kogan, M. and McGrath, D. (1993) Integrated pest management: present dilemmas and future challenges. In: *Anais 14o Congreso Brasileiro de Entomologia*. SEB, Piracicaba, SP, Brasil, Sociedade Entomologica do Brasil.

Kogan, M. and Ortman E.F. (1978) Antixenosis – a new term proposed to define Painter's 'nonpreference' modality of resistance. *Bulletin of the Entomological Society of America* 24, 175–176.

Kolodny-Hirsch, D.M. and Harrison, F.P. (1986) Yield loss relationships of tobacco and tomato hornworms (Lepidoptera: Sphingidae) at several growth stages of Maryland tobacco. *Journal of Economic Entomology* 79, 731–735.

Kolodny-Hirsch, D.M. and Schwalbe, C.P. (1990) Use of disparlure in the management of the gypsy moth. In: Ridgway, R.L., Silverstein, R.M. and Inscoe, M.N. (eds) *Behaviour-modifying Chemicals for Insect Management. Applications of Pheromones and other Attractants*. Marcel Dekker, New York and Basel, pp. 363–385.

König, A. (1998) Comparison of the EU and the U.S. regulatory framework for the environmental safety assessment of transgenic crops – an industry perspective. In: Kerry, B.R. (ed.) *British Crop Protection Council Symposium Proceedings No. 71. Biotechnology in Crop Protection: Facts and Fallacies*. British Crop Protection Council, Farnham, UK, pp. 101–108.

Kötze, A.C. and Reynolds, S.E. (1989) Actions of cyromaine on insect cuticle. In: McFarlane, N.R. (ed.) *British Crop Protection Council Monograph No. 43. Progress and Prospects in Insect Control*. British Crop Protection Council, Farnham, UK, pp. 265.

Kouskolekas, C. and Decker, G.C. (1968) A quantitative evaluation of factors affecting alfalfa yield reduction caused by the potato leafhopper attack. *Journal of Economic Entomology* 61, 921–927.

Koyama, J., Teruya, T. and Tanaka, K. (1984) Eradication of the oriental fruit fly (Diptera: Tephritidae) from the Okinawa Islands by male annihilation. *Journal of Economic Entomology* 77, 468–472.

Koziel, M.G., Beland, G.L., Bowman, C., Carozzi, N.B., Crenshaw, R., Crossland, L., Dawson, J., Desai, N., Hill, M., Kadwell, S., Lauris, K., Lewis, K., Maddox, D., McPherson, K., Meghji, M.R., Merlin, E., Rhodes, R., Warren, G.W., Wrights, M. and Erola, S.T. (1993) Field performance of elite transgenic maize plants expressing an insecticidal protein derived from *Bacillus thuringiensis*. *Bio/Technology* 11, 194–200.

Kulman, H.M. (1971) Effects of insect defoliation on growth and mortality of trees. *Annual Review of Entomology* 16, 289–324.

Kumar, R. (1984) *Insect Pest Control with Special Reference to African Agriculture*. Edward Arnold, London.

Kunz, S.E., Drummon, R.O. and Weintraub, J. (1984) A pilot test to study the use of the sterile insect technique for eradication of cattle grubs. *Preventative Veterinary Medicine* 2, 523.

Kunz, S.E., Graham, M.R., Hogan, B.F. and Eschle, J.L. (1974) Effect of releases of sterile horn flies into a native population of horn flies. *Environmental Entomology* 3, 159.

Kuperstein, M.L. (1974) Utilisation of the precipitin test of the quantitative estimation of the influence of *Pterostichus crenuliger* (Coleoptera: Carabidae) upon the population dynamics of *Eurygaster integriceps* (Hemiptera: Scutelleridae). *Zoologicheski Zhurnal* 53, 557–562.

Kuperstein, M.L. (1979) Estimating carabid effectiveness in reducing the sunn pest, *Eurygaster integriceps* Puton (Heteroptera: Scutelleridae) in the USSR. Serology in insect predator–prey studies. *Entomological Society of America, Miscellaneous Publication* 11, 80–84.

La Brecque, G.C. and Meifert, D.W. (1966) Control of house flies (Diptera: Muscidae) in poultry houses with chemosterilants. *Journal of Medical Entomology* 3, 232–236.

La Brecque, G.C., Meifert, D.W. and Fye, R.L. (1963) A field study on the control of house flies with chemosterilant techniques. *Journal of Economic Entomology* 56, 159.

La Brecque, G.C., Smith, C.N. and Meifert, D.W. (1962) A field experiment in the control of houseflies with chemosterilant baits. *Journal of Economic Entomology* 55, 449–451.

Laing, J.E. and Eden, G.M. (1990) Mass-production of *Trichogramma minutum* Riley on factitious host eggs. *Memoirs of the Entomological Society of Canada* 153, 10–24.

Landsmann, J. (1998) The regulatory framework in Europe. The genetechnology law in Germany. In: Richter, J., Huber, J. and Schuler, B. (eds) *Biotechnology for Crop Protection – its Potential for Developing Countries*. Seutsche Stiftung für Internationale Entwicklung, Feldafing, Germany, pp. 121–127.

Langley, P.A. (1998) Target technologies for insect pest control. *Pesticide Outlook* 9, 6–12.

Langley, P.A. and Weidhaas, D. (1986) Trapping as means of controlling tsetse, *Glossina* spp. (Diptera: Glossinidae): the relative merits of killing and sterilisation. *Bulletin of Entomological Research* 76, 89–95.

Lanier, G.N. (1990) Principles of attraction-annihilation: mass trapping and other means. In: Ridgway, R.L., Silverstein, R.M. and Inscoe, M.N. (eds) *Behaviour-modifying Chemicals for Insect Management. Applications of Pheromones and other Attractants*. Marcel Dekker, New York and Basel, pp. 25–45.

Larson, L.L. (1997) Spinodad, the first member of a new class of insect control products, the naturalytes. *IBC's 2nd Annual Conference on Biopesticides and Transgenic Plants*. International Business Communications, Southborough, MA, USA.

Lashomb, J.H. and Ng, Y.-S. (1984) Colonisation by Colorado potato beetles, *Leptinotarsa decemlineata* (Say) Coleoptera: Chrysomelidae), in rotated and nonrotated potato fields. *Environmental Entomology* 13, 1352–1356.

Lawrence, W.J.C. (1968) *Plant Breeding*. Institute of Biology, Studies in Biology No. 12. Edward Arnold, London.

Lazzeri, P.A. (1998) Techniques for the development of transgenic crops in crop protection. In: Kerry, B.R. (ed.) *British Crop Protection Council Symposium Proceedings No. 71. Biotechnology in Crop Protection: Facts and Fallacies*. British Crop Protection Council, Farnham, UK, pp. 3–10.

Leather, S.R. (1993) Influence of site factor modification on the population development of the pine beauty moth (*Panolis flammea*) on a Scottish lodgepole pine (*Pinus contorta*) plantation. *Forest Ecology and Management* 59, 207–223.

Leather, S.R., Carter, N., Walters, K.F.A., Chroston, J.R., Thornback, N., Gardner, S.M. and Watson,

S.J. (1984) Epidemiology of cereal aphids on winter wheat in Norfolk, 1979–1981. *Journal of Applied Ecology* 21, 103–114.

Leather, S.R., Watt, A.D. and Barbour, D.A. (1985) The effect of host-plant and delayed mating on the fecundity and lifespan of the pine beauty moth, *Panolis flammea* (Denis and Shiffermuller) (Lepidoptera: Noctuidae): their influence on population dynamics and relevance to pest management. *Bulletin of Entomological Research* 75, 641–651.

LeClerg, E.L. (1971) Field experiments for assessment of crop losses. In: Chiarappa, L. (ed.) *Crop Loss Assessment Methods, FAO Manual on the Evaluation and Prevention of Losses by Pests, Disease and Weeds*. Commonwealth Agricultural Bureaux, Slough, UK, pp. 2.1/1–2.1/9.

Lee, G., Stevens, D.J., Stokes, S. and Wratten, S.D. (1981) Duration of cereal aphid populations and the effects on wheat yield and bread making quality. *Annals of Applied Biology* 98, 169–178.

Leffelaar, P.A. and Ferrari, T.J. (1989) Some elements of dynamic simulation. In: Rabbinge, R., Ward, S.A. and Laar, H.H. van (eds) *Simulation and Systems Management in Crop Protection*. Pudoc, Wageningen, pp. 19–45.

Legner, E.F. (1986) Importation of exotic natural enemies. In: Franz, J.M. (ed.) *Biological Plant and Health Protection. Biological Control of Plant Pests and of Vectors of Human and Animal Diseases*. Gustav Fischer Verlag, Stuttgart and New York, pp. 19–30.

Leibee, G.L. (1985) Effects of storage at 1.1°C on the mortality of *Liriomyza trifolii* (Burgess) (Diptera: Agromyzidae) life stages in celery. *Journal of Economic Entomology* 78, 407–411.

Leius, K. (1967a) Influence of wild flowers on parasitism of the tent caterpillar and codling moth. *Canadian Entomologist* 99, 444–446.

Leius, K. (1967b) Food sources and preferences of adults of a parasite, *Scambus buolianae* (Hym.: Ichn.), and their consequences. *Canadian Entomologist* 99, 865–871.

Lenz, M. and Runko, S. (1994) Protection of buildings, other structures and materials in ground contact from attack by subterranean termites (Isoptera) with a physical barrier – a fine mesh of high grade stainless steel. *Sociobiology* 24, 1–16.

Lewis, T. (1981) Pest monitoring to aid insecticide use. *Philosophical Transactions of the Royal Society London B* 295, 153–162.

Lewis, T. (1998) Commitment to long-term agricultural research: a message for science, sponsors and industry. In: Lewis, T. (ed.) *The Bawden Memorial Lectures*. British Crop Protection Council, Farnham, UK, pp. 263–292.

Lewis, T. and Macauley, E.D.M. (1976) Design and elevation of sex-attractant traps for pea moth, *Cydia nigricana* (Steph.) and the effect of plume shape on catches. *Ecological Entomology* 1, 175–187.

Liber, H. and Niccoli, A. (1988) Observations on the effectiveness of an attractant food spray in increasing chrysopid predation on *Prays oleae* (Bern.) eggs. *Redia* 71, 467–482.

Lilley, R., Hardie, J. and Wadhams, L.J. (1997) Manipulation of *Praon* populations with synthetic aphid sex pheromones for the control of cereal aphids. *Boletin de la Asociacion Española de Entomologia*. Supl. 21, 23–29.

Linblad, M. and Solbreck, C. (1998) Predicting *Oscinella frit* population densities from suction trap catches and weather data. *Journal of Applied Ecology* 35, 871–181.

Lindgren, D.I. and Vincent, L.E. (1970) Effect of atmospheric gases alone or in combination on the mortality of granary and rice weevils. *Journal of Economic Entomology* 63, 1926–1929.

Lindquist, D.A. and Busch-Petersen, E. (1987) Applied insect genetics and IPM. In: Delucchi, V. (ed.) *Integrated Pest Management, Protection Integrée: Quo Vadis? An International Perspective*. Parasitis 86, Geneva, pp. 237–255.

Lindquist, D.A., Abusowa, M. and Hall, M.J.R. (1992) The New World screwworm fly in Libya: a review of their introduction and eradication. *Medical and Veterinary Entomology* 6, 2–8.

Lindroth, C.H. (1957) *The Faunal Connections between Europe and North America*. John Wiley & Sons, New York.

Lingren, P.D., Westbrook, J.K., Bryant, V.M. Jr., Raulston, J.R., Esquivel, J.F. and Jones G.D. (1994) Origin of corn earworm (Lepidoptera: Noctuidae) migrants as determined by *Citrus* pollen markers and synoptic weather systems. *Environmental Entomology* 23(3), 562–570.

Logan, J.A. (1989) Exposure of operators and bystanders to pesticides in the UK. In: McFarlane, N.R. (ed.) *British Crop Protection Council Monograph No. 43. Progress and Prospects in Insect Control*. British Crop Protection Council, Farnham, UK, pp. 231–242.

Lösel, P.M., Lindemann, M., Scherkenbeck, J., Campbell, C.A.M., Hardie, J., Pickett, J.A. and Wadhams, L.J. (1996) Effect of primary-host dairomones on the attractiveness of the hop-aphid sex pheromone to *Phorodon humuli* males and gynoparae. *Entomologia Experimentalis et Applicata* 80, 79–82.

Lövei, G.L., Hodgson, D.J., MacLeod, A. and Wratten, S.D. (1993) Attractiveness of some novel crops for flower-visiting hoverflies (Diptera: Syrphidae): comparisons from two continents. In: Corey, S.A., Dall, D.J. and Milne, W.M. (eds) *Pest Control and Sustainable Agriculture*. CSIRO Publications, Canberra, Australia, pp. 368–370.

Lowe, H.J.B. (1973) Resistance to aphids in sugar beet. *Annual Report of Plant Breeding Institute 1972*, pp. 150.

Lucey, S., Parker, C., Campion, S. and Tatchell, M. (1997) A framework to improve the uptake of computer models within the UK horticulture industry. *MAFF Project Report: HH9913T*.

Luck, R.F. (1990) Evaluation of natural enemies for biological control: a behavioural approach. *Trends in Ecology and Evolution* 5, 196–202.

Luck, R.F., Shepard, B.M. and Kenmore, P.E. (1988) Experimental methods for evaluating arthropod natural enemies. *Annual Review of Entomology* 33, 367–391.

Luijk, R., Lefferts, L.Y. and Groth, E. (1998) The importance of food safety issues from the public perspective. Public perception and the consumer's interest in pesticide residues. *The 1998 Brighton Conference: Pests and Diseases* Volume 1. British Crop Protection Council, Farnham, UK, pp. 475–482.

Lukefahr, M.J., Houghtaling, J.E. and Graham, H.M. (1971) Suppression of *Heliothis* populations with glabrous cotton strains. *Journal of Economic Entomology* 64, 486–488.

Lukefahr, M.J., Houghtaling, J.E. and Cruhm, D.G. (1975) Suppression of *Heliothis* spp. with cotton containing combinations of resistant characters. *Journal of Economic Entomology* 68, 743–746.

Lund, A.E. (1985) Insecticides: effects on the nervous system. In: Kerkut, G.A. and Gilbert, L.I. (eds) *Comprehensive Insect Physiology Biochemistry and Pharmacology*. Pergamon, Oxford, pp. 9–56.

Luttrell, R.G. (1994) Cotton pest management: Part 2. A US perspective. *Annual Review of Entomology* 39, 527–542.

Luttrell, R.G., Fitt, G.P., Ramalho, F.S. and Sugonyaev, E.S. (1994) Cotton pest management Part 1. A worldwide perspective. *Annual Review of Entomology* 39, 517–526.

MacArthur, R.H. and Wilson, E.O. (1967) *The Theory of Island Biogeography*. Princeton University Press, Princeton, New Jersey, USA.

McCarthy, M.T., Shepard, M. and Turnipseed, S.G. (1980) Identification of predacious arthropods in soybeans by using autoradiography. *Environmental Entomology* 9, 199–203.

McCown, B.H., McCabe, D.E., Russell, D.R., Robinson, D.J., Barton, K.A. and Raffa, K.F. (1991) Stable transformation of *Populus* and incorporation of pest resistance by electric discharge particle acceleration. *Plant Cell Report* 9, 590–594.

McCoy, C.W., Samson, R.A. and Boucias, D.G. (1986) Entomogenous fungi. In: Ignoffo, C.M. and Mandava, N.B. (eds) *Handbook of Natural Pesticides. Volume 5: Microbial Pesticides*. CRC Press, Boca Raton, USA, pp. 151–236.

McEwen, P.K., Jervis, M.A. and Kidd, N.A.C. (1993) The effect on olive moth (*Prays oleae*) population levels, of applying artificial food to olive trees. *A.N.P.P 3rd International Conference on Pests in Agriculture, Montpellier*, pp. 361–368.

McEwen, P.K., Jervis, M.A. and Kidd, N.A.C. (1994) Use of a sprayed L-tryptophan solution to concentrate numbers of the green lacewing *Chrysoperla carnea* in olive tree canopy. *Entomologia Experimentalis et Applicata* 70, 97–99.

Mack, T.P. (1992) Implementing innovative insect management systems in soybean in the southeastern US. In: Copping, L.G., Green, M.B. and Rees, R.T. (eds) *Pest Management in Soybean*. Elsevier Applied Science, London and New York, pp. 36–45.

Mackauer, M. (1972) Genetic aspects of insect production. *Entomophaga* 17, 27–48.

McKenzie, R.I.H. and Lambert, J.W. (1961) Comparison of F_3 lines and their related F_6 lines in two barley crosses. *Crop Science* 1, 246–249.

McKibben, G.H., Smith, J.W. and McGovern, W.L. (1990) Design of an attract-and-kill device for boll weevil (Coleoptera: Curculionidae). *Journal of Entomological Science* 25, 581–586.

McKinion, J.M. (1989) Modelling and economics. In: McFarlane, N.R. (ed.) *British Crop Protection*

Council Monograph No. 43. Progress and Prospects in Insect Control. British Crop Protection Council, Farnham, UK, pp. 203–214.

MacLeod, J. and Donnelly, J. (1961) Failure to reduce an isolated blowfly population by the sterile males release method. *Entomologia Experimentalis et Applicata* 4, 101.

MacLeod, N.D. and Norton, G.A. (1996) Economic considerations for the assessment of losses due to pasture diseases. In: *Pasture and Forage Crop Pathology.* American Society of Agronomy, Crop Science Society of America and Soil Science Society of America, Madison, USA.

McManus, M.T., White, D.W.R. and McGregor, P.G. (1994) Accumulation of a chymotrypsin inhibitor in transgenic tobacco can affect the growth of insect pests. *Transgenic Research* 3, 50–58.

McNaughton, S.J. (1983) Compensatory plant growth as a response to herbivory. *Oikos* 40, 329–336.

McPherson, R.M. and Newsom, L.D. (1984) Trap crops for control of stink bugs in soybean. *Journal of the Georgia. Entomological Society* 19, 470–480.

Manjunath, T.M. (1998) Mass production and application of biocontrol agents. In: Ananthakrishnan, T.N. (ed.) *Technology in Biological Control.* Science Publishers Inc, Enfield, New Hampshire, USA, pp. 112–120.

Manly, B.F.J. (1974) Estimation of stage-specific survival rates and other parameters for insect populations passing through stages. *Oecologia (Berl.)* 15, 277–285.

Manly, B.F.J. (1977) The determination of key factors from life table data. *Oecologia* 31, 111–117.

Manly, B.F.J. (1990) *Stage Structured Populations: Sampling Analysis and Simulation.* Chapman & Hall, London.

Manly, B.F.J. and Parr, M.J. (1968) A new method for estimating population size, survivorship and birth rate from capture-recapture data. *Transactions of the Society for British Entomology* 18, 81–89.

Markland, R.E. and Sweigart, J.R. (1987) *Quantitative Methods: Application to Managerial Decision making.* John Wiley & Sons, Chichester, UK.

Marks, R.J. (1977) Laboratory studies of plant searching behaviour by *Coccinella septempunctata* L. larvae. *Bulletin of Entomological Research* 67, 235–241.

Marris, G.C., Weaver, R., Olieff, S.M., Mosson, H.J. and Edwards, J.P. (1996) Ectoparasitoid venom as a regulator of Lepidopteran host development and moulting. *Brighton Crop Protection Conference 1996: Pests and Diseases* Volume 3. British Crop Protection Council, Farnham, UK, pp. 1021–1028.

Martin, P. and Bateson, P.P.G. (1993) *Measuring Behaviour. An Introductory Guide,* 2nd edn. Cambridge University Press, Cambridge, UK.

Matteson, P.C., Gallagher, K.D. and Kenmore, P.E. (1994) Extension of integrated pest management for planthoppers in Asian irrigated rice: empowering the user. In: Denno, R.F. and Perfect, T.J. (eds) *Planthoppers Their Ecology and Management.* Chapman & Hall, New York and London, pp.656–685.

Matthews, G.A. (1979) *Pesticide Application Methods.* English Language Book Society/Longman, Harlow, UK.

Matthews, G.A. (1981) Improved systems of pesticide application. *Philosophical Transactions of the Royal Society B* 295, 163–173.

Matthews, G.A. (1983) Suitability of pesticide application equipment take over for small-scale farmers in tropical countries. *10th International Congress of Plant Protection, Plant Protection for Human Welfare, Brighton,* Volume 2, 517.

Matthews, G.A. (1984) *Pest Management.* Longman, London and New York.

Matthews, G.A. (1985) Application from the ground. In: Haskell, P.T. (ed.) *Pesticide Application: Principles and Practice.* Clarendon Press, Oxford, pp. 95–117.

Matthews, G.A. (1992) *Pesticide Application Methods,* 2nd edn. Longman Scientific & Technical, Harlow, UK/John Wiley & Sons, New York, USA.

Matthews, G.A. (1997a) Techniques to evaluate insecticide efficacy. In: Dent, D.R. and Walton, M.P. (eds) *Methods in Ecological and Agricultural Entomology.* CAB International, Wallingford, UK, pp. 243–270.

Matthews, G.A. (1997b) Implementing cotton integrated pest management. *Experimental Agriculture* 33, 1–14.

Matthews, G.A. and Clayphon, J.E. (1973) Safety precautions for pesticide application in the tropics. *PANS* 19, 1–12.

Mattson, W.J. and Haack, R.A. (1987) The role of drought stress in provoking outbreaks of phytophagous insects. In: Barbosa, P. and Schultz, J.C. (eds) *Insect Outbreaks.* Academic Press, San Diego and London, pp. 365–407.

May, R.M. (1978) Host-parasitoid systems in patchy environments: a phenomenological model. *Journal of Animal Ecology* 47, 833–843.

May, R.M. and Hassell, M.P. (1981) The dynamics of multiparasitoid-host interactions. *American Naturalist* 117, 234–261.

May, R.M. and Hassell, M.P. (1988) Population dynamics and biological control. *PTRB* 318, 129–169.

Maynard-Smith, J. (1974) *Models in Ecology.* Cambridge University Press, Cambridge.

Mayo, O. (1987) *The Theory of Plant Breeding.* Clarendon Press, Oxford.

Mayse, M.A. (1978) Effects of spacing between rows on soybean arthropod populations. *Journal of Applied Ecology* 15, 439–450.

Medawar, P.B. (1981) *Advice to a Young Scientist.* Pan Books, London, Sydney.

Medawar, P.B. (1986) *The Limits of Science.* Oxford University Press, Oxford, UK.

Mellado, L. (1971) La tecnica de machos esteriles en el control de la mosca Mediterrano. Programas realizados en Espana. In: *Sterility Principle for Insect Control or Eradication.* IAEA/FAO, Vienna, pp. 49.

Menn, J.J., Raina, A.K. and Edwards, J.P. (1989) Juvenoids and neuropeptides as insect control agents: retrospect and prospects. In: McFarlane, N.R. (ed.) *British Crop Protection Council Monograph No. 43. Progress and Prospects in Insect Control.* British Crop Protection Council, Farnham, UK, pp. 87–106.

Merlin, J., Lemaitre, O. and Grégoire, J.-C. (1996) Oviposition in *Cryptolaemus montrouzieri* stimulated by wax filaments of its prey. *Entomologia Experimentalis et Applicata* 79, 141–146.

Merritt, C.R. (1998) The commercialisation of transgenic crops – the *Bt* experience. In: Kerry, B.R. (ed.) *British Crop Protection Council Symposium Proceedings No. 71. Biotechnology in Crop Protection: Facts and Fallacies.* British Crop Protection Council, Farnham, UK, pp. 79–86.

Messenger, P.S. and van den Bsoch, R. (1971) The adaptability of introduced biological control agents. In: Huffaker, C.G. (ed.) *Biological Control.* Plenum Press, New York and London, pp. 16–67.

Metcalf, R.L. (1980) Changing role of insecticides in crop protection. *Annual Review of Entomology* 25, 219–256.

Metcalf, R.L. (1986) Coevolutionary adaptations of rootworm beetles (Coleoptera: Chrysomelidae) to cucurbitacins. *Journal of Chemical Ecology* 12, 1109–1124.

Metcalf, R.L. and Metcalf, E.R. (1992) *Plant Kairomones in Insect Ecology and Control.* Chapman and Hall, New York.

Metcalf, R.L. and Metcalf, R.A. (1993) *Destructive and Useful Insects,* 5th edn. McGraw-Hill, New York, USA.

Metcalf, R.L., Rhodes, A.M., Metcalf, R.A., Ferguson, J., Metcalf, E.R. and Lu, P. (1982) Cucurbitacin contents and diabroticite (Coleoptera: Chrysomelidae) feeding upon *Cucurbita* spp. *Environmental Entomology* 11, 931–937.

Meyer, J.A., Lancaster, J.L. Jr. and Simco, J.S. (1982) Comparison of habitat modification, animal control and standard spraying for control of the lone star tick. *Journal of Economic Entomology* 75, 724–729.

Michael, P.J. (1989) Importation and establishment of new natural enemies of *Heliothis* spp. (Lepidoptera: Noctuidae) in Australia. In: King, E.G. and Jackson, R.D. (eds) *Proceedings of the Workshop on Biological Control of* Heliothis*: Increasing the Effectiveness of Natural Enemies.* Far Eastern Regional Research Office, US Department of Agriculture, New Delhi, India, pp. 364–373.

Mihm, J.A. (1989) Evaluating maize for resistance to tropical stem borers, armyworms, and earworms. In: Mihm, J.A., Wiseman, B.R. and Davis, F.M. (eds) *Toward Insect Resistant Maize for the Third World.* International Wheat and Maize Improvement Center (CIMMYT), El Batan, Mexico, pp. 109–121.

Miller, J.R. and Cowles, R.S. (1990) Stimulo-deterrent diversion: a concept and its possible application to onion maggot control. *Journal of Chemical Ecology* 16, 3197–3212.

Mills, N. (1997) Techniques to evaluate the efficacy of natural enemies. In: Dent, D.R. and Walton, M.P. (eds) *Methods in Ecological and Agricultural Entomology.* CAB International, Wallingford, UK, pp. 271–291.

Milne, W.M. and Bishop, A.L. (1987) The role of predators and parasites in the natural regulation of lucerne aphids in eastern Australia. *Journal of Applied Ecology* 24, 893–905.

Minks, A.K. and Jong, D.J. de (1975) Determination of spraying dates for *Adoxophyes orana* by sex pheromone traps and temperature recordings. *Journal of Economic Entomology* 68, 729–732.

Mishoe, J.W., Jones, J.W., Swaney, D.P. and Wilkerson, G.G. (1984) Using crop and pest models for management applications. *Agricultural Systems* 15, 153–170.

Mollema, C., Steenhuis, M.M., Inggamer, H. and Soria, C. (1993) Evaluating the resistance of *Frankliniella occidentalis* in cucumber: methods, genotypic variation and effects upon thrips biology. *Bulletin IOBC/WPRS* 16, 113–116.

Mols, P.J.M., Booy, C.J.H. and de Visser, P. (1992) "Gaby" a computerized advisory system for IPM in apple orchards. *Acta Phytopathological Entomology* 27, 461–464.

Molthan, J. and Rupert, V. (1988) Zur bedeutung bluhender wildkrauter in felrainen und ackern für blutenbesuchende nutzinsekten. *Mitteilungen aus der Biologischen Budnesanstalt für Land-und Fortwirtschaft* 247, 85–99.

Montiel, A. (1992) The influence of the project Éclair 209 on the development of the new olivecul-ture. In: Haskell, P.T. (ed.) *Research Collaboration in European IPM Systems*. British Crop Protection Council Monograph No. 52, British Crop Protection Council, Farnham, UK, pp. 77–80.

Morewood, W.D. (1992) Cold storage of *Phytoseiulus persimilis* (Phytoseiidae). *Experimental and Applied Acarology* 13, 231–236.

Morgan, D., Walters, K.F.A., Oakley, J.N. and Lane, A. (1998) An internet-based decision support sys-tem for the rational management of oilseed rape invertebrate pests. *The 1998 Brighton Conference: Pests and Diseases* Volume 1. British Crop Protection Council, Farnham, UK, pp. 259–264.

Morlan, H.B., McCray, E.M. and Kilpatrick, J.W. (1962) Field tests with sexually sterile males for con-trol of *Aedes aegypti*. *Mosquito News* 22, 295.

Morrison, R.E. and King, E.G. (1977) Mass production of natural enemies. In: Ridgway, R.L. and Vinson, S.B. (eds) *Biological Control by Augmentation of Natural Enemies. Insect and Mite Control with Parasites and Predators*. Plenum Press, New York and London, pp. 183–217.

Morse, S. and Buhler, W. (1997) IPM in developing countries: the danger of an ideal. *Integrated Pest Management Reviews* 2, 175–186.

Moscardi, F. and Sosa-Gomez, D.R. (1992) Use of viruses against soybean caterpillars in Brazil. In: Copping, L.G., Green, M.B. and Rees, R.T. (eds) *Pest Management in Soybean*. Elsevier Applied Science, London and New York, pp. 98–109.

Mulaa, M.A. (1995) Evaluation of factors leading to rational pesticide use for the control of the maize stalk borer *Busseola fusca* (Lepidoptera: Noctuidae) in Trans-Nzoia district, Kenya. Unpublished PhD Thesis, University of Wales, UK.

Mulder, P.G. and Showers, W.B. (1986) Defoliation by the armyworm (Lepidoptera: Noctuidae) on field corn in Iowa. *Journal of Economic Entomology* 79, 368–373.

Müller, H.J. (1958) The behaviour of *Aphis fabae* in selecting its host plants, especially different vari-eties of *Vicia faba*. *Entomologia Experimentalis et Applicata* 1, 66–72.

Mullin, C.A. and Croft, B.A. (1985) An update on development of selective pesticides favouring arthropod natural enemies. In: Hoy, M.A. and Herzog, D.C. (eds) *Biological Control in Agricultural IPM Systems*. Academic Press, Orlando and London, pp. 123–150.

Mullins, L.J. (1989) *Management and Organisational Behaviour*, 2nd edn. Pitman Publishing, London, UK.

Mumford, J.D. and Knight, J.D. (1997) Injury, damage and threshold concepts. In: Dent, D.R. and Walton, M.P. (eds) *Methods in Ecological and Agricultural Entomology*. CAB International, Wallingford, UK, pp. 203–220.

Mumford, J.D. and Norton, G.A. (1984) Economics of decision making in pest management. *Annual Review of Entomology* 29, 157–174.

Mumford, J.D. and Norton, G.A. (1987) Economic aspects of integrated pest management. In: Delucchi, V. (ed.) *Integrated Pest Management, Protection Intégrée: Quo Vadis? An International Perspective*. Parasitis 86, Geneva, pp. 397–408.

Mumford, J.D. and Norton, G.A. (1989) Expert systems in pest management implementation on an international basis. *AI Applications* 3, 67–69.

Mumford, J.D. and Norton, G.A. (1991) Economics of integrated pest control. In: Teng, P.S. (ed.) *Crop Loss Assessment and Pest Management*. APS Press, St Paul, Minnesota, pp. 191–200.

Mumford, J.D. and Norton, G.A. (1993) Survey and knowledge acquisition techniques. In: Norton, G.A. and Mumford, J.D. (eds) *Decision Tools for Pest Management*. CAB International, Wallingford, UK, pp. 79–88.

Munthali, D.C. (1981) Biological efficiency of small pesticide deposits. Unpublished PhD thesis, University of London.

Munthali, D.C. and Scopes, N.E.A. (1982) A technique for studying the biological efficiency of small droplets of pesticide solutions and a consideration of the implications. *Pesticide Science* 13, 60–62.

Murchie, A.K., Williams, I.H. and Smart, L.E. (1995) Responses of brassica pod midge (*Dasineura brassicae*) and its parasitoid (*Platygaster* sp.) to isothiocyanates. *Proceedings of the 9th International Rapeseed Congress 1995: Rapeseed Today and Tomorrow*. Groupe Consultatif International de Recherche sur le Colza.

Murdoch, W.W. (1979) Predation and dynamics of prey populations. *Fortschritte der Zoologie* 25, 295–310.

Murdoch, W.W. and Oaten, A.A. (1975) Predation and population stability. *Advances in Ecological Research* 9, 1–131.

Murdoch, W.W., Chesson, J. and Chesson, P.L. (1985) Biological control in theory and practice. *American Naturalist* 125, 344–366.

Murray, R.A. and Solomon, M.G. (1978) A rapid technique for analysing diets of invertebrate predators by electrophoresis. *Annals of Applied Biology* 90, 7–10.

Musgrave, A.E. (1980) Kuhn's second thoughts. In: Cutting, G. (ed.) *Paradigms and Revolutions*. University of Notre Dame Press, Notre Dame, pp. 39–53.

Myers, J.H., Higgins, C. and Kovacs, E. (1988) How many insect species are necessary for the biological control of insects? *Environmental Entomology* 18, 541–547.

Nadel, H. and van Alphen, J.J.M. (1986) The role of host- and host-plant odours in the attraction of the parasitoid *Epidinocarsis lopezi* to its host, the cassava mealybug, *Phenacoccus manihoti*. *Mededlingen van de Faculteit Landbouwwetenschappen Rijksuniversiteit (Gent)* 51, 1079–1086.

Nadel, H. and van Alphen, J.J.M. (1987) The role of host and host plant odours in the attraction of a parasitoid, *Epidinocarsis lopezi* to the habitat of its host, the cassava mealybug, *Phenacoccus manihoti*. *Entomologia Experimentalis et Applicata* 45, 181–186.

Nakamura, K. (1976) The active space of the pheromone of *Spodoptera litura* and the attraction of adult males to the pheromone source. *Proceedings of a Symposium on Insect Pheromones and their Applications, Nagaoka and Tokyo 1976*, pp. 145–155.

Nakamura, M. and Nakamura, K. (1977) Population dynamics of the chestnut gall wasp, *Dryocosmus kuriphilus* Yasumatsu (Hymenoptera: Cynipidae). *Oecologia* 27, 97–116.

Neethling, D.C. and Dent, D.R. (1998) *Metarhizium anisopliae*, isolate IMI 330189: A mycoinsecticide for locust and grasshopper control. *The 1998 Brighton Conference: Pests and Diseases*, Volume 1. British Crop Protection Council, Farnham, UK, pp. 37–42.

Nei, M., Maruyama, T. and Chakraborty, R. (1975) The bottleneck effect and genetic variability in populations. *Evolution* 29, 1–10.

Nelson-Smith, D. (1998) 1995 Food or famine: politics, economics and science in the world's food supply. In: Lewis, T. (ed.) *The Bawden Memorial Lectures*. British Crop Protection Council, Farnham, UK, pp. 279–291.

Neuenschwander, P. and Herren, H.R. (1988) Biological control of the cassava mealybug, *Phenacoccus manihoti*, by the exotic parasitoid *Epidinocarsis lopezi* in Africa. *Philosophical Transactions of the Royal Society B* 318, 319–333.

Neuenschwander, P., Schulthess, F. and Madojemu, E. (1986) Experimental evaluation of the efficiency of *Epidinocarsis lopezi*, a parasitoid introduced into Africa against the mealybug, *Phenacoccus manihoti*. *Entomologia Experimentalis et Applicata* 42, 133–138.

Neuman, U. (1990) Commercial development: mating disruption of the European grape berry moth. In: Ridgway, R.L., Silverstein, R.M. and Inscoe, M.N. (eds) *Behaviour-modifying Chemicals for Insect Management. Applications of Pheromones and other Attractants*. Marcel Dekker, New York and Basel, pp. 539–546.

Newton, P.J. and Odendaal, W.J. (1990) Commercial inundative releases of *Trichogrammatoidea cryp-*

tophlebiae (Hymenoptera: Trichogrammatidae) against *Cryptophlebia leucotreta* (Lepidoptera: Tortricidae) in citrus. *Entomophaga* 35, 545–556.

Ng, S.S., Davis, F.M. and Williams, W.P. (1990) Ovipositional response to southwestern corn borer (Lepidoptera: Pyralidae) and fall armyworm (Lepidoptera: Noctuidae) to selected maize hybrids. *Journal of Economic Entomology* 83, 1575–1577.

Nicholson, A.G. (1992) Compensatory growth of potatoes (*Solanum tuberosum* L.) in response to defoliation. Thesis, Cornell University, USA.

Nicholson, A.J. and Bailey, V.A. (1935) The balance of animal populations. *Proceeding of the Zoological Society of London* 1935, 551–598.

Nicol, D. and Wratten, S.D. (1997) The effect of hydroxamic acid concentration at late growth stages of wheat on the performance of the aphid *Sitobion avenae*. *Annals of Applied Biology* 130, 387–396.

Nielson, M.W. and Lehman, W.F. (1980) Breeding approaches in alfalfa. In: Maxwell, F.G. and Jennings, P.R. (eds) *Breeding Plants Resistant to Insects*. John Wiley and Sons, Chichester, pp. 277–312.

Niles, G.A. (1980) Breeding cotton for resistance to insect pests. In: Maxwell, F.G. and Jennings, P.R. (eds) *Breeding Plants Resistant to Insects*. John Wiley and Sons, Chichester, pp. 337–370.

Nordlund, D.A. (1981) Semiochemicals: a review of terminology. In: Nordlund, D.A., Jones, R.L. and Lewis, W.J. (eds) *Semiochemicals, Their Role in Pest Control*. John Wiley and Sons, Chichester, UK, pp. 13–23.

Norris, D.M. and Kogan, M. (1980) Biochemical and morphological bases of resistance. In: Maxwell, F.G. and Jennings, P.R. (eds) *Breeding Plants Resistant to Insects*. John Wiley and Sons, Chichester, pp. 23–61.

North, C. (1979) *Plant Breeding and Genetics in Horticulture*. Macmillan Press Limited, London.

Norton, G.A. (1976) Analysis of decision making in crop protection. *Agroecosystems* 3, 27–44.

Norton, G.A. (1982a) Crop protection decision making – an overview. *British Crop Protection Council Monograph No. 25. Decision Making in the Practice of Crop Protection*. British Crop Protection Council, Farnham, UK, pp. 3–11.

Norton, G.A. (1982b) A decision-analysis approach to integrated pest control. *Crop Protection* 1, 147–164.

Norton, G.A. (1985) Economics of pest control. In: Haskell, P.T. (ed.) *Pesticide Application: Principles and Practice*. Clarendon Press, Oxford, pp. 175–189.

Norton, G.A. (1987) Pest management and world agriculture – policy: research and extension. *Papers in Science, Technology and Pubic Policy*, No. 13. Imperial College of Science and Technology University of London, The Science Policy Research Unit University of Sussex, The Technical Change Centre, London.

Norton, G.A. (1990) Decision tools for pest management: their role in IPM design and delivery. *FAO/UNEP/USSR Workshop on Integrated Pest Management*. Kishinev, Moldavia 1990.

Norton, G.A. (1993) Philosophy, concepts and techniques. In: Norton, G.A. and Mumford, J.D. (eds) *Decision Tools for Pest Management*. CAB International, Wallingford, UK, pp. 1–22.

Norton, G.A. and Evans, D.E. (1974) The economics of controlling froghopper (*Aeneolamia varia saccharina* (Dist.) (Hom., Cercopidae)) on sugar-cane in Trinidad. *Bulletin of Entomological Research* 63, 619–627.

Norton, G.A. and Mumford, J.D. (1983) Decision making in pest control. In: Coaker, T.H. (ed.) *Advances in Applied Biology, Vol. VIII*. Academic Press, London, pp. 87–119.

Norton, G.A. and Mumford, J.D. (eds) (1993) *Decision Tools for Pest Management*. CAB International, Wallingford, UK.

Nottingham, S.F., Hardie, J., Dawson, G.W., Hick, A.J., Pickett, J.A., Wadhams, L.J. and Woodcock, C.M. (1991) Behavioural and electrophysiological response of aphids to host and nonhost plant volatiles. *Journal of Chemical Ecology* 17, 1231–1242.

Nwanze, K.F. (1989) Insect pests of pearl millet in Sahelian West Africa I. *Acigona ignefusalis* (Pyralidae, Lepidoptera): distribution, population dynamics and assessment of crop damage. *Tropical Pest Management* 35, 137–142.

Nyffeler, M., Dean, D.A. and Sterling, W.L. (1987) Evaluation of the importance of the striped lynx spider, *Oxyopes salticus* (Araneae: Oxyopidae), as a predator in Texas cotton. *Environmental Entomology* 16, 1114–1123.

Oakley, J.N., Walters, K.F.A., Ellis, S.A. and Young, J.E.B. (1998) The economic impact and evalua-
    tion of control strategies for the reduced-rate use of aphicides against winter wheat aphids in the
    UK. *The 1998 Brighton Conference: Pests and Diseases* Volume 3. British Crop Protection
    Council, Farnham, UK, pp. 1083–1088.
OECD (1993) *Safety Evaluation of Foods Produced by Modern Biotechnology: Concepts and
    Principles*. OECD, Paris.
Onstad, D.W., Shoemaker, C.A. and Hansen, B.C. (1984) Management of potato leafhopper, *Empoasca
    fabae* (Homoptera: Cicadellidae), on alfalfa with the aid of systems analysis. *Environmental
    Entomology* 13, 1046–1058.
Ooi, P.A.C. (1986) Insecticides disrupt natural biological control of *Nilaparvata lugens* in Sekinchan,
    Malaysia. In: Hussein, M.Y. and Ibrahim, A.G. (eds) *Biological Control in the Tropics*. Universiti
    Pertanian, Serdang, Malaysia, pp. 109–120.
Oomen, P.A. (1998) Risk assessment and risk management of pesticide effects on non-target arthro-
    pods in Europe. *The 1998 Brighton Conference: Pests and Diseases* Volume 2. British Crop
    Protection Council, Farnham, UK, pp. 591–598.
Ortega, A., Vasal, S.K., Mihim, J. and Hershey, C. (1980) Breeding for insect resistance in maize. In:
    Maxwell, F.G. and Jennings, P.R. (eds) *Breeding Plants Resistant to Insects*. John Wiley and Sons,
    Chichester, pp. 371–420.
Ostlie, K.R. (1997) Context, performance and management issues of B.t. corn. *IBC's 2nd Annual
    Conference on Biopesticides and Transgenic Plants*. International Business Communications,
    Southborough, MA, USA.
Ozgur, A.F. and Sekeroglu, E. (1986) Population development of *Bemisia tabaci* (Homoptera:
    Aleurodidae) on various cotton cultivars in Cukurova, Turkey. *Agriculture Ecosystems and
    Environment* 17, 83–88.
Pace, M.E. and MacKenzie, D.R. (1987) Modelling of crop growth and yield for loss assessment. In:
    Teng, P.S. (ed.) *Crop Loss Assessment and Pest Management*. APS Press, St Paul, Minnesota,
    pp. 30–36.
Painter, R.H. (1951) *Insect Resistance in Crop Plants*. The Macmillan Company, New York.
Palmer, D.F., Windels, M.B. and Hiang, H.C. (1979) Artificial infestation of corn with western corn
    rootworm eggs in agar-water. *Journal of Economic Entomology* 70, 277–278.
Panda, N. and Khush, G.S. (1995) *Host Plant Resistance to Insects*. CAB International, Wallingford,
    UK, IRRI, Manila, Phillipines.
Parker, R.D. and Nilakhe, S.S. (1990) Evaluation of predators and parasites and chemical grain pro-
    tectants on insect pests of sorghum stored in commercial bins. *Proceedings of 3rd National
    Stored Grain Pest Management Training Conference*. Kansas City, MO, USA, pp. 229–239.
Parker, W.E. and Biddle, A.J. (1998) Assessing the damage caused by black bean aphid (*Aphis fabae*)
    on spring beans. *Proceedings 1998 Crop Protection Conference – Pests and Diseases*. British
    Crop Protection Council, Farnham, UK, pp. 1077–1082.
Parnell, F.R. (1935) Origin and development of the U4 cotton. Empire cotton growing. *Crop Review*
    12, 177–182.
Parr, W.J., Gould, H.J., Jessop, N.H. and Ludlam, F.A.B. (1976) Progress towards a biological control
    programme for glasshouse whitefly (*Trialeurodes vaporariorum*) on tomatoes. *Annals of Applied
    Biology* 83, 349–363.
Pathak, P.K., Saxena, R.C. and Heinrichs, E.A. (1982) Parafilm sachet for measuring honeydew excre-
    tion by *Nilaparvata lugens* on rice. *Journal of Economic Entomology* 75, 194–195.
Pawar, C.S. (1986) Ultra low volume spraying for pest control in pigeonpea. *Indian Journal of Plant
    Protection* 14, 37–41.
Payne, C.C. (1982) Insect viruses as control agents. *Parasitology* 84, 35–77.
Payne, C.C. (1988) Pathogens for the control of insects: where next? *Philosophical Transactions of
    the Royal Society B* 318, 225–248.
Peakall, D.B. (1993) DDE-induced eggshell thinning: an environmental detective story.
    Environmental Review 1, 13–20.
Pearl, R. (1928) *The Rate of Living*. Knopf, New York.
Pedgley, D.E., Reynolds, D.R. and Tatchell, G.M. (1995) Long-range insect migration in relation to cli-
    mate and weather: Africa and Europe. In: Drake, V.A. and Gatehouse, A.G. (eds) *Insect
    Migration: Tracking Resources through Space and Time*. Cambridge University Press, pp. 3–29.

Pedigo, L.P. (1996) *Entomology and Pest Management*, 2nd edn. Prentice Hall, Englewood Cliffs, NJ, USA.

Pedigo, L.P. and Higley, L.G. (1992) The economic injury level concept and environmental quality. A new perspective. *American Entomologist* Spring 92, 12–21.

Pedigo, L.P. and Zeiss, M.R. (1996) Effect of soybean planting date on bean leaf beetle (Coleoptera Chrysomelidae) abundance and pod injury. *Journal of Economic Entomology* 89, 183–188.

Pedigo, L.P., Hutchins, S.H. and Higley, L.G. (1986) Economic injury levels in theory and practice. *Annual Review of Entomology* 31, 341–368.

Pena, J.E., Pohronezny, K., Waddill, V.H. and Stimac, J. (1986) Tomato pinworm (Lepidoptera: Gelechiidae) artificial infestation: effect on foliar and fruit injury of ground tomatoes. *Journal of Economic Entomology* 79, 957–960.

Penman, D.R., Rohitha, B.H., White, J.G.H. and Smallfield, B.M. (1979) Control of bluegreen lucerne aphid by grazing management. *Proceedings 32nd New Zealand Weed and Pest Control Conference*, pp. 186–191.

Percy-Smith, A. and Philpsen, H. (1992) Implementation of carrot fly monitoring in Denmark. In: Finch, S. and Freuler, J. (eds) *Proceedings of Working Group Meeting: Integrated Control in Field Vegetable Crops*. IOBC/WPRS pp. 36–42.

Perkins, J.H. (1982) *Insects, Experts and the Insecticide Crisis: the Quest for New Pest Management Strategies*. Plenum, New York, USA.

Perlack, F.J., Deaton, R.W., Armstrong, T.A., Fuchs, R.L., Sims, S.R., Greenplate, J.T. and Fischhoff, D.A. (1990) Insect resistant cotton plants. *Bio/Technology* 8, 939–943.

Perrin, R.M. (1995) Synthetic pyrethroids success story. In: Best, G.A. and Ruthven, A.D. (eds) *Pesticides – Developments, Impacts, and Controls*. The Royal Society of Chemistry, London, UK, pp. 19–27.

Perrin, R.M. and Phillips, M.L. (1978) Some effects of mixed cropping on the population dynamics of insect pests. *Entomologia Experimentalis et Applicata* 24, 385–393.

Perry, J.N. (1997) Statistical aspects of field experiments. In: Dent, D.R. and Walton, M.P. (eds) *Methods in Ecological and Agricultural Entomology*. CAB International, Wallingford, UK, pp. 171–202.

Perry, J.N., Wall, C. and Greenway, A.R. (1980) Latin Square designs in field experiments involving insect sex attractants. *Ecological Entomology* 5, 385–396.

Petrie, H.G. (1976) Do you see what I see: the epistemology of interdisciplinary inquiry. *Journal of Aesthetic Education* 10, 29–43.

Petterson, J., Pickett, J.A., Pye, B.J., Quiroz, A., Smart, L.E., Wadhams, L.J. and Woodcock, C.M. (1994) Winter host component reduces colonization by bird–cherry–oak aphid *Rhapalosiphum padi* (L.) (Homoptera, Aphidae) and other aphids in cereal fields. *Journal of Chemical Ecology* 20, 2565–2574.

Pfeiffer, D.G., Kaakeh, W., Killian, J.C., Lachance, M.W. and Kirsch, P. (1993) Mating disruption for control of damage by codling moth in Virginia apple orchards. *Entomologia Experimentalis et Applicata* 67, 57–64.

Pfeiffer, D.G., Killian, J.C., Rajotte, E.G., Hull, L.A. and Snow, J.W. (1991) Mating disruption for reduction of damage by lesser peach tree borer (Lepidoptera: Sesiidae) in Virginia and Pennsylvania peach orchards. *Journal of Economic Entomology* 84, 218–223.

Pickett, C.H., Wilson, L.T. and Flaherty, D.L. (1990) The role of refuges in crop protection, with reference to plantings of French prune trees in a grape agroecosystem. In: Bostanian, N.J., Wilson, L.T. and Dennehy, T.J. (eds) *Monitoring and Integrated Management of Arthropod Pests of Small Fruit Crops*. Intercept, Andover, UK, pp. 151–165.

Pickett, J.A. (1988) The future of semiochemicals in pest control. *Aspects of Applied Biology No. 17, Environmental Aspects of Applied Biology*. The Association of Applied Biologists, Wellesbourne, UK, pp. 397–406.

Pickett, J.A. and Woodcock, C.M. (1993) Chemical ecology of plants and insects, helping crops to help themselves. *Interdisciplinary Science Reviews* 18, 68–72.

Pickett, J.A., Dawson, G.W., Griffiths, D.C., Hassanali, A., Merritt, L.A., Mudd, A., Smith, M.C., Wadhams, L.J., Woodcock, C.M. and Zhang, Z.-N. (1987) Development of plant-derived antifeedants for crop protection. In: Greenhalgh, R. and Roberts, T.R. (eds) *Pesticide Science and Biotechnology*. Blackwell Scientific, Oxford, pp. 125–128.

Pickett, J.A., Wadhams, L.J. and Woodcock, C.M. (1995) Exploiting chemical ecology for sustainable pest control. *British Crop Protection Monograph No. 63, Integrated Crop Protection: Towards Sustainabilities?* British Crop Protection Council, Farnham, pp. 353–362.

Pickup, J. and Brewer, A.M. (1994) The use of aphid suction-trap data in forecasting the incidence of potato leafroll virus in Scottish seed potatoes. *Proceedings 1994 British Crop Protection Conference – Pests and Diseases.* British Crop Protection Council, Farnham, UK, pp. 351–358

Pike, K.S. and Glazer, M. (1982) Strip rotary tillage: a management method for reducing *Fumibotys fumalis* (Lepidoptera: Pyralidae) in peppermint. *Journal of Economic Entomology* 75, 1136–1139.

Pimentel, D. (1963) Introducing parasites and predators to control native pests. *Canadian Entomologist* 95, 785–792.

Pimentel, D., Acquay, H., Biltonen, M., Rice, P., Silva, M., Nelson, J., Lipner, V., Giordana, S., Horowitz, A. and D'Amore, M. (1993) Assessment of environmental and economic impacts of pesticide use. In: Pimentel, D. and Lehman, H. (eds) *The Pesticide Question, Environment, Economics and Ethics.* Chapman & Hall, New York, pp. 47–84.

Pincus, J., Waibel, H. and Jungbluth, F. (1999) Pesticide policy: an international perspective. In: Poapongsakorn, N., Meenakanit, L., Waibel, H. and Jungbluth, F. (eds) *Approaches to Pesticide Policy Reform – Building Consensus for Future Action.* Publication Series No. 7. Pesticide Policy Project, Hannover, Germany, pp. 4–22.

Pinnschmidt, H.O., Batchelor, W.D. and Teng, P.S. (1995) Simulation of multiple species pest damage in rice using CERES-rice. *Agricultural Systems* 48, 193–222.

Pitcairn, M.J., Zalom, F.G. and Rice, R.E. (1992) Degree-day forecasting of generation time of *Cydia pomonella* (Lepidoptera: Tortricidae) populations in California. *Environmental Entomology* 21(3), 441–446.

Pitre, H.N. (ed.) (1983) *Natural Enemies of Arthropod Pests in Soybean.* South Carolina Agricultural Experimental Station Southern Cooperative Series Bulletin 285. Clemson, SC, USA.

Podoler, H. and Rogers, D. (1975) A new method for the identification of key factors from life-table data. *Journal of Animal Ecology* 44, 85–114.

Poinar, G.O. (1983) Recent developments in the use of nematodes in the control of insect pests. *10th International Congress of Plant Protection 1983, Plant Protection for Human Welfare, Brighton.* Volume 3, pp. 751–758.

Pollard, E. (1971) Hedges. IV. Habitat diversity and crop pests: a study of *Brevicoryne brassicae* and its syrphid predator. *Journal of Applied Ecology* 8, 751–780.

Ponti, O.M.B. de (1982) Resistance to insects: a challenge to plant breeders and entomologists. In: Visser, J.H. and Minks, A.K. (eds) *Proceedings of the 5th International Symposium on Insect–Plant Relationships.* Pudoc, Wageningen, The Netherlands, pp. 337–348.

Ponti, O.M.B. de (1983) Resistance to insects promotes the stability of integrated pest control. In: Lamberti, F., Waller, J.M. and Graff, N.A. van der (eds) *Durable Resistance in Crops.* Plenum Press, New York, pp. 211–223.

Ponti, O.M.B. de, Romanov, L.F. and Berlinger, M.J. (1990) Whitefly-plant relationships: plant resistance. In: Gerling, D. (ed.) *Whiteflies: their Bionomics, Pest Status and Management.* Intercept, Andover, pp. 91–106.

Poston, F.L., Pedigo, L.P. and Welch, S.M. (1983) Economic injury levels: reality and practicality. *Bulletin of the Entomological Society of America* 29, 49–53.

Potter, D.A. and Timmons, G.M. (1983) Forecasting emergence and flight of the lilac borer (Lepidoptera: Sessiidae) based on pheromone trapping and degree-day accumulations. *Environmental Entomology* 12, 400–403.

Powell, G., Hardie, J. and Pickett, J.A. (1997) Laboratory evaluation of antifeedant compounds for inhibiting settling by cereal aphids. *Entomologia Experimentalis et Applicata* 84, 189–193.

Powell, J.E. (1989) Importation and establishment of predators and parasitoids of *Heliothis* into the USA. In: King, E.G. and Jackson, R.D. (eds) *Proceedings of the Workshop on Biological Control of Heliothis: Increasing the Effectiveness of Natural Enemies.* Far Eastern Regional Research Office, US Department of Agriculture, New Delhi, India, pp. 387–395.

Powell, W. (1986) Enhancing parasitoid activity in crops. In: Waage, J. and Greathead, D. (eds) *Insect Parasitoids.* Academic Press, London, pp. 319–340.

Powell, W. and Walton, M.P. (1995) Populations and communities. In: Jervis, M.A. and N.A.C. Kidd (eds) *Insect Natural Enemies.* Chapman & Hall, London, pp. 223–292.

Powell, W. and Wright, A.F. (1988) The abilities of the aphid parasitoids *Aphidius ervi* Haliday and *A. rhopalosiphi* De Stefani Perez (Hymenoptera: Braconidae) to transfer between different known host species and the implications for the use of alternative hosts in pest control strategies. *Bulletin of Entomological Research* 78, 683–693.

Powell, W., Walton, M.P. and Jervis, M.A. (1996) Populations and communities. In: Jervis, M.A. and Kidd, N.A.C. (eds) *Insect Natural Enemies*. Chapman & Hall, London, pp. 223–292.

Price, P.W. (1987) Pathogen-induced cycling of outbreak insect populations. In: Barbosa, P. and Schultz, J.C. (eds) *Insect Outbreaks*. Academic Press, San Diego and London, pp. 269–285.

Prinsley, R.T. (ed.) (1987) *Integrated Crop Protection for Small Scale Farms in East Africa*. Commonwealth Science Council, London, UK.

Prior, C. (1997) Susceptibility of target acridoids and non-target organisms to *Metarhizium anisopliae* and *Metarhizium flavoviride*. In: Krall, S., Peveling, R. and BaDiallo, D. (eds) *New Strategies in Locust Control*. Berkhauser Verlag, Basel, Switzerland, pp. 369–375.

Prior, C., Carey, M., Abraham, Y.J., Moore, D. and Bateman, R.P. (1996) Development of a bioassay method for the selection of entomopathogenic fungi virulent to the desert locust *Schistocerca gregaria* (Forskal). *Journal of Applied Ecology* 119, 567–573.

Prokopy, R.J., Collier, R.H. and Finch, S. (1983) Visual detection of host plants by cabbage root flies. *Entomologia Experimentalis et Applicata* 34, 85–89.

Pruess, K.P. (1983) Day-degree methods for pest management. *Environmental Entomology* 12, 613–619.

Pschorn-Walcher, H. (1977) Biological control of forest insects. *Annual Review of Entomology* 22, 1–22.

Puntener, W. (1981) *Manual for Field Trials in Plant Protection*, 2nd edn. Ciba Geigy, Basle, Switzerland.

Puterka, G.J., Black, W.C. (IV), Steiner, W.M. and Burton, R.L. (1993) Genetic variation and phylogenetic relationships amongst world-wide collections of Russian wheat aphid, *Diuraphis noxia* (Mordvilko), inferred from allozyme and RAPD-PCR markers. *Heredity* 70, 604–618.

Quicke, D. (1988) Spiders bite their way towards safer insecticides. *New Scientist* 120(1640), 38–41.

Quisumbing, A.R. and Kydonieus, A.F. (1990) Controlled-release technologies for pest management. In: Wilkins, R.M. (ed.) *Controlled Delivery of Crop-protection Agents*. Taylor & Francis, London, UK, pp. 43–63.

Rabbinge, R. (1989) Population models for fruit-tree spider mite and predatory mites. In: Rabbinge, R., Ward, S.A. and Laar, H.H. van (eds) *Simulation and Systems Management in Crop Protection*. Pudoc, Wageningen, pp. 131–144.

Rabbinge, R. and Wit, C.T. de (1989) Systems, models and simulation. In: Rabbinge, R., Ward, S.A. and Laar, H.H. van (eds) *Simulation and Systems Management in Crop Protection*. Pudoc, Wageningen, pp. 3–15.

Rabbinge, R., Ankersmit, G.W. and Pak, G.A. (1979) Epidemiology and simulation of population development of *Sitobion avenae* in winter wheat. *Netherlands Journal of Plant Pathology* 85, 197–200.

Rabbinge, R., Ward, S.A. and van Laar, H.H. (1989) *Simulation and Systems Management in Crop Protection*. Pudoc, Wageningen, The Netherlands.

Raheja, A.K. (1995) Practice of IPM in South and Southeast Asia. In: Mengech, A.N., Saxena, K.N. and Gopalan, H.N.B. (eds) *Integrated Pest Management in the Tropics*. John Wiley & Sons, Chichester, UK, pp. 69–120.

Rai, B.K. (1977) Damage to coconut palms by *Azteca* sp. (Hymenoptera: Formicidae) and insecticidal control with bait, in Guyana. *Bulletin of Entomological Research* 67, 175–183.

Ramalho, F.S. (1994) Cotton pest management: Part 4. A Brazilian perspective. *Annual Review of Entomology* 39, 563–578.

Ratcliffe, D.A. (1967) Decrease in eggshell weight in certain birds of prey. *Nature* 215, 208–210.

Raty, L., Drumont, A., de Windt, N and Grégoire, J.-C. (1995) Mass trapping of the spruce bark beetle *Ips typographus* L.: traps or trap trees? *Forest Ecology and Management* 78, 191–205.

Readshaw, J.L. (1973) The numerical response of predators to prey density. *Journal of Applied Ecology* 10, 342–351.

Readshaw, J.L. and Van Gerwen, A.C.M. (1983) Age-specific survival, fecundity and fertility of the adult blowfly, *Lucilia cuprina*, in relation to crowding, protein food and population cycles. *Journal of Animal Ecology* 52(3), 879–887.

Redfearn, A. and Pimm, S.L. (1987) Insect outbreaks and community structure. In: Barbosa, P. and Schultz, J.C. (eds) *Insect Outbreaks*. Academic Press, San Diego and London, pp. 99–133.

Reed, W. (1983) Crop losses caused by insect pests in the developing world. *10th International Congress of Plant Protection 1983, Plant Protection for Human Welfare*, Brighton, Volume 1, pp. 74–80.

Reed, W., Davies, J.C. and Green, S. (1985) Field experimentation. In: Haskell, P.T. (ed.) *Pesticide Application: Principles and Practice*. Clarendon Press, Oxford, pp. 153–174.

Reeve, J.D. (1988) Environmental variability, migration and persistence in host-parasitoid systems. *American Naturalist* 132, 810–836.

Reeve, J.D. (1990) Stability, variability and persistence in host-parasitoid systems. *Ecology* 71, 422–426.

Regev, U., Shalit, H. and Gutierrez, A.P. (1976) Economic conflicts in plant protection: the problems of pesticide resistance, theory and application to the Egyptian alfalfa weevil. In: Norton, G.A. and Holling, C.S. (eds) *Proceeding of the Conference on Pest Management*. IIASA, Laxenburg, Austria.

Reichelderfer, K.H. (1989) Economic contributions of pest management to agricultural development. *Tropical Pest Management* 35, 248–251.

Reichelderfer, K.H., Carlson, G.A. and Norton, G.A. (1984) *Economic Guidelines for Crop Pest Control*. FAO Plant Production and Protection Paper 58. Food and Agriculture Organisation of the United Nations, Rome.

Renn, N. (1995) Mortality of immature houseflies (*Musca domestica* L.) in artificial diet and chicken manure after exposure to encapsulated entomopathogenic nematodes (Rhabditida: Steinernematidae, Heterorhabditidae). *Biocontrol Science and Technology* 5, 349–359.

Reuter, O.M. (1913) *Lebensgewahnheiten und Instinkte der Inseckten*. Friedlander, Berlin.

Reynolds, D.R., Riley, J.R., Armes, N.J., Cooter, R.J., Tucker, M.R. and Colvin, J. (1997) Techniques for quantifying insect migration. In: Dent, D.R. and Walton, M.P. (eds) *Methods in Ecological and Agricultural Entomology*. CAB International, Wallingford, UK, pp. 111–146.

Reynolds, S.E. (1989) The integument as a target for insecticides: disruption of cuticle chemistry, structure and function. In: McFarlane, N.R. (ed.) *British Crop Protection Council Monograph No. 43. Progress and Prospects in Insect Control*. British Crop Protection Council, Farnham, UK, pp. 73–87.

Rice, M.E. and Wilde, G.E. (1988) Experimental evaluation of predators and parasitoids in suppressing greenbugs (Homoptera: Aphididae) in sorghum and wheat. *Environmental Entomology* 17, 836–841.

Richter, A.R. and Fuxa, J.R. (1984) Timing, formulation and persistence of a nuclear polyhedrosis virus and a microsporidium for control of the velvetbean caterpillar (Lepidoptera: Noctuidae) in soybeans. *Journal of Economic Entomology* 77, 1299–1306.

Riechert, S.E. and Bishop, L. (1990) Prey control by an assemblage of generalist predators: spiders in garden test systems. *Ecology* 71, 1441–1450.

Riechert, S.E. and Lockley, T. (1984) Spiders as biological control agents. *Annual Review of Entomology* 29, 299–320.

Riley, J.R. (1974) Radar observations of individual desert locusts (*Schistocerca gregaria*) (Forsk.). *Bulletin of Entomological Research* 64, 19–32.

Riley, J.R. (1993) Flying insects in the field. In: Wratten, S.D. (ed.) *Video Techniques in Animal Ecology and Behaviour*. Chapman and Hall, London, pp. 1–15.

Riley, J.R. and Reynolds, D.R. (1995) Monitoring the movement of migratory insect pests. *Proceedings of the 3rd International Conference on Tropical Entomology, Nairobi, 1994*.

Risch, S.J. (1981) Insect herbivore abundance in tropical monocultures and polycultures: an experimental test of two hypotheses. *Ecology* 62, 1325–1340.

Risch, S.J. (1987) Agricultural ecology and insect outbreaks. In: Barbosa, P. and Schultz, J.C. (eds) *Insect Outbreaks*. Academic Press, San Diego and London, pp. 217–238.

Risch, S.J., Andow, D. and Altieri, M.A. (1983) Agroecosystem diversity and pest control: data, tentative conclusions, and new research directions. *Environmental Entomology* 12, 625–629.

Robinson, J., Fischer, M. and Hoisington, D. (1993) Molecular characterisation of *Diuraphis* spp. using random amplified polymorphic DNA. *Southwestern Entomologist* 18, 121–127.

Robinson, R.A. (1976) *Plant Pathosystems*. Springer-Verlag, Berlin.

Robinson, R.A. (1979) Permanent and impermanent resistance to crop parasites: a re-examination of the pathosystem concept with special reference to rice blast. *Zeitschrift für Pflanzenzüchtung* 83, 1–39.

Robinson, R.A. (1980a) The pathosystem concept. In: Maxwell, F.G. and Jennings, P.R. (eds) *Breeding Plants Resistant to Insects*. John Wiley & sons, Chichester, Brisbane and Toronto, pp. 157–182.

Robinson, R.A. (1980b) New concepts in breeding for disease resistance. *Annual Review of Phytopathology* 18, 189–210.

Robinson, R.A. (1983) Theoretical resistance models. In: Lamberti, F., Waller, J.M. and Graff, N.A. van der (eds) *Durable Resistance in Crops*. Plenum Press, New York, pp. 45–55.

Robinson, R.A. (1987) *Host Management in Crop Pathosystems*. Macmillan Publishing Company, New York and Collier-Macmillan Limited, London.

Robinson, R.A. (1991) The controversy concerning vertical and horizontal resistance. *Revista Mexicana de Fitopatologia* 9, 57–63.

Robinson, R.A. (1996) *Return to Resistance: Breeding Crops to Reduce Pesticide Dependence*. Ottawa: agAccess, Davis, California and International Development and Research Centre.

Robinson, S.H., Wolfenbarger, D.A. and Dilday, R.H. (1980) Antixenosis of smooth leaf cotton to the ovipositional response of tobacco budworm. *Crop Science* 20, 646–649.

Rodenhouse, N.L., Barrett, G.W., Zimmerman, D.M. and Kemp, J.C. (1992) Effects of uncultivated corridors on arthropod abundances and crop yields in soybean agroecosystems. *Agriculture, Ecosystems and Environment* 38, 179–191.

Rogers, C.E. (1985) Cultural management of *Dectes texanus* (Coleoptera: Cerambycidae) in sunflower. *Journal of Economic Entomology* 78, 1145–1148.

Rogers, D.J. (1972) Random search and insect population models. *Journal of Animal Ecology* 41, 369–383.

Römbke, J. and Moltman, J.F. (1996) *Applied Ecotoxicology*. CRC Press, Boca Raton, USA.

Root, R.B. (1973) Organisation of a plant-arthropod association in simple and diverse habitats: the fauna of collards (*Brassica oleracea*). *Ecological Monographs* 43, 95–124.

Rosier, M. (1992) The importance of soybean as a market for agricultural chemicals. In: Copping, L.G., Green, M.B. and Rees, R.T. (eds) *Pest Management in Soybean*. Elsevier Applied Science, London and New York, pp. 1–9.

Rossing, W.A.H. (1989) Application of operations research techniques in crop protection. In: Rabbinge, R., Ward, S.A. and Laar, H.H. van (eds) *Simulation and Systems Management in Crop Protection*. Pudoc, Wageningen, pp. 278–298.

Rossing, W.A.H. and Heong, K.L. (1997) Opportunities for using systems approaches in pest management. *Field Crops Research* 51, 83–100.

Rossing, W.A.H. and Wiel, L.A.J. M. van de (1990) Simulation of damage in winter wheat caused by the grain aphid *Sitobion avenae*. I. Quantification of the effects of honeydew on gas exchange of leaves and aphid populations of different size on crop growth. *Netherlands Journal of Plant Pathology* 96, 343–364.

Rossini, F.A., Porter, A.L., Kelly, P. and Chubin, D.E. (1978) Frameworks and factors affecting integration within technology assessments. *Department of Social Science Final Technical Report, NSF Grnt ERS 76–04474*. Georgia Institute of Technology.

Rotem, J., Bashi, E. and Kranz, J. (1983a) Studies of crop loss in potato blight caused by *Phytophthora infestans*. *Plant Pathology* 32, 117–122.

Rotem, J., Kranz, J. and Bashi, E. (1983b) Measurement of healthy and diseased haulm area for assessing late blight epidemics in potatoes. *Plant Pathology* 32, 109–115.

Rothschild, G.H.L. (1966) A study of a natural population of *Conomelus anceps* (Germar) (Homoptera: Delphacidae) including observations on predation using the precipitin test. *Journal of Applied Ecology* 35, 413–434.

Roush, D.K. and McKenzie, J.A. (1987) Ecological genetics of insecticide and acaricide resistance. *Annual Review of Entomology* 32, 361–380.

Rubia, E.G., Heong, K.L., Zalucki, M., Gonzlaes, B. and Norton, G. (1996) Mechanisms of compensation of rice plants to stem borer injury. *Crop Protection* 15, 335–340.

Russell, D.A. (1987) A simple method for improving estimates of percentage parasitism by insect parasitoids from field sampling of hosts. *New Zealand Entomologist* 10, 38–40.

Russell, E.P. (1989) Enemies hypothesis: a review of the effect of vegetational diversity on predatory insects and parasitoids. *Environmental Entomology* 18, 590–599.

Russell, G.E. (1978) *Plant Breeding for Pest and Disease Resistance.* Butterworth, London and Boston.

Sadras, V.O. (1996) Cotton responses to simulated insect damage: radiation-use efficiency, canopy architecture and leaf nitrogen content as affected by loss of reproductive organs. *Field Crops Research* 48, 199–208.

Sadras, V.O. and Wilson, L.J. (1997a) Growth analysis of cotton crops infested with spider mites: I. Light interception and radiation-use efficiency. *Crop Science* 37, 481–491.

Sadras, V.O. and Wilson, L.J. (1997b) Growth analysis of cotton crops infested with spider mites: II. Partitioning of dry matter. *Crop Science* 37, 492–497.

Sadras, V.O. and Wilson, L.J. (1997c) Nitrogen accumulation and partitioning in shoots of cotton plants infested with two-spotted spider mites. *Australian Journal of Agricultural Research* 48, 525–533.

Saiki, R.K., Gelfand, D.H., Stoffel, S., Scharf, S.J., Higuchi, R., Horn, G.T., Mullis, K.B. and Erlich, H.A. (1988) Primer-directed enzymatic amplification of DNA with a thermostable DNA polymerase. *Science* 293, 487–491.

Sailer, R.I. (1983) History of insect introduction. In: Wilson, C.I. and Graham, C.I. (eds) *Exotic Plant Pests and North American Agriculture.* Academic Press, New York, USA, pp. 15–38.

Salifu, A.B., Singh, R. and Hodgson, C.J. (1988) Mechanism of resistance in cowpea (*Vigna unguiculata*) genotype TVx3236, to the bean flower thrips *Megalurothrips sjostedt* (Thysanoptera: Thripidae) 2. Non-preference and antibiosis. *Tropical Pest Management* 34, 185–188.

Salt, D.W. and Ford, M.G. (1984) The kinetics of insecticide action. Part III. The use of stochastic modelling to investigate the pick-up of insecticides from ULV-treated surfaces by larvae of *Spodoptera littoralis. Pesticide Science* 15, 382–410.

Samways, M.J. (1981) *Biological Control of Pests and Weeds.* Institute of Biology, Studies in Biology No. 132. Edward Arnold, London.

Santoso, T., Sunjaya, Dharmapurtra, O.S., Halid, H. and Hodges, R.J. (1996) Pest management in psocids in milled rice stores in the humid tropics. *International Journal of Pest Management* 42, 189–197.

Sattaur, O. (1989) Genes on deposit: saving for the future. *New Scientist* 123, 37–41.

Saud, S., Saadon, M., Esa, Y.M. and Taharim, N. (1985) Implementation of integrated pest management programme for paddy in the Tanjong Karang irrigation scheme. In: Lee, B.S., Loke, W.H. and Heong, K.L. (eds) *Integrated Pest Management in Malaysia.* The Malaysian Plant Protection Society, Kuala Lumpur, pp. 283–294.

Sawaki, R.M. and Denholm, I. (1989) Insect resistance management revisited. In: McFarlane, N.R. (ed.) *British Crop Protection Council Monograph No. 43. Progress and Prospects in Insect Control.* British Crop Protection Council, Farnham, UK, pp. 191–203.

Schillinger, J.A. and Gallun, R.L. (1968) Leaf pubescence of wheat as a deterrent to the cereal leaf beetle, *Oulema melanopus. Annals of the Entomological Society of America* 61, 900–903.

Schneider, F. (1948) Beitrag zar Kenntris Generationsverhaltnisse und Diapause rauberischer Schwebfliegen. *Mitteilungen der Schweizerischen Entomologischen Gelleschaft* 13, 1–24.

Schoneveld, J.A. and Ester, A. (1994) The introduction of systems for the supervised control of the carrot fly (*Psila rosae* F.) in the Netherlands. In: Finch, S. and Brunel, E. (eds) *International Control in Field Vegetable Crops.* IOBC wprs Bulletin Vol. 17, pp. 55–66.

Schooley, D.A. and Edwards, J.P. (1996) Anti juvenile hormones: from precocenes to peptides. *Brighton Crop Protection Conference: Pests and Diseases,* Volume 3. British Crop Protection Council, Farnham, UK, pp. 1029–1038.

Schroeder, H.E., Gollasch, S., Moore, A., Tabe, L.M., Craig, S., Hardie, D.C., Chrispeels, M.J., Spencer, D. and Higgins, T.J.V. (1995) Bean amylase inhibitor resistance to the pea weevil (*Bruchus pisorum*) in transgenic peas (*Pisum sativum* L.). *Plant Physiology* 107, 1233–1239.

Schuster, M.F. (1980) Insect resistance in cotton. In: Harris, M.K. (ed.) *Biology and Breeding for Resistance to Arthropods and Pathogens in Agricultural Plants.* Texas A & M University, pp. 101–112.

Scopes, N.E.A. (1969) The economics of mass rearing *Encarsia formosa*, a parasite of the whitefly *Trialeurodes vaporariorum*, for use in commercial horticulture. *Plant Pathology* 18, 130–132.

Scopes, N.E.A. and Biggerstaff, S.M. (1971) The production, handling and distribution of the whitefly *Trialeurodes vaporariorum* and its parasite *Encarsia formosa* for use in biological control programmes in glasshouses. *Plant Pathology* 20, 111–116.

Scriven, R. and Meloan, C.E. (1986) The importance of particle size and bait position to the effectiveness of boric acid and sodium borate as blatticides for the American cockroach, *Periplaneta americana. International Pest Control* 28, 9–10.

Seaby, D.A. and Mowat, D.J. (1993) Growth changes in 20-year-old Sitka spruce *Picea sitchensis* after attack by the green spruce aphid *Elatobium abietinum. Forestry* 66, 371–379.

Seckler, D.M. and Cox, G. (1994) Population projections by the United Nations and the World Bank: zero growth in 40 years. *Discussion Paper No. 21.* Winrock International Institute, Arlington VA.

Seeworthrum, S.L., Permalloo, S., Gungah, B., Soonnoo, A.R. and Alleck, M. (1998) Eradication of an exotic fruit fly from Mauritius. *Proceedings of the 5th International Symposium on Fruitflies of Economic Importance*, Penang, Malaysia.

Sekhon, S.S. and Sajjan, S.S. (1987) Antibiosis in maize (*Zea mays* L.) to maize borer, *Chilo partellus* (Swinhoe) (Pyralidae: Lepidoptera) in India. *Tropical Pest Management* 33, 55–60.

Sengonca, C. and Frings, B. (1989) Enhancement of the green lacewing *Chrysoperla carnea* (Stephens), by providing artificial facilities for hibernation. *Turkiye Entomoloji Dergisis* 13, 245–250.

Sengonca, C., Hoffman, A. and Kleinhenz, B. (1994) Investigations on development, survival and fertility of the cereal aphids *Sitobion avenae* and *Rhopalosiphum padi* at different low temperatures. *Zeitschrift fur Angewandte Entomologie* 117, 224–233.

Senn, R., Hofer, D., Hoppe, T., Anst, M., Wyss, P., Brandl, F., Maienfisch, P., Zang, L. and White, S. (1998) CGA 293'343: a novel broad-spectrum insecticide supporting sustainable agriculture worldwide. *The 1998 Brighton Conference: Pests and Diseases*, Volume 1. British Crop Protection Council, Farnham, UK, pp. 27–36.

Service, M.W. (1993) *Mosquito Ecology: Field Sampling Methods*, 2nd edn. Elsevier Applied Science, London, UK.

Settle, W.H., Ariawan, H., Astuti, E.T., Cahyana, W., Hakim, A.L., Hindayan, D., Lestari, A.S., Pajarningsih and Saartanto (1996) Managing tropical rice pests through conservation of generalist natural enemies and alternative prey. *Ecology* 77, 1975–1988.

Shaner, W.W., Philipp, P.F. and Schmehl, W.R. (1982) *Farming Systems Research and Development: Guidelines for Developing Countries.* Westview Press, Boulder, Colorado, USA.

Shanower, T.G., Guiterrez, A.P. and Wightman, J.A. (1993) Effect of temperature on development rates, fecundity and longevity of the groundnut leaf miner, *Aproaerema modicella* (Lepidoptera: Gelechiidae), in India. *Bulletin of Entomological Research* 83, 413–419.

Shapere, D. (1980) The structure of scientific revolution. In: Cutting, G. (ed.) *Paradigms and Revolutions.* University of Notre Dame Press, Notre Dame, pp. 27–38.

Shapiro, M. and Argauer, R. (1995) Effects of pH, temperature, and ultraviolet radiation on the activity of an optical brightener as a viral enhancer for the gypsy moth (Lepidoptera: Lymantriidae) baculovirus. *Journal of Economic Entomology* 88, 1602–1606.

Shapiro, M. and Argauer, R. (1997) Components of the stilbene optical brightener Tinopal LPW as enchancers for the gypsy moth (Lepidoptera: Lymantriidae) baculovirus. *Journal of Economic Entomology* 90, 899–904.

Sharp, J.L. (1986) Hot-water treatment for control of *Anastrepha suspensa* (Diptera: Tephritidae) in mangos. *Journal of Economic Entomology* 79, 706–708.

Sharp, J.L., Ouye, M.T., Hart, W., Ingle, S., Hallman, G., Gould, W. and Chew, V. (1989) Immersion of Florida mangos in hot water as a quarantine treatment for Caribbean fruit fly (Diptera: Tephritidae). *Journal of Economic Entomology* 82, 186–188.

Sheppard, C.A. and Smith, E.H. (1997) Entomological heritage. *American Entomologist* 142–146.

Shoemaker, C.A. (1979) Optimal timing of multiple applications of pesticides with residual toxicity. *Biometrics* 35, 803–812.

Shoemaker, C.A. (1982) Optimal integrated control of univoltine pest populations with age structure. *Operations Research* 30, 40–61.

Shoemaker, C.A. (1984) The optimal timing of multiple applications of residual pesticides: deterministic and stochastic analyses. In: Conway, G.R. (ed.) *Pest and Pathogen Control: Strategic, Tactical and Policy Models.* John Wiley and Sons, Chichester, pp. 290–310.

Shorey, H.H. (1976) *Animal Communication by Pheromones*. Academic Press, New York.

Shorey, H.H. (1977) Concepts and methodology involved in pheromone control of Lepidoptera by disruption of premating communication. In: McFarlane, N.R. (ed.) *Crop Protection Agents – Their Biological Evaluation*. Academic Press, London, New York and San Francisco, pp. 187–200.

Shu-Sheng, L. (1985) Aspects of the numerical and functional responses of the aphid parasite *Aphidius sonchi* in the laboratory. *Entomologia Experimentalis et Applicata* 37, 247–256.

Siddiqui, W.H., Barlow, C.A. and Randolph, P.A. (1973) Effects of some constant and alternating temperatures on population growth of the pea aphid, *Acyrthosiphon pisum* (Homoptera: Aphididae). *Canadian Entomologist* 105, 145–156.

Simmonds, F.J. (1968) Economics of biological control. *PANS* 14, 207–215.

Simmonds, N.W. (1979) *Principles of Crop Improvement*. Longman, London and New York.

Simmons, A.M. and Yeargan, K.V. (1990) Effect of combined injuries from defoliation and green stink bug (Hemiptera: Pentatomidae) and influence of field cages on soybean yield and seed quality. *Journal of Economic Entomology* 83, 599–609.

Sinclair, J.B. and Backman, P.A. (eds) (1989) *Compendium of Soybean Diseases*, 3rd edn. APS Press, St. Paul, MN, USA.

Sinclair, J.B., Kogan, M. and McGlamery, M.D. (1997) *Guidelines for the Integrated Management of Soybean Pests*. National Soybean Research Laboratory Publication No. 2, College of Agricultural, Consumer and Environmental Sciences, University of Illinois at Urbana-Champaign, Urbana, Illinois, USA.

Singh, B.B., Hadley, H.H. and Bernard, R.L. (1971) Morphology of pubescence in soybean and its relationship to plant vigour. *Crop Science* 11, 13–16.

Singh, D.P. (1986) *Breeding for Resistance to Diseases and Insect Pests*. Springer-Verlag, New York and Heidelberg.

Singh, K.G. (1988) New pest threats and plant quarantine development in the Southeast Asia Region. *PLANTI News* 7, 3–7.

Singh, R. and Ellis, P.R. (1993) Sources, mechanisms and bases of resistance in cruciferae to the cabbage aphid, *Brevicoryne brassicae*. *IOBC/WPRS Bulletin* 16, 2–35.

Singh, R., Ellis, P.R., Pink, D.A.C. and Phelps, K. (1994) An investigation of the resistance to cabbage aphid in brassica species. *Annals of Applied Biology* 125, 457–465.

Sivasubramaniam, W., Wratten, S.D. and Klimaszewski, J. (1997) Species composition, abundance, and activity of predatory arthropods in carrot fields, Canterbury, New Zealand. *New Zealand Journal of Zoology* 24, 205–212.

Sloderbeck, P.E. and Yeargan, K.V. (1983) Green cloverworm (Lepidoptera: Noctuidae) populations in conventional and double-crop, no-till soybeans. *Journal of Economic Entomology* 76, 785–791.

Slovic, P. (1987) Perception of risk. *Science* 236, 280–285.

Smart, E. and Stevenson, J.H. (1982) Laboratory estimation of toxicity of pyrethroid insecticides to honey bees: relevance to hazard in the field. *Bee World* 63, 150–152.

Smart, L.E., Pickett, J.A. and Powell, W. (1997) "Push-Pull" strategies for pest control. *Grain Legumes* 15, 14–15.

Smelser, R.B. and Pedigo, L.P. (1992) Soybean seed yield and quality reduction by bean leaf beetle (Coleoptera: Chrysomelidae) pod injury. *Journal of Economic Entomology* 85, 2401–2403.

Smith, E.H. (1978) Integrated pest management needs – teaching, research, and extension. In: Smith, E.H. and Pimentel, D. (eds) *Pest Control Strategies*. Academic Press, New York, USA, pp. 309–329.

Smith, H.G. and Hallsworth, P.B. (1990) The effects of yellowing viruses on yield of sugar beet in field trials, 1985 and 1987. *Annals of Applied Biology* 116, 503–511.

Smith, H.S. (1980) The effect of 10 generations of selection for oil and protein content in corn. *Bulletin of Illinois Agricultural Experimental Station* 1907, 533.

Smith, J.G. (1969) Some effects of crop background on populations of aphids and their natural enemies on Brussels sprouts. *Annals of Applied Biology* 63, 326–330.

Smith, J.W., McKibben, G.H., Villavasao, J., McGovern, W.L. and Jones, R.G. (1994) Management of the cotton boll weevil with attract-and-kill devices. In: Constable, G.A. and Forester, N.W. (eds) *Proceedings of the World Cotton Research Conference, No. 1*. CSIRO, Brisbane, Australia, pp. 480–484.

Smith, P. (1983) Farming – partly an art, partly a science. *10th International Congress of Plant Protection 1983, Plant Protection for Human Welfare*, Brighton, Volume 1, pp. 56–60.

Smith, P. (1995) Pesticides in foodstuffs. In: Best, G.A. and Ruthven, A.D. (eds) *Pesticides – Developments, Impacts, and Controls*. The Royal Society of Chemistry, London, UK, pp. 62–73.

Smith, R.F. and Reynolds, H.T. (1966) Principles, definitions and scope of integrated pest control. *Proceedings of FAO (United Nations Food and Agriculture Organisation) Symposium on Integrated Pest Control* 1, 11–17.

Smits, P.H. (1997) Insect pathogens: their suitability as biopesticides. In: Evans, H.F. (ed.) *British Crop Protection Council Symposium Proceedings No. 68. Microbial Insecticides: Novelty or Necessity?*. British Crop Protection Council, Farnham, UK, pp. 21–28.

Snedecor, G.W. and Cochran, W.G. (1978) *Statistical Methods*. Iowa State University Press, Ames, Iowa.

Snyder, R.G. (1993) The U.S. greenhouse vegetable industry update and greenhouse vegetables in Mississippi. In: Carpenter, T.D. and Gibbs, C.R. (eds) *Proceedings of American Greenhouse Vegetable Growers Conference and Trade Show*. Denver, CO, USA.

Soderland, D.M. and Bloomquist, J.R. (1990) Molecular mechanisms of insecticide resistance. In: Roush, R.T. and Tabashnik, B.R. (eds) *Pesticide Resistance in Arthropods*. Chapman and Hall, New York and London, pp. 58–196.

Soderstrom, E.L. and Brandl, D.G. (1982) Antifeeding effect of modified atmospheres on larvae of the navel orangeworm and Indianmeal moths (Lepidoptera: Pyralidae). *Journal of Economic Entomology* 75, 704–705.

Soderstrom, E.L., Gardner, P.D., Baritelle, J.L., Nolan de Lozano, K. and Brandl, D.G. (1984) Economic cost evaluation of a generated low-oxygen atmosphere as an alternative fumigant in the bulk storage of raisins. *Journal of Economic Entomology* 77, 457–461.

Solomon, M.G. (1981) Windbreaks as a source of orchard pests and predators. In: Thresh, J.M. (ed.) *Pests, Pathogens and Vegetation*. Pitman, London, pp. 273–283.

Solomon, M.G. (1989) The role of natural enemies in top fruit. *Proceedings of the Symposium on Insect Control Strategies and the Environment*. Amsterdam, pp. 70–85.

Solomon, M.G., Fitzgerald, J.D. and Murray, R.A. (1996) Electrophoretic approaches to predator–prey interactions. In: Symondson, W.O.C. and Liddell, J.E. (eds) *The Ecology of Agricultural Pests*. Chapman & Hall, London, pp. 457–468.

Song, Y.H., Heong, K.L., Lazaro, A.A. and Yeun, K.S. (1994) Use of geographical information systems to analyse large area movement and dispersal of rice insects in South Korea. In: Teng, P.S., Heong, K.L. and Moody, K. (eds) *Rice Pest Science and Management*. International Rice Research Institute, Los Banos, Philippines, pp. 31–42.

Sopp, P.I. and Sunderland, K.D. (1989) Some factors affecting the detection period of aphid remains in predators using ELISA. *Entomologia Experimentalis et Applicata* 51, 11–20.

Sopp, P.I., Sunderland, K.D., Fenlon, J.S. and Wratten, S.D. (1992) An improved quantitative method for estimating invertebrate predation in the field using an enzyme-linked immunosorbent assay (ELISA). *Journal of Applied Ecology* 79, 295–302.

Sorenson, W.C. (1995) *Brethren of the Net, American Entomology 1840–1880*. University of Alabama Press, Tuscaloosa, AL, USA.

Soroka, J.J. and Mackay, P.A. (1991) Antibiosis and antixenosis to pea aphid (Homoptera: Aphididae) in cultivars of field peas. *Journal of Economic Entomology* 84(6), 195–196.

Southwood, T.R.E. (1975) The dynamics of insect populations. In: Pimentel, D. (ed.) *Insects, Science and Society*. Academic Press, New York, USA, pp. 151–199.

Southwood, T.R.E. (1977a) The relevance of population dynamic theory to pest status. In: Cherrett, J.M. and Sagar, G.R. (eds) *Origin of Pest, Parasite, Disease and Weed Problems*. Blackwell Scientific Publications, Oxford, pp. 35–54.

Southwood, T.R.E. (1977b) Habitat, the templet for ecological strategies? *Journal of Animal Ecology* 46, 337–365.

Southwood, T.R.E. (1978) *Ecological Methods, with Particular Reference to the Study of Insect Populations*, 2nd edn. Chapman & Hall, London.

Southwood, T.R.E. and Comins, H.N. (1976) A synoptic population model. *Journal of Animal Ecology* 45, 949–965.

Southwood, T.R.E. and Jepson, W.F. (1962) Studies on the populations of *Oscinella frit* L. (Diptera: Chloropidae) in the oat crop. *Journal of Animal Ecology* 31, 481–495.

Southwood, T.R.E. and Norton, G.A. (1973) Economic aspects of pest management strategies and decisions. *Memoirs of the Ecological Society of Australia* 1, 168–184.

Soyatech Inc. (1995) *Soya Bluebook*. Soyatech Inc., Bar Harbor, ME, USA.

Spedding, C.R.W. (1996) *Agriculture and the Citizen*. Chapman & Hall, London, UK.

Spedding, C.R.W. (1998) 1992 Modern Agriculture: the role and impact of technology, legislation and public opinion. In: Lewis, T. (ed.) *The Bawden Memorial Lectures*. British Crop Protection Council, Farnham, UK, pp. 227–248.

Speight, M.R. and Wainhouse, D. (1989) *Ecology and Management of Forest Insects*. Oxford Science Publishers, Clarendon Press, Oxford.

Spencer, J. and Dent, D.R. (1991) Walking speed as a variable in knapsack sprayer operation: perception of speed and the effect of training. *Tropical Pest Management* 37, 321–323.

Spitters, C.J.T. (1990) Crop growth models: their usefulness and limitations. *Acta Horticulture* 267, 349–368.

Srivastava, S.K. (1974) Some field problems related to the use of pesticides in India. *PANS* 20, 323–326.

Stapleton, J.J., Summers, C.G., Duncan, R.A. and Newton, A.S. (1994) Using reflectorised spray mulch to reduce yield losses of zucchini squash due to aphid-vectored virus diseases. *Proceedings of the 9th Congress of the Mediterranean Phytopathological Union*, Turkey, pp. 343–346.

Stapleton, J.J., Summers, C.G., Turini, T.A. and Duncan, R.A. (1993) Effect of reflective polyethylene and spray mulches on aphid populations and vegetative growth of bell pepper. In: Lamont, W.J. Jr. (ed.) *Proceedings of the 24th National Agricultural Plastics Congress*, Kansas, pp. 117–122.

Star, J. and Estes, J. (1990) *Geographic Information System – An Introduction*. Prentice Hall, Englewood Cliffs, New Jersey.

Starks, K.J., Burton, R.L., Wilson, R.L. and Davis, F.M. (1982) Southwestern corn borer: influence of planting dates and times of infestation on damage to corn, pearl millet and sorghum. *Journal of Economic Entomology* 75, 57–60.

Steenwyk, R.A. van, Oatman, E.R. and Wyman, J.A. (1983) Density treatment level for tomato pinworm (Lepidoptera: Gelechiidae) based on pheromone trap catches. *Journal of Economic Entomology* 76, 440–445.

Steiner, L.F., Mitchell, W.C., Harris, E.J., Kozuma, T.T. and Fujimoto, M.S. (1965) Oriental fruit fly eradication by male annihilation. *Journal of Economic Entomology* 58, 961–964.

Stenseth, N.C. (1981) How to control pest species: application of models from the theory of island biogeography in formulating pest control strategies. *Journal of Applied Ecology* 18, 773–94.

Stenseth, N.C. (1987) Evolutionary processes and insect outbreaks. In: Barbosa, P. and Schultz, J.C. (eds) *Insect Outbreaks*. Academic Press, San Diego and London, pp. 533–564.

Stephens, S.G. (1957) Sources of resistance of cotton strains to the boll weevil and their possible utilization. *Journal of Economic Entomology* 50, 415–418.

Stern, V.M., Smith, R.F., Bosch, R. van den and Hagen, K.S. (1959) The integrated control concept. *Hilgardia* 29, 81–101.

Stevens, R.E., Gutierrez, A.P. and Butler, G.P. (1994) An algorithm for temperature-dependent growth rate simulation. *Canadian Entomologist* 106, 519–524.

Stewart, S., Reed, J., Luttrell, R. and Harris, F.A. (1998) Cotton insect control strategy project: comparing Bt with conventional cotton management and plant bug control strategies at five locations in Mississippi, 1995–7. *Proceedings 1998 Beltwide Cotton Conference*, pp. 807–1672.

Stinner, B.R. and House, G.J. (1990) Arthropods and other invertebrates in conservation-tillage agriculture. *Annual Review of Entomology* 35, 299–318.

Stinner, R.E., Gutierrez, A.P. and Butler, G.P. (1974) An algorithm for temperature-dependent growth rate simulation. *Canadian Entomologist* 106, 519–524.

Stone, N.D. and Gutierrez, A.P. (1986a) Pink bollworm control in Southwestern desert cotton. I. A field oriented simulation of pink bollworm in south western desert cotton. *Hilgardia* 54, 1–24.

Stone, N.D. and Gutierrez, A.P. (1986b) Pink bollworm control in Southwestern desert cotton. II. A strategic management model. *Hilgardia* 54, 1–24.

Stone, R. (1992) Researchers score victory over pesticides – and pests – in Asia. *Science* 256, 1272–1273.

Stower, W.J. and Greathead, D.J. (1969) Numerical changes in a population of the desert locust, with special reference to factors responsible for mortality. *Journal of Applied Ecology* 6, 203–235.

Strickberger, M.W. (1976) *Genetics* 2nd edn. Macmillan Publishing Company, New York and Collier Macmillan Publishers, London.

Strong, D.R., Lawton, J.H. and Southwood, R. (1984) *Insects on Plants – Community Patterns and Mechanisms.* Blackwell Scientific Publications, Oxford, UK.

Stübler, H. and Kern, M. (1998) Biotechnology in crop protection and crop production: the view from industry. In: Richter, J., Huber, J. and Schuler, B. (eds) *Biotechnology for Crop Protection – its Potential for Developing Countries.* Seutsche Stiftung für Internationale Entwicklung, Feldafing, Germany, pp. 35–49.

Su, N.-Y. (1994) Field evaluation of hexaflumuron bait for population suppression of subterranean termites (Isoptera: Rhinotermitidae). *Journal of Economic Entomology* 87, 389–397.

Su, N.-Y. and Scheffrahn, R.H. (1998) A review of subterranean termite control practices and prospects for integrated pest management programmes. *Integrated Pest Management Reviews* 3, 1–14.

Su, N.-Y., Tamashiro, M. and Yates, J.R. (1982) *Trials on the Field Control of the Formosan Subterranean Termite with Amdro Bait.* The International Research Group on Wood Preservation. Document No. IRG/WP/1163, Stockholm, Sweden.

Su, N.-Y, Scheffrahn, R.H. and Ban, P.M. (1995) Effects of sulfluramid-treated bait blocks on field colonies of the formosan subterranean termite (Isoptera: Rhinotermitidae). *Journal of Economic Entomology* 88, 1343–1348.

Subramanyan, B. and Hagstrum, D.W. (1991) Quantitative analysis of temperature, relative humidity, and diet influencing development of the larger grain borer, *Prostephanus truncatus* (Horn) (Coleoptera Bostrichidae). *Tropical Pest Management* 37, 195–202.

Sumberg, J. and Okali, C. (1989) Farmers, on-farm research and new technology. In: Chambers, R., Pacey, A. and Thrupp, L.A. (eds) *Farmer First. Farmer Innovation and Agricultural Research.* Intermediate Technology Publications, London, UK, pp. 189–207.

Summers, C.G. (1992) Integrated pest management in the alfalfa agroecosystem. *Proceedings of the 11th Nevada Alfalfa Symposium, 5–6 February 1992, Sparks, Nevada.* University of Nevada Cooperative Extension Service, Reno, Nevada, pp. 119–126.

Summers, C.G., Stapleton, J.J., Newton, A.S., Duncan, R.A. and Hart, D. (1995) Comparison of sprayable and film mulches in delaying the onset of aphid-transmitted virus diseases in zucchini squash. *Plant Disease* 79, 1126–1131.

Sunderland, K.D. (1975) The diet of some predatory arthropods in cereal crops. *Journal of Applied Ecology* 12, 507–515.

Sunderland, K.D., de Snoo, G.R., Dinter, A., Hance, T., Helenius, J., Jepson, P., Kromp, B., Samu, F., Sotherton, N.W., Toft, S. and Ulber, B. (1995) Density estimation for invertebrate predators in agroecosystems. *Acta Jutlandica* 70, 133–164.

Sunderland, K.D., Fraser, A.M. and Dixon, A.F.G. (1986) Field and laboratory studies on money spiders (linyphiidae) as predators of cereal aphids. *Journal of Applied Ecology* 23, 433–447.

Sutter, G.R. and Branson, T.F. (1980) A procedure for artificially infesting field plot with corn rootworm eggs. *Journal of Economic Entomology* 73, 135–137.

Swanson, E.R. (1979) Working with other disciplines. *American Journal of Agricultural Economics* 61, 849–859.

Symondson, W.O.C. and Hemingway, J. (1997) Biochemical and molecular techniques. In: Dent, D.R. and Walton, M.P. (eds) *Methods in Ecological and Agricultural Entomology.* CAB International, Wallingford, UK, pp. 293–350.

Symondson, W.O.C. and Liddell, J.E. (1995) Decay rates for slug antigens within the carabid predator *Pterostichus melanarius* monitored with a monoclonal antibody. *Entomologia Experimentalis et Applicata* 75, 245–250.

Symondson, W.O.C. and Liddell, J.E. (eds) (1996) *The Ecology of Agricultural Pests.* Chapman & Hall, London, UK.

Szmedra, P.I., Wetzstein, M.E. and McClendon, R.W. (1990) Economic threshold under risk: a case study of soybean production. *Journal of Economic Entomology* 83, 641–646.

Tabashnik, B.E. and Mau, R.F.L. (1986) Suppression of diamondback moth (Lepidoptera: Plutellidae) oviposition by overhead irrigation. *Journal of Economic Entomology* 79, 189–191.

Tait, E.J. (1987) Planning an integrated pest management system. In: Burn, A.J., Coaker, T.H. and Jepson, P.C. (eds) *Integrated Pest Management*. Academic Press, London, pp. 189–207.

Taneja, S.L. and Jayaswal, A.P. (1981) Capture threshold of pink bollworm moths on *Hirsutum* cotton. *Tropical Pest Management* 27, 318–324.

Tang, J.Y., Cheng, J.A. and Norton, G.A. (1994) HOPPER – an expert system for forecasting the risk of white-backed planthopper attack in the first crop season in China. *Crop Protection* 13(6), 463–474.

Tanigoshi, L.K., Fargerlund, J., Nishio-Wong, J.Y. and Griffiths, H.J. (1985) Biological control of citrus thrips, *Scirtothrips citri* (Thysanoptera: Thripidae), in southern California citrus groves. *Environmental Entomology* 14, 733–741.

Tanton, M.T. (1962) The effect of leaf 'toughness' on the feeding of larvae of the mustard beetle *Phaedon cochleariae* Fab. *Entomologia Experimentalis et Applicata* 5, 74–78.

Taylor, A.D. (1988) Parasitoid competition and the dynamics of host-parasitoid models. *American Naturalist* 132, 417–436.

Taylor, A.D. (1990) Metapopulations, dispersal, and predator-prey dynamics: an overview. *Ecology* 71, 429–433.

Taylor, A.D. (1991) Studying metapopulation effects in predator-prey systems. *Biological Journal of the Linnean Society* 72, 305–323.

Taylor, L.R. (1961) Aggregation, variance and the mean. *Nature* 189, 732–735.

Taylor, L.R. (1962a) The absolute efficiency of insect suction traps. *Annals of Applied Biology* 50, 405–421.

Taylor, L.R. (1962b) The efficiency of cylindrical sticky insect traps and suspended nets. *Annals of Applied Biology* 50, 681–685.

Taylor, L.R. (1963) Analysis of the effect of temperature on insects in flight. *Journal of Animal Ecology* 32, 99–117.

Taylor, L.R. (1974) Monitoring change in the distribution and abundance of insects. *Report of Rothamsted Experimental Station for 1973, Part* 2, 202–239.

Taylor, W.E. and Bardner, R. (1968) Leaf injury and food consumption by larvae of *Phaedon cochleariae* (Coleoptera: Chrysomelidae) and *Plutella maculipennis* (Lepidoptera: Plutellidae) feeding on turnip and radish. *Entomologia Experimentalis et Applicata* 11, 177–184.

Teetes, G.L. (1980) Breeding sorghums resistant to insects. In: Maxwell, F.G. and Jennings, P.R. (eds) *Breeding Plants Resistant to Insects*. John Wiley and Sons, Chichester, pp. 457–485.

Teetes, G.L. (1991) The environmental control of insects using planting times and plant spacing. In: Pimentel, D. (ed.) *Handbook of Pest Management in Agriculture*, Volume 1, 2nd edn. CRC, Boca Raton, Florida, USA, pp. 169–182.

Teng, P.S. (1990) *Integrated Pest Management in Rice: An Analysis of the Status Quo with Recommendations for Action*. Department of Plant Pathology, IRRI, Manila, Philippines.

Theunissen, J. and Den Ouden, H. (1980) Effects of intercropping with *Spergula arvensis* on pests of Brussels sprouts. *Entomologia Experimentalis et Applicata* 27, 260–268.

Thomas, D.J.I., Morgan, J.A.W., Whipps, J.M. and Saunders, J.R. (1997) Plasmid transfer by the insect pathogen *Bacillus thuringiensis* in the environment. In: Evans, H.F. (ed.) *British Crop Protection Council Symposium Proceedings No. 68. Microbial Insecticides: Novelty or Necessity?*. British Crop Protection Council, Farnham, UK, pp. 361–365.

Thomas, M. and Waage, J. (1996) *Integration of Biological Control and Host Plant Resistance*. Technical Centre for Agricultural and Rural Cooperation, Wageningen, The Netherlands; CAB International, Wallingford, UK.

Thomas, M.B., Blanford, S., Gbongboui, C. and Lomer, C.J. (1998) Experimental studies to evaluate spray applications of a mycoinsecticide against the rice grasshopper, *Hieroglyphus daganensis*, in northern Benin. *Entomologia Experimentalis et Applicata* 87, 93–102.

Thomas, M.B., Wratten, S.D. and Sotherton, N.W. (1991) Creation of "island" habitats in farmland to manipulate populations of beneficial arthropods: predator densities and emigration. *Journal of Applied Ecology* 28, 906–917.

Thomas, M.B., Wratten, S.D. and Sotherton, N.W. (1992) Creation of "island" habitats in farmland to manipulate populations of beneficial arthropods: predator densities and species composition. *Journal of Applied Ecology* 29, 524–531.

Thomas, R.C. and Miller, H.G. (1994) The interaction of green spruce aphid and fertilizer applications on the growth of Sitka spruce. *Forestry* 67, 329–342.

Thompson, G.D. (1997) Industry and individual company's perspective on resistance management. *Resistant Pest Management* 9, 6–8.

Thompson, K.F. (1963) Resistance to the cabbage aphid (*Brevicoryne brassicae*) in brassica plants. *Nature* 198, 209.

Thorne, G.N. (1986) Confessions of a narrow-minded applied biologist, or why do interdisciplinary research? *Annals of Applied Biology* 108, 205–217.

Thurston, R., Smith, W.T. and Cooper, B. (1966) Alkaloid secretion by trichomes of *Nicotiana* species and resistance to aphids. *Entomologia Experimentalis et Applicata* 9, 428–432.

Tingey, W.M. and Lamont, W.J. Jr. (1988) Insect abundance in field beans altered by intercropping. *Bulletin of Entomological Research* 78, 527–535.

Tisdell, C. (1990) Economic impact of biological control of weeds and insects. In: Mackauer, M., Ehler, L.E. and Roland, J. (eds) *Critical Issues in Biological Control*. Intercept, Andover, pp. 301–316.

Titmarsh, I.J. (1992) Mortality of immature Lepidoptera, a case study with *Heliothis* species. (Lepidoptera: Noctuidae) in agricultural crops on the Darling Downs. PhD thesis, University of Queensland, Brisbane, Australia.

Toba, H.H., Kishaba, A.N., Wolf, W.W. and Gisbon, T. (1970) Spacings of screen traps baited with synthetic sex-pheromone of the cabbage looper. *Journal of Economic Entomology* 63, 197–200.

Todd, J.W. and Schumann, F.W. (1988) Combination of insecticide applications with trap crops of early maturing soybean and southern peas for population management of *Nezara viridula* in soybean (Hemiptera: Pentatomidae). *Journal of Entomological Science* 23, 192–199.

Todd, J.W., McPherson, R.M. and Boethel, D. (1994) Management tactics for soybean insects. In: Higley, L.G. and Boethel, D.J. (eds) *Handbook of Soybean Insect Pests*. Entomological Society of America, Lanham, MD, USA, pp. 115–117.

Townley-Smith, T.F. and Hurd, E.A. (1973) Use of moving means in wheat yield trials. *Canadian Journal of Plant Science* 53, 447–450.

Travasso, M.I., Caldiz, D.O. and Saluzzo, J.A. (1986) Yield prediction using the SUBSTOR-potato model under Argentinian conditions. *Potato Research* 39, 305–312.

Troxclair, N.N. Jr. and Boethel, D.J. (1984) Influence of tillage practices and row spacing on soybean insect populations in Louisiana. *Journal of Economic Entomology* 77, 1571–1579.

Tucker, M.R. (1984a) Possible sources of outbreaks of the armyworm, *Spodoptera exempta* (Walker) (Lepidoptera: Noctuidae), in East Africa at the beginning of the season. *Bulletin of Entomological Research* 74, 599–607.

Tucker, M.R. (1984b) Forecasting the severity of armyworm seasons in East Africa from early season rainfall. *Insect Science Application* 5(1), 51–55.

Tucker, M.R. and Pedgeley, D.E. (1983) Rainfall and outbreaks of the African armyworm, *Spodoptera exempta* (Walker) (Lepidoptera: Noctuidae). *Bulletin of Entomological Research* 73, 195–199.

Tucker, M.R., Mwandoto, S. and Pedgley, D.E. (1982) Further evidence for windborne movement of armyworm moths, *Spodoptera exempta*, in East Africa. *Ecological Entomology* 7, 763–773.

Tukahirwa, E.M. and Coaker, T.H. (1982) Effect of mixed cropping on some insect pests of brassicas; reduced *Brevicoryne brassicae* infestations and influences on epigeal predators and the disturbance of oviposition behaviour in *Delia brassicae*. *Entomologia Experimentalis et Applicata* 32, 129–140.

Turchin, P., Odendaal, F.J. and Rausher, M.D. (1991) Quantifying insect movement in the field. *Environmental Entomology* 20, 955–963.

Turnbull, A.L. and Chant, D.A. (1961) The practice and theory of biological control of insects in Canada. *Canadian Journal of Zoology* 39, 697–753.

Turnock, W.J. (1972) Geographical historical variability in population patterns and life systems of the larch sawfly. *Canadian Entomologist* 104, 1883–1900.

Unterstenhofen, G. (1976) The basic principles of crop protection field trials. *Pflanzenschultz-Nachrichten Bayer* 29, 85.

Usherwood, P.N.R. and Blagborough, I.S. (1989) Amino-acid synapses and receptors. In: McFarlane, N.R. (ed.) *British Crop Protection Council Monograph No. 43. Progress and Prospects in Insect Control*. British Crop Protection Council, Farnham, UK, pp. 45–58.

Usherwood, P.N.R., Duce, I.R. and Boden, P. (1984) Slowly reversible block of glutamate receptor channels by venoms of *Argiope trifasciata* and *araneus gemma*. *Journale de Physiologie (Paris)* 79, 169–171.

Uthamasamy, S. (1986) Studies on the resistance in okra, *Abelmoschus esculentus* (L.) Moench. to the leafhopper *Amrasca devastans* (Dist.). *Tropical Pest Management* 32, 146–147.

Uvah, I.I.I. and Coaker, T.H. (1984) Effect of mixed cropping on some insect pests of carrots and onions. *Entomologia Experimentalis et Applicata* 36, 159–167.

Vaeck, M., Reynaerts, A., Hofte, H., Jansens, S., DeBeuckeleer, M., Dean, C., Zabeau, M., Montagu, M. van and Leemans, J. (1987) Transgenic plants protected from insect attack. *Nature* 328, 33–37.

Vale, G.A. (1982) The trap-orientated behaviour tsetse flies (Glossinidae) and other Diptera. *Bulletin of Entomological Research* 72, 71–93.

Vale, G.A., Bursell, E. and Hargrove, J.W. (1985) Catching-out the tsetse fly. *Parasitology Today* 1, 106–110.

Valentine, B.J., Gurr, G.M. and Thwaite, W.G. (1996) Efficacy of the insect growth regulators tebufenozide and fenoxycarb for lepidopteran pest control in apples, and their compatibility with biological control for integrated pest management. *Australian Journal of Experimental Agriculture* 36, 501–506.

Valentine, W.J., Newton, C.M. and Talevio, R.L. (1976) Compatible systems and decision models for pest management. *Environmental Entomology* 5, 891–900.

van de Fliert, E. (1992) *Risk and Reward of IPM Implementation and Good Farmers and IPM Farmers*. Internal Report, National Programme for Development and Training of Integrated Pest Management in Rice-Based Cropping Systems, Indonesia.

van den Bosch, J. and Messenger, P.S. (1973) *Biological Control*. International Textbook Company, Limited, Aylesbury, UK.

van den Bosch, R. (1978) *The Pesticide Conspiracy*. Doubleday, Garden City, NY, USA.

van den Bos, J. and Rabbinge, R. (1976) *Simulation of the Fluctuations of the Grey Larch Bud Moth*. Pudoc, Wageningen, The Netherlands.

van den Ende, E., Blommers, L. and Trapman, M. (1996) Gaby: a computer-based decision support system for integrated pest management in Dutch apple orchards. *Integrated Pest Management Reviews* 1, 147–162.

van den Ende, J.E. (1994a) Practical use of a computer based advisory system for IPM in apple. *Mededelingen Faculteit Landbouwkundige en Toegepaste Biologische Wetenschappen Universiteit Gent* 5913, 1241–1245.

van den Ende, J.E. (1994b) Op maat gesneden voorlichting via de computer. *Fruitteelt* 15, 22–23.

van der Kraan, C. and Deventer, G.C. (1982) Range of action and interaction of pheromone traps for the summerfruit tortrix moth *Adoxophyes orana* (F.v.R.). *Journal of Chemical Ecology* 8, 1251–1262.

van der Plank, J.E. (1963) *Plant Diseases: Epidemics and Control*. Academic Press, New York.

van Driesche, R.G. (1983) The meaning of percent parasitism in studies of insect parasitoids. *Environmental Entomology* 12, 1611–1622.

van Driesche, R.G. and Bellows, T.S. Jr. (1996) *Biological Control*. Chapman & Hall, London, UK.

van Duyn, J.W., Turnipseed, S.G. and Maxwell, H.D. (1972) Resistance in soybeans to the Mexican bean beetle. II. Reactions to the beetle to resistant plants. *Crop Science* 12, 561–562.

van Emden, H.F. and Peakall, D.B. (1996) *Beyond Silent Spring*. Chapman & Hall, London, UK.

Van Emden, H.F., Hughes, R.D., Eastop, V.R. and Way, M.J. (1969) The ecology of *Myzus persicaei*. *Annals of Applied Biology* 14, 197–270.

Van Hamburg, H. and Hassell, M.P. (1984) Density dependence and augmentative release of egg parasitoids against graminaceous stalkborers. *Ecological Entomology* 9, 101–108.

van Lenteren, J.C. (1986) Parasitoids in the greenhouse: successes with seasonal inoculative release systems. In: Waage, J. and Greathead, D. (eds) *Insect Parasitoids*. Academic Press, London, pp. 341–374.

van Lenteren, J.C. (1987) Environmental manipulation advantageous to natural enemies of pests. In: Delucchi, V. (ed.) *Integrated Pest Management, Protection Integrée: Quo Vadis? An International Perspective*. Parasitis 86, Geneva, pp. 123–163.

van Lenteren, J.C. (1989) Implementation and commercialization of biological control in western Europe. *Proceedings and Abstracts. International Symposium of Biological Control Implementation. North American Plant Protection Bulletin* 6, 50–70.

van Lenteren, J.C. (1992a) Insect invasions: origins and effects. In: *Ecological Effects of Genetically Modified Organisms*. Netherlands Ecological Society, Amsterdam, The Netherlands, pp. 59–90.

van Lenteren, J.C. (1992b) Biological control in protected crops: where do we go? *Pesticide Science* 36, 321–327.

van Lenteren, J.C. (1993) Biological control and integrated pest management in glasshouses – a commercial success. *Pesticide Science*, 37, 430–432.

van Lenteren, J.C. (1995) Integrated pest management in protected crops. In: Dent, D.R. (ed.) *Integrated Pest Management*. Chapman & Hall, London.

van Lenteren, J.C. (1997) Benefits and risks of introducing exotic macro-biological control agents into Europe. *Bulletin OEPP/EPPO Bulletin* 27, 15–27.

van Lenteren, J.C. and Woets, J. (1988) Biological and integrated pest control in greenhouses. *Annual Review of Entomology* 33, 239–269.

van Lenteren, J.C., Roskam, M.M. and Timmer, R. (1997) Commercial mass production and pricing of organisms for biological control of pests in Europe. *Biological Control* 10, 143–149.

van Tol, N.B. and Lentz, G.L. (1998) Influence of *Bt* cotton on beneficial arthropod populations. *Proceedings of the 1998 Beltwide Cotton Conference*, 1052–1054.

Vandermeer, J. (1989) *The Ecology of Intercropping*. Cambridge University Press, Cambridge.

Vandermeer, J. and Andow, D.A. (1986) Prophylactic and responsive components of an integrated pest management program. *Journal of Economic Entomology* 79, 299–302.

Vansteenkiste, G.C. (1984) Formulation, quantification and control of pest-crop processes. In: Cavalloro, R. (ed.) *Statistical and Mathematical Methods in Population Dynamics and Pest Control*, Proceedings of a Meeting of the EC Experts' Group/Parma 26–28 October 1983. A.A. Balkema, Rotterdam and Boston, pp. 136–146.

Varley, G.C. (1970) The need for life tables for parasites and predators. In: Rabb, R.L. and Guthrie, F.E. (eds) *Concepts of Pest Management*. Proceedings of a Conference held at North Carolina State University, 25–27 March 1970, pp. 59–70.

Varley, G.C. and Gradwell, G. R (1960) Key factors in population ecology. *Journal of Animal Ecology* 29, 399–401.

Varley, G.C., Gradwell, G.R. and Hassell, M.P. (1973) *Insect Population Ecology: An Analytical Approach*. Blackwell Scientific Publications, Oxford.

Vaughn, T.T. and Hoy, C.W. (1993) Effects of leaf age, injury, morphology, and cultivars on feeding behaviour of *Phylloteta cruciferae* (Coleoptera: Chrysomelidae). *Environmental Entomology* 22, 418–424.

Verhalen, L.M., Baker, J.L. and McNew, R.W. (1975) Gardner's grid system and plant selection efficiency in cotton. *Crop Science* 57, 863–864.

Verma, B.R., Lal, B. and Wadhi, S.R. (1988) A quarantine note on *Prostephantus truncatus* – a pest that needs to be watched. *Current Science* 57, 863–864.

Vet, L.E.M., van Lenteren, J.C. and Woets, J. (1980) The parasite-host relationship between *Encarsia formosa* (Hymenoptera: Aphelinidae) and *Trialeurodes vaporariorum* (Homoptera: Aleyrodidae). *Zeitschrift für angewandte Entomologie* 90, 26–51.

Vickerman, G.P. (1982) Distribution and abundance of cereal aphid parasitoids (*Aphidius* spp.) on grassland and winter wheat. *Annals of Applied Biology* 101, 185–190.

Vickerman, G.P. (1988) Farm scale evaluation of the long-term effects of different pesticide regimes on the arthropod fauna of winter wheat. In: Greeves, M.P., Grieg-Smith, P.W. and Smith, B.D. (eds) *Field Methods for the Environmental Study of the Effects of Pesticides*. BCPC Monograph No. 40. British Crop Protection Council, Farnham, UK, pp. 127–135.

Vickerman, G.P. and Wratten, S.D. (1979) The biology and pest status of cereal aphids (Hemiptera: Aphididae) in Europe: a review. *Bulletin of Entomological Research* 69, 1–32.

Villaloboss, F.J., Hall, A.J., Ritchie, J.T. and Orgaz, F. (1996) OILCROP-SUN: a development, growth and yield model of the sunflower crop. *Agronomy Journal* 88, 403–415.

Vlug, H.J. and Paul, H. (1986) Sampling of leatherjackets. *Mededelingen van de Faculteit Landbouwwetenschappen Rijksuniversiteit Gent* 51, 939–942.

von Klinger, K. (1987) Effects of margin-strips along a winter wheat field on predatory arthropods and the infestation by cereal aphids. *Journal of Applied Entomology* 104, 47–58.

Waage, J.K. (1983) Aggregation in field parasitoid populations: foraging time allocation by a population of *Diadegma* (Hymenoptera: Ichneumonidae). *Ecological Entomology* 8, 447–453.

Waage, J.K. (1984) Sperm competition and the evolution of odonate mating systems. In: Smith, R.L. (ed.) *Sperm Competition and the Evolution of Animal Mating Systems*. Academic Press, New York, USA, pp. 257–290.

Waage, J.K. (1989) Ecological theory and the selection of biological control agents. In: Mackauer, M., Ehler, L.E. and Roland, J. (eds) *Critical Issues in Biological Control*. Intercept, Andover, UK, pp. 1–41.

Waage, J.K. (1990) Ecological theory and the selection of biological control agents. In: Mackauer, M., Ehler, L.E. and Roland, J. (eds) *Critical Issues in Biological Control*. Intercept, Andover, UK, pp. 135–58.

Waage, J.K. (1993) Making IPM work: developing country experiences and prospects. In: Srivastava, J.P. and Alderman, H. (eds) *Agriculture and Environmental Challenges. Proceedings of the Thirteenth Agricultural Sector Symposium*. World Bank, Washington DC, USA, pp. 119–134.

Waage, J.K. (1996) Introduction to the Symposium. In: Waage, J.K. (ed.) *Biological Control Introductions – Opportunities for Improved Crop Production*, BCPC Proceedings No. 67. British Crop Protection Council, Farnham, UK, pp. 3–11.

Waage, J.K. (1997) Biopesticides at the cross-roads: IPM products or chemical clones. In: Evans, H.F. (ed.) *British Crop Protection Council Symposium Proceedings No. 68. Microbial Insecticides: Novelty or Necessity?*. British Crop Protection Council, Farnham, UK, pp. 11–19.

Waage, J.K. and Greathead, D.J. (eds) (1986) *Insect Parasitoids*. Academic Press, London, UK.

Waage, J.K. and Greathead, D.J. (1988) Biological control: challenges and opportunities. *Philosophical Transactions of the Royal Society B* 318, 111–128.

Waage, J.K. and Hassell, M.P. (1982) Parasitoids as biological control agents – a fundamental approach. *Parasitology* 84, 241–268.

Waage, J.K. and Mills, N.J. (1992) Biological control. In: Crawley, M.J. (ed.) *Natural Enemies: The Population Biology of Predators, Parasites and Diseases*. Blackwell Scientific Publications, Oxford, UK, pp. 412–430.

Waddill, H. van, Pohronezny, K., McSorley, R. and Bryan, H.H. (1984) Effect of manual defoliation on pole bean yield. *Journal of Economic Entomology* 77, 1019–1023.

Waggoner, P.E. and Berger, R.D. (1987) Defoliation, disease and growth. *Phytopathology* 77, 393–98.

Waibel, H. (1987) Farmers' practices and recommended economic threshold levels in irrigated rice in the Philippines. In: Tait, J. and Napompeth, B. (eds) *Management of Pests and Pesticides, Farmers' Perceptions and Practises*. Westview Press, Boulder and London, pp. 191–197.

Walgate, R. (1990) *Miracle or Menace? Biotechnology and the Third World*. The Panos Institute, London, UK.

Walker, C.H., Hopkin, S.P., Sibly, R.M. and Peakall, D.B. (1996) *Principles of Ecotoxicology*. Taylor & Francis, London.

Walker, P.T. (1981) Methods for studying the pest/yield relationship and for appraising losses due to insects and mites. In: Chiarappa, L. (ed.) *Crop Loss Assessment Methods, FAO Manual on the Evaluation and Prevention of Losses by Pest, Disease and Weeds*. Commonwealth Agricultural Bureaux, Slough, UK, pp. 73–78.

Walker, P.T. (1987) Losses in yield due to pests in tropical crops and their value in policy decision-making. *Insect Science and its Application* 8, 665–671.

Wall, C. (1990) Principles of monitoring. In: Ridgway, R.L., Silverstein, R.M. and Inscoe, M.N. (eds) *Behaviour-modifying Chemicals for Insect Management. Applications of Pheromones and other Attractants*. Marcel Dekker, New York and Basel, pp. 9–23.

Wall, R. (1995) Fatal attraction: the disruption of mating and fertilisation for insect control. In: Leather, S.R. and Hardie, J. (eds) *Insect Reproduction*. CRC Press, Cambridge, pp. 109–128.

Wall, R. and Howard, J. (1994) Autosterilisation for control of the housefly *Musca domestica. Journal of Theoretical Biology* 171, 431–437.

Wall, R. and Shearer, D. (1997) *Veterinary Entomology*. Chapman & Hall, London.

Wall, R., Green, C.H., French, N. and Morgan, K.L. (1992) Development of an attractive target for the sheep blowfly *Lucilia sericata. Medical and Veterinary Entomology* 6, 67–74.

Walsh, D.B., Zalom, F.G., Welch, N.C., Pickel, C. and Shaw, D.V. (1997) Pretransplant cold storage of strawberries: effects on plant vigor, yield and spider mite (Acari: Tetranychidae) abundance. *Journal of Economic Entomology* 90, 818–823.

Walsh, G.R. (1985) *An Introduction to Linear Programming*, 2nd edn. Wiley-Interscience Publications, John Wiley & Sons, Chichester, UK.

Walton, M.P. and Dent, D.R. (1997) Introduction. In: Dent, D.R. and Walton, M.P. (eds) *Methods in Ecological and Agricultural Entomology*. CAB International, Wallingford, UK, pp. 1–4.

Walton, M.P., Loxdale, H.D. and Allen-Williams, L. (1990a) Electrophoretic 'keys' for the identification of parasitoids (Hymenoptera: Braconidae: Aphelinidae) attacking *Sitobion avenae* (F.) (Hemiptera: Aphididae). *Biological Journal of the Linnean Society* 40, 333–346.

Walton, M.P., Powell, W., Loxdale, H.D. and Allen-Williams, L (1990b) Electrophoresis as a tool for estimating levels of hymenopterous parasitism in field populations of the cereal aphid, *Sitobion avenae*. *Entomologia Experimentalis et Applicata* 54, 271–279.

Walton, S.B. (1998) MORPH: expediting the production and distribution of decision support systems to the horticultural industry. *The 1998 Brighton Conference: Pests and Diseases,* Volume 1. British Crop Protection Council, Farnham, UK, pp. 253–258.

Ward, S.A., Rabbinge, R. and Mantel, W.P. (1985) The use of incidence counts for estimation of aphid populations. 1. Minimum sample size for required accuracy. *Netherlands Journal of Plant Pathology* 91, 93–99.

Wardlow, L.R. (1992) The role of extension services in integrated pest management in glasshouse crops in England. In: van Lenteren, J.C., Minks, A.K. and de Ponti, O.M.B. (eds) *Biological Control and Integrated Crop Protection: Towards Environmentally Safer Agriculture*. Pudoc Scientific Publishers, Wageningen, The Netherlands, pp. 193–199.

Watanabe, T., Yamamoto, H. and Sogawa, K. (1994) Growth and yield analysis of rice plants infested with long-distance migratory rice planthoppers. II. Measurement of recovery of vegetative growth of rice plants infested with the white-backed planthopper, *Sogatella furcifera* Horvathe (Homoptera: Delphacidae), by spectral reflectivity. *Japanese Journal of Applied Entomology and Zoology* 38, 169–175.

Watt, A.D. (1983) The influence of forecasting on cereal aphid control strategies. *Crop Protection* 2, 417–429.

Watt, K.E.F. (1965) Community stability and the strategy of biological control. *Canadian Entomologist* 97, 887–895.

Way, M.J. and Banks, C.J. (1968) Population studies on the active stages of the black bean aphid, *Aphis fabae* Scop., on its winter host *Euonymus europaeus* L. *Annals of Applied Biology* 62, 177–197.

Way, M.J. and Murdie, G. (1965) An example of varietal variations in resistance of Brussels sprouts. *Annals of Applied Biology* 56, 326–328.

Way, M.J., Cammell, M.E., Taylor, L.R. and Woiwod, I.P. (1981) The use of egg counts and suction-trap samples to forecast the infestation of spring-sown field beans, *Vicia faba*, by the black bean aphid, *Aphis fabae*. *Annals of Applied Biology* 98, 21–34.

Wearing, C.H. (1988) Evaluating the IPM implementation process. *Annual Review of Entomology* 33, 17–38.

Weaver, R.J., Marris, G.C., Olieff, S., Mosson, J.H. and Edwards, J.P. (1997) Role of ecotparasitoid venom in the regulation of haemolymph ecdysteroid titres in a host noctuid moth. *Archives of Insect Biochemistry and Physiology* 35, 169–178.

Webster, C.C. and Wilson, P.N. (1980) *Agriculture in the Tropics*, 2nd edn. English Language Book Society/Longman.

Webster, J.A., Smith, D.H., Rathke, H. and Cress, C.E. (1975) Resistance to cereal leaf beetle in wheat: density and length of leaf-surface pubescence in four wheat lines. *Crop Science* 12, 561–562.

Webster, J.P.G. and Bowles, R.G. (1996) Estimating the economic costs and benefits of pesticide use in apples. *Brighton Crop Protection Conference: Pests and Diseases,* Volume 1. British Crop Protection Council, Farnham, UK, pp. 325–330.

Weidhaas, D.E., Schmidt, E.L. and Seabrook, E.L. (1962) Field studies on the release of sterile males for the control of *Anopheles quadrimaculatus*. *Mosquito News* 22, 283.

Welsh, J. and McClelland, M. (1990) Fingerprinting genomes using PCR with arbitrary primers (RAPDs). *Nucleic Acids Research* 18, 7213–7218.

West, L.S. (1951) *The Housefly, its Natural History, Medical Importance and Control.* Comstock Publishing Company, New York.

Westbrook, J.K., Esquivel, J.D., López, J.D. Jr., Jones, G.D., Wolf, W.W. and Raultson, J.R. (1997) Validation of bollworm migration across south-central Texas in 1994–1996. *1997 Proceedings: Beltwide Cotton Conference.* National Cotton Council of America, Memphis, USA.

Westbrook, J.K., Raulston, J.R., Wolf, W.W., Pair, S.D., Eyster, R.S. and Lingren, P.D. (1995a) Field observations and simulations of atmospheric transport of Noctuids from northeastern Mexico and the south-central US. *Southwestern Entomologist* 18, 25–44.

Westbrook, J.K., Raulston, J.R., Wolf, W.W., Pair, S.D., Eyster, R.S. and Lingren, P.D. (1995b) Migration pathways of corn earworm (Lepidoptera: Noctuidae) indicated by tetroon trajectories. *Agricultural and Forest Meteorology* 73, 67–87.

Westbrook, J.K., Wolf, W.W., Lingren, P.D. and Raulston, J.R. (1994) Tracking tetroons to evaulate tobacco budworm and bollworm migration. *Cotton Insect Research and Control Conference 1994.* Beltwide Cotton Conferences.

Wetzler, R.E. and Risch, S.J. (1984) Experimental studies of beetle diffusion in simple and complex crop habitats. *Journal of Animal Ecology* 53, 1–19.

Whalon, M.E. and Smilowitz, Z. (1979) The interaction of temperature and biotype on development of the green peach aphid *Myzus persicae. American Potato Journal* 56, 591–596.

Wheatley, A.R.D., Wightman, J.A., Williams, J.H. and Wheatley, S.J. (1989) The influence of drought stress on the distribution of insects on four groundnut genotypes grown near Hyderabad, India. *Bulletin of Entomological Research* 79, 567–577.

Whitten, M.J. and McKenzie, J.A. (1982) The genetic basis for pesticide resistance. In: Lee, K.E. (ed.) *Proceedings of the 3rd Australian Conference on Grassland Invertebrate Ecology.* South Australian Government Printer, Adelaide, Australia, pp. 1–16.

WHO (1991) Strategies for Assessing the Safety of Foods Produced by Biotechnology. *Report of a Joint FAO/WHO Consultation, WHO/FNU/FOS/91.6.* World Health Organisation, Geneva, Switzerland.

WHO (1995) Application of the principles of substantial equivalence to the safety evaluation of foods and food components from plants derived by modern biotechnology. *Report of a WHO Workshop, WHO/FNU/FOS/95.1.* World Health Organisation, Geneva, Switzerland.

Widstrom, N.W. and Burton, R.L. (1970) Artificial infestation of corn with suspensions of corn earworm eggs. *Journal of Economic Entomology* 63, 443–446.

Widstrom, N.W., Williams, W.P., Wiseman, B.R. and Davies, F.M. (1992) Recurrent selection for resistance to leaf feeding fall armyworm on maize. *Crop Science* 32, 1171–1174.

Wiktelius, S. and Pettersson, J. (1985) Simulations of bird-cherry-oat aphid population dynamics: a tool for developing strategies for breeding aphid-resistant plants. *Agriculture, Ecosystems and Environment* 14, 159–170.

Williams, D.G. and McDonald, G. (1982) The duration and number of immature stages of codling moth *Cydia pomonella* (L.) (Tortricidae: Lepidoptera). *Journal of Australian Entomological Society* 21, 1–4.

Wilson, L.T. (1985) Estimating the abundance and impact of arthropod natural enemies in IPM systems. In: Hoy, M.A. and Herzog, D.C. (eds) *Biological Control in Agricultural IPM Systems.* Academic Press, Orlando and London, pp. 303–322.

Wilson, M.F. (1995) Monitoring and adapting to the changes in pesticide use profiles that occur in response to modern pest control and environmental requirements. In: Best, G.A. and Ruthven, A.D. (eds) *Pesticides – Developments, Impacts, and Controls.* The Royal Society of Chemistry, London, UK, pp. 1–7.

Wing, K.D. and Ramsay, J.R. (1989) Non-nervous system targets for chemicals: other hormonal agents: ecdysone agonists. In: McFarlane, N.R. (ed.) *British Crop Protection Council Monograph No. 43. Progress and Prospects in Insect Control.* British Crop Protection Council, Farnham, UK, pp. 105–116.

Winstanley, D., Jarrett, P.J. and Morgan, J.A.W. (1998) The use of transgenic biological control agents to improve their performance in the management of pests. In: Kerry, B.R. (ed.) *British Crop Protection Council Symposium Proceedings No. 71. Biotechnology in Crop Protection: Facts and Fallacies.* British Crop Protection Council, Farnham, UK, pp. 37–44.

Wiseman, B.R. (1989) Technological advances for determining resistance in maize to *Heliothis zea*. In: Mihm, J.A., Wiseman, B.R. and Davis, F.M. (eds) *Toward Insect Resistant Maize for the Third World*. International Wheat and Maize Improvement Center (CIMMYT), El Batan, Mexico, pp. 94–100.

Wiseman, B.R., Widstrom, N.W. and McMilian, W.W. (1974) Methods of application and numbers of eggs of the corn earworm required to infest ears of corn artificially. *Journal of Economic Entomology* 67, 74–76.

Wittwer, S.H. and Castilla, N. (1995) Protected cultivation of horticultural crops world-wide. *HortTechnology* 5, 6–23.

Wolf, W.W., Westbrook, J.K., Raulston, J.R., Pair, S.D. and Lingren, P.D. (1993) Radar detection of ascent of *Helicoverpa zea* (Lepidoptera: Noctuidae) moths from corn in the lower Rio Grande valley of Texas. *Proceedings of the 13th International Congress of Biometeorology*. Calgary, Alberta, Canada.

Wood MacKenzie Consultants Limited (1995) Agrochemical monitor. *Agricultural Biotechnology Report* No. 120. Edinburgh, UK.

Woodworth, C.M., Long, E.R. and Jugenheimer, R.W. (1952) Fifty generations of selection for protein and oil in corn. *Agronomy Journal* 44, 60–65.

Wool, D., van Emden, H.F. and Bunting, S.W. (1978) Electrophoretic detection of the internal parasite, *Aphidius matricariae* in *Myzus persicae*. *Annals of Applied Biology* 90, 21–26.

Wootten, N.W. and Sawyer, K.F. (1954) The pick-up of spray droplets by flying locusts. *Bulletin of Entomological Research* 45, 177–197.

Wratten, S.D. (1975) The nature of the effects of the aphids *Sitobion avenae* and *Metopolophium dirhodum* on the growth of wheat. *Annals of Applied Biology* 79, 27–34.

Wratten, S.D. (1987) The effectiveness of native natural enemies. In: Burn, A.J., Coaker, T.H. and Jepson, P.C. (eds) *Integrated Pest Management*. Academic Press, London, pp. 89–112.

Wratten, S.D. (1988) The role of field boundaries as reservoirs of beneficial insects. In: Park, J.R. (ed.) *Environmental Management in Agriculutre: European Perspectives*. EEC/Pinter Publishers Ltd, London, UK, pp. 144–150.

Wratten, S.D. and Pearson, J. (1982) Predation of sugar beet aphids in New Zealand. *Annals of Applied Biology* 101, 143–203.

Wratten, S.D. and van Emden, H.F. (1995) Habitat management for enhanced activity of natural enemies of pests. In: Glen, D.M., Greaves, M.P. and Anderson, H. M (eds) *Proceedings of the 13th Long Ashton Symposium, England: Ecology and Integrated Farming Systems*. John Wiley & Sons, London, UK, pp. 117–145.

Wratten, S.D., Lee, G. and Stevens, D.J. (1979) Duration of cereal aphid populations and the effects on wheat yield and quality. *Proceedings 1979 British Crop Protection Conference – Pests and Diseases*. British Crop Protection Council, Farnham, UK, pp. 1–8.

Wratten, S.D., Watt, A.D., Carter, N. and Entwistle, J.C. (1990) Economic consequences of pesticide use for grain aphid control in winter wheat in 1984 in England. *Crop Protection* 9, 73–78.

Wright, D.W., Hughes, R.D. and Worrall, J. (1960) The effect of certain predators on the numbers of cabbage root fly (*Erioishia brassicae* (Bouche)) and on the subsequent damage caused by the pest. *Annals of Applied Biology* 48, 756–763.

Wright, J.W. (1976) *Introduction to Forest Genetics*. Academic Press, New York.

Wright, R.J. (1984) Evaluation of crop rotation for control of Colorado potato beetles (Coleoptera: Chrysomelidae) in commercial potato fields on Long Island. *Journal of Economic Entomology* 77, 1254–1259.

Wu, G., Wu, Z.F., Zhao, S.X. and Xu, C.J. (1993) The effects of resistance of rice varieties on carboxylesterase and phosphatase activity of the white backed rice planthopper. *Acta Phytophylacica Sinica* 20, 139–142.

Wyatt, T.D. (1997) Methods in studying insect behaviour. In: Dent, D.R. and Walton, M.P. (eds) *Methods in Ecological and Agricultural Entomology*. CAB International, Wallingford, UK, pp. 27–56.

Yang, P.J., Carey, J.R. and Dowell, R.V. (1994) Temperature influences on the development and demography of *Bactrocera dorsalis* in China. *Environmental Entomology* 23(4), 971–974.

Yasumatsu, K. (1976) Rice stem-borers. In: Delucchi, V.L. (ed.) *Studies in Biological Control*. Cambridge University Press, pp. 121–137.

Yencho, G.C., Getzin, L.W. and Long, G.E. (1986) Economic injury level, action threshold, and a yield-loss model for the pea aphid, *Acyrthosiphon pisum* (Homoptera: Aphididae), on green peas, *Pisum sativum. Journal of Economic Entomology* 79, 1681–1687.

Yokoyama, V.Y. and Miller, G.T. (1993) Pest-free period for walnut husk fly (Diptera: Tephritidae) and host status of stone fruits for export to New Zealand. *Journal of Economic Entomology* 86, 1766–1772.

Yokoyama, V.Y. and Miller, G.T. (1996) Response of walnut husk fly (Diptera: Tephritidae) to low temperature, irrigation and pest-free period for exported stone fruits. *Journal of Economic Entomology* 89, 1186–1191.

Yokoyama, V.Y., Hatchett, J.H. and Miller, G.T. (1993a) Hessian fly (Diptera: Cecidomyiidae) control by hydrogen phosphide fumigation and compression of hay for export to Japan. *Journal of Economic Entomology* 86, 76–85.

Yokoyama, V.Y., Hatchett, J.H. and Miller, G.T. (1993b) Hessian fly (Diptera: Cecidomyiidae) control by compression of hay for export to Japan. *Journal of Economic Entomology* 86, 803–808.

Yokoyama, V.Y., Miller, G.T. and Hartsell, P.L. (1992) Pest-free period and methyl bromide fumigation for control of walnut husk fly (Diptera: Tephritidae) in stone fruits exported to New Zealand. *Journal of Economic Entomology* 85, 150–156.

Yokoyama, V.Y., Hatchett, J.H., Miller, G.T. and Hartsell, P.L. (1994) Hydrogen phosphide residues and efficacy to control hessian fly (Diptera: Cecidomyiidae) in compressed hay for export to Japan. *Journal of Economic Entomology* 87, 1272–1277.

Yu, X., Heong, K.L., Hu, C. and Barrion, A.T. (1996) Role of non-rice habitats for conserving egg parasitoids of rice planthoppers and leafhoppers. In: Hokyo, N. and Norton, G. (eds) *Proceedings of the International Workshop on Pest Management Strategies in Asian Monsoon Agrocecosystems.* Kyushu National Agricultural Experimental Station, Kumamoto, Japan, pp. 63–77.

Zadoks, J.C. (1991) Rationale and concepts of crop loss assessment for improving pest management and crop protection. In: Teng, P.S. (ed.) *Crop Loss Assessment and Pest Management.* APS Press, St Paul, Minnesota, pp. 1–5.

Zadoks, J.C., Chang, T.T. and Konzak, C.F. (1974) A decimal code for the growth stages of cereals. *Weed Research* 14, 415–421.

Zalom, F. (1993) Reorganising to facilitate the development and use of integrated pest management. *Agriculture, Ecosystems and Environment* 46, 245–256.

Zalom, F.G. and Jones, A. (1994) Insect fragments in processed tomatoes. *Horticultural Entomology* 87, 181–186.

Zalom, F.G., Castañé, C. and Gabarra, R. (1996) Effects of chilling of *Bemisia argentifolii* (Homoptera: Aleyrodidae) infesting cabbage. *Journal of Entomological Science* 31, 39–51.

Zalucki, M.P., Daglish, G., Firempong, S. and Twine, P. (1986) The biology and ecology of *Heliothis armigera* (Hübner) and *H. punctigera* Wallengren (Lepidoptera: Noctuidae) in Australia: what do we know? *Australian Journal of Zoology* 34, 779–814.

Zandstra, B.H. and Motooka, P.S. (1978) Beneficial effects of weeds in pest management – a review. *PANS* 24, 333–238.

Zeiss, M.R. and Pedigo, L.P. (1996) Timing of food plant availability: effect on survival and oviposition of the bean leaf beetle (Coleoptera: Chrysomelidae). *Environmental Entomology* 25, 295–302.

Zethner, O. (1995) Practice of integrated pest management in tropical and sub-tropical Africa: an overview of two decades (1970–1990). In: Mengech, A.N., Saxena, K.N. and Gopalan, H.N.B. (eds) *Integrated Pest Management in the Tropics.* John Wiley & Sons, Chichester, UK, pp. 1–68.

Zimmerman, N.L. (1990) Coleoptera found in imported stored-food products entering southern California and Arizona between December 1984 through December 1987. *Coleopterists Bulletin* 44, 235–240.

Index